WAVES AND RAYS IN SEISMOLOGY

Answers to Unasked Questions

Second Edition

Other Related Titles from World Scientific

Waves and Rays in Elastic Continua
3rd Edition
by Michael A. Slawinski
ISBN: 978-981-4641-75-3
ISBN: 978-981-4641-76-0 (pbk)

Wavefronts and Rays as Characteristics and Asymptotics
2nd Edition
by Andrej Bóna and Michael A. Slawinski
ISBN: 978-981-4644-78-5
ISBN: 978-981-4644-79-2 (pbk)

On Foundations of Seismology: Bringing Idealizations Down to Earth
by James Robert Brown and Michael A. Slawinski
ISBN: 978-981-4329-49-1

WAVES AND RAYS IN SEISMOLOGY

Answers to Unasked Questions

Second Edition

Michael A. Slawinski

World Scientific

NEW JERSEY • LONDON • SINGAPORE • BEIJING • SHANGHAI • HONG KONG • TAIPEI • CHENNAI

Published by

World Scientific Publishing Co. Pte. Ltd.
5 Toh Tuck Link, Singapore 596224
USA office: 27 Warren Street, Suite 401-402, Hackensack, NJ 07601
UK office: 57 Shelton Street, Covent Garden, London WC2H 9HE

British Library Cataloguing-in-Publication Data
A catalogue record for this book is available from the British Library.

WAVES AND RAYS IN SEISMOLOGY
Answers to Unasked Questions
2nd Edition

ISBN 978-981-3239-87-6
ISBN 978-981-3270-96-1 (pbk)

For any available supplementary material, please visit
http://www.worldscientific.com/worldscibooks/10.1142/10979#t=suppl

Desk Editor: Ng Kah Fee

This book is dedicated to those who, together with Hermann Weyl (1949), appreciate that

> *exact natural science, if not the most important, is the most distinctive feature of our culture in comparison to other cultures. Philosophy has the task to understand this feature in its peculiarity and singularity.*

And who, together with Mario Bunge (1967),

> *are concerned with finding out the status of concepts, statements and systems of hypotheses as well as with clarifying and rearranging them in a logical order.*

Acknowledgments

The author wishes to acknowledge improvements of this book that resulted from collaborations and discussion with Len Bos, Jim Brown, Chris Chapman, David Dalton, Tomasz Danek, Freeman Dyson, Klaus Helbig, Misha Kochetov, Ken Larner, Michael Rochester, Raphaël Slawinski, Piotr Stachura, Theodore Stanoev and Maurizio Vianello. Graphics were created, and photographs were taken, by Elena Patarini; computations were realized by Rafael Abreu, Colin Brisco and Tomasz Danek; editorial work was done by David Dalton and Theodore Stanoev.

The second edition benefited from collaborations with Len Bos, David Dalton, Andrey Melnikov, Piotr Stachura and Theodore Stanoev; from discussions with Filip Adamus, Hilary Chang, Chris Chapman, Marcelo Epstein, Martino Ischia, Izabela Kudela, Misha Kochetov, Thomas Meehan, Jacob Newman, Giuseppe Orlando and Yves Rogister; from proofreading and scientific editing of Md Abu Sayed, David Dalton, Thomas Meehan, Andrey Melnikov, Piotr Stachura and Theodore Stanoev; and from editorial support of, and graphics created by, Elena Patarini. Kah-Fee Ng ensured the adjustment of final details prior to printing.

Contents

Acknowledgments vii

List of Figures xix

List of Tables xxiii

Foreword to First Edition xxv

Changes from First Edition xxxi

1. Science of seismology 1

Preliminary remarks 1
1.1 Purpose and methodology: Historical sketch 2
1.2 Classification 10
Closing remarks 11
1.3 Exercises 14

2. Seismology and continuum mechanics 19

Preliminary remarks 19
2.1 On axiomatic formulation 21
2.2 Kinematic descriptions 23
2.2.1 Spacetime 23
2.2.2 Motion 25
2.2.3 Coordinates 27
2.3 Field equations 29
2.3.1 Balance equations 29
2.3.2 Balance of mass: Continuity equation 30
2.3.3 Balance of momentum: Equation of motion 31

Closing remarks 33
2.4 Exercises 34

3. Hookean solid: Material symmetry 41

Preliminary remarks 41
3.1 Hookean solids 42
3.2 Material symmetry 45
3.2.1 On symmetries 45
3.2.2 On tensor rotations 48
3.2.3 Finite and infinitesimal elasticities 49
3.2.3.1 Deformation gradient 49
3.2.3.2 Elasticity tensor 52
3.2.3.3 Prestressed linearly elastic materials . . . 54
3.2.3.4 Material symmetry: Finite elasticity . . . 58
3.2.3.5 Material symmetry: Relation between finite and infinitesimal elasticities 59
3.2.4 Symmetry classes 62
3.2.4.1 Material-symmetry conditions 62
3.2.4.2 Hooke's law in $\mathbb{R}^3$ and $\mathbb{R}^6$ 65
3.2.4.3 Index symmetries 66
3.2.4.4 Kelvin notation 68
3.2.4.5 Monoclinic tensor 70
3.2.4.6 On tensorial properties of monoclinic tensor 70
3.2.4.7 Orthotropic tensor 77
3.2.4.8 Tetragonal tensor 78
3.2.4.9 On tensorial properties of tetragonal tensor 78
3.2.4.10 Transversely isotropic tensor 81
3.2.4.11 Trigonal tensor 81
3.2.4.12 Cubic tensor 82
3.2.4.13 Isotropic tensor 82
3.2.4.14 On isotropic parameters 82
3.2.4.15 Relations among elasticity parameters . . 83
3.2.4.16 Diclinic solids 88
3.2.4.17 Hexagonal solids 90
Closing remarks 92
3.3 Exercises 93

4. Hookean solid: Effective symmetry and equivalent medium 103

Preliminary remarks . . . 103
4.1 Effective symmetries . . . 104
4.1.1 On accuracy . . . 104
4.1.2 Fixed orientation of coordinate system . . . 111
4.1.2.1 Introduction . . . 111
4.1.2.2 Monoclinic tensor . . . 112
4.1.2.3 Orthotropic tensor . . . 113
4.1.2.4 Tetragonal tensor . . . 114
4.1.2.5 Transversely isotropic tensor . . . 115
4.1.2.6 Trigonal tensor . . . 115
4.1.2.7 Cubic tensor . . . 117
4.1.2.8 Isotropic tensor . . . 117
4.1.3 Optimal orientation of coordinate system . . . 118
4.2 Equivalent media . . . 121
4.2.1 Introduction . . . 121
4.2.2 Equivalence parameters for isotropic layers . . . 124
4.2.2.1 Formulæ . . . 124
4.2.2.2 Justification . . . 127
4.2.2.3 Interpretation . . . 137
4.2.2.4 Possible issues with Backus averaging . . 138
4.2.3 Equivalence parameters for TI layers . . . 140
4.3 Relation between effective and equivalent formulations . . 142
4.3.1 Introduction . . . 142
4.3.2 Analytical formulation . . . 144
4.3.2.1 Anisotropic layers, monoclinic medium . 144
4.3.2.2 Monoclinic layers, orthotropic medium . 147
4.3.2.3 Orthotropic layers, tetragonal medium . 150
4.3.2.4 Transversely isotropic layers . . . 151
4.3.2.5 Combining rotational averages . . . 152
4.3.3 Discussion . . . 152
Closing remarks . . . 152
4.4 Exercises . . . 155

5. Body waves 181

Preliminary remarks . . . 181
5.1 Wave equations . . . 182
5.1.1 Assumptions and formulation . . . 182
5.1.2 Particular case: Isotropy and homogeneity . . . 183
5.1.3 Particular case: Inhomogeneous string . . . 190

5.1.4 Particular case: String with friction 193
5.2 Solutions of wave equation 194
5.2.1 Introduction 194
5.2.2 Product solution 194
5.2.3 d'Alembert solution 196
5.2.3.1 d'Alembert's approach 196
5.2.3.2 Euler's approach 197
5.2.3.3 Spherical-symmetry approach 199
5.2.4 Fourier-transform solution 201
5.2.5 Green's-function solution 204
5.3 On approximations 208
Closing remarks 210
5.4 Exercises 211

6. Surface, guided and interface waves 221

Preliminary remarks 221
6.1 Introduction 222
6.2 Surface waves: Homogeneous elastic halfspace 224
6.3 Guided waves: Homogeneous layer above halfspace 234
6.3.1 Elastic layer above rigid halfspace 234
6.3.2 Elastic layer above elastic halfspace 237
6.3.3 Love wave as superposition of SH waves 240
6.3.4 On Rayleigh and Love waves in Earth's crust . . . 241
6.4 Existence of surface and guided waves 242
6.4.1 Introduction 242
6.4.2 Elasticity parameters and mass densities 242
6.4.3 On Love waves in homogeneous halfspace 243
6.4.4 On P waves in homogeneous halfspace 243
6.5 Interface waves: Homogeneous halfspaces 245
6.5.1 Introduction 245
6.5.2 Elastic and liquid halfspaces 246
6.5.3 Liquid halfspaces 257
6.6 Existence of interface waves 261
6.6.1 Introduction 261
6.6.2 Elasticity parameters and mass densities 262
6.6.3 On SH waves as interface waves 262
Closing remarks 264
6.7 Exercises 265

7. Variational principles in seismology 267

Preliminary remarks 267
7.1 Historical comments 268
7.2 Fermat's principle 269
7.2.1 Isotropic layered medium 269
7.2.2 Isotropic continuously inhomogeneous medium . . 272
7.2.3 Global optimization and causality 275
7.2.4 Stationarity versus minimization 277
7.2.5 Mathematical justification 278
7.2.5.1 Fermat's principle 278
7.2.5.2 Head waves 280
7.2.6 Physical interpretation 283
7.2.6.1 Macroscopic interpretation 283
7.2.6.2 Microscopic interpretation 285
7.2.6.3 Phase consideration 286
7.2.7 On teleology of Fermat's principle 287
7.3 Hamilton's principle 290
7.3.1 Action 290
7.3.2 Wave equation 292
7.3.3 Mathematical justification 292
7.3.4 Physical interpretation 293
7.4 Conserved quantities 294
7.4.1 Introduction 294
7.4.2 Ray parameter 294
7.4.2.1 Isotropy 294
7.4.2.2 Anisotropy 296
7.4.3 Hamiltonian and Lagrangian 298
7.4.3.1 Ray theory 298
7.4.3.2 Classical mechanics 300
Closing remarks 301
7.5 Exercises 302

8. Gravitational and thermal effects in seismology 309

Preliminary remarks 309
8.1 Gravitation 310
8.1.1 Body forces 310
8.1.2 Wave speeds 313
8.2 On weak gravitational waves 317

8.3 Temperature 324
8.3.1 Propagation and diffusion 324
8.3.2 Isothermal and adiabatic formulations 325
8.3.2.1 Lamé parameters 325
8.3.2.2 Bulk moduli 327
Closing remarks 328
8.4 Exercises 329

9. Seismology as science 335
Preliminary remarks 335
9.1 Hypotheticodeductive formulation 336
9.1.1 Hypotheses 336
9.1.2 Deductive argumentation 338
9.2 Theory versus data 341
9.2.1 Introduction 341
9.2.2 Theory-ladenness of data 341
9.2.3 Underdetermination of theory by data 342
9.3 Bayesian inference 343
9.4 Predictions versus explanations 346
9.4.1 Introduction 346
9.4.2 Covering-law model 347
9.4.3 Inference to best explanation 349
9.5 Realistic approach versus instrumental approach 349
9.6 Coherence theory of justification 351
Closing remarks 352
9.7 Exercises 354

Appendix A On strains 359
Preliminary remarks 359
A.1 Displacement-vector field 360
A.2 Deformation-gradient tensor 363
A.3 Compatibility of deformation gradient 373
A.4 Cauchy-Green strain tensors 377
A.4.1 Right Cauchy-Green strain tensor 377
A.4.2 Left Cauchy-Green strain tensor 380
A.4.3 Relations between right and left tensors 382
A.5 Green-Lagrange and Almansi strain tensors 383
A.6 Deformation of volume element 386

A.7 Deformation of oriented surface element 388
A.8 Stretch ratio 389
A.9 Stretch and rotation tensors 390
A.9.1 Polar decomposition 390
A.9.2 Right Stretch Tensor 391
A.9.3 Left Stretch Tensor 396
A.9.4 Relations between right and left Stretch Tensors . 397
A.9.5 Local and global rotations 399
A.10 Infinitesimal strain and rotation tensors 401
A.11 Objective tensors 405
A.12 On terminology 411
Closing remarks 413

Appendix B On stresses 415

Preliminary remarks 415
B.1 Cauchy stress tensor 416
B.2 Piola-Kirchhoff stress tensors 420
B.2.1 First Piola-Kirchhoff stress tensor 420
B.2.2 Second Piola-Kirchhoff stress tensor 422
B.3 Stress tensors for infinitesimal strains 424
B.4 Objective stress tensors 425
Closing remarks 425

Appendix C On thermoelasticity 427

Preliminary remarks 427
C.1 First law of thermodynamics 428
C.2 Second law of thermodynamics 431
C.3 Heat conduction 437
C.4 Thermodynamic processes and potentials 438
C.4.1 Processes 438
C.4.2 Helmholtz free energy as potential 440
C.4.3 Internal energy as potential 440
C.4.4 Gibbs free energy as potential 441
C.4.5 Enthalpy as potential 441
C.5 Stress tensors in terms of potentials 442
C.6 Isothermal infinitesimal strains 447
C.6.1 Introduction 447
C.6.2 Single spatial dimension 448

C.6.3 Three spatial dimensions 449
Closing remarks 450

Appendix D On hyperelasticity 453

Preliminary remarks 453
D.1 Strain-energy function 454
D.2 Hyperelastic materials 455
D.2.1 Introduction 455
D.2.2 Finite strains 455
D.2.3 Infinitesimal strains 457
D.3 Material symmetry 459
D.3.1 Rotation invariance 459
D.3.2 Isotropy 461
D.3.3 Infinitesimal strain 461
Closing remarks 463

Appendix E On covariant and contravariant transformations 465

Preliminary remarks 465
E.1 Contravariant transformations 466
E.2 Covariant transformations 468
E.3 Mixed transformations 471
Closing remarks 472

Appendix F On covariant derivatives 475

Preliminary remarks 475
F.1 Metric tensor 476
F.2 Christoffel symbol 480
F.3 Covariant derivative 482
Closing remarks 489

Appendix G List of symbols 493

G.1 Mathematical relations and operations 493
G.2 Physical quantities 494
G.2.1 Greek letters 494
G.2.2 Roman letters 495

Bibliography 499

Subject Index 511

Name Index 539

About the Author 544

List of Figures

1.1 Rudzki's grave 7
1.2 Seismic trace 14

2.1 Aristotelian spacetime as product manifold 24
2.2 Spatial and material coordinates 26
2.3 Uniaxial stretch 35
2.4 Simple shear 37

3.1 Monument of Piola 55
3.2 City square named after Piola 56
3.3 Order relation of material-symmetry classes of elasticity tensors 63
3.4 Order relation of index-symmetry classes of fourth-rank tensors 67
3.5 Orientations of thirteen-parameter monoclinic tensor 71
3.6 Orientations of twelve-parameter monoclinic tensor 72
3.7 Values of coordinate coefficients of monoclinic tensor 73
3.8 Values of coordinate coefficients of monoclinic tensor 76
3.9 Orientations of six-parameter tetragonal tensor 80
3.10 Values of $2c_{2313}$ under rotations of monoclinic tensor 84
3.11 Values of $\sqrt{2}\hat{c}_{1112}$ under rotations of tetragonal tensor 86
3.12 Frobenius norm of tetragonal tensor 87
3.13 Values of $\sqrt{2}\hat{c}_{1113}$ and $\sqrt{2}\hat{c}_{1123}$ under rotations of trigonal tensor 89
3.14 Mohr's circle 96

4.1 Effective transversely isotropic Hookean solid 107
4.2 Effective orthotropic Hookean solid 109
4.3 Case of failure for product approximation in Backus's average . 139
4.4 Backus's moving average 169

4.5 Standard normal distribution as weight function 170
4.6 Backus average as convolution 172
4.7 Backus average as function of its convolution operator 173
4.8 Backus average as function of inhomogeneity 178

5.1 Vertical displacement of vibrating string 191
5.2 Green's mill 205

6.1 Vertical displacement of Rayleigh waves 233
6.2 Solubility of Scholte-wave determinant 254

7.1 Snell's law as a consequence of Huygens's principle 270
7.2 Snell's law as a consequence of Fermat's principle 271
7.3 Monument of Lagrange 273
7.4 Streets named after Lagrange 274
7.5 Stationarity of Fermat's principle 278
7.6 Direct, reflected and refracted signals 281
7.7 Head wave from plane interface 282
7.8 Head wave from curved interface 283
7.9 Traveltime and probability amplitudes 284
7.10 Street named after Beltrami 296

8.1 Gravitational effect on P-wave speed 315

9.1 Direct and refracted signals in forward and inverse problems . . 348

A.1 Reference and current configurations 360
A.2 Reference and current configurations with coinciding axes . . . 361
A.3 Inhomogeneous deformation 367
A.4 Deformation of spherical shell 372
A.5 Medal in honour of Levi-Civita and Ricci-Curbastro 374
A.6 Deformation gradient 377
A.7 Polar decomposition 392
A.8 Polar decomposition of simple shear 400
A.9 Rotation due to simple shear 401
A.10 Objectivity of orthogonal transformations 405
A.11 Objectivity of two-point second-rank tensor 406
A.12 Active and passive rotations 410

B.1 Traction vectors for Piola-Kirchhoff and Cauchy stresses 420

D.1 Material symmetry as invariance in reference configurations . . 460

E.1 Ray velocity and wavefront slowness as duals 470

F.1 Radially symmetric vector field 476
F.2 Riemann's grave 478

List of Tables

4.1 Elasticity parameters of stack of isotropic layers 168
4.2 Traveltimes through stack of layers and for equivalent medium 177

6.1 Surface, interface and guided waves 262

A.1 Strain tensors 413

B.1 Stress tensors 425

Foreword to First Edition

Foundational research is not popular: most philosophers shrink at its technicalities and most scientists feel the urge to go forward rather than retrace their steps, understand what they have gone through, analyze it critically and thereby be in a better position to plan what to do next.

Mario Bunge (1967)

Seismology, similarly to astrophysics, relies on remote sensing. Inferring material properties of the Earth's interior is akin to inferring the composition of a distant star. In both cases, scientists rely on matching theoretical predictions or explanations with observations. Also, obtaining a sample of a material from the interior of our planet might not be less difficult than obtaining a sample from the surface of a distant celestial object.

Waves and rays in seismology: Answers to unasked questions is an examination of fundamental issues in physical sciences in the sense of methodologies for quantitative descriptions of Nature. The motivation for this volume, as suggested by its subtitle, is to pose questions that are not commonly raised in applied sciences. In the following quotation, Charles Sanders Peirce (1955) warns us about the loss of an inquisitive attitude.

> Persons who know science chiefly by its results—that is to say, have no acquaintance with it as a living inquiry—are apt to acquire the notion that the universe is now entirely explained in its leading features; and that it is only here and there that the fabric of scientific knowledge betrays any rents.

The loss of an inquisitive attitude is also due to the fact that many applications of physics do not require a deep understanding of its foundations. Hence, there are questions we might never ask, even though the answers to these questions would allow us to reach alternative interpretations and understand fundamental limitations.[1]

Paraphrasing the first paragraph of Bunge (1967), this book is not an introduction to seismology, but an examination of its scientific formulation. It can be an accompaniment to a quantitative study of seismology presented by Slawinski (2015) and Bóna and Slawinski (2015). However, this book is a stand-alone examination of quantitative scientific theories. Seismology, which uses abstract mathematical tools to examine tangible phenomena, is just an insightful example.

For instance, in mathematical physics, we must be aware of potential circular arguments that result from our knowledge of the conclusions. Let us consider the following explanation of Brush (1980), which is stated in a seismological context.

> Contemporary books and articles on the earth sciences rarely acknowledge that it was Jeffreys who established the fluidity of the core. Whatever one may think about the importance of rectifying historical injustices, geophysicists should at least recognize that a scientific error is contained in the usual statement that fluidity is implied by the failure to observe S waves propagated through the core. That failure was well known before 1926, but was discounted because the evidence for rigidity seemed stronger. The most convincing argument for the fluidity of the core was not the direct one, that it must have very low rigidity because it does not transmit S waves, but rather the indirect one, that it must have a very low rigidity because the mantle has much higher rigidity than the average rigidity of the entire Earth. [...] the transmission or nontransmission of shear waves has not been a decisive test for solidity or fluidity, contrary to frequent statements in the literature.

The failure to observe S waves is the absence of evidence, not the evidence of the absence. However, arguing for an accepted result, one might be tempted to avoid the necessary rigor. In doing so, one might succumb to

[1]Also, as Marcel Proust (1992) comments insightfully,

> Nous croyons savoir exactement les choses [...] pour la simple raison que nous ne nous en soucions pas. Mais dès que nous avons le désir de savoir [...] c'est un vertigineux kaléidoscope, où nous ne distinguons plus rien.

> We think we know exactly how things stand [...] for the simple reason that we are not preoccupied by it. But as soon as we wish to know [...] things become a dizzying kaleidoscope in which we no longer distinguish anything.

logical fallacies such as confusing consistency with a deductive necessity or *epistemology* with *ontology*. Robert Hooke's not answering my letters is consistent with his being dead since 1703, but his lack of response is not a valid reason to conclude his death.[2]

To familiarize the reader with the personality of Sir Harold Jeffreys, whose name appears in the above quote and who, during his long and distinguished career, contributed greatly to the science of geophysics, I would like to quote Freeman Dyson (email comm., 2014).

> I knew Jeffreys well when I was a student at Cambridge. That was in the middle of the war when there were very few students. Jeffreys gave a course on fluid dynamics, Mondays, Wednesdays and Fridays at 9 am. He always appeared formally dressed in academic gown and gave the lecture in a formal style, although I was the entire audience. I wondered whether, if I did not appear, he would still give the lecture to an empty room. The course was a good one, and taught me most of what I know about oceans and tides. It is remarkable that he put so much effort into teaching a single student.

Understanding of technicalities associated with arguments presented in this book requires the mathematics and physics backgrounds of a senior undergraduate student of science or engineering. However, in many cases, the explanatory style of presentation allows the reader to understand the crux of issues discussed even without focussing on technicalities.

The issues discussed in this book belong to subjects presented by Slawinski (2015) and Bóna and Slawinski (2015). Footnotes with references to these two volumes are for convenience of a reader interested in further details. These volumes contain derivations that are omitted in the present book, whose focus is on evaluating the assumptions of these derivations and on interpreting their results.

This book consists of nine chapters and three appendices. Each chapter commences with *Preliminary remarks* and concludes with *Closing remarks*. In these sections, we comment on the importance of discussed topics and emphasize the role of a given chapter within the book.

In Chapter 1, we present a historical sketch of seismology in the context of its purpose and methodology. Arguments presented in this chapter are common in justifying the existence of distinct branches of a physical science whose division is associated with such criteria as pragmatic purposes and

[2]Readers interested in the life and achievements of Hooke might refer to Chapman (2005).

tools of inquiry with different levels of abstraction and generality. We conclude this chapter by describing two broad fields of seismology, wave propagation and normal modes, to narrow, for the remainder of the book, our attention on the former.

In Chapter 2, we set the background of continuum mechanics, which is the scientific realm of quantitative seismology. Therein—in anticipation of the mathematical concept of seismic waves as analogies for propagation of physical disturbances—we discuss the equations that allow us to examine motion within a continuum.

In Chapter 3, we discuss seismological applications of continuum mechanics. In particular, we discuss the theory of elasticity and Hookean solids within it. In discussing material symmetries of these solids, we examine their analogies for the Earth's materials. We emphasize limitations of these analogies, such as—in accordance with Herman's theorem—impossibility of distinguishing among material symmetries that are greater than fourfold, even though the Earth contains materials exhibiting hexagonal and octagonal symmetries.

We discuss assumptions and their consequences, such as linearity, which—limiting the study to small deformations—entails the convenience of the superposition principle, which implies, for instance, that a result caused by several events occurring together is equivalent to the sum of the results of each event happening independently. Since concepts used to examine seismic phenomena, such as the mathematical structure of continuum mechanics, models, symmetry, linearity and the superposition principle, are ubiquitous in physical sciences, the issues exemplified in this chapter are pertinent to problems of physics, in general.

In Chapter 4, to achieve a closer quantitative agreement between the aforementioned analogies and empirical studies, we discuss effective tensors and equivalent media within Hookean solids. These two concepts allow us to incorporate—in the process of modelling—limitations of both the accuracy of analogies and the resolution of data.

In Chapter 5, we consider waves in Hookean solids. By representing the Earth as such a solid, we consider equations of motion in Hookean solids and examine wave equations in an unbounded medium, where they are referred to as body waves. We proceed to discuss solutions of wave equations in such media. Also, we examine several methods to obtain a general solution of the wave equation. For each method, we obtain different results, which exemplify intrinsic limitations associated with each approach. Again—apart from their seismological meaning—these are examples of consequences of the choice of a method in solving partial differential equations, which is an important issue in mathematical physics.

In Chapter 6, we restrict the wave propagation to bounded media, which is tantamount to imposing side conditions on the wave equation. We discuss surface, guided and interface waves, which—apart from their seismological meaning—are examples of the importance of side conditions in solutions of differential equations. In a seismological context, these waves, as mathematical analogies, allow us to infer quantitative information about the layered structure of the Earth.

In Chapter 7, we discuss the variational principles of Fermat and Hamilton in the context of the causality of wave propagation. We show that no teleological interpretation of Fermat's or Hamilton's principles is required. We complete the chapter by studying the conserved quantities that appear in seismology: the conservation of energy and of linear momentum. The concept of causality is the key tenet of classical physics. The calculus of variations has important physical applications and an interpretation as a search for a global optimization rather than a consequence of causality. Seismological examples presented in this chapter allow us to discuss this tenet at different levels of explanation—from the macroscopic level of continuum mechanics to the microscopic one of quantum mechanics. Conserved quantities can be viewed as pillars of physical sciences. Again, seismological examples provide us with a subject for insightful comments.

In Chapter 8, we examine the arguments that allow seismologists to view gravitational and thermal effects as negligible when studying seismic waves. Also, we comment on ignoring the Earth's rotation in such studies.

Furthermore, we discuss another effect that might, albeit weakly, affect seismic waves: the gravitational waves. We comment on the possibility of detection of gravitational waves by seismic experiments.

Any physical theory must ignore certain concepts, and the argument of what to ignore and what not to ignore is one of the key issues of physical sciences. Seismological examples presented in this chapter allow us to discuss the criteria used in neglecting certain concepts and to evaluate the consequences of these choices.

In Chapter 9, we examine seismology as a hypotheticodeductive science in the context of continuum mechanics. While, in Chapter 1, we present subjects that belong to seismology, in Chapter 9 we examine features that qualify seismology as a science. In other words, we examine the concept of science not as a list of disciplines but as a set of criteria that makes these disciplines scientific. Also, we use seismology to comment on such concepts as realism and empiricism in natural sciences, and to discuss the issues of explanations and predictions, as well as experimental verifications and refutations.

In Appendix A (Appendix E, in the second edition), to allow quantitative formulations of seismology in the context of curvilinear coordinates, we discuss the resulting two types of transformation, and hence, present the distinction between a vector and a one-form.

In Appendix B (Appendix F, in the second edition), we discuss an extension of the partial derivative that allows us to apply its operation beyond Cartesian coordinates. The need for such an extension appears in quantitative seismology in the description of deformable materials and their deformations.

To facilitate the use of this book, we provide a list of symbols and their definitions; they appear in Appendix C (Appendix G, in the second edition). Also, there is a detailed, three-level index.

The book contains many footnotes, which play an important—but secondary—part; herein, *secondary* means that the continuity of initial reading need not be interrupted by referring to a footnote, which can be consulted upon deeper study of the material. Since this book is the third volume of the series that contains Slawinski (2015) and Bóna and Slawinski (2015), the footnotes marked by *see also* direct the reader to specific locations with further details in these two volumes. Other footnotes refer to pertinent work whose reference might broaden the reader's perspective and facilitate the understanding of material.

Let us conclude this preface by wishing that discussions of foundations of seismology might allow the reader to appreciate also the following statement of George Santayana (1983).

> A theory is not an unemotional thing. If music can be full of passion, merely by giving form to a single sense, how much more beauty or terror may not a vision be pregnant with which brings order and method into everything that we know.

Changes from First Edition

The usefulness of asymptotic,[1] *as opposed to convergent, expansions is that when we approximate a function $f(x)$ in practice, we will do so with a finite series of N terms. We thus replace the concept of fixing x and letting N tend to infinity with that of fixing N and considering asymptotic limits as $x \to x_0$. Almost invariably, this concept is of more practical utility.*

Andrew Cadle Fowler (1997)

Changes in the second edition are a result of lecturing from this book in the Department of Earth Sciences at Memorial University of Newfoundland and in the Department of Mathematics at Politecnico di Milano. A summary of these changes, and the reason for their inclusion, are given below.

The main addition consists of four new appendices, namely, Appendices A, B, C and D, which also include many new figures, examples and remarks. These appendices, accompanied by Appendices E and F, can serve—on their own—as a continuum-mechanics course on finite elasticity. As such they broaden the scope of elasticity theory commonly considered in seismology. Appendices E and F (Appendices A and B, in the first edition) are also modified for the coherency of such a course, with inclusion of new items, such as Examples E.1, E.2, F.5 and Remark F.1. This is also the case of Appendix G (Appendix C, in the first edition).

Let us describe briefly Appendices A–F. This description could be an outline of a finite-elasticity course.

In Appendix A, we consider—in terms of strain tensors—the kinematics of deformation, without limiting it to infinitesimal deformations. Implicitly, by introducing the distinction between the current and the reference configurations, we consider deformations of elastic solids, which—in their

[1] *see also*: Bóna and Slawinski (2015)

current configuration—preserve a memory of their configuration prior to deformation, which commonly serves as a reference.

In Appendix B, we examine—in terms of stress tensors—forces that are associated with deformations discussed in Appendix A. Thus, in general, these tensors correspond to finite strains.

In Appendix C, we relate the stress and strain tensors. We do so by obtaining the former as a derivative of the energy function, with respect to the deformation gradient and temperature, and, using the fact that, as shown in Appendix A, strain tensors are formulated in terms of the deformation gradient. The main subject of Appendix C is thermoelasticity, and therein we discuss the first and second laws of thermodynamics.

In Appendix D, we consider a special case of materials discussed in Appendix C, namely, the ones whose stress tensors are derived from an energy function that depends only on the deformation gradient. The thermal effect is ignored. As discussed in Chapter 8, resulting models are pertinent to a wide range of applications in engineering and seismology.

In Appendices E and F, we discuss tensorial operations, which appear in Appendices A–D, but we express them in curvilinear coordinates to bring the reader's attention to such formulations. Even though, in Appendices A–D, we use mainly Cartesian coordinates, curvilinear coordinates might be preferable in studies of deformable solids.

Let us also summarize modifications of the main text. Sections 3.2.4.14, 4.2.2.4, 4.3 and 6.3.3 are new. So are Remarks 2.3, 3.1, 3.2, 3.3, 4.10, 6.1, 6.2, 8.2, Exercises 3.1, 3.2, 3.3, 3.4, 3.9, 6.2, 8.4, and Figures 3.14, 4.3, 4.8. Furthermore, there are extensions and clarifications of Sections 2.2, 3.2.1, 3.2.3, 3.2.4, 5.2.2, 5.2.3.1, 6.2, Remark 4.3, Exercise 4.2, Figures 3.5, 3.6, 3.9. Also, there are new references, in particular to the fourth volume, Brown and Slawinski (2017), which was published after the first edition of this volume. To facilitate the referencing of material within the book, its index is split—in the second edition—into *Subject Index* and *Name Index*. In the former, the page numbers that contain defining or principal discussions of a given term are marked in bold font.

Chapter 1

Science of seismology

[T]he rich available stock of structures that pure mathematics studies with great generality not only furnishes physicists with a practically inexhaustible source of revolutionary concepts, but it also facilitates the comparison between new theories and their predecessors.[1]

Roberto Toretti (1999)

Preliminary remarks

In this chapter, we discuss disciplines that, in a variety of ways, contribute to the science of seismology, such as mathematical physics, continuum mechanics, computational science and electromagnetism. Also, we discuss branches of seismology, such as earthquake seismology, exploration seismology, global seismology and theoretical seismology. Even such a provisional, and admittedly arbitrary, list, with different methods and research strategies for each discipline or branch, indicates the multidisciplinarity of

[1]Readers interested in relations between mathematics and physics might refer to Dyson (1964, 2011), where the author writes, respectively,

> For a physicist mathematics is not just a tool by means of which phenomena can be calculated, it is the main source of concepts and principles by means of which new theories can be created.

and

> *Mathematics as metaphor* is a good slogan for birds. It means that the deepest concepts in mathematics are those which link one world of ideas with another. [Birds fly high in the air and survey broad vistas of mathematics out to the far horizon. They delight in concepts that unify our thinking and bring together diverse problems from different parts of the landscape.]

seismology. The purpose of this chapter is to examine such a list. This purpose is distinct from the purpose of Chapter 9, where we discuss features that make seismology a science. Therein, the question is not which disciplines contribute to seismology, but whether or not seismology is a science. Herein, instead of discussing criteria required by seismology to be a science, we examine common threads for seismological subjects.

We begin this chapter with a historical sketch. We conclude with a classification of seismological subjects, and select those pertaining to this book.

1.1 Purpose and methodology: Historical sketch

In this section, we examine scientific pursuits that constitute seismology and the place of seismology within a hierarchy of science. We pursue this discussion within the context of a history of seismology.

Seismology could be classified as a discipline belonging to geophysics and, in general, to Earth Sciences, due to commonality of purpose, *sensu lato*, among the geosciences. Such a classification, however, ignores distinct methodologies, even within geophysics itself. A geophysicist studying earthquakes, to whom we refer as an earthquake seismologist, invokes largely different physical principles from those used by a geophysicist examining the Earth's magnetic field. Even fewer commonalities of method appear if we compare an earthquake seismologist with a palæobotanist.

In view of commonality of method, seismology—or at least its theoretical underpinnings—could be classified as a branch of continuum mechanics. In an analogous manner, a study of the Earth's magnetic field could be classified as a branch of electromagnetism. As discussed on page 3, below, both continuum mechanics and electromagnetism are basic sciences, and both introduce an abstract concept of a field as their underlying entity.

The word *seismology* is composed of Greek terms. Its first part refers to earthquakes and the second to an enquiry. It appeared in the middle of the nineteenth century. The word *geophysics* is also composed of Greek terms. Its first part refers to the Earth and the second to the science of natural things. It appeared towards the end of the nineteenth century.

Thus, for instance, Edmond Halley, whom we could now describe as an astronomer, mathematician, physicist, meteorologist and geophysicist, would not have been described as a geophysicist in his lifetime. Philosophically, Halley would have objected to the distinction between a physicist and a geophysicist, since there are no fundamental principles of physics that distinguish physics, in general, from physics of the Earth. Removal of such a distinction—which stemmed from Aristotelian physics that distinguished between the changeable terrestrial realm and the permanent celestial realm—was one of the main achievements of the Scientific Revolution, which provided a research platform for Halley. The present distinction between geophysics and physics is based on the focus of enquiry, not on different physical principles.

A difficulty for the taxonomy of seismology and geophysics can be, at least in part, explained by considering the foundational work of Mario Bunge (1967), a philosopher of physics, according to whom neither geophysics nor seismology are *basic sciences*. According to the nomenclature of Bunge (1967), they are *based sciences*, since they rely on other disciplines, which themselves are basic, like continuum and quantum mechanics. Neither continuum nor quantum mechanics relies on other disciplines, even though only the latter enquires into fundamental properties of materials; the former is, to a certain degree, a selfsufficient black-box theory.

If we view quantitative seismology as a subject of continuum mechanics, we accept—using the terminology of Bunge (1967)—its *hypotheticodeductive* formulation. For instance, the P and S waves, which are commonly—and rather hastily—viewed as physical phenomena within the Earth, are neither exclusive properties of seismology nor—perhaps more importantly—does their concept originate within empirical studies of Earth Sciences: they belong to the mathematical field of mechanics, where they are abstract entities whose properties serve as quantitative analogies for experimental measurements.

Under Robert Hooke's seventeenth-century assumption of linear dependence between force and deformation, and the principle of balance of linear momentum—together with the stress tensor, formulated in the nineteenth century by the French mathematician Augustin-Louis Cauchy—we derive wave equations in an isotropic continuum. The form and properties of these equations were already known to Jean-Baptiste le Rond d'Alembert, a French mathematician, mechanician, physicist, philosopher and music theorist. Hooke, himself, was a natural philosopher, architect and polymath. Thus, as a subject of continuum mechanics, quantitative seismology is a hypotheticodeductive discipline, where hypotheses are Hooke's law and balance of momentum, and deduction is the mathematical derivation of the wave equation.

As described by Love (1892, Historical introduction, p. 25), based on these assumptions and using a deductive argument,

> [t]he theory of the propagation of waves in an unlimited isotropic elastic medium was first considered by Poisson, who, in his memoir of 1828, shewed that there are two kinds of waves, one waves of compression, and the other waves of distortion, and that these are propagated independently with different velocities.

Thus, in 1828, the mathematical existence of the P and S waves was recognized in the realm of Hookean solids. Strictly speaking, their existence is limited to that realm. Yet, their empirical counterparts can be considered in the physical world.

Richard Dixon Oldham, a British geologist, is credited with having first interpreted a seismogram in the context of separate arrivals of the P and S waves. Such an interpretation, however, does not imply that these waves are intrinsic physical phenomena within the Earth. As illustrated in Exercises 1.1 and 1.2, such an interpretation is tantamount to extracting, from many events that appear on a seismic record, the two events that are in a reasonable agreement with the quantitative predictions resulting from deduction. In interpreting recorded data, Oldham pursued an empirical science, but he achieved this by invoking analogies between physical observations and mathematical entities; the latter resulted from the work of Hooke, d'Alembert, Cauchy and others. In doing so, Oldham demonstrated that Hooke's assumption and subsequent derivations result in pertinent analogies for physical observations. He justified a deductive continuum-mechanics formulation in the context of experimental seismology.

To understand the meaning of analogy and interpretation in Oldham's work, let us emphasize that the P and S waves do not have an independent physical existence in a manner akin to Mars or Venus, which exist as planets, regardless of any model of the solar system. Again, as described by Love (1892, Historical introduction, pp. 25-26), following a deductive argument, in 1842,

> Green considered the propagation of plane waves in an æolotropic medium and concluded that there are three kinds of waves which are propagated with different velocities.

As in the case of P and S waves, an interpretation of seismological data in terms of three kinds of waves, which we denote by qP, qS_1 and qS_2, where q stands for "quasi",[2] was not feasible, from a pragmatic viewpoint,

[2] *see also*: Slawinski (2015, Section 9.2.3: *Displacement directions*)

until sufficiently accurate seismograms were developed. However, fundamentally, such an interpretation was impossible prior to the recognition of the mathematical existence of these waves.

The requirement of a mathematical model to precede a specific physical interpretation can be illustrated as follows. In general, each of the three waves in an anisotropic solid exhibits a different velocity than the speed of either of the two waves in an isotropic one. Thus—depending on the choice of the model—a given event on the seismogram is associated with a wave from the model under consideration. In spite of this apparent inconsistency—where the same event can be associated with two distinct waves, say, P and qP—both models lend themselves to interpretations of a seismogram.

With a certain abuse of terminology, we could say that the ontology of a particular seismic wave is derived from its epistemology: the physical existence of a given wave is a function of the method for its enquiry. In other words, among a plethora of mechanical disturbances recorded by a seismograph, one model allows us to identify disturbances whose physical properties exhibit analogies that are pertinent to the behaviour of waves contained within that model, and another model allows us to identify the same disturbances with different waves, which are contained within the second model. Even though one of the models might be more accurate, it cannot be endowed with more physical reality, since this accuracy is a quality of a model, which remains in the abstract realm. The increase of accuracy means only that a given model might provide a better analogy. However, as discussed in Chapter 4, it is not necessarily so; in particular, if data are contaminated with measurement errors.

In the context of various pursuits pertaining to the Earth, it is of interest to note that, based on his seismic interpretation, Oldham is also credited with the first convincing inference that the Earth has a central core, which had been already suggested by Halley as a hypothesis to account for magnetic measurements (Brush, 1980).

To examine seismograms, seismologists require an experimental apparatus. In 1906, Boris Borisovich Golitsyn, a Russian physicist, invented the first electromagnetic seismograph. Historically, this invention would not have been possible prior to the second half of the nineteenth century, since it required theoretical concepts of electromagnetism formulated by James Clerk Maxwell, a Scottish mathematical physicist; this is another example of the multidisciplinary aspect of geophysics, and—of course—interrelations among scientific pursuits in general.

Relating observations to theoretical predictions is a *modus operandi* of seismology. Thus, to facilitate measurements of disturbances propagating in the Earth, a modified seismograph—sensitive to horizontal displacements, which are commonly observed on the Earth's surface following an earthquake—was designed a couple of years after Golitsyn's invention by Emil Johann Wiechert, a German mathematician and geophysicist. It was used in 1909 by Andrija Mohorovičić, a Croatian meteorologist and seismologist, best known for the discovery of the boundary between the Earth's crust and its mantle. This boundary, known in his honour as the *Moho discontinuity*, is one of the key properties of the layered Earth.

Wiechert made many contributions to both theoretical physics and geophysics. He was among the first to discover the electron and to present a verifiable model of a layered structure of the Earth. Even though his doctoral degree was in mathematics, he became, at the University of Göttingen, one of the first professors of geophysics, which at that time was becoming a distinct discipline; this does not mean that it became, or could ever become, an independent science. It means only that the questions posed within its realm—either in terms of answers themselves or methods to reach them—became sufficiently complex to justify a substantial engagement of scientists involved.

The first Chair of Geophysics was established for Maurycy Pius Rudzki. He held this chair at the Jagiellonian University in Kraków, where, in 1895, he established the Institute of Geophysics, and where he died in 1916, as commemorated by his gravestone shown in Figure 1.1. In contrast to fading inscriptions on this gravestone, his research subject continues.

Rudzki's research specialty was seismic anisotropy (Rudzki, 1911, 1912, 1913, 2000, 2003), which—within the realm of continuum mechanics—required a formulation of Hooke's law in three dimensions for which he relied on results provided by Woldemar Voigt (1910), a German physicist. The work on anisotropic Hookean solids resulted in derivation of the qP, qS_1 and qS_2 waves, in the realm of differential equations stemming from the balance of linear momentum. Half-a-century later, these waves were convincingly associated with three separate arrivals on a seismogram, thus justifying—in a manner analogous to Oldham's work, albeit with more sensitive equipment—a continuum-mechanics formulation in the context of experimental seismology.

A commonality of terrestrial interests, under the auspices of geophysics, allows for inferences about our planet by distinct approaches. An example of such a result is the confirmation of the liquid outer core, which can be

photo: Elena Patarini

Fig. 1.1: Inscription of Maurycy Pius Rudzki's grave at the Rakowicki cemetery in Kraków, Poland; in translation:

Dr Maurycy Prus [sic] Rudzki of the Prawdzic Coat of Arms, Professor of the Jagiellonian University, lived 54 years, died 20 July 1916

The fading inscriptions on the gravestone are at odds with the continuous importance of Rudzki's work.

independently inferred from geodynamics, as an interpretation of wobbles of the Earth, and from global seismology, as an interpretation of the S-wave shadow zone on its surface. Also, the commonality of terrestrial interests can motivate research within distinct fields.

Plate tectonics is an example of a coherent theory capable of predicting, retrodicting and explaining many observed properties and phenomena, including shapes of continents, distribution and formation of mountain chains, heat flow, earthquakes and volcanism. It relies on many *noncompeting sciences*, such as paleontology and fluid mechanics, to establish a convincing scenario of processes and their results.

Agreements among noncompeting sciences are an important criterion for the validity of a broad theory. Consistent inferences from noncompeting sciences support ontological claims, such as the liquid outer core inferred from both geodynamics and global seismology. On the other hand, we accept that *competing sciences*, such as classical and quantum mechanics, disagree on their predictions or explanations.

In the context of plate tectonics, the contribution of several branches of noncompeting science might have its origins in a single hypothesis of Francis Bacon, an English philosopher. While examining world maps—whose accuracy, motivated by navigation, increased significantly at the beginning of the seventeenth century—he commented on the fit of Africa and America, and speculated that these two continents might have been connected. A reconstruction, with no Atlantic Ocean, was published in 1858 by a geographer, Antonio Snider-Pellegrini. These were hypotheses based on shapes of continents, akin to puzzle arrangements. They motivated comparison of rocks on either side of the Atlantic; prominent among these investigators was Alexander du Toit, a geologist from South Africa, and an early supporter of Alfred Wegener's theory of *continental drift*. Wegener, a German polar researcher, known during his lifetime for his achievements in meteorology, is remembered today for advancing, in 1912, the concept of continental drift, which was a precursor to the present theory of plate tectonics; he compiled many geological, biological and glaciological observations, which he presented in his book (Wegener, 1929). These observations, even though suggestive according to the criteria of descriptive sciences, were insufficient to formulate a physical theory.

From a physics point of view, a completion of the theory—beyond convincing symptoms of fitting shapes, consistent geology and an insightful interpretation of paleomagnetic strips on the ocean floor of the Atlantic—required an explanatory mechanism for the movement of continents. This required further research by geophysicists, and, presently, such a mechanism is represented by a deep mantle thermal convection system that moves lithospheric plates upon deforming plastic asthenosphere.

The commonality of purpose that allows the formulation of broad theories such as plate tectonics does not imply a commonality of approaches, methodologies or research strategies. The variety of possible approaches, together with distinct areas of competence within each approach, produces a richness of intellectual resources. Direct collaboration among scientists across different subjects is hardly possible; for instance, the intricacies of the *Curie point*, which is crucial in interpreting the magnetic strips, might be foreign to a geologist for whom subtleties of stratigraphic knowledge are required to correlate lithological information on both sides of the Atlantic. Hence, many scientific issues that arise across disciplines need to be discussed at the level pertaining to results, independent of methodologies. A geodynamicist and a seismologist would agree on the existence of the liquid outer core. However, their common understanding of methods involved in such an inference, which for either of them is reached by different means, might be lacunary. This is hardly a weakness, since an agreement reached by independent means strengthens the hypothesis in question.

Commonality of purpose, such as understanding mechanical processes within, and properties of, the Earth, motivates fruitful interactions among seismologists and creation of common scientific meetings, institutes, departments. Their individual work, however, can belong to distinct realms, such as mathematics to derive equations of motion in complex media, mechanics to consider dissipation processes, statistics to examine reliability of data-based inferences, computer science to perform complicated computations, geology to associate theoretical predictions with observations, etc. Thus, even in what might appear as a narrow field, commonality of purpose does not imply commonality of methodology or competence.

For instance, the book by Sir Harold Jeffreys (1924), which he dedicated to Sir George Howard Darwin as "the founder of modern cosmogony and geophysics", is readable only by seismologists fluent in the language of mathematics. Also, only seismologists with an understanding of such a language can share Jeffreys's interest in Bayesian analysis to evaluate seismic measurements. The same is true for aspects of the work of Keith Edward Bullen, a New-Zealand-born professor of applied mathematics at the University of Sydney in Australia (Bullen and Bolt, 1987), noted for his seismological interpretation of the deep structure of the Earth's mantle and core. As the reader can see, this book also relies on a mathematical literacy that might not be shared by all seismologists.

Another example of commonality of purpose is the extraction of natural resources. For instance, both Ludger Mintrop and Bruno Pontecorvo could be described as applied geophysicists; the former could be, even more specifically, described as an applied seismologist. In 1908, in Göttingen, Mintrop participated in creating the first artificial earthquake, which consisted of dropping a four-thousand-kilogram weight from a height of fourteen metres. During World War I, he used a portable seismograph to locate the positions of heavy artillery. After the war, Mintrop moved to the United

States, where—setting off explosions and measuring traveltimes between sources and receivers—he estimated mechanical properties of the subsurface by comparing field measurements to mathematical predictions based on assumptions of continuum mechanics. Pontecorvo was an Italian nuclear physicist and an early assistant of Enrico Fermi. After World War II, while working at an oil company in Tulsa, Oklahoma, he developed a technology and an instrument for well logging based on the properties of neutrons. This technology is the first application of slow neutrons discovered by *I ragazzi di Via Panisperna*[3]: a group of young scientists led by Fermi, which included Pontecorvo. Today, the methods formulated by Mintrop and by Pontecorvo, stemming from distinct branches of physics, are standard techniques in exploration geophysics.

In conclusion, let us define geophysics as a subject that combines theories and methods that are pertinent to physical properties of, and processes within, the Earth, and seismology as a branch of geophysics whose theories and methods focus on mechanical properties and processes within the Earth. Neither of these subjects is a discipline independent from other sciences. Theoretical advancement of these subjects is possible only by building theories with concepts originating from physics, with the methodological scaffolding offered by mathematics, statistics and computational sciences. Empirical examination of resulting theories is possible only with observations, which in turn rely on experimental equipment and subsequent data analysis and interpretation, each of them relying on competence within different scientific or technological pursuits.

The practice of science is a social activity. Thus, it is a choice of its practitioners to see the multifaceted aspects of seismology either as an intellectual asset or as an inconvenience preventing direct interaction among scientists involved in its different aspects.

1.2 Classification

Let us comment on the classification of seismology as it pertains to this book. The criteria for this classification are based on methodology, which in turn depends on mathematical subjects and assumptions invoked within a given method. Methods of quantitative seismology are methods of continuum mechanics, whose purpose is a description of mechanical properties of deformable materials. The origin of such materials, which could be terrestrial, jovian or other, does not affect this classification.

[3] *The boys of Panisperna street*

Broadly speaking, in view of mathematical methods and assumptions, we might divide seismology into two branches: examination of waves propagating within the Earth and examination of vibrations of the entire planet. The boundary between these two branches is not sharp and depends on assumptions used in either of them. If we consider a nonrotating Earth model, these branches are, respectively, examinations of the propagating and standing waves.

Both wave propagation and planetary vibrations are treated by Dahlen and Tromp (1998). Therein, vibrations are presented as the study of *normal modes* for the nonrotating and rotating hydrostatic Earth models.

Herein, we restrict our discussions to wave propagation. Propagation is presented as the study of equations of motion, where—unlike for vibrations of the entire planet, and as discussed in Section 8.1—we can assume that the effect of gravitation is negligible. This formulation leads to the wave equation in the context of *body waves*, *surface waves*, *guided waves* and *interface waves*; these distinctions depend on the side conditions of that equation.

Perhaps the classifications that are based on methodological criteria are more objective than their counterparts, such as *global seismology* versus *exploration seismology* or *theoretical seismology* versus *applied seismology*. Also, their criteria might be flawed, since, as stated by Kurt Lewin (1952),

> [t]here is nothing more practical than a good theory.

Be that as it may, distinctions between subjects are a matter of emphasis. Geophysics can be viewed as a study of physics to understand Earth and terrestrial planets or as a study of terrestrial properties to understand underlying physics. To avoid hasty classifications, it might be worth noting that one could object to limiting the scope to terrestrial planets, since there is an interest in *helioseismology*: the study of waves in the Sun. There could also be *zenoseismology*: the study of waves in Jupiter, and in other nonterrestrial planets.

Closing remarks

In seismology we use remote measurements—such as geophones on the surface responding to disturbances within the interior—so the inferences from the measurements to the properties of the interior must be mediated by a theory. For seismology, this theory is continuum mechanics, together with its mathematical structure and methods.

For better or worse, seismologists are confined to having an intermediate step between measurements and information about properties of the subsurface. In inferring material properties from mathematical models, the best we can do is to achieve consistency between model predictions and observations. As emphasized by Tarantola (2006), we can infer properties from models by induction, not deduction, even though the models themselves are deductive products of underlying assumptions.[4]

Such models—with their assumptions, properties and limitations—constitute the topic of this book.

In discussing such models, which—broadly speaking—is a comparison of observations with theoretical predictions, we focus our attention on formulation of these predictions, and examine consequences of including or neglecting certain quantities, such as the effect of gravitation on seismic waves. We do not discuss properties of features within our planet, such as, say, anisotropy of the inner core, nor do we discuss subtleties of experimental techniques to justify the inclusion of particular model parameters. This book is an examination of methodology within quantitative seismology, which—in the last chapter—allows us to discuss criteria that classify it as a scientific pursuit.

This book might help the reader in addressing issues of *scientific truth*. Does a quantitative accuracy of predictions or *retrodictions* of mathematical models—compared to experiments—suggest that these models represent the physical essence of objects or phenomena? In what manner is a continuum-mechanics model that provides us with such an accuracy realistic? After all, it assumes a continuous medium contrary to discreteness of atomic structure, whose existence—according to other scientific enquiries—we accept as a scientific truth. For instance, why would we accept the existence of the outer core and its liquid state, which are inferred from continuum-mechanics modelling, not only as a mathematical model to account for observed mechanical behaviours of our planet but also as a scientific truth?

All in all, scientific enquiry—even though its path abounds in confusions, errors and prejudices, as well as in biases stemming from hope or

[4]Also, as emphasized by Popper (1963, Chapter 1, Section III, Science: Conjectures and refutations),

> No scientific theory can ever be deduced from observation statements,
> or be described as a truth-function of observation statements.

fear—proceeds towards a truth. Arguments such as that we look for truth because we do not know it, and hence, if we find it, we cannot recognize it are *sophisms*, which are clever but fallacious and deceptive statements. Objections to such statements require criteria for acceptance of a scientific truth, which is a subject of epistemology, whose vastness is symptomatic of the variety of issues to be addressed. The variety and complexity of these issues necessarily foster an opposition in terms of sophistic reasoning.

Scientific revolutions, as described by Thomas Kuhn (1996), and their *paradigm shifts*, as well as—in general—the history of science and its changing foundations should render us dubious about ever reaching the final answers in search of scientific truths or criteria for their acceptance. However, they should not prevent us from striving towards a progressive enhancement of our understanding.

Be that as it may, in addressing these issues, we shall heed Lord Byron's statement,

> [t]hose who will not reason, are bigots, those who cannot, are fools, and those who dare not, are slaves,

while remaining aware of its rhetorical aspect, since—even though Byron was a fellow of the Royal Society, the oldest society for science still in existence—his criteria of reason and enquiry were representative of the romantic ethos and its opposition to science. Science requires rational thinking but rational thinking, in general, need not be scientific.

Also, while inspired by Byron's poetic phrase, we need to be mindful of the warning of Karl Popper (1963, Introduction: On the sources of knowledge and of ignorance), an Austrian-British philosopher famous for his *critical rationalism*,

> [t]he theory that truth is manifest—that it is there for everyone to see, if only he wants to see it—this theory is the basis of almost every kind of fanaticism. For only the most depraved wickedness can refuse to see the manifest truth; only those who have reason to fear truth conspire to suppress it.

1.3 Exercises

Exercise 1.1. Given the seismogram trace shown in Figure 1.2, discuss its interpretation in terms of Hookean solids.

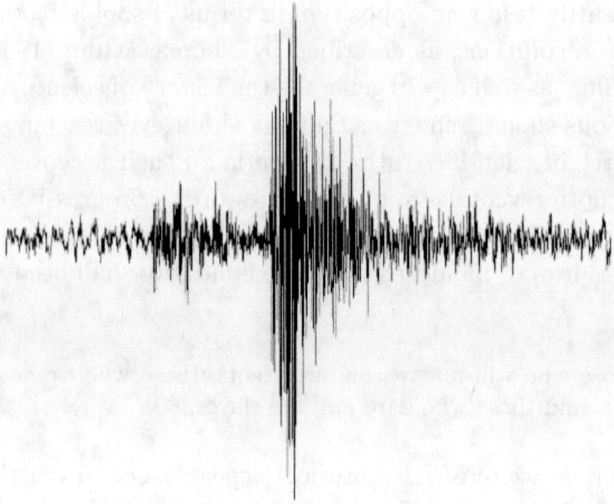

Fig. 1.2: Seismic trace recorded at a particular location. The horizontal dimension stands for time and the vertical dimension for amplitudes of disturbances recorded at that location. Depending on the theory invoked to mediate between observations and retrodictions of physical phenomena, this trace can be interpreted in terms of a variety of waves.

Solution. Figure 1.2 is a trace recorded on an earthquake seismograph. This trace represents the disturbance measured in a particular location on the Earth's surface. We interpret this disturbance as a result of a distant earthquake whose displacements propagate from its focus.

To interpret this seismic record, we must invoke a theoretical model. Herein, we assume that the Earth can be modelled by a Hookean solid, which is a mathematical object, whose properties can be used as quantitative analogies for a mechanical behaviour.

If we assume an isotropic inhomogeneous Hookean solid, we associate events on this seismic trace with the body waves, P and S,[5,6] and with

[5] *see also*: Slawinski (2015, Section 6.1.2)
[6] *see also*: Slawinski (2015, Section 6.1.3)

surface waves, which are discussed in Section 6.2, below. Assuming that these three types of waves are generated at the same instant and at the same location, we can express their traveltimes—from their focus at 0 to the measurement point at X—as

$$t_P = \int_0^X \frac{\mathrm{d}\xi}{v_P(\xi)}, \qquad t_S = \int_0^X \frac{\mathrm{d}\zeta}{v_S(\zeta)}, \qquad t_G = \int_0^X \frac{\mathrm{d}\eta}{v_G(\eta)},$$

where—in accordance with Fermat's principle,[7] formulated in 1662 by Pierre de Fermat—they follow different trajectories due to their different speeds of propagation;[8] herein, a dependence of $v(\cdot)$ on position denotes inhomogeneity. Given $v(\cdot)$, each trajectory could be obtain by solving *Lagrange's ray equations*,[9] whose solutions are $\xi(\tau)$, $\zeta(\tau)$ and $\eta(\tau)$, where τ is a parameter along a curve.

If we assume an anisotropic inhomogeneous Hookean solid, we associate events—on the same seismic trace—with the qP waves, qS_1 and qS_2 waves,[10] and surface waves. We can express their four traveltimes as

$$t_{qP} = \int_0^X \frac{\mathrm{d}\xi}{V_{qP}(\xi,\xi')}, \quad t_{qS1} = \int_0^X \frac{\mathrm{d}\zeta}{V_{qS_1}(\zeta,\zeta')}, \quad t_{qS2} = \int_0^X \frac{\mathrm{d}\nu}{V_{qS_2}(\nu,\nu')}, \quad t_G = \int_0^X \frac{\mathrm{d}\eta}{v_G(\eta)},$$

where $'$ denotes the directional dependence of propagation speed, which is anisotropy. V stands for the *ray velocity*, which is the signal velocity in anisotropic media; surface waves, which are constrained to a single surface are assumed to be isotropic.

In either case—within the context of Hookean solids—we might interpret the first arrivals, recorded in Figure 1.2, as P or qP waves, the second arrivals as S or qS_1 and qS_2 waves, followed by the arrivals of the surface waves, whose amplitude is the largest, which is the reason for most earthquake damage being due to surface waves. Also, it follows from mathematical requirements within Hookean solids that $v_P > v_S > v_G$, as shown—for body waves—in Exercise 5.3, and as discussed—for surface wave—in Section 6.2, below. In the anisotropic case, the qP waves are faster than both

[7] *see also*: Slawinski (2015, Section 13.1)
[8] *see also*: Slawinski (2015, Sections 6.1.2 and 6.1.3)
[9] *see also*: Slawinski (2015, Chapter 11)
[10] *see also*: Slawinski (2015, Section 9.2.3: *Phase velocities in arbitrary directions*)

qS waves, for all directions;[11,12] however, the superiority of speed for the qS_1 versus qS_2 waves depends on direction of propagation.[13]

For either case, the information consists of the measurement location, X, and the arrival time of a given wave depends on the chosen model. The arrival-time information, however, is subject to interpretation, since the measurements themselves are part of the physical world and do not belong to the realm of Hookean solids. Consequently, this information has to be extracted from a plethora of physical events and their ambient noise, which are illustrated in Figure 1.2.

The unknowns contain the location of the earthquake and the time of its occurrence, as well as mechanical properties of the medium through which disturbances propagate and which determine speeds of their propagations. Hence, the trajectory of each wave is unknown, even though—given these properties—they could be calculated by solving Lagrange's ray equations, which are statements of Fermat's principle.

A simplification in the formulation of the problem could be achieved by assuming that the model of the Earth be spherically symmetric; the model consists of concentric layers. Consequently, by Noether's theorem, each seismic ray exhibits a conserved quantity along the ray, which is unique to that ray, and is referred to as its *ray parameter*. This parameter is

$$p = \frac{r \sin \vartheta}{v(\vartheta, r)}, \tag{1.1}$$

where v is the wavefront propagation speed, which is a function of the radius, r, and of the propagation direction; ϑ is the angle between the wavefront normal and the radius. Ray parameters are discussed in Section 7.4.2, below.[14]

If we also assume that the concentric layers are isotropic, the problem is simplified further. The wavefront propagation speed becomes equivalent to ray velocity and the wavefront angle to ray angle,[15] which are entities related directly to the traveltime.

Exercise 1.2. Assume a homogeneous Hookean solid, where $v_P = 5000$ and $v_S = 3000$. Given that $t_S - t_P = 60$, calculate the distance to the earthquake focus. All quantities are in the *SI* units.

[11] *see also*: Slawinski (2015, Section 9.3.2)

[12] Readers interested in an exception to that statement might refer to Bucataru and Slawinski (2009b).

[13] *see also*: Slawinski (2015, Section 9.3.3)

[14] *see also*: Slawinski (2015, Chapter 14)

[15] *see also*: Slawinski (2015, Section 8.4)

Solution. Since the medium is homogeneous, both P-wave and S-wave trajectories are the straight line connecting the focus with the measurement point; in general, the distance travelled by a signal is greater than the distance between the focus and the measurement point. Herein, traveltimes are $t_P = d/v_P$ and $t_S = d/v_S$, where d stands for distance between these two locations. Hence, $t_S - t_P := \Delta t = d(1/v_S - 1/v_P)$. Solving for d, we get

$$d = \frac{v_P\, v_S}{v_P - v_S}\, \Delta t\,. \tag{1.2}$$

Thus, the distance is 450 kilometres.

Remark 1.1. Comparing Exercises 1.1 and 1.2, we note several simplifying assumptions in the latter. These simplifications might render the accuracy of its results insufficient for a satisfactory retrodiction of the examined earthquake. On the other hand, the former—even though unavoidably relying on mathematical simplifications, such as the concept of a Hookean solid—might require input information that could be unavailable with sufficient accuracy, and hence, the precision of its results would not be justifiable.

In facing such *conundra*, it might be comforting to quote Richard Feynman from one of his last interviews.

> I can live with doubt, and uncertainty, and not knowing. I think it's much more interesting to live not knowing than to have answers which might be wrong. I have approximate answers, and possible beliefs, and different degrees of certainty about different things, but I'm not absolutely sure of anything.

Chapter 2

Seismology and continuum mechanics

Just as in all experimental sciences, theoretical and observational aspects of seismology must be considered. The first are based on the principles of mechanics of continuous media with the assumption that the Earth is an imperfectly elastic body in which vibrations are produced by earthquakes.[1]

Agustín Udías (1999)

Preliminary remarks

Continuum mechanics, unlike quantum mechanics, does not attempt to investigate the microstructure of matter; its formulation is not rooted in constituents of matter. To study quantitatively behaviours, particularly deformations, of materials without dealing with individual constituents, continuum mechanics—similarly to James Clerk Maxwell's electromagnetic-field theory—invokes the concept of *continuum*, which is a smooth *field*. Necessarily, by doing so, it limits its realm of enquiry to macroscopic phenomena, which means also that properties postulated for, and inferred from, continuum mechanics are close to our everyday experience. We experience water as a continuum, not as a material composed of discrete particles or of H_2O molecules.

Furthermore, macroscopic properties are not a simple consequence of the microscopic ones, even though they *emerge* from them. For instance, as stated by Holland (2014, Chapter 6),

[1]A similar statement is made by Dahlen and Tromp (1998, p. 25).

> In seismology and geodynamics we regard the Earth as a *continuum*; that is, as a continuous distribution of matter which can interact through both short-range and long-range forces.

> the characteristic of "wetness" cannot be reasonably assigned to individual H_2O molecules, so we see that the wetness of water is not obtained by summing up the wetness of the constituent molecules—wetness *emerges* from the interaction between molecules.

Emergence,[2] which is defined by Bunge (2014) as a qualitative novelty and a property of systems, appears also in such properties as rigidity and material symmetry. Rigidity of an iron bar is not the sum or average of rigidities of its *Fe* atoms, which by themselves might not even exhibit rigidity as a defined physical property. *Material symmetry* of a salt dome is not simply inherited from the symmetry of its individual crystals; both material and crystal symmetries are well-established properties, but they are distinct. The latter is the symmetry of a lattice; the former is the symmetry of a tensor: the invariance of its components under rotations of a coordinate system.

Historically, particle mechanics was developed by Isaac Newton in the second half of the seventeenth century, prior to continuum mechanics, whose development is closely associated with the work of Augustin-Louis Cauchy in the first half of the nineteenth century, with important advancements in the second half of the twentieth century, as described by Maugin (2013, 2014).

This historical sequence is also followed heuristically, since continuum mechanics is commonly introduced as a means for dealing with a large number of particles by describing their ensemble by such properties as viscosity, rigidity and compressibility. However, as stated by Noll (1974a) in the quote on page 181, such a formulation cannot be achieved deductively, even though it might be used as a motivation for a didactic introduction.

Furthermore, there are aspects of the theory of continuum mechanics that have no counterparts in particle mechanics. *Constitutive equations*, such as *Hooke's law*, which are analogous to *equations of state* in such field theories as electromagnetism, have no counterparts in the theory of particle mechanics. The necessity for these equations is a consequence of considering macroscopic properties only, whose general laws, such as *conservation of mass*, do not form a system that suffices to obtain solutions to be compared with physical observations. This system must be supplemented by

[2]Readers interested in a discussion of emergence versus *reduction* might refer to Popper (1979, Chapter 8).

descriptions of material responses known as *constitutive laws*, which are expressed by constitutive equations.[3]

Thus, continuum mechanics is divided into two complementary parts: the general theory, which is valid for all materials and consists of field equations, and particular theories, which refer to specific materials, such as *theory of elasticity* and *fluid mechanics*. Particular theories are distinguished from each other by their constitutive equations, which need not agree among each other. For instance, for elastic solids, such an equation can be solely a function of deformation and, for a liquid, solely a function of rate of deformation.

We begin this chapter, in which we focus on the general theory, with a comment on the axiomatic formulation and with a discussion on deformations within continua.[4] In Chapter 3, we focus on a particular theory: the theory of elasticity.

2.1 On axiomatic formulation

To interpret seismological observations, we invoke a theory that allows us to both quantify observable properties of, and infer further information about, the Earth. Such a theory establishes the concept of a dynamical process for seismic disturbances traveling through the Earth and provides quantitative predictions of measurable processes.

This theory is continuum mechanics, which is a mathematical formulation whose essence is a description of behaviours of a body represented by continuous media subject to fundamental laws, such as the balance of linear momentum. These laws are axioms and hypotheses from which behaviours of continua are deduced.

Hence, as stated by Bunge (1967) and mentioned in Chapter 1, continuum mechanics is a hypotheticodeductive theory. Its axiomatic approach[5] allows us to examine the veracity of conclusions and to avoid paradoxical statements, which might pass unnoticed in a heuristic description, wherein they might appeal misleadingly to our intuition.

As stated by Coleman and Noll (1974), in an axiomatic approach to continuum mechanics, a dynamical process is a triad consisting of a body, a motion and a system of forces acting within it.

[3] *see also*: Slawinski (2015, Chapters 2 and 3)

[4] Readers interested in a primer to continuum mechanics might refer to Epstein (2010, Appendix A) and Roberts (1994).

[5] *see also*: Slawinski (2015, Section 1.2)

> The basic physical concepts of classical continuum mechanics are *body*, *configuration* of a body, and *force system* acting on a body. [...] Once these concepts are made precise one can proceed to the statement of *general principles*, such as the principle of objectivity or the law of balance of linear momentum, and the statement of specific *constitutive assumptions* [...] While the general principles are the same for all work in classical continuum mechanics, the constitutive assumptions vary with the application in mind and serve to define the *material* under consideration.

In this book, a material body is represented by a *Hookean solid*. Motion, which is a family of configurations of a body parametrized by time, results—within Hookean solids—in the *elastodynamic equation*. A system of forces, which is composed of the contact and body forces, is a family of vector-value functions brought to the elastodynamic equation by *traction* and gravitation.

The key purpose of introducing the concept of a continuum to mechanics is to study quantitatively deformations of materials. The very assumption that renders this study feasible is to view materials not as composed of molecules, atoms or subatomic particles, but as being continuous.

A consequence of this assumption is a convenient use of calculus together with the explicit limitation of a theory to macroscopic properties of a material, since its bulk properties are assumed to represent this material at any scale. However, even though ignoring multitudes of interactions among discrete constituents of matter results in simplifications, the resulting theory necessitates—in part, by introduction of deformability—a sophisticated mathematical apparatus.

In view of the apparent contradiction with accepted physics, we emphasize that this simplifying assumption does not imply any ontological property of Nature. It is a convenient epistemological property of the formulation for an enquiry into these properties.

We do not, as Aristotle did, argue against the indivisibles of Democritus; nor do we argue for them. We accept that the reality of continuum mechanics resides in the mathematical, not the physical, realm. We use continuum mechanics, which is a *semiphenomenological*, not a *representational*, theory of matter. It allows us to describe phenomena associated with matter without examining its composition. The term *semi* implies that this theory contains entities that are not observable, such as the stress tensor, discussed below.[6]

2.2 Kinematic descriptions

2.2.1 *Spacetime*

[7]Let us consider the kinematics within continua, which takes place in the spacetime. Let us examine *Galilean spacetime* and its particular case: *Aristotelian spacetime*. Also, to allow a quantitative description of geophysical phenomena, we introduce the concept of a manifold, which is a generalization of a surface to higher dimensions.

Physical phenomena occur in spacetime, which is represented mathematically by a *differentiable manifold*. Unlike classical particle mechanics, which is constrained to rigid-body motions, continuum mechanics requires also a *material manifold*. Herein, we discuss the concept of spacetime. In Section 2.2.2, below, we introduce the spatial and material manifolds, and examine the relation between them.

A common view of spacetime is that a *physical event* occurs at an identifiable time and in an identifiable location. Remaining in classical physics, which insists on distinguishing between time and space, geometrically, we consider time to be one-dimensional and space to be three-dimensional. Such geometry leads us to a four-dimensional manifold.

Let us discuss three common spacetime structures that result in three distinct manifolds. Given two physical events in *Aristotelian spacetime*, we can assert whether or not they occur simultaneously and identify their locations. Such an assertion requires *absolute time* and *absolute space*. In mathematical language, we must have a *product manifold*, $t \times S$, where t stands for time, S for space and $\times$ is the *Cartesian product*, as illustrated in Figure 2.1.

A key example of the Cartesian product is the *Cartesian plane* in analytic geometry, where—to each point—René Descartes, a French philosopher, mathematician and scientist, assigned a pair of real numbers, $\mathbb{R}$,

[6]Readers interested in a semantical and methodological status of physical entities might refer to Bunge (1967, Chapter 2.3), which begins with the statement that

> [n]ot all the symbols occurring in physics refer to some physical object.

[7]This section stems from discussions and collaborations with Marcelo Epstein, Albert Tarantola and Jeroen Tromp at Princeton University.

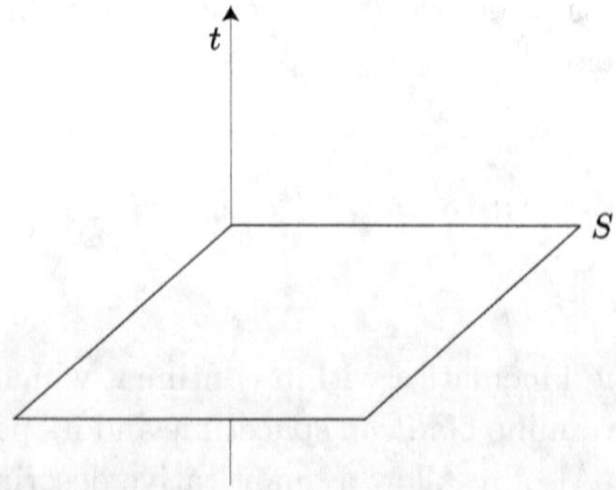

Fig. 2.1: An illustration of the physical three-dimensional space, S, and time, t, which is Aristotelian spacetime. Each event can be identified by Cartesian product, $t \times S$. Also, this diagram can be a local representation of Galilean spacetime or an inertial frame. *Modified from Penrose (2004)*

which are coordinates of that point. The set of all such pairs, $\mathbb{R} \times \mathbb{R}$, is assigned to the set of all points in the plane. Each pair identifies unambiguously the location of a point on the plane, and *vice versa*, the location of each point is unambiguously identified.

In Einstein's relativity, the spacetime structure is simplified to a single four-dimensional manifold, without distinguishing between space and time. In other words, time is not endowed with a particular status, and hence neither simultaneity nor location can be asserted unambiguously.

Unlike Aristotelian spacetime, its generalization, namely, Galilean spacetime, does not accept absolute space, since—according to Newton's first law—the natural state of an object is uniform motion, not rest. Points in space do not retain their identity, since we cannot assert whether or not a given material point is now at the same location as before. There is no immovable background space in which physical events take place. No physical experiment allows us to distinguish between moving uniformly and not moving at all. Herein lies the importance of an *inertial reference frame*, within which we describe geophysical phenomena. However, unlike in Einsteinian spacetime, in Galilean spacetime, one assumes absolute time.

Commonly, seismological studies take place in a given inertial reference frame, which is tantamount to Aristotelian spacetime; the physical events of interest take place within a product manifold of time and space. Mathematical operations, such as differentiation, are associated with given inertial frames.

2.2.2 *Motion*

An intrinsic issue of continuum mechanics is the fact that any description of motion within it must address the lack of discrete elements, which is the essence of the notion of a continuum. On the one hand, it allows us—due to resulting smoothness—to use calculus in describing physical phenomena therein. On the other hand, it results in a description of motion that is more complicated than the one for discrete objects.

Let us begin with a description of motion in continuum mechanics that is analogous to the one in particle mechanics. To formulate an analogy of discrete points within a continuum, we introduce the concept of *material points*, $\mathcal{M}$. We assume that these points are identifiable individually and can be followed during deformations. One may think of them as chalk marks on a deforming solid. Quantitative descriptions of a deforming solid—as seen from a chalk-mark viewpoint—require methods that belong to the calculus on manifolds, discussed briefly in Appendix F. We can view a *material body* itself as a set of material points.

The *motion* within a continuous medium is described completely if the location, $\mathcal{S}$, of a material point, $\mathcal{M}$, is given for any time, t, by

$$\mathcal{S} = \phi(\mathcal{M}, t)\,, \tag{2.1}$$

where ϕ is a continuous map. This is akin to considering the motion within a particular inertial frame, which is endowed with absolute space and time. As stated by Noll (1974a), the concept of a reference frame, in general, is

> a set of objects whose mutual distances change very little in time, like the walls of a laboratory, the fixed stars or the wooden horses on the merry-go-around.

Herein, the inertial reference frame is represented by the walls and by the stars, and the noninertial one by the merry-go-around. Following these analogies, one would consider our planet, which rotates in a manner akin to the merry-go-around, as a noninertial frame. However, in the context of many seismic phenomena discussed in this book—as opposed to vibrations of the entire planet—the Earth can be viewed locally as an inertial frame due to the negligible effect of its rotation upon these phenomena. We do not include effects of the Earth's rotation in examining seismic-wave propagation.

Given inertial frames whose velocities with respect to each other are known, we can relate events happening within these frames by a *Galilean transformation*. However, we do not invoke this transformation in our discussions.

To obtain another description of motion in continuum mechanics, which is distinct from the one in particle mechanics, we use the continuity of motion. Thus—for a particular time—map ϕ in expression (2.1) is an *isomorphism*, which is a bijection such that both ϕ and its inverse are structure-preserving maps; hence,

$$\mathcal{M} = \Phi(\mathcal{S}, t)\,, \tag{2.2}$$

where Φ and ϕ are the inverses of one another. In this case, the preservation of structure means the smoothness of both Φ and ϕ.

To understand the qualifier of a particular time, let us examine Figure 2.2. In general, ϕ does not have an inverse, as illustrated by comparing distinct spatial coordinates of $\mathcal{S}_1$ and $\mathcal{S}_2$, at t_1 and t_2, respectively, with the same material coordinates of $\mathcal{M}$; that is, in general, the map is not one-to-one.

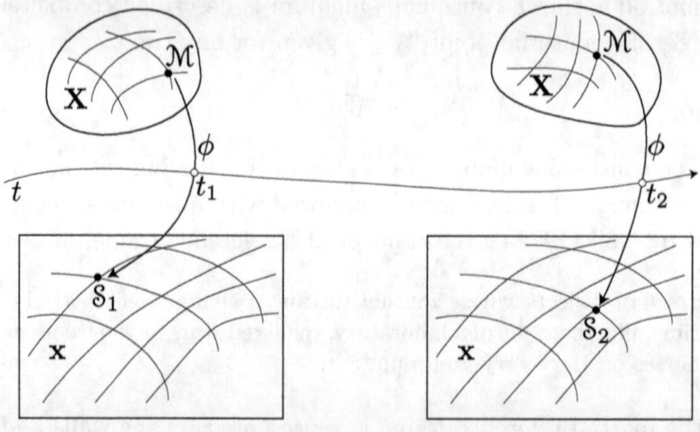

Fig. 2.2: Material body subject to deformation and its inertial reference frame, shown at two instances, t_1 and t_2. $\mathcal{M}$ is the same material point at two distinct spatial locations, $\mathcal{S}_1$ and $\mathcal{S}_2$. $\mathbf{X}$ and $\mathbf{x}$ are the comoving material coordinates and the spatial coordinates, respectively; they are related by ϕ. Fixed, as opposed to comoving, material coordinates are illustrated in Figures A.1 and A.2 of Appendix A.

Restricting our description to an inertial frame, we can view expressions (2.1) and (2.2) as descriptions of motion therein. They enable us to describe an event by focusing, respectively, on a motion of a single material point or on a flow of material points through a spatial location. Either description requires a corresponding coordinate system; the existence of such systems allows us to endow the Galilean spacetime with two physical interpretations: a *spatial manifold* and a *material manifold*, even though, mathematically—due to diffeomorphisms (2.1) and (2.2), together with abstractness of the notion of manifolds—it might be natural to view it as a single manifold.

2.2.3 *Coordinates*

For a quantitative analysis of motion, we need to introduce coordinates to specify both spatial and material points.[8] We consider two types of coordinate systems: one to identify spatial points, another to identify material points, and we refer to these systems as *spatial coordinates* and *material coordinates*, respectively. In view of their transformation properties, discussed in Appendix E, both coordinates are vectors, as illustrated in Figures A.1 and A.2 of Appendix A.

Notation 2.1. In this chapter—in accordance with Appendix E—the indices of contravariant tensors, including vectors, are placed as superscripts. Also, in this chapter, bold letters represent vectors.

To identify points $\mathcal{S}$, we assign a set of *spatial coordinates*,

$$\mathbf{x} = \left(x^1, x^2, x^3\right) , \tag{2.3}$$

which allow us to specify each point by its coordinates and to assign the coordinates to each point,

$$\mathcal{S} = \mathcal{S}(\mathbf{x}) , \qquad x^i = x^i(\mathcal{S}) , \quad i = 1, 2, 3 . \tag{2.4}$$

Coordinates (2.3) refer to a single inertial reference frame. They are not affected by the motion of a continuum, as illustrated in the lower part of Figure 2.2. It is also common to refer to these coordinates as *Eulerian coordinates*.

Herein, to identify points $\mathcal{M}$, we assign a set of *comoving material coordinates*, as opposed to fixed material coordinates discussed in Appendix A,

$$\mathbf{X} = \left(X^1, X^2, X^3\right) , \tag{2.5}$$

such that

$$\mathcal{M} = \mathcal{M}(\mathbf{X}) , \qquad X^I = X^I(\mathcal{M}) , \quad I = 1, 2, 3 . \tag{2.6}$$

[8] *see also*: Slawinski (2015, Section 1.3)

This comoving coordinate system deforms with the continuum, as illustrated in the upper part of Figure 2.2 and as commented in Appendix A.1. In contrast to fixed coordinates, the comoving material coordinates are not inertial; they can accelerate together with the material to which they are attached. In general, it is common to refer to any material coordinates as *Lagrangian coordinates.*

Let us restate the expressions of motion in terms of coordinates. Expression (2.1) requires

$$x^i = \phi^i(\mathbf{X}, t)\,, \quad i = 1, 2, 3\,, \tag{2.7}$$

and expression (2.2) requires

$$X^I = \Phi^I(\mathbf{x}, t)\,, \quad I = 1, 2, 3\,. \tag{2.8}$$

Expressions (2.7) and (2.8) offer us distinct perspectives to study quantitatively the motion within a continuum. These expressions are mappings between material and spatial coordinates.

To study deformations, we describe flow of the continuum with respect to its inertial reference frame. For this purpose, we introduce the *deformation matrix*, which is defined in terms of partial derivatives of expression (2.7), with respect to the fixed, as opposed to comoving, material coordinates,

$$\begin{bmatrix} \dfrac{\partial\phi^1(\mathbf{X},t)}{\partial X^1} & \dfrac{\partial\phi^1(\mathbf{X},t)}{\partial X^2} & \dfrac{\partial\phi^1(\mathbf{X},t)}{\partial X^3} \\[2ex] \dfrac{\partial\phi^2(\mathbf{X},t)}{\partial X^1} & \dfrac{\partial\phi^2(\mathbf{X},t)}{\partial X^2} & \dfrac{\partial\phi^2(\mathbf{X},t)}{\partial X^3} \\[2ex] \dfrac{\partial\phi^3(\mathbf{X},t)}{\partial X^1} & \dfrac{\partial\phi^3(\mathbf{X},t)}{\partial X^2} & \dfrac{\partial\phi^3(\mathbf{X},t)}{\partial X^3} \end{bmatrix}. \tag{2.9}$$

We can view this 3×3 matrix as a transformation between material and spatial coordinates. Formally, with a certain abuse of notation, we can write its entries as $\partial x^i / \partial X^I$, where $i, I = 1, 2, 3$. In continuum mechanics, matrix (2.9) is also referred to as a *deformation gradient*, discussed in Appendix A.2.

Matrix (2.9) is also referred to as a *Jacobi matrix*, in honour of the German mathematician Carl Gustav Jacob Jacobi. It plays an important role as the generalization of the gradient operator acting beyond the scalar-valued fields. As illustrated in Exercises 2.1 and 2.2 and as discussed in Remark 2.3, matrix (2.9) allows us to quantify the change of respective positions of material points, which is tantamount to a deformation.

Assuming the existence of the inverse of ϕ, we can write concisely the inverse of the deformation matrix as

$$\left[\frac{\partial \Phi^I(\mathbf{x},t)}{\partial x^i}\right], \quad I,i=1,2,3, \tag{2.10}$$

whose entries we can rewrite as $\partial X^I/\partial x^i$.

2.3 Field equations

2.3.1 *Balance equations*

Let us comment on the term *field* in the title of this section. In the context of continuum mechanics, the term *field equations* is indicative of its relation with other field theories, such as electromagnetism, where the field equations are *Maxwell's equations*.[9] A field, which can be expressed in term of scalars, vectors or higher-rank tensors, allows us to quantify macroscopic physical phenomena, without examining microscopic interaction of matter.

The field equations discussed in this section belong to the fundamental laws of continuum mechanics and express balance of physical quantities; herein, we consider the balance of mass and linear momentum. In general, these laws include also the *balance of energy* and *entropy*. However, for the purpose of this chapter, as well as Chapters 3–7, where we focus our attention on Hookean solids, these laws do not offer any constraints that are not provided by the balance of mass and of the linear and angular momenta.

Remark 2.1. The balance of angular momentum is not an independent principle in continuum mechanics, but a consequence of the conservation of mass and balance of linear momentum. Under the strong form of Newton's third law, the balance of angular momentum entails the symmetry of the Cauchy stress tensor.[10]

As discussed by Romano and Marasco (2014, Section 5.1), all balance equations have a similar form. First, the amount of a certain physical quantity, Q, which is contained in volume, V, enclosed in surface S, changes with time, t; in other words,

$$\frac{\mathrm{d}}{\mathrm{d}t}\iiint\limits_V Q\,\mathrm{d}V\,. \tag{2.11}$$

[9] *see also*: Bóna and Slawinski (2015, Appendix C)
[10] *see also*: Slawinski (2015, Section 2.7)

Secondly, this change can be balanced by the flux, $\Phi \cdot \mathbf{n}$, through surface S, where $\mathbf{n}$ is the outward unit normal to S. It can be also balanced by the amount that is generated or destroyed within V. In other words,

$$\iint\limits_S \Phi \cdot \mathbf{n} \, \mathrm{d}S + \iiint\limits_V f \, \mathrm{d}V . \tag{2.12}$$

A balance equation is the equality between expressions (2.11) and (2.12).

Let us conclude this section with a quote from Epstein (2010, p. 114).

> Many important objects in Continuum Mechanics are made to be integrated, since only then do they acquire a clear physical meaning. Mass density, internal energy density, body forces, surface tractions, and heat influx are just a few important examples.

2.3.2 *Balance of mass: Continuity equation*

Let us consider the balance of mass under the assumption that mass is neither created nor destroyed, but can flow through space. This means, in view of expression (2.12), that the volume integral is zero. Thus, we write this balance principle as[11]

$$\frac{\mathrm{d}}{\mathrm{d}t} \iiint\limits_V \rho(\mathbf{x}, t) \, \mathrm{d}V = - \iint\limits_S \rho(\mathbf{x}, t) \, \mathbf{v} \cdot \mathbf{n} \, \mathrm{d}S, \tag{2.13}$$

where $\mathbf{v}$ is the velocity of the flow and $\mathbf{n}$ is the unit normal to the surface; if $\mathbf{v} \perp \mathbf{n}$, there is no flow across the surface. Since $\mathbf{n}$ is an outward normal, the negative sign corresponds to the mass entering the volume.

Assuming that the volume is fixed in space, which means that V is not a function of time, we can interchange the integration and differentiation on the left-hand side of equation (2.13). Also, invoking the Divergence Theorem,[12] illustrated in Exercise 2.3, we express the double integral as a triple integral. Thus—under certain restricting assumptions—we combine the resulting expressions to get

$$\iiint\limits_V \left(\frac{\partial}{\partial t} \rho(\mathbf{x}, t) + \nabla \cdot (\rho(\mathbf{x}, t) \, \mathbf{v}) \right) \mathrm{d}V = 0 .$$

Since this equation must be valid for an arbitrary volume, we require the integrand itself to be zero,

$$\frac{\partial \rho(\mathbf{x}, t)}{\partial t} + \nabla \cdot \left(\rho(\mathbf{x}, t) \frac{\partial \mathbf{u}(\mathbf{x}, t)}{\partial t} \right) = 0 . \tag{2.14}$$

[11] *see also*: Slawinski (2015, Section 2.1)

[12] *see also*: Bóna and Slawinski (2015, Appendix A.1)

Example 2.1. [13]To exemplify the need for this requirement to ensure the validity of the equation for arbitrary integration limits, let us consider the case for which the equality, $\int_a^b f(x)\,\mathrm{d}x = 0$, is a function of the integration limits and does not depend solely on $f(x)$.

Let $f(x) = \cos x$. Then, if $b - a = 2\pi$, it follows that $\int_a^b f(x)\,\mathrm{d}x = 0$; in general, $\int_a^b f(x)\,\mathrm{d}x \neq 0$. If $f(x) = 0$, however, then $\int_a^b f(x)\,\mathrm{d}x = 0$, for all a and b.

In equation (2.14), $\mathbf{u}$ denotes the displacement, whose time derivative is the velocity. This is the equation of continuity. We can rewrite that equation as

$$\left(\frac{\partial}{\partial t}, \frac{\partial}{\partial x^1}, \frac{\partial}{\partial x^2}, \frac{\partial}{\partial x^3}\right) \cdot \left(\rho\left(1, \frac{\partial u^1}{\partial t}, \frac{\partial u^2}{\partial t}, \frac{\partial u^3}{\partial t}\right)\right) = 0,$$

which is a typical form of a balance equation: the spacetime divergence is zero.

Equation (2.13), which is stated in terms of integrals, is more general than equation (2.14); it allows for discrete changes of position, as long as the total amount within the volume is conserved. The requirements imposed by the derivation of a convenient differential equation restricts our scope of examination, since the mass density must vary continuously.

Let us complete this section with an analogy from electromagnetism. The equation of continuity therein has a form similar to expression (2.14); it is $\partial\rho/\partial t + \nabla \cdot J = 0$, where ρ and J are the electric-charge and electric-current densities, respectively.[14] Notably, the Minkowski metric, which allows us to express concisely the continuity equation in spacetime and is used in Section 8.2 to examine gravitational waves, was originally introduced by Hermann Minkowski in the context of the electromagnetic field.

2.3.3 *Balance of momentum: Equation of motion*

Cauchy's equations of motion—which contain quantitative information, such as speeds of the P and S waves, used as analogies to study seismic disturbances—are rooted in the balance of linear momentum, which states that the forces acting on a body are equal to the rate of change of the linear momentum of this body.[15] In accordance with expression (2.11),

---

[13] *see also*: Slawinski (2015, Exercise 2.1)

[14] *see also*: Bóna and Slawinski (2015, Appendix C.1.1)

[15] *see also*: Slawinski (2015, Sections 2.4-2.6)

the temporal change of linear momentum is

$$\frac{\mathrm{d}}{\mathrm{d}t}\iiint\limits_{V(t)}\rho(\mathbf{x},t)\frac{\mathrm{d}\,\mathbf{u}(\mathbf{x},t)}{\mathrm{d}\,t}\,\mathrm{d}V\,,$$

where ρ stands for mass density and $\mathbf{u}$ is the displacement vector. Herein, we allow the volume to change with time. In accordance with expression (2.12), we state the balance of this change as

$$\iint\limits_{S(t)}\mathbf{T}(\mathbf{x},t)\,\mathrm{d}S+\iiint\limits_{V(t)}\mathbf{f}(\mathbf{x},t)\,\mathrm{d}V\,,$$

where $\mathbf{T}$ and $\mathbf{f}$ are the contact-force and body-force densities; symbol $\mathbf{T}$ denotes traction. Equating these two expressions, we obtain the integral equation for the balance of linear momentum.

To derive the corresponding differential equation, we need to invoke the Divergence Theorem. To do so, we must introduce two important concepts. First—to bring the temporal derivative inside the triple integral, whose integration limits are time-dependent—we invoke the material time derivative, which is illustrated in Exercises 2.4 and 2.5, to obtain, under certain restricting assumptions,[16,17,18]

$$\iiint\limits_{V(t)}\rho(\mathbf{x})\frac{\partial^2\,\mathbf{u}\,(\mathbf{x},t)}{\partial\,t^2}\,\mathrm{d}V\,.$$

Secondly, to express the integrand of the double integral in the required form, we introduce the Cauchy stress tensor, σ , which linearly relates $\mathbf{T}$ to the orientation of the surface on which it acts,

$$\mathbf{T}=\sigma\,\mathbf{n}\,; \tag{2.15}$$

herein, $\mathbf{n}$ is the unit normal to the surface; σ is a second-rank symmetric tensor, $\sigma^{ij}=\sigma^{ji}$, where $i,j=1,2,3$.[19] No properties other than orientation, such as curvature, can be considered in formulating σ , as discussed by Noll (1974a, Theorem IV) and Romano and Marasco (2014, Theorem 5.1).

Combining the resulting expressions and writing them in terms of components, we obtain

$$\iiint\limits_{V(t)}\left(\rho\,(\mathbf{x})\,\frac{\partial^2 u^i\,(\mathbf{x},t)}{\partial t^2}-\sum_{j=1}^{3}\frac{\partial\sigma^{ij}(\mathbf{x})}{\partial x^j}-f^i(\mathbf{x},t)\right)\mathrm{d}V=0\,,\qquad i=1,2,3\,,$$

[16] *see also*: Slawinski (2015, Section 2.2)

[17] *see also*: Bóna and Slawinski (2015, Appendix B.1)

[18] Readers interested in an insightful discussion on the material time derivative might refer to Epstein (2010, Appendix A.5).

[19] *see also*: Slawinski (2015, Section 2.7.3)

where the summation represents the divergence of σ, which results from the Divergence Theorem. To ensure the validity for an arbitrary volume, we require that the integrand be zero,

$$\rho(\mathbf{x}) \frac{\partial^2 u^i(\mathbf{x},t)}{\partial t^2} = \sum_{j=1}^{3} \frac{\partial \sigma^{ij}(\mathbf{x})}{\partial x^j} + f^i(\mathbf{x},t), \qquad i = 1,2,3, \tag{2.16}$$

which are Cauchy's equations of motion. Below, in Section 5.1, we use the fact that—in the context of isotropic elastic solids, discussed in Section 3.1—these equations entail the existence of both P and S waves.

Let us comment on the derivation of equation (2.16). The requirements of differentiability and assumptions associated with the introduction of the Cauchy stress tensor render the differential equation less general than the integral equation from which it originates. We could suggest that such a derivation is a consequence of the history of mathematics. The techniques for studying differential equations were developed earlier than for studying integral equations. Perhaps, had computers been invented earlier, our ancestors would have a preference for integral equation, where the process of infinitesimal summation lends itself to computer operations. Be that as it may, the introduction of the stress tensor—at least implicitly, for the purpose of invoking the Divergence Theorem—has played the key role in the development of continuum mechanics.

Closing remarks

The purpose of continuum mechanics is a quantitative examination of macroscopically observable physical phenomena, without examining details of the constitution of matter. As stated by James (2015, p. 446),

> Everything is discrete and, at the finest level, indivisible. It is quite surprising, then, that continuum mechanics, possibly the most successful theory of general use in applied mathematics, does not explicitly recognize the existence of atoms.

The criterion to evaluate this success, and the justification of mathematical entities used in continuum mechanics, is the agreement between macroscopic observations and results entailed by a mathematical model; it is an *a posteriori* justification of a hypotheticodeductive formulation whose underlying entities need not carry intrinsic physical meanings.

The concept of an *a posteriori* justification of continuum mechanics has an important consequence. The validity of continuum mechanics does not rely on reducing it to microphysical concepts. Desirable and insightful as such a reduction might be, it is not a criterion of validity of continuum mechanics, which is a basic science. Hence, while inspired by the essay of Eugene Wigner (1960) on "The unreasonable effectiveness of mathematics in the natural sciences," we might question the reason for, but not the usefulness of, an abstractness of a vector field to provide an accurate quantitative description of observed phenomena. Notably, this very abstractness is the reason for the actual and potential fruitfulness of connections between mathematics and physics not to be missed, as emphasized by Dyson (1996a) in his lecture entitled "Missed opportunities."

In accordance with the quote of Udías (1999) at the beginning of this chapter, theoretical seismology is a science based on continuum mechanics, since to argue for a seismological formulation, we invoke concepts of continuum mechanics. However, in view of limitations of continuum mechanics, we must stop our direct analysis at the level of mathematical entities. As pointed out by Epstein and Slawinski (1998), we cannot avoid the issue of

> conceptual commonality between all pursuits dealing with media which, for better or worse, have been assumed *ab initio* to be continuous.

Inferences of properties of physical materials, such as information about grain orientations, rely on interpretations of results obtained by methods of continuum mechanics; however, such inferences do not belong to continuum mechanics itself.

2.4 Exercises

Exercise 2.1. Consider a uniaxial stretch parallel to the x^1-axis; its expression (2.7) is

$$\begin{aligned} x^1 &= \phi^1(\mathbf{X},t) = \alpha(t)\,X^1\,, \\ x^2 &= \phi^2(\mathbf{X},t) = X^2\,, \\ x^3 &= \phi^3(\mathbf{X},t) = X^3\,, \end{aligned} \tag{2.17}$$

where α is a time-dependent stretch factor. Prior to the stretch, $\alpha(0) = 1$; hence, material and spatial coordinates coincide.

Obtain matrix (2.9) and its inverse. Using the fact that, geometrically, the determinant of matrix (2.9) is tantamount to the change in volume, comment on the effect of a uniaxial stretch on the change in volume. Illustrate and discuss the effect of deformation for $\alpha(t) = 1 + t$.

Remark 2.2. Herein, with a certain abuse of notation, we consider t and $\alpha(t)$ as unitless quantities.

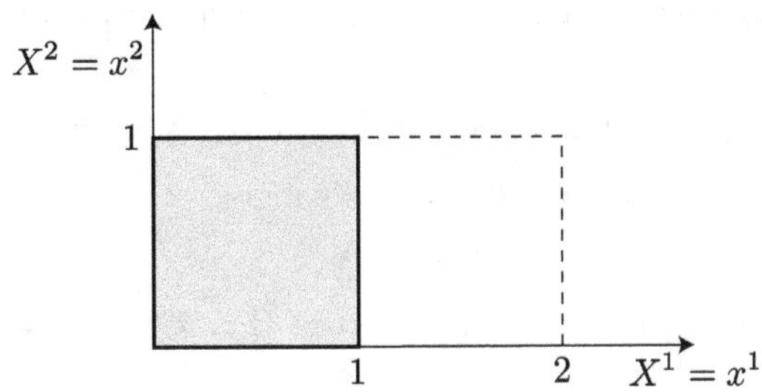

Fig. 2.3: A unit square at $t = 0$, shown in gray; the dashed line is its incremental deformation due to a uniaxial stretch, given in expression (2.20); herein, $t = 1$.

Solution. Matrix (2.9) and its inverse are

$$\begin{bmatrix} \dfrac{\partial\phi^1(\mathbf{X},t)}{\partial X^1} & \dfrac{\partial\phi^1(\mathbf{X},t)}{\partial X^2} & \dfrac{\partial\phi^1(\mathbf{X},t)}{\partial X^3} \\ \dfrac{\partial\phi^2(\mathbf{X},t)}{\partial X^1} & \dfrac{\partial\phi^2(\mathbf{X},t)}{\partial X^2} & \dfrac{\partial\phi^2(\mathbf{X},t)}{\partial X^3} \\ \dfrac{\partial\phi^3(\mathbf{X},t)}{\partial X^1} & \dfrac{\partial\phi^3(\mathbf{X},t)}{\partial X^2} & \dfrac{\partial\phi^3(\mathbf{X},t)}{\partial X^3} \end{bmatrix} = \begin{bmatrix} \alpha(t) & 0 & 0 \\ 0 & 1 & 0 \\ 0 & 0 & 1 \end{bmatrix} \tag{2.18}$$

and

$$\begin{bmatrix} \alpha^{-1} & 0 & 0 \\ 0 & 1 & 0 \\ 0 & 0 & 1 \end{bmatrix}, \tag{2.19}$$

respectively. Their determinants, which are α and α^{-1}, respectively, show that the change in volume is directly proportional to stretch, as expected.

Let us consider the x^1x^2-plane. For $\alpha(t) = 1 + t$, expression (2.17) becomes

$$\begin{aligned} x^1 &= (1+t)\,X^1\,, \\ x^2 &= X^2\,. \end{aligned} \tag{2.20}$$

At $t = 0$, $x^i = X^I$, as required; the four vertices are $(x^1, x^2) = (X^1, X^2) = (0,0)$, $(1,0)$, $(1,1)$ and $(0,1)$. The location of a given vertex, (x^1, x^2), at time t is obtained by inserting the corresponding pair, (X^1, X^2), and the value of t into expressions (2.20). Considering the four vertices, at $t = 1$, we obtain the deformation illustrated in Figure 2.3.[20] Mohr's circle, which illustrates the corresponding state of stress, is discussed in Exercise 3.4, on page 94.

Exercise 2.2. Consider a *simple shear* that acts parallel to the x^1-axis;[21] its expression (2.7) is

$$\begin{aligned} x^1 &= \phi^1(\mathbf{X}, t) = X^1 + \alpha(t)\, X^2\,, \\ x^2 &= \phi^2(\mathbf{X}, t) = X^2\,, \\ x^3 &= \phi^3(\mathbf{X}, t) = X^3\,, \end{aligned} \tag{2.21}$$

where α is a time-dependent shear factor. Prior to the shear, $\alpha(0) = 0$; hence, material and spatial coordinates coincide.

Obtain matrix (2.9) and its inverse. Using the fact that, geometrically, the determinant of matrix (2.9) is tantamount to the change in volume, comment on the effect of a simple shear on the change in volume. Illustrate and discuss the effect of deformation for $\alpha(t) = t$.

Solution. Matrix (2.9) and its inverse are

$$\begin{bmatrix} \dfrac{\partial \phi^1(\mathbf{X},t)}{\partial X^1} & \dfrac{\partial \phi^1(\mathbf{X},t)}{\partial X^2} & \dfrac{\partial \phi^1(\mathbf{X},t)}{\partial X^3} \\ \dfrac{\partial \phi^2(\mathbf{X},t)}{\partial X^1} & \dfrac{\partial \phi^2(\mathbf{X},t)}{\partial X^2} & \dfrac{\partial \phi^2(\mathbf{X},t)}{\partial X^3} \\ \dfrac{\partial \phi^3(\mathbf{X},t)}{\partial X^1} & \dfrac{\partial \phi^3(\mathbf{X},t)}{\partial X^2} & \dfrac{\partial \phi^3(\mathbf{X},t)}{\partial X^3} \end{bmatrix} = \begin{bmatrix} 1 & \alpha & 0 \\ 0 & 1 & 0 \\ 0 & 0 & 1 \end{bmatrix} \tag{2.22}$$

and

$$\begin{bmatrix} 1 & -\alpha & 0 \\ 0 & 1 & 0 \\ 0 & 0 & 1 \end{bmatrix},$$

respectively. Their determinants are equal to unity, which means that a simple shear results in no change in volume; it is an *equivoluminal* deformation, which is also referred to as *isochoric*.

[20] *see also*: Slawinski (2015, Figure 1.4.1)

[21] Readers interested in a simple shear in the context of a deformation gradient, discussed in Appendix A.2, might refer to Ogden (1997, pp. 101–104).

Let us consider the x^1x^2-plane. For $\alpha(t) = t$, expression (2.21) becomes

$$\begin{aligned} x^1 &= X^1 + t\,X^2\,, \\ x^2 &= X^2\,. \end{aligned} \tag{2.23}$$

At $t = 0$, $x^i = X^I$, as required; the four vertices are $(x^1, x^2) = (X^1, X^2) = (0,0)$, $(1,0)$, $(1,1)$ and $(0,1)$. The location of a given vertex, (x^1, x^2), at time t is obtained by inserting the corresponding pair, (X^1, X^2), and the value of t into expressions (2.23). Considering the four vertices, at $t = 1$, we obtain the deformation illustrated in Figure 2.4. The deformation at $t = 2$ is illustrated in Figure A.8 on page 400.[22]

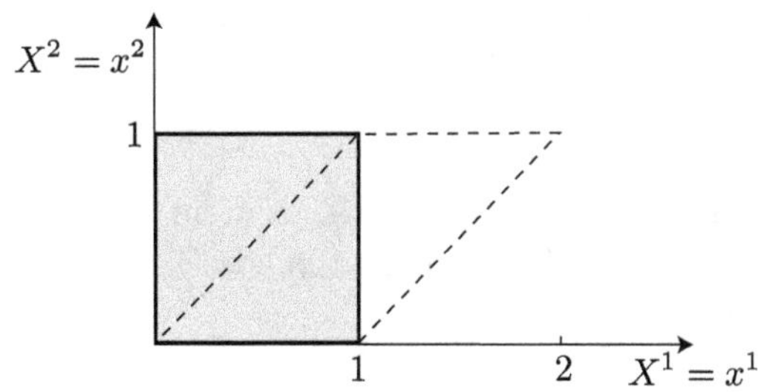

Fig. 2.4: A unit square at $t = 0$, shown in gray; the dashed line is the deformed square due to a simple shear, given in expression (2.23); herein, $t = 1$.

Remark 2.3. Exercises 2.1 and 2.2 illustrate geometrical properties of matrices (2.9) and (2.10), which, respectively, are the deformation matrix and its inverse. The main diagonal entries specify deformations for which the shape remains unaltered, in the sense that a rectangular parallelepiped remains a rectangular parallelepiped. These entries are related only to the change in size. The other entries specify deformations that lead to a change in shape, in the sense that a rectangular parallelepiped loses the rectangularity of its faces.

The value of the determinant of matrix (2.9) indicates the change in volume, as discussed in Appendix A.6. As illustrated in Exercise 2.1, if the determinant of matrix (2.9) is equal to unity at $t = 0$ and is greater than unity for $t > 0$, the volume is increased; if—for $t > 0$—the value of this determinant were between zero and unity, it would be decreased. The opposite is true for the determinant of matrix (2.10), since these two matrices are inverses of one another and, hence, their determinants are

[22] *see also*: Slawinski (2015, Figure 1.4.2)

reciprocals of one another. If the determinants remain equal to unity, the volume is preserved in spite of deformation, as illustrated in Exercise 2.2, as well as in Example A.2, on page 368. Geometrically, the meaning of matrices (2.9) and (2.10) must be the same. Thus, for Exercise 2.1, we can view the determinant of matrix (2.9) being greater than unity as the spatial coordinates being expanded with respect to the material ones, and being smaller than unity as the spatial coordinates being contracted with respect to the material ones.

The determinant is assumed to be nonzero to ensure the existence of the inverse of matrix (2.9), which results in matrix (2.10). This requirement is necessary for a proper transformation between the material and spatial coordinates. If the determinant is zero, there is no one-to-one relation between ϕ and Φ, due to the loss of dimensions, as is the case if a volume becomes a surface.

Exercise 2.3. Illustrate the Divergence Theorem using a vector field given by the gradient of $r^3/3$. Comment on the applicability of this theorem to the gradient of $1/r$.

Theorem 2.1. *Divergence Theorem: The surface integral of a differentiable vector field along a closed surface can be expressed as the integral of the divergence of this field over the volume enclosed by this surface,*

$$\iint\limits_S \mathbf{F}\cdot\mathbf{n}\,\mathrm{d}S = \iiint\limits_V \nabla\cdot\mathbf{F}\,\mathrm{d}V\,, \tag{2.24}$$

where $\mathbf{n}$ *is the unit outward normal vector on surface* S. $\mathbf{F}$ *must be differentiable within the domains of integration,* S *and* V.

Solution. Using the spherical symmetry of r^3, we invoke the gradient operator of a spherically symmetric function to write the vector field as $\mathbf{F} = (\partial/\partial r)r^3/3\,\mathbf{r} = r^2\,\mathbf{r}$, where $\mathbf{r}$ is the radial unit vector. Since, in spherical coordinates, the element of surface is $\mathrm{d}S = R^2\sin\alpha\,\mathrm{d}\alpha\,\mathrm{d}\beta$, where R is the magnitude of the radius corresponding to S, the left-hand side of equation (2.24) becomes

$$R^2\int\limits_0^{2\pi}\int\limits_0^{\pi} R^2\sin\alpha\,\mathrm{d}\alpha\,\mathrm{d}\beta = 4\pi R^4\,,$$

where the vector field is evaluated on S, where $r = R$.

Invoking the divergence operator of a spherically symmetric function, we obtain

$$\frac{1}{r^2}\frac{\partial}{\partial r}\left(r^2\, r^2\right) = 4r\,.$$

Since the element of volume is $\mathrm{d}V = r^2 \sin\alpha\,\mathrm{d}r\,\mathrm{d}\alpha\,\mathrm{d}\beta$, the right-hand side of equation (2.24) becomes

$$4\int\limits_0^{2\pi}\int\limits_0^{\pi}\int\limits_0^{R} r^3 \sin\alpha\,\mathrm{d}r\,\mathrm{d}\alpha\,\mathrm{d}\beta = 4\pi R^4\,.$$

As expected from Theorem 2.1, the left-hand and right-hand sides are equal to one another.

Herein, Theorem 2.1 applies since the vector field is differentiable everywhere. In particular, for the left-hand side, it is differentiable at $r = R$ for $\alpha \in \{0, \pi\}$ and $\beta \in \{0, 2\pi\}$. For the right-hand side, it is differentiable for $r \in \{0, R\}$, $\alpha \in \{0, \pi\}$ and $\beta \in \{0, 2\pi\}$. However, if we let $\mathbf{F}$ be the gradient of $1/r$, then the left-hand side is -4π and the right-hand side is zero. Theorem 2.1 does not apply due to nondifferentiability at $r = 0$, which affects the right-hand side of equation (2.24).[23]

Exercise 2.4. Consider an object falling in a gravitational field, along the y-axis. Assume the initial speed of this object to be zero. Use the material-time-derivative operator,

$$\left(\frac{\partial}{\partial t} + \mathbf{v}\cdot\nabla\right), \tag{2.25}$$

to show that the acceleration of this object is constant and that its value is the same regardless of the speed being expressed in the spatial or material coordinates. Let the y-axis point downwards.

Solution. In the spatial coordinates, the position of the object is

$$y(t) = \frac{1}{2}gt^2\,;$$

its speed and acceleration are

$$v(t) = \frac{\partial y(t)}{\partial t} = gt$$

and

$$a(t) = \frac{\partial^2 y(t)}{\partial t^2} = g\,,$$

respectively; the latter is a constant.

[23] *see also*: Bóna and Slawinski (2015, Example A.4)

In material coordinates, the speed is

$$v(y)=\sqrt{2gy}\,. \tag{2.26}$$

Applying operator (2.25) and using the fact that we are dealing with a problem in a single spatial dimension, we obtain

$$a(y)=\frac{\partial v(y)}{\partial t}+v(y)\frac{\partial v(y)}{\partial y}=g\,, \tag{2.27}$$

where—to get the final result—we use expression (2.26); $v(y)$ has no temporal dependence: the speed is always the same at a given position. Thus, we conclude that $a(t)=a(y)=g$, as required.

Exercise 2.5. Using $Q(\mathbf{x}(t),t)$, show that operator (2.25), in Exercise 2.4, is tantamount to the chain rule. Discuss the physical meaning of this operator.

Solution. Using the chain rule, let us write

$$\frac{\mathrm{d}}{\mathrm{d}t}Q(\mathbf{x}(t),t)=\frac{\partial Q}{\partial x^1}\frac{\mathrm{d}x^1}{\mathrm{d}t}+\frac{\partial Q}{\partial x^2}\frac{\mathrm{d}x^2}{\mathrm{d}t}+\frac{\partial Q}{\partial x^3}\frac{\mathrm{d}x^3}{\mathrm{d}t}+\frac{\partial Q}{\partial t}\frac{\mathrm{d}t}{\mathrm{d}t}\,.$$

Since $\mathrm{d}x^i/\mathrm{d}t$ are components of the velocity vector, $\mathbf{v}$, and $\mathrm{d}t/\mathrm{d}t=1$, we write

$$\frac{\mathrm{d}}{\mathrm{d}t}Q(\mathbf{x}(t),t)=\frac{\partial Q}{\partial x^1}v^1+\frac{\partial Q}{\partial x^2}v^2+\frac{\partial Q}{\partial x^3}v^3+\frac{\partial Q}{\partial t}\,.$$

Since the first three summands are the scalar product of two vectors, we write

$$\frac{\mathrm{d}}{\mathrm{d}t}Q(\mathbf{x}(t),t)=\left(\frac{\partial Q}{\partial x^1},\frac{\partial Q}{\partial x^2},\frac{\partial Q}{\partial x^3}\right)\cdot\left(v^1,v^2,v^3\right)+\frac{\partial Q}{\partial t}\,. \tag{2.28}$$

Recognizing that $\partial Q/\partial x^i$ are components of the gradient, ∇Q, we write

$$\frac{\mathrm{d}}{\mathrm{d}t}Q(\mathbf{x}(t),t)=\nabla Q\cdot\mathbf{v}+\frac{\partial Q}{\partial t}\equiv\left(\frac{\partial}{\partial t}+\mathbf{v}\cdot\nabla\right)Q\,,$$

where we use linearity of the required differential operator, and commutativity of the scalar product.

In considering the temporal rate of change, operator (2.25) accounts for the velocity with which the object in question is moving. Let us recall Exercise 2.4. In equation (2.27), the acceleration itself, $\partial v/\partial t$ is zero; however, the additional term, which corresponds to $\mathbf{v}\cdot\nabla$, accounts for the motion, and allows us to obtain the correct answer.

Chapter 3

Hookean solid: Material symmetry

Chodzi, najogólniej mówiąc, o materiały, które pod wpływem działań (sił) zewnętrznych doznają odkształceń, a po ich ustaniu powracają do stanu wyjściowego. [...] Chociaż efekty spreżyste wystepują równiez w płynach, to teoria spreżystości zajmuje się głównie zjawiskami takimi w ciałach stałych. [...] Najwiekszą jego własnością odrózniającą go od płynów jest sztywność postaciowa. Wynika ona z silnych więzów pomiędzy elementami struktury, które ograniczają znacznie ich ruchy. Dlatego też w ciałach stałych energia wewnętrzna wzajemnych oddziaływań przewyższa znacznie energię kinetyczną ruchów cząsteczek lub atomów. Jest to więc sytuacja odwrotna niż w płynach, gdzie najczęściej dominjąca jest energia kinetyczna elementów strukturalnych, co prowadzi do braku sztywności postaciowej płynów.[1]

Czesław Rymarz (1993)

Preliminary remarks

In this chapter, we discuss a mathematical analogy of physical materials used to examine quantitatively phenomena of interest to seismologists. To do so, we invoke a particular theory within continuum mechanics: the *theory of elasticity*. Within this theory, we examine mathematical symmetries of elastic continua as analogies for mechanical symmetries of physical materials.

We begin this chapter by introducing Hooke's law: a constitutive equation for elastic solids. Subsequently, we discuss mathematical symmetries of solids in both finite and infinitesimal elasticities. Since the latter—which correspond to Hookean solids—are inherited from the former, we conclude this chapter by examining the material-symmetry classes of Hookean solids.

3.1 Hookean solids

A Hookean solid is a mathematical entity. It is defined as an object, $c_{ijk\ell}$, that relates stress, σ_{ij}, and strain, $\varepsilon_{k\ell}$, in a linear fashion. This linear relationship is called Hooke's law.

As a mathematical entity, $c_{ijk\ell}$ shares the Platonic realm with its other occupants, such as a point, a sphere, an equation. Also, it provides analogies to examine quantitatively physical phenomena.

Invoking Platonic analogies, which is a common approach in physics, we should be aware of their limitations. For instance, to say that P and S waves propagate in the Earth is a shorthand for saying that if we use an isotropic Hookean solid as an analogy for the material of our planet, we can associate the behaviour of disturbances in the Earth with the behaviour of wave equations in such a solid. These disturbances result in detectable phenomena, and the wave equations have solutions, which might be used as quantitative analogies to study these phenomena.

The subject of this chapter is a connection between detectable phenomena and mathematical entities used as their analogies. As stated in the quote of Udías (1999), at the beginning of Chapter 2, theoretical and observational aspects of seismology, which deal with physical phenomena at a macroscopic level, are based on the principles of mechanics of continuous media. To set the stage for most of this book, we limit our focus to elastic continua, which are referred to as *Hookean solids* in honour of Robert Hooke.

[1] *We are, most broadly speaking, dealing with materials that upon influence of external forces are subject to deformations, but upon disappearance of forces return to their original form.* [...] *Even though elastic effects occur also in liquids, theory of elasticity is concerned mainly with such phenomena in solids.* [...] *The most important property distinguishing solids from liquids is the rigidity. It results from strong bonds among structural elements, which restrict significantly their movements. Also, for that reason, internal energy among elements in solids surpasses significantly kinetic energy of particles or atoms. Therefore, it is the opposite of liquids, where most commonly the kinetic energy of structural elements is dominant, which leads to the lack of rigidity of liquids.*

To gain further insight into this quote, which implicitly relates a field theory to microscopic properties of matter, let us examine a quote of Berdichevsky (2009),

> In the system of enclosed gas forced by the piston, a thermodynamic description is possible if the piston velocity is much smaller than the average molecule velocity. If the velocity of the piston is on the order of the average molecule velocity, thermodynamic description fails: the relation between the force and the gas volume becomes dependent on the details of the molecule motion.

Let us consider

$$\sigma_{ij} = \sum_{k=1}^{3}\sum_{\ell=1}^{3} c_{ijk\ell}\varepsilon_{k\ell}\,, \qquad i,j = 1,2,3\,, \tag{3.1}$$

where σ_{ij} are the stress-tensor components, $\varepsilon_{k\ell}$ are the strain-tensor components, and $c_{ijk\ell}$ are the components of the elasticity tensor, which relates linearly the stress and strain tensors. The assumption of the linear relation between loads and deformations is the crux of *Hooke's law*. The tensorial form of expression (3.1) is symptomatic of the fact that, as stated by James (2015, p. 447),

> The conventional language of continuum mechanics is tensor analysis.

Let us use the terminology presented in *Preliminary remarks* of Chapter 2. Expression (3.1) is the constitutive relation that—within the theory of elasticity—constrains our studies to Hookean solids. Specifically, expression (3.1) is the material response that relates deformations, $\varepsilon_{k\ell}$, to applied loads, σ_{ij}. Mathematically, $c_{ijk\ell}$ is the proportionality constant. Mechanically—under restricting assumptions of index symmetries and positive definiteness,[2] $c_{ijk\ell}(\mathbf{x})$, together with a mass density—is a Hookean solid: a continuum defined by a *tensor field* of a fourth-rank tensor, which exhibits properties of anisotropy and inhomogeneity in $\mathbf{x} \in \mathbb{R}^3$.

The linearity of the relation between loads and deformations is implicit in the original statements of Robert Hooke who—in *De potentia restitutiva, Or of spring: Explaining the power of spring bodies*, published in London in 1678—stated the law that now bears his name as *ut tensio sic vis*: as the extension so the force, and alternatively as *ut pondus sic tensio*: as the weight so the extension.[3] For Hooke, these two statements were equivalent to one another, since, as stated by Chapman (2005, p. 175), at that time

> scientists had not yet developed a coherent concept of "energy", and often spoke of *pondus*, or weight, force, pressure and such when trying to define the powers of nature.

In stating expression (3.1), we assume that the deformation described by $\varepsilon_{k\ell}$ refers to the difference between the undeformed and deformed states

[2] *see also*: Slawinski (2015, Sections 3.2.2 and 4.2.1)

[3] Readers interested in the historical context of the formulation of Hooke's law might refer to Chapman (2005, Chapter 10). Notably, the essence of this law might have been grasped by Hooke a couple of decades prior to the publication of *De potentia restitutiva*, which led to a priority conflict with Christiaan Huygens.

or, implicitly, between two states separated by an incremental difference. Explicitly, we examine the solid in its stressed and unstressed states; no strain, $\varepsilon_{k\ell} = 0$, results in no stress and *vice versa*, as implied by

$$\varepsilon_{ij} = \sum_{k=1}^{3}\sum_{\ell=1}^{3} s_{ijk\ell}\sigma_{k\ell}\,, \qquad i,j = 1,2,3\,, \tag{3.2}$$

where $s_{ijk\ell}$ are the components of the inverse of the elasticity tensor. Commonly, $c_{ijk\ell}$ and $s_{ijk\ell}$ are referred to as *stiffness tensor* and *compliance tensor*, respectively.

Within the realm of mathematics, equation (3.1) is exact. However, if viewed as a physical law, it is an approximation. The elasticity tensor is a mathematical object whose properties are not properties of a physical material, even though they might serve as their analogies. As indicated in Section 3.2, below, examination of physical phenomena using symmetries of this tensor allows us to gain an insight into material properties of objects, such as directions of layers and fractures.

In the context of seismology, the material through which disturbances might propagate is subject to preexisting stresses and strains, to which we refer as preloads. These preloads are due to such phenomena as self-gravitation of the Earth and its rotation. Propagation generates additional stresses. Hooke's law allows us to examine incremental stresses. The validity of such an approach is dependent on validity of the superposition principle, which is an assumption according to which the result of several deformations described by distinct values of $\varepsilon_{k\ell}$ and corresponding to different instances is the same as the sum of $\varepsilon_{k\ell}$ occurring together at the same instant; also, the result remains the same independently of the sequence in which deformations occur.

If we use seismic waves to study propagation of disturbances within the Earth, we can ignore the effects of selfgravitation, rotation and other sources of preload on propagation. Again, the argument for validity of this approach relies on the linearity and, hence, on the superposition principle, which means that the relation between the incremental stress and deformation is independent of preload, even though, in general, the values of the components, $c_{ijk\ell}$, are not. The selfgravitation issue is addressed in Chapter 8, below. The justification for ignoring its effects relies on comparing the estimates of magnitudes of such properties as propagation speed with and without selfgravitation.[4]

[4]Readers interested in such fundamentals of elasticity as the thermodynamics of deformation, deformations with temperature change and contacts among solids might refer to Landau and Lifshitz (1986, Chapter 1).

In concluding this section, let us quote Dyson (2011), who writes,

> Before quantum mechanics was invented, classical physics was always nonlinear, and linear models were only approximately valid. After quantum mechanics, nature itself suddenly became linear.

In the spirit emphasized by Stebbing (1937), we distinguish between *nature* and its description. Behaviours of Hookean solids, which are linear models for classical physics, are approximations within the physical world. Nonlinear models, which might result in more accurate approximations, could also be considered. Theoretically, behaviours of continua might be examined further within condensed matter physics and, fundamentally, within quantum mechanics, as discussed by Bunge (1967). Nevertheless, they all remain within the mathematical realm.

Thus, we view a Hookean solid as a mathematical analogy for physical materials, not as their approximation. An approximation, *sensu stricto*, can be achieved within the same realm only.

3.2 Material symmetry

3.2.1 *On symmetries*

The concept of symmetry is one of the most important notions of mathematics and physics. For instance, as shown by Emmy Noether (1918) a century ago, all conservation laws are expressions of symmetry. Notably, the ray parameter, discussed below in Section 7.4.2, is a conserved quantity that is a consequence of a translational symmetry within parallel layers, and an example of Noether's theorem. In this section, we illustrate rotational symmetry of tensors.

In a manner similar to the everyday meaning of the word, mathematical symmetries mean that we can perform operations on an object without modifying its appearance. A sphere can be rotated by any amount about any axis without changing its appearance. Also, wave-propagation properties within an isotropic Hookean solid are independent of its orientation, giving them spherical symmetry.

As an aside, we expect that isotropy is a good analogy for wave propagation in granites, which exhibit a random arrangement of quartz, mica and feldspar, but not in shales, where properties of disturbances propagating along laminations might be different from properties of disturbances propagating obliquely to laminations.

Even though, heuristically, a Hookean solid could be illustrated as an idealized physical object, it is no more and no less than a tensor, which exists in the realm of mathematics only. Hence, its symmetry is only a mathematical property: an operation that leaves the components of this tensor unchanged.

We can exemplify symmetry of tensors by rotating a second-rank tensor in two dimensions, whose components, a_{ij}, can be written as a 2×2 matrix. Herein, a symmetry of a tensor means that it is not affected by a given rotation. In other words, the condition of symmetry is

$$\begin{bmatrix} \cos\theta & -\sin\theta \\ \sin\theta & \cos\theta \end{bmatrix} \begin{bmatrix} a_{11} & a_{12} \\ a_{21} & a_{22} \end{bmatrix} \begin{bmatrix} \cos\theta & -\sin\theta \\ \sin\theta & \cos\theta \end{bmatrix}^T = \begin{bmatrix} a_{11} & a_{12} \\ a_{21} & a_{22} \end{bmatrix}, \tag{3.3}$$

where T denotes transpose. In other words, we ask the values of a_{ij} to remain unchanged for a given rotation angle, θ. It can be shown that if $a_{11} = a_{22}$ and $a_{12} = a_{21} = 0$, the tensor is isotropic: its components remain unchanged for all angles.

The matrix containing trigonometric functions is the rotation matrix in $\mathbb{R}^2$. In contrast to deformation matrices discussed on page 28, the properties of rotation matrices ensure that there is no deformation. These properties consist of the determinant being equal to unity, the magnitude of each column and each row being equal to unity, and the scalar product of a column with another column and any row with another row being equal to zero. Hence, such matrices are referred to as orthonormal matrices. An examination of these properties for a rotation matrix in $\mathbb{R}^6$ is shown in Exercise 3.8. One of the consequences of these properties is the invariance of the length of a rotated vector, as illustrated in Exercise 3.1.

In expression (3.3), if $\theta = \pi$, the rotated matrix remains unchanged for any values of a_{ij}. This is a consequence of the so-called point symmetry, which is invariance under the action of the negative identity, $-I$. Geometrically, it means that properties are sensitive to direction but not to the sense; it does not matter if we go NW or SE as long as we remain on a NW-SE line. Notably, all even-rank tensors, which includes Hookean solids, exhibit that property. Used in seismology, this property implies that velocity might depend on direction but cannot depend on the sense.

Herein, the point symmetry is also—according to Theorem 3.2,[5] formulated by Herman (1945) and discussed on page 91, below—the highest

[5] *see also*: Slawinski (2015, Theorem 5.10.1 and Exercise 5.8)

discrete symmetry that a second-rank tensor can have, namely, the twofold symmetry. Hookean solids, which are fourth-rank tensors, $c_{ijk\ell}$, can reach the fourfold symmetry, but not beyond.

Let us note that not all tensors can exhibit anisotropy. The zero-rank tensors, which are represented by scalars, cannot; they are rotation-invariant. Also, not all tensors can have an isotropic form. As illustrated in Exercises 3.2 and 3.3, the first-rank tensors, which are represented by vectors, cannot.

Let us use expression (3.3) to gain further insight into limitations of tensors as analogies for quantifying physical properties. Using a_{ij}, we cannot distinguish among objects whose discrete symmetries are twofold, threefold, fourfold, etc., since invariance of a second-rank tensor under such rotations implies its isotropy. Consider, for instance, a fourfold rotation, which means that $\theta = \pi/2$. In such a case, the symmetry condition becomes

$$\begin{bmatrix} a_{22} & -a_{21} \\ -a_{12} & a_{11} \end{bmatrix} = \begin{bmatrix} a_{11} & a_{12} \\ a_{21} & a_{22} \end{bmatrix}.$$

To satisfy this condition, we require that $a_{11} = a_{22}$ and $a_{12} = a_{21} = 0$, which, as stated above, is tantamount to isotropy. Also—due to point symmetry—we cannot distinguish phenomena whose properties depend on left versus right along a given direction. Thus, analogies provided by a second-rank tensor in two dimensions are limited.

However, since Hookean solids are fourth-rank tensors, $c_{ijk\ell}$, in three dimensions, the analogies they provide are richer than analogies provided by the second-rank tensors, a_{ij}, in two dimensions. Nevertheless, they are limited. Let us state a few such limitations.

As discussed in Section 3.2.4.16, below, the fourth-rank tensors cannot have only two orthogonal symmetry planes due to their even rank. They can have either a single symmetry plane or three symmetry planes. This statement does not mean that physical materials cannot exhibit only two sets of planar features that are orthogonal to one another. It means, however, that such an arrangement does not have a symmetry-class analogy among Hookean solids.

As discussed in Section 3.2.4.17, below, another limitation is the fact that, since Hookean solids are fourth-rank tensors, we cannot distinguish among discrete rotational symmetries beyond fourfold; they all appear as a complete rotational invariance. Hence, there are no pentagonal or hexagonal classes for Hookean solids. As proven by Herman (1945),[6] all rotational symmetries higher than fourfold result in transverse isotropy.

[6] *see also*: Slawinski (2015, Theorem 5.10.1 and Exercise 5.8)

Let us comment on terminology used for material symmetries. The importance of crystallography in studies of symmetries led to borrowing of its nomenclature by other disciplines, which resulted in misleading terminology for material symmetries of elastic continua. For instance, the hexagonal symmetry is an inappropriate term for Hookean solids, even though we might infer such a symmetry from tessellation, as indicated in Section 3.2.4.17.

Also, since symmetries of tensors are distinct from lattice symmetries, we must not be hasty with comparisons between symmetries of Hookean solids and symmetries of crystals. For instance, the former allow for continuous symmetry groups, namely, isotropy and transverse isotropy, and the latter allow for discrete symmetries only. The above distinction stems from the essence of continuum mechanics, where we describe properties of materials by continuous functions, not by discrete structures. This statement does not contradict the fact that symmetries of continua serve as analogies for such structures.

Let us complete this section by emphasizing that—while using seismic information—we must remember that seismology is a subject of continuum mechanics. As stated by Aki and Richards (2002),

> there is a conjecture that two sets of small motions may be superimposed without interfering with each other in a nonlinear fashion. [...] These conjectures, and many others that are generally assumed by seismologists to be true, are properties of infinitesimal motion in classical continuum mechanics.

3.2.2 *On tensor rotations*

As an aside, to familiarize the reader with aspects of tensorial transformations and the study of tensor symmetries, let us revisit expression (3.3) and write it as

$$\begin{bmatrix} \hat{a}_{11} & \hat{a}_{12} \\ \hat{a}_{21} & \hat{a}_{22} \end{bmatrix} = \begin{bmatrix} \cos\theta & -\sin\theta \\ \sin\theta & \cos\theta \end{bmatrix} \begin{bmatrix} a_{11} & a_{12} \\ a_{21} & a_{22} \end{bmatrix} \begin{bmatrix} \cos\theta & -\sin\theta \\ \sin\theta & \cos\theta \end{bmatrix}^T ,$$

where a_{ij} and $\hat{a}_{ij}$ are components of a second-rank tensor in $\mathbb{R}^2$ expressed in two coordinate systems rotated by angle θ with respect to one another. Performing the multiplications and using trigonometric identities, we can write expressions of the components after the rotation as

$$\hat{a}_{11} = \tfrac{1}{2}\left(a_{11} + a_{22} + (a_{11} - a_{22})\cos(2\theta) - (a_{12} + a_{21})\sin(2\theta)\right) , \tag{3.4}$$

$$\hat{a}_{12} = \tfrac{1}{2}\left(a_{12} - a_{21} + (a_{12} + a_{21})\cos(2\theta) + (a_{11} - a_{22})\sin(2\theta)\right) , \tag{3.5}$$

$$\hat{a}_{21} = \tfrac{1}{2}\left(a_{21} - a_{12} + (a_{12} + a_{21})\cos(2\theta) + (a_{11} - a_{22})\sin(2\theta)\right) ,$$

$$\hat{a}_{22} = \tfrac{1}{2}\left(a_{11} + a_{22} + (a_{22} - a_{11})\cos(2\theta) + (a_{12} + a_{21})\sin(2\theta)\right) .$$

Examining these expressions, we see that a rotation of coordinates by θ results in double-angle expressions. This is a symptom of a deeper property, which appears in other contexts.

One such context is the *Mohr's circle*, illustrated in Exercise 3.4, which was formulated in 1882 by a German civil engineer Christian Otto Mohr to study the stress and strain tensors. It offers, for symmetric second-rank tensors in $\mathbb{R}^2$, their graphical illustration independent of a coordinate system, thus emphasizing this independence, which is an intrinsic property of any tensor. Also, Mohr's circle offers to symmetric second-rank tensors in $\mathbb{R}^2$ an image analogous to an arrow for vectors.

Proceeding with the mathematical abstraction, we can say that the appearance of the double angle results from the double cover of the *special orthogonal group*, $SO(3)$, by the *special unitary group*, $SU(2)$.[7] This property provides us with a convenient method to study tensor symmetries.[8] It stems from the use of complex numbers for rotations as a multiplication by $\exp \iota\theta$, where $\iota := \sqrt{-1}$. It is also related to the *Lie groups* and *Lie algebras*.[9] Also, it follows from the Clebsch-Gordan theorem, that all even-rank tensors in $\mathbb{R}^2$ exhibit this double-angle property.

3.2.3 *Finite and infinitesimal elasticities*

3.2.3.1 *Deformation gradient*

[10] To motivate the examination of material symmetries of Hookean solids in Section 3.2.4, which correspond to infinitesimal deformations, let us consider a more general case, namely, finite deformations, discussed in Appendices A, B and D. As we see below, and as discussed in Appendix D.3, material symmetries of the former are inherited from the latter; information

[7]Readers interested in the orthogonal and unitary groups might refer to Schutz (1980, Section 3.15).

[8]Readers interested in such an approach to study material symmetry might refer to Bóna *et al.* (2004a), where symmetries are characterized using $SU(2)$.

[9]Readers interested in Lie theory, named in the honour of a Norwegian mathematician, Sophus Lie, and in its relations to tensor operations might refer to Stillwell (2008). Readers interested in the biography of Sophus Lie might refer to Stubhaug (2002).

[10]Section 3.2.3 is based on the lecture by Maurizio Vianello of Politecnico di Milano, delivered at Memorial University in 2014.

of directional dependence of deformations to orientations of applied loads is contained within the elasticity tensor.

The theory of finite elasticity contains two assumptions. First, as implied by its adjective, the displacements are allowed to be finite, not only infinitesimal. Secondly, the stress tensor, at any point within a continuum, is a function of the gradient of the displacement field, discussed in Appendix A.1, evaluated at the same point.

Notation 3.1. In this section, in contrast to Appendix A, we do not use upper and lower cases to distinguish between the material and spatial coordinates.

Let us use the Cartesian coordinates to write $f_i(x,t)$ for the coordinates of the motion that places the element of continuum—originally located in the reference configuration at x—at the point whose coordinates at time t are given by f_i, in the current configuration. The correspondence between these two configurations is assumed to be one-to-one. Moreover, the second-rank tensor, F_{ik}, whose components are defined by $\partial f_i/\partial x_k$, and which is referred to as the *deformation gradient*, discussed in Appendix A.2, is required to have a positive determinant at each point and at each instant.

Thus, the displacement components of a point are given by

$$u_i(x,t) = f_i(x,t) - x_i\,, \qquad i,k = 1,2,3\,,$$

where x_i are the components of its reference position. This expression corresponds to expression (A.3) on page 361. Taking the derivative with respect to the reference coordinates, we get

$$\frac{\partial u_i}{\partial x_k} = \frac{\partial f_i}{\partial x_k} - \frac{\partial x_i}{\partial x_k}\,, \qquad i,k = 1,2,3\,,$$

which is the relation between the displacement-vector gradient and the deformation gradient,

$$\frac{\partial u_i}{\partial x_k} = F_{ik} - \delta_{ik}\,, \qquad i,k = 1,2,3\,; \tag{3.6}$$

this expression corresponds to expression (A.5), on page 363.

Formally, the constitutive equation of an elastic solid can be stated as the Cauchy stress tensor, σ_{ij}, being a function of the deformation gradient,

$$\sigma_{ij} = \hat{\sigma}_{ij}(F_{k\ell})\,, \qquad i,j,k,\ell = 1,2,3\,, \tag{3.7}$$

where ˆ distinguishes the function, $\hat{\sigma}_{ij}$, from its values, σ_{ij}. Notice that, for simplicity, we do not assume any explicit dependence of $\hat{\sigma}_{ij}$ on the material point, x_k, which is tantamount to the hypothesis that the body is homogeneous. Still, equation (3.7) is nonlinear; also, its right-hand side can involve coefficients that are components of a tensor of any rank.

To consider material symmetries, let A_{ih} be the Cartesian components of an orthogonal tensor: a tensor that represents a transformation belonging to $O(3)$, which contains—as a subgroup—the set of rotations, $SO(3)$. For any deformation gradient F_{ik}, we can think of $F^*_{hk} = \sum_{i=1}^3 A_{hi}F_{ik}$ as the gradient of the deformation followed by an orthogonal transformation of the current configuration about the point whose coordinates are given by $f_i(x,t)$. The *principle of material frame indifference*,[11] which in this context is universally accepted, states that the stress tensor, σ_{ij}, associated with F_{ik}, and stress $\sigma^*_{\ell m}$ that is associated with F^*_{hk}, must be related by

$$\sigma^*_{\ell m} = \sum_{i=1}^{3}\sum_{j=1}^{3} A_{\ell i}A_{mj}\sigma_{ij}\,, \qquad \ell, m = 1,2,3\,, \tag{3.8}$$

where $*$ distinguishes between the stress tensors and deformation gradients in two configurations. This principle is a restriction to be placed *a priori* on any constitutive equation.[12]

Let us justify equation (3.8), for finite elasticity. Following equation (3.7), we have

$$\hat{\sigma}_{\ell m}\left(\sum_{i=1}^{3} A_{hi}F_{ik}\right) = \sum_{i=1}^{3}\sum_{j=1}^{3} A_{\ell i}A_{mj}\hat{\sigma}_{ij}(F_{hk})\,, \qquad \ell, m, h, k = 1,2,3\,. \tag{3.9}$$

An intuitive interpretation of expression (3.9) can be given by considering a surface, $\mathcal{S}$ which, at point $f_i(x,t)$, has a unit normal, n. According to Cauchy's theorem, the components of traction, T_i, on this oriented surface, are

$$T_i = \sum_{k=1}^{3} \sigma_{ik}n_k\,, \qquad i = 1,2,3\,, \tag{3.10}$$

which is expression (2.15). If we consider an orthogonal transformation, A_{ij}, acting in space, $\mathcal{S}$ is transformed into $\mathcal{S}^*$, whose unit normal is $n^*_i = \sum_{j=1}^3 A_{ij}n_j$, where $i = 1,2,3$. The traction acting on $\mathcal{S}^*$ is $T^*_i = \sum_{k=1}^3 \sigma^*_{ik}n^*_k$, where σ^*_{ik} are the components of the stress tensor associated with the deformation gradient, $F^*_{hk} = \sum_{i=1}^3 A_{hi}F_{ik}$. In this context,

[11]Readers interested in this principle might refer to Truesdell (1966, Lecture I). Such a principle, *sensu lato*, was discussed earlier by Oldroyd (1950).

[12]Readers interested in the condition imposed on the stress tensor by the frame-indifference principle might refer to Truesdell and Noll (2004, Section 84).

the principle of frame indifference is expressed by the natural requirement that

$$T_i^* = \sum_{k=1}^{3} A_{ik} T_k \,, \qquad i = 1, 2, 3 \,,$$

which—in view of expression (3.10)—can be written as

$$\sum_{k=1}^{3} \sigma_{ik}^* n_k^* = \sum_{k=1}^{3} A_{ik} \sum_{j=1}^{3} \sigma_{kj} n_j \,, \qquad i = 1, 2, 3 \,,$$

and, following an algebraic manipulation, as

$$\sum_{k=1}^{3} \sigma_{ik}^* \sum_{j=1}^{3} A_{kj} n_j = \sum_{k=1}^{3} A_{ik} \sum_{j=1}^{3} \sigma_{kj} n_j \,, \qquad i = 1, 2, 3 \,.$$

In view of the arbitrariness of the unit vector, n_j, this is equivalent to

$$\sum_{k=1}^{3} \sigma_{ik}^* A_{kj} = \sum_{k=1}^{3} A_{ik} \sigma_{kj} \,, \qquad i, j = 1, 2, 3 \,,$$

which we multiply on both sides by $A_{\ell j}$ and sum over j. Since $\sum_{j=1}^{3} A_{kj} A_{\ell j} = \delta_{k\ell}$, the final relation is

$$\sigma_{i\ell}^* = \sum_{k=1}^{3} \sum_{j=1}^{3} A_{ik} A_{\ell j} \sigma_{kj} \,, \qquad i, \ell = 1, 2, 3 \,,$$

which, upon renaming the indices is equation (3.8), as required.

3.2.3.2 *Elasticity tensor: Inheritance from finite elasticity*

The elasticity tensor, $c_{ijk\ell}$, which in the constitutive equation given in expression (3.1) provides a linear relation between two second-rank tensors, is defined here—within finite elasticity—as the derivative of the Cauchy stress tensor with respect to the deformation gradient, evaluated at δ_{hk},

$$c_{ijhk} = \left. \frac{\partial \hat{\sigma}_{ij}}{\partial F_{hk}} \right|_{F_{hk} = \delta_{hk}} \,, \qquad i, j, h, k = 1, 2, 3 \,. \tag{3.11}$$

The motivation for this formulation is quite natural. If we multiply a finite displacement, $\bar{u}_h$, by ϵ, to get $u_h = \epsilon \, \bar{u}_h$, in view of expression (3.6), we have

$$\sigma_{ij} = \hat{\sigma}_{ij} \left(\delta_{hk} + \epsilon \frac{\partial \bar{u}_h}{\partial x_k} \right) , \qquad i, j, h, k = 1, 2, 3 \,,$$

which can be approximated by

$$\sigma_{ij} = \hat{\sigma}_{ij}(\delta_{hk}) + \sum_{h=1}^{3}\sum_{k=1}^{3} \left.\frac{\partial \hat{\sigma}_{ij}}{\partial F_{hk}}\right|_{F_{hk}=\delta_{hk}} \epsilon \frac{\partial \bar{u}_h}{\partial x_k} + o(\epsilon)\,, \qquad i,j = 1,2,3\,,$$

where the first term on the right-hand side is the stress tensor in the reference configuration and $o(\epsilon)$ is a Landau symbol, named in honour of a Russian physicist Lev Davidovich Landau, and also refereed to as the little o.[13,14] Invoking expression (3.11) in the second term on the right-hand side, we write

$$\sigma_{ij} = \hat{\sigma}_{ij}(\delta_{hk}) + \sum_{h=1}^{3}\sum_{k=1}^{3} c_{ijhk} \frac{\partial u_h}{\partial x_k} + o(\epsilon)\,, \qquad i,j = 1,2,3\,. \tag{3.12}$$

Thus, if we assume—from now on—that the stress tensor in the reference configuration is zero, and if we consider the last term on the right-hand side to be negligible, the stress-strain equation becomes

$$\sigma_{ij} = \sum_{h=1}^{3}\sum_{k=1}^{3} c_{ijhk} \frac{\partial u_h}{\partial x_k}\,, \qquad i,j = 1,2,3\,, \tag{3.13}$$

which is similar to Hooke's law stated in expression (3.1), except that the strain tensor

$$\varepsilon_{hk} := \frac{1}{2}\left(\frac{\partial u_h}{\partial x_k} + \frac{\partial u_k}{\partial x_h}\right)\,, \tag{3.14}$$

is only the symmetric part of $\partial u_h/\partial x_k$, which also contains rotations.

Note that even if the *residual stress*, $\hat{\sigma}_{ij}(\delta_{hk})$, is not assumed to be zero, we still obtain an infinitesimal elasticity theory but a nonlinear one. Herein, with $\hat{\sigma}_{ij}(\delta_{hk}) = 0$, we do not make any distinction in terminology between the infinitesimal and linear elasticity.

Every tensor can be written as the sum of its symmetric and antisymmetric parts.[15] The justification for the absence of the antisymmetric part results from the principle of frame indifference.

We need to use the property that states that, for any antisymmetric tensor, $\xi_{hk} = -\xi_{kh}$, there is a function, $A_{hk}(s)$, defined on a real interval containing zero and with values in $SO(3)$, such that $A_{hk}(0) = \delta_{hk}$ and $A'_{hk}(0) = \xi_{hk}$, where $'$ denotes the derivative with respect to s. In the context of differential geometry, this implies that the space of antisymmetric

[13] *see also*: Bóna and Slawinski (2015, Section 4.3.1)

[14] Readers interested in the little o notation might refer to Fowler (1997, Section 3.1)

[15] *see also*: Slawinski (2015, Section 1.5)

tensors is the tangent space to the group of rotations at the identity; a proof of this result is shown in Exercise 3.5.[16]

Since we assume that $\hat{\sigma}_{ij}(\delta_{hk}) = 0$, it follows—from the principle of frame indifference—that, for any rotation, $\hat{\sigma}_{ij}(A_{ih}\delta_{hk}) = \hat{\sigma}_{ij}(A_{ik}) = 0$. In other words, any configuration obtained through a rotation of the unstressed reference configuration is also unstressed.

To proceed, we fix an antisymmetric tensor, ξ_{hk}, and let $A_{hk}(s) \in SO(3)$ be such that $A_{hk}(0) = \delta_{hk}$ and $A'_{hk}(0) = \xi_{hk}$. Taking the derivative of $\hat{\sigma}_{ij}(A_{hk}(s)) = 0$ with respect to s, we obtain

$$\frac{\mathrm{d}}{\mathrm{d}s}\,\hat{\sigma}_{ij}(A_{hk}(s))\big|_{s=0} = \sum_{h=1}^{3}\sum_{k=1}^{3}\left.\frac{\partial\hat{\sigma}_{ij}}{\partial F_{hk}}\right|_{F_{hk}=\delta_{hk}} A'_{hk}(0) = \sum_{h=1}^{3}\sum_{k=1}^{3} c_{ijhk}\xi_{hk} = 0\,, \tag{3.15}$$

where $i,j = 1,2,3$. This expression implies that if we write the displacement gradient as the sum of its symmetric and antisymmetric parts, $\varepsilon_{hk} + \xi_{hk}$, it follows—from expressions (3.13) and (3.15)—that

$$\sigma_{ij} = \sum_{h=1}^{3}\sum_{k=1}^{3} c_{ijhk}\varepsilon_{hk}\,, \qquad i,j = 1,2,3\,, \tag{3.16}$$

which is expression (3.1), as expected. Thus, the elasticity tensor defined by expression (3.11) exhibits the index symmetries given by $c_{ijhk} = c_{jihk} = c_{ijkh}$, which are known as the *minor symmetries* and—within the infinitesimal theory of elasticity—are usually assumed *a priori*. The other index symmetries, $c_{ijhk} = c_{hkij}$, which are known as the *major symmetries*, can be deduced under the assumption of the existence of the strain-energy function, from which equation (3.7) can be obtained.[17] To derive major symmetries in finite elasticity, we would need to invoke the Piola-Kirchhoff stress tensor, which is discussed in Appendix B; herein, the first name corresponds to Gabrio Piola, whose monument and street name are shown in Figures 3.1 and 3.2, respectively.[18]

3.2.3.3 *Prestressed linearly elastic materials*

It should be emphasized that expression (3.16) is derived under the assumption that the stress tensor in the reference configuration is zero. This

[16]Readers interested in standard proofs might refer to textbooks on geometry of the Lie Groups, for instance to Warner (1983, Chapter 3, Section 3.37); another proof, whose formulation is particularly pertinent to our discussion, can be found in Gurtin (1981, Appendix, Section 36).

[17]*see also*: Slawinski (2015, Chapters 3 and 4)

[18]Readers interested in the life and work of Gabrio Piola might refer to Ruta (2016).

photo: Elena Patarini

Fig. 3.1: The monument of Gabrio Piola in Milan, by Vincenzo Vela

is a significant restriction: it excludes situations in which the material is *prestressed*,[19] which can be found in a wide range of physical applications.[20]

One might be tempted to think that a correct constitutive relation

[19]Readers interested in mathematical formulations that include such situations might refer to Hoger (1986).

[20]Readers interested in geophysical applications that include the residual stress might refer to Dahlen and Tromp (1998, Chapter 3).

photo: Elena Patarini

Fig. 3.2: City square in Milan named after Gabrio Piola; according to encyclopedic information, the date of birth is 1794.

can be obtained just by adding one term to the right-hand side of expression (3.16), namely, $\sigma_{ij}^0 = \hat{\sigma}_{ij}(\delta_{hk})$, which is the residual stress: the Cauchy stress tensor in the reference configuration. However, this is not the case. As we show below—in the presence of a residual stress—another modification is needed.

Indeed, differentiating the constitutive relation for the Cauchy stress tensor, given by expression (3.7), we obtain expression (3.12) which—neglecting, as before, $o(\epsilon)$, but under the new assumption that the residual stress is not zero—we can write as

$$\sigma_{ij} = \sigma_{ij}^0 + \sum_{h=1}^{3}\sum_{k=1}^{3} c_{ijhk}\frac{\partial u_h}{\partial x_k}, \qquad i,j = 1,2,3. \tag{3.17}$$

We separate the displacement gradient into its symmetric and antisymmetric parts, ε_{hk} and ξ_{hk}, and deduce that

$$\sigma_{ij} = \sigma_{ij}^0 + \sum_{h=1}^{3}\sum_{k=1}^{3} c_{ijhk}\xi_{hk} + \sum_{h=1}^{3}\sum_{k=1}^{3} c_{ijhk}\varepsilon_{hk}, \qquad i,j = 1,2,3. \tag{3.18}$$

As we proceed to show, in general—with $\sigma_{ij}^0 \neq 0$—the first double sum is not zero.

Let us invoke the principle of frame indifference stated by expression (3.9). If $F_{ik} = \delta_{ik}$, we obtain

$$\hat{\sigma}_{\ell m}(A_{hk}) = \sum_{i=1}^{3}\sum_{j=1}^{3} A_{\ell i}A_{mj}\sigma_{ij}^0, \qquad \ell,m,h,k = 1,2,3, \tag{3.19}$$

where $\sigma_{ij}^0 = \hat{\sigma}_{ij}(\delta_{hk})$ is the residual stress.

In a manner presented on page 54—for an arbitrary antisymmetric tensor, ξ_{hk}—we let $A_{hk}(s) \in SO(3)$ be such that $A_{hk}(0) = \delta_{hk}$ and $A'_{hk}(0) = \xi_{hk}$. In view of the definition of $c_{\ell mhk}$ stated in expression (3.11), after substitution of $A_{hk}(s)$ in expression (3.19) and differentiation with respect to s evaluated at $s = 0$, we obtain

$$\sum_{h=1}^{3}\sum_{k=1}^{3} c_{\ell mhk}\xi_{hk} = \sum_{i=1}^{3}\sum_{j=1}^{3} \xi_{\ell i}\delta_{mj}\sigma_{ij}^{0} + \sum_{i=1}^{3}\sum_{j=1}^{3} \delta_{\ell i}\xi_{mj}\sigma_{ij}^{0}$$

$$= \sum_{i=1}^{3} \xi_{\ell i}\sigma_{im}^{0} + \sum_{j=1}^{3} \xi_{mj}\sigma_{\ell j}^{0}, \qquad \ell, m = 1,2,3,$$

which—due to the antisymmetry of ξ_{hk}, and upon renaming an index—can be written as

$$\sum_{h=1}^{3}\sum_{k=1}^{3} c_{\ell mhk}\xi_{hk} = \sum_{i=1}^{3} \xi_{\ell i}\sigma_{im}^{0} - \sum_{i=1}^{3} \sigma_{\ell i}^{0}\xi_{im}, \qquad \ell, m = 1,2,3.$$

We conclude that relation (3.18) for the Cauchy stress in a linearly elastic material with a given prestress is

$$\sigma_{\ell m} = \sigma_{\ell m}^{0} + \sum_{i=1}^{3} \xi_{\ell i}\sigma_{im}^{0} - \sum_{i=1}^{3} \sigma_{\ell i}^{0}\xi_{im} + \sum_{h=1}^{3}\sum_{k=1}^{3} c_{\ell mhk}\varepsilon_{hk}, \qquad i,j = 1,2,3. \tag{3.20}$$

The two middle summands,

$$\sum_{i=1}^{3} \xi_{\ell i}\sigma_{im}^{0} - \sum_{i=1}^{3} \sigma_{\ell i}^{0}\xi_{im}, \qquad \ell, m = 1,2,3, \tag{3.21}$$

which contain the effect of the antisymmetric part of the displacement gradient, disappear if the residual stress is isotropic, which means that $\sigma_{ij}^{0} = a\,\delta_{ij}$, where a is a scalar. It is also possible to argue that if the residual stress is infinitesimal, comparable in a certain manner to the displacement gradient, then these two summands could be considered negligible, being infinitesimal of a higher order.

In general, however, the constitutive relation for the Cauchy stress tensor should be written as in expression (3.20), not forgetting the additional—and perhaps unexpected—terms given in expression (3.21).

It is also important to point out that in such a case the Piola-Kirchhoff stress tensors and the Cauchy stress tensor do not coincide, even to the first order, as is the case if $\sigma_{ij}^{0} = 0$. This implies that the linearized constitutive relation for the Piola-Kirchhoff stress tensors is different from expression (3.20), and, as a consequence, makes the theory of prestressed linearly elastic materials more challenging. This topic, however, is beyond the aim of the present text.

3.2.3.4 *Material symmetry: Finite elasticity*

By definition, the material-symmetry group for finite elasticity is the symmetry group of function $\hat{\sigma}_{ij}$, given in expression (3.7); it is

$$\hat{G}^{\mathrm{sym}} = \left\{A_{k\ell} \in O(3) \colon \hat{\sigma}_{ij}\left(\sum_{m=1}^{3} F_{km}A_{m\ell}\right) = \hat{\sigma}_{ij}(F_{k\ell})\right\}, \tag{3.22}$$

for all $F_{k\ell}$, where $i,j,k,\ell = 1,2,3$. The terms in parentheses are the arguments of $\hat{\sigma}_{ij}$. In other words, an orthogonal transformation, $A_{k\ell} \in O(3)$, belongs to $\hat{G}^{\mathrm{sym}}$ if the stress tensor remains unchanged if the continuum—prior to an arbitrary deformation—is transformed by $A_{k\ell}$ about point x_k. This means that $A_{k\ell} \in \hat{G}^{\mathrm{sym}}$ is undetectable by any mechanical experiment, since the stress induced by any deformation is not changed if the continuum is subject to $A_{k\ell}$ before being deformed. It is not difficult to check that $\hat{G}^{\mathrm{sym}}$ is a group in the mathematical sense of the term, since it satisfies all requirements.

There are two extreme cases, with many others in between. The symmetry group can be $\hat{G}^{\mathrm{iso}} = O(3)$, which means that the stress tensor is invariant under all orthogonal transformations; in such a case the continuum is isotropic. The other extreme, $\hat{G}^{\mathrm{aniso}} = \{\delta_{ik}\}$, means that any orthogonal transformation—different from the identity—is detectible, and does indeed change the stress tensor by its action; the body is completely anisotropic.

If the orthogonal group remains invariant under rotation about a given direction, we say that the body is transversely isotropic, and its symmetry group is $\hat{G}^{\mathrm{TI}} = O(2)$. If the group coincides with the set of all orthogonal transformations that leave a given cube invariant, the body is said to be of cubic material symmetry, and its group is $\hat{G}^{\mathrm{cubic}}$. However, it is important to emphasize that material symmetry is distinct from the geometric symmetry: unlike for crystal symmetries, the cubic symmetry in the sense of $\hat{G}^{\mathrm{cubic}}$ does not imply and is not implied by a cubic shape.

Below, we proceed to discuss the relationship between material symmetry for the finite and infinitesimal elasticities.

3.2.3.5 *Material symmetry: Relation between finite and infinitesimal elasticities*

The symmetry group of the elasticity tensor itself, which is tantamount to the symmetry group for the infinitesimal elasticity, is defined by

$$G^{\mathrm{sym}} = \left\{ A_{ij} \in O(3) \colon c_{ijk\ell} = \sum_{m=1}^{3}\sum_{n=1}^{3}\sum_{o=1}^{3}\sum_{p=1}^{3} A_{im}A_{jn}A_{ko}A_{\ell p}c_{mnop} \right\}, \tag{3.23}$$

where $i, j, k, \ell = 1, 2, 3$; it is possible—without loss of generality, for elastic continua—to use the rotation subgroup of $O(3)$, which is $SO(3)$, as discussed by Bóna *et al.* (2004b, Section 2.4).

Remark 3.1. $O(3)$ contains both reflections and rotations. $SO(3)$ contains only rotations; thus, $SO(3) \subset O(3)$. Rotations constitute a group, since the product of two rotations is also a rotation. Reflections, however, do not constitute a group, since the product of two reflections is a rotation. For rotations, its transpose is its inverse; a reflection is its own inverse, which herein implies that reflection is a symmetric tensor.

Definition (3.23) is used following expression (3.31), below. However, prior to using this definition, let us derive it from the finite-elasticity theory, since—if we begin within the context of the infinitesimal elasticity itself—it is not easy to motivate definition (3.23). A convincing approach comes from the relationship between the material-symmetry groups for finite and infinitesimal elasticities, since definition (3.22) is grounded in physical intuition.

Indeed, we now prove that $\hat{G}^{\mathrm{sym}} \subset G^{\mathrm{sym}}$ by showing that each symmetry transformation of the material-symmetry group of finite elasticity induces the same symmetry transformation for the elasticity tensor. In other words, any symmetry transformation of the elasticity tensor, $c_{ijk\ell}$, originates as a symmetry transformation of function $\hat{\sigma}_{ij}$; in other words, as we see below, properties of a function are inherited by its derivative with respect to the deformation gradient.

To consider the deformation gradient for an infinitesimal deformation, let $H_{\ell m}$ be a symmetric tensor and let

$$F_{\ell m}(s) = \delta_{\ell m} + sH_{\ell m}\,, \qquad \ell, m = 1, 2, 3\,, \tag{3.24}$$

for s in a small interval containing zero. Notice that, by reason of continuity, since $\det(\delta_{\ell m}) = 1$, we know that $\det(F_{\ell m}(s)) > 0$, for sufficiently small values of s.

Consider the property of frame indifference stated in expression (3.9) and assume that a given orthogonal transformation belongs to the material-symmetry group of finite elasticity, $A_{hk} \in \hat{G}^{\rm sym}$. In view of definition (3.22), it follows that

$$\hat{\sigma}_{ij}\left(\sum_{\ell=1}^{3}\sum_{m=1}^{3}A_{h\ell}A_{km}F_{\ell m}\right)=\sum_{p=1}^{3}\sum_{r=1}^{3}A_{ip}A_{jr}\hat{\sigma}_{pr}(F_{\ell m})\,,\qquad i,j=1,2,3\,,$$

for any deformation gradient, $F_{\ell m}$. Next, we insert expression (3.24) into the above equation to obtain

$$\hat{\sigma}_{ij}\left(\sum_{\ell=1}^{3}\sum_{m=1}^{3}A_{h\ell}A_{km}(\delta_{\ell m}+sH_{\ell m})\right)=\sum_{p=1}^{3}\sum_{r=1}^{3}A_{ip}A_{jr}\hat{\sigma}_{pr}(\delta_{\ell m}+sH_{\ell m})\,,$$

where $i,j=1,2,3$. Since $\sum_{\ell=1}^{3}\sum_{m=1}^{3}A_{h\ell}A_{km}\delta_{\ell m}=\sum_{\ell=1}^{3}A_{h\ell}A_{k\ell}=\delta_{hk}$, it follows that

$$\hat{\sigma}_{ij}(\delta_{hk}+s\sum_{\ell=1}^{3}\sum_{m=1}^{3}A_{h\ell}A_{km}H_{\ell m})=\sum_{p=1}^{3}\sum_{r=1}^{3}A_{ip}A_{jr}\hat{\sigma}_{pr}(\delta_{\ell m}+sH_{\ell m})\,,\tag{3.25}$$

where $i,j=1,2,3$. Using expression (3.24), let us find the derivative of equation (3.25), with respect to s and evaluated at $s=0$.

Considering the left-hand side, we write

$$\sum_{hk\ell m=1}^{3}\left.\frac{\partial\hat{\sigma}_{ij}}{\partial F_{hk}}\right|_{F_{hk}=\delta_{hk}+s\sum_{\ell m}A_{h\ell}A_{km}H_{\ell m}}A_{h\ell}A_{km}H_{\ell m}\,,$$

where, for conciseness, we state the four summations as a single summation symbol. This expression—evaluated at $s=0$ and in view of definition (3.11)—becomes

$$\sum_{hk\ell m=1}^{3}\left.\frac{\partial\hat{\sigma}_{ij}}{\partial F_{hk}}\right|_{F_{hk}=\delta_{hk}}A_{h\ell}A_{km}H_{\ell m}=\sum_{hk\ell m=1}^{3}c_{ijhk}A_{h\ell}A_{km}H_{\ell m}\,.\tag{3.26}$$

Considering the derivative of the right-hand side of equation (3.25), we write

$$\sum_{\ell mpr=1}^{3}A_{ip}A_{jr}\left.\frac{\partial\hat{\sigma}_{pr}}{\partial F_{\ell m}}\right|_{F_{\ell m}=\delta_{\ell m}+sH_{\ell m}}H_{\ell m}\,,\qquad i,j=1,2,3\,,$$

which—for $s=0$ and in view of definition (3.11)—becomes

$$\sum_{\ell mpr=1}^{3}A_{ip}A_{jr}\left.\frac{\partial\hat{\sigma}_{pr}}{\partial F_{\ell m}}\right|_{F_{\ell m}=\delta_{\ell m}}H_{\ell m}=\sum_{\ell mpr=1}^{3}A_{ip}A_{jr}c_{pr\ell m}H_{\ell m}\,.\tag{3.27}$$

Equating derivatives (3.26) and (3.27), we obtain

$$\sum_{hk\ell m=1}^{3} c_{ijhk} A_{h\ell} A_{km} H_{\ell m} = \sum_{\ell mpr=1}^{3} A_{ip} A_{jr} c_{pr\ell m} H_{\ell m} \,. \tag{3.28}$$

This equation is valid for any $A_{km} \in \hat{G}^{\mathrm{sym}}$ and for any symmetric tensor, $H_{\ell m}$. Since $H_{\ell m}$ is arbitrary, we deduce that equation (3.28) holds for any $A_{km} \in G^{\mathrm{sym}}$ if and only if

$$\sum_{h=1}^{3}\sum_{k=1}^{3} c_{ijhk} A_{h\ell} A_{km} = \sum_{p=1}^{3}\sum_{r=1}^{3} A_{ip} A_{jr} c_{pr\ell m} \,, \qquad i,j,\ell,m = 1,2,3 \,. \tag{3.29}$$

Since $\sum_{\ell=1}^{3} A_{o\ell} A_{h\ell} = \delta_{oh}$ and $\sum_{m=1}^{3} A_{sm} A_{km} = \delta_{sk}$, we multiply both sides by $A_{o\ell} A_{sm}$ and sum over ℓ and m to obtain

$$c_{ijos} = \sum_{\ell=1}^{3}\sum_{m=1}^{3}\sum_{p=1}^{3}\sum_{r=1}^{3} A_{ip} A_{jr} A_{o\ell} A_{sm} c_{pr\ell m} \,, \qquad i,j,o,s = 1,2,3 \,,$$

which—compared with expression (3.23), upon renaming of indices—shows that $A_{ik} \in G^{\mathrm{sym}}$, as required.

Thus, each orthogonal transformation that belongs to $\hat{G}^{\mathrm{sym}}$, which is the symmetry group for finite elasticity, belongs to G^{sym}, which is the symmetry group for the infinitesimal-elasticity tensor: $\hat{G}^{\mathrm{sym}} \subset G^{\mathrm{sym}}$, as required. In other words, any symmetry transformation that exists within the finite elasticity generates a symmetry transformation within the infinitesimal elasticity. This is not to say that the finite and infinitesimal elasticities have the same number of symmetry groups, since the same symmetry transformations can be arranged differently within their respective groups. For instance, the finite-elasticity rotational symmetries greater than fourfold about a given axis, which could form distinct groups therein, such as hexagonal and octagonal symmetries, generate—due to Theorem 3.2 of Herman,[21] discussed in Section 3.2.4.17, below—the rotational invariance about that axis, within infinitesimal elasticity; hence, they are all contained within the single group, which is transverse isotropy.

A different number of symmetry classes in the finite and infinitesimal elasticities is a consequence of the fact that symmetry groups of the former express the invariance of a second-rank tensor-valued function in $\mathbb{R}^3$, whose general form is given by expression (3.7), and, of the latter, express the invariance of a fourth-rank tensor, $c_{ijk\ell}$, with required index symmetries. Herein, *class* stands for a symmetry group that has a specific geometrical

[21] *see also*: Slawinski (2015, Theorem 5.10.1 and Exercise 5.8)

property, even though it can be a subgroup of another group; for instance, the cubic class and isotropic class exhibit distinct geometrical properties, even though the former is a subgroup of the latter, which consists of $O(3)$.

Thus, only several among infinitely many subgroups of $O(3)$ belong to the material-symmetry groups of infinitesimally elastic continua; for instance, again due to the Herman (1945) theorem, there are no discrete symmetries greater than fourfold rotations about a given axis. If the major index symmetries hold, there are eight symmetry classes, listed in Figure 3.3, below,[22] which is the case we discuss in the remainder of this chapter, if they do not, there are ten classes. Many more—and, depending on complexity of function (3.7), possibly all—subgroups of $O(3)$ could be symmetry classes for finitely elastic continua; yet, it is not known how many among these groups are mathematical analogies for physical materials.

3.2.4 *Symmetry classes*

3.2.4.1 *Material-symmetry conditions*

Once we have chosen a Hookean solid as the mathematical concept to represent the Earth in examining propagation of disturbances therein, we have to be aware of properties of that solid, with emphasis on possible analogies between its properties and properties of materials composing our planet. Since properties of these materials exhibit certain symmetries of mechanical behaviour, let us consider symmetries of Hookean solids, which are mathematical properties.

There are eight material-symmetry classes of a Hookean solid; this is a theorem, as shown by Forte and Vianello (1996) and Bóna *et al.* (2004b). One cannot invent a symmetry class beyond these eight to accommodate a particular material. Notably, as discussed below, in Sections 3.2.4.16 and 3.2.4.17, respectively, there are no Hookean solids with only two perpendicular symmetry planes and no solids whose discrete rotation symmetry is greater than fourfold, even though there are materials that exhibit a hexagonal symmetry, for instance, crystals.

The most symmetric Hookean solid is *isotropic*; it exhibits no directional properties. The least symmetric is *generally anisotropic*; its only directional property is no distinction of direction along any line of a given orientation. The remaining six classes, which exhibit different directional properties, are organized by partial ordering shown in Figure 3.3.

---

[22] *see also*: Slawinski (2015, Chapter 5)

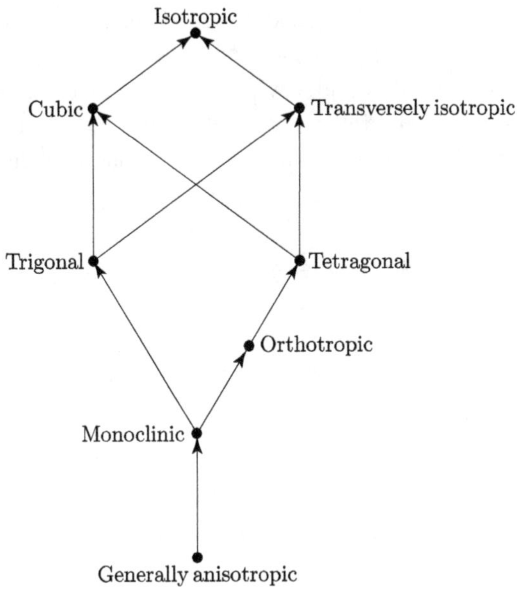

Fig. 3.3: Order relation of material-symmetry classes of elasticity tensors: Arrows indicate subgroups in this partial ordering. For instance, monoclinic is a subgroup of all nontrivial symmetries, in particular, of both orthotropic and trigonal; orthotropic is not a subgroup of trigonal or *vice versa.*

The eight symmetry classes are isotropy, cubic symmetry, transverse isotropy (TI), tetragonal symmetry, trigonal symmetry, orthotropic symmetry, monoclinic symmetry and general anisotropy. Let us examine the definition of symmetry.

The material symmetry of tensor $\mathbf{c}$ is its invariance to orthogonal transformations,

$$\mathbf{c} = g \circ \mathbf{c}; \tag{3.30}$$

herein, g is an orthogonal transformation that belongs to the symmetry group of $\mathbf{c}$, where a group is an algebraic concept of the geometric idea of symmetry. Elasticity tensors that satisfy condition (3.30), for each g of a given symmetry group, are denoted by $\mathbf{c}^{\mathrm{sym}}$, where sym stands for the symmetry class corresponding to that group. As stated in definition (3.23), expressing $\mathbf{c}$ in an orthonormal basis in $\mathbb{R}^3$, we can write—in components—its symmetry with respect to a given element of the symmetry group as

$$c_{ijk\ell} = \sum_{m=1}^{3}\sum_{n=1}^{3}\sum_{o=1}^{3}\sum_{p=1}^{3} A_{im}A_{jn}A_{ko}A_{\ell p}c_{mnop}\,, \qquad A \in O(3)\,, \tag{3.31}$$

which means that $\mathbf{c}$ is invariant under the action of transformation A.

As discussed by Bóna *et al.* (2004b) and Diner *et al.* (2011b), the symmetry group of any elasticity tensor, $G^{\mathrm{sym}} = \{g_1, \ldots, g_n\}$, where g_i are the orthogonal transformations under whose action a given tensor is invariant, is conjugate to one of the following eight groups, which is tantamount to saying that any G^{sym} belongs to at least one of these groups.

$$G^{\mathrm{iso}} = O(3)\,, \tag{3.32}$$

$$G^{\mathrm{cubic}} = \{A \in O(3) \mid A(e_i) = \pm e_j\}\,, \tag{3.33}$$

for any $i, j = 1, 2, 3$,

$$G^{\mathrm{TI}} = \{\pm I, \pm R_{\theta, \hat{\mathbf{e}}_3}, \pm M_{\mathbf{v}} \mid \theta \in [0, 2\pi)\}\,, \tag{3.34}$$

where $\mathbf{v}$ is in the $\hat{\mathbf{e}}_1\hat{\mathbf{e}}_2$-plane,

$$\begin{aligned} G^{\mathrm{tetra}} = \{ &\pm I, \pm R_{\frac{\pi}{2}, \hat{\mathbf{e}}_3}, \pm R_{-\frac{\pi}{2}, \hat{\mathbf{e}}_3}, \\ &\pm M_{\hat{\mathbf{e}}_1}, \pm M_{\hat{\mathbf{e}}_2}, \pm M_{\hat{\mathbf{e}}_3}, \pm M_{(1,1,0)}, \pm M_{(1,-1,0)}\}\,, \end{aligned} \tag{3.35}$$

$$\begin{aligned} G^{\mathrm{trig}} = \{ &\pm I, \pm R_{\frac{2\pi}{3}, \hat{\mathbf{e}}_3}, \pm R_{-\frac{2\pi}{3}, \hat{\mathbf{e}}_3}, \\ &\pm M_{\hat{\mathbf{e}}_1}, \pm M_{(\cos\frac{\pi}{3}, \sin\frac{\pi}{3}, 0)}, \pm M_{(\cos\frac{2\pi}{3}, \sin\frac{2\pi}{3}, 0)}\}\,, \end{aligned} \tag{3.36}$$

$$G^{\mathrm{ortho}} = \{\pm I, \pm M_{\hat{\mathbf{e}}_1}, \pm M_{\hat{\mathbf{e}}_2}, \pm M_{\hat{\mathbf{e}}_3}\}\,, \tag{3.37}$$

$$G^{\mathrm{mono}} = \{\pm I, \pm M_{\hat{\mathbf{e}}_3}\}\,, \tag{3.38}$$

$$G^{\mathrm{aniso}} = \{\pm I\}, \tag{3.39}$$

where $R_{\theta, \hat{\mathbf{e}}_i}$ denotes the rotation around $\hat{\mathbf{e}}_i$ by θ and $M_{\mathbf{v}}$ denotes the reflection about the plane whose normal is $\mathbf{v}$.

The fact that any G^{sym} belongs to *at least* one of these groups is illustrated by the property that, for example, if $\mathbf{c}$ exhibits the symmetry of G^{ortho}, it also exhibits symmetries of G^{mono} and G^{aniso}, since it possesses the required symmetry-group elements. By reciprocity, any element of G^{aniso} belongs to G^{mono}, and any element of G^{mono} belongs to G^{ortho}.

The order relation among these groups is shown in Figure 3.3, where the vertices are connected by a line if the symmetry group corresponding to the vertex below is contained in the symmetry group corresponding to the vertex above. The partial ordering allows us to infer physical properties. For instance, it is impossible to obtain an orthotropic continuum from a trigonal one or *vice versa*, since it is not enough to remove group elements of one continuum to obtain the other. In the context of inferring material properties, this suggests, for instance, an intrinsic distinction between oblique and orthogonal fractures.

In the context of symmetry groups, conjugation means that two elasticity tensors, $\mathbf{c}_1$ and $\mathbf{c}_2$, belong to the same symmetry class if their symmetry groups, $G^{\mathbf{c}_1}$ and $G^{\mathbf{c}_2}$, are conjugate to one another, which means that there exists an orthogonal transformation, $A \in O(3)$, such that $G^{\mathbf{c}_2} = A\, G^{\mathbf{c}_1} A^T$. Not including the condition for conjugation in the definition might result in counting twice the same symmetry group due to distinct coordinate systems in which it is expressed.

Since $\mathbf{c}$ is an even-rank tensor, $-I$ belongs to each of its symmetry groups. Hence, if $\mathbf{c}$ is invariant under A, it is also invariant under $-A$. Thus, without loss of generality, we can consider only rotations for transformations in expressions (3.32)–(3.38), which is tantamount to replacing $O(3)$ by $SO(3)$.

3.2.4.2 *Hooke's law in* $\mathbb{R}^3$ *and* $\mathbb{R}^6$

[23]According to its definition, the elasticity tensor possesses the following index symmetries:

$$c_{ijk\ell} = c_{jik\ell} = c_{k\ell ij}\,, \tag{3.40}$$

which are distinct from material symmetries, and result from properties of the stress and strain tensors as well as from the existence of the strain-energy function. Hence, we can write its components as entries of a symmetric 6×6 matrix whose index $m = 1, \ldots, 6$, is related to $i, j = 1, 2, 3$, by $m = i\,\delta_{ij} + (1 - \delta_{ij})(9 - i - j)$, with δ_{ij} being the Kronecker delta, named in honour of the German mathematician Leopold Kronecker; indices n, k, ℓ are related by the same formula.

Thus, equation (3.1) can be written as

$$\begin{bmatrix} \sigma_{11} \\ \sigma_{22} \\ \sigma_{33} \\ \sqrt{2}\sigma_{23} \\ \sqrt{2}\sigma_{13} \\ \sqrt{2}\sigma_{12} \end{bmatrix} = \begin{bmatrix} c_{1111} & c_{1122} & c_{1133} & \sqrt{2}c_{1123} & \sqrt{2}c_{1113} & \sqrt{2}c_{1112} \\ c_{1122} & c_{2222} & c_{2233} & \sqrt{2}c_{2223} & \sqrt{2}c_{2213} & \sqrt{2}c_{2212} \\ c_{1133} & c_{2233} & c_{3333} & \sqrt{2}c_{3323} & \sqrt{2}c_{3313} & \sqrt{2}c_{3312} \\ \sqrt{2}c_{1123} & \sqrt{2}c_{2223} & \sqrt{2}c_{3323} & 2c_{2323} & 2c_{2313} & 2c_{2312} \\ \sqrt{2}c_{1113} & \sqrt{2}c_{2213} & \sqrt{2}c_{3313} & 2c_{2313} & 2c_{1313} & 2c_{1312} \\ \sqrt{2}c_{1112} & \sqrt{2}c_{2212} & \sqrt{2}c_{3312} & 2c_{2312} & 2c_{1312} & 2c_{1212} \end{bmatrix} \begin{bmatrix} \varepsilon_{11} \\ \varepsilon_{22} \\ \varepsilon_{33} \\ \sqrt{2}\varepsilon_{23} \\ \sqrt{2}\varepsilon_{13} \\ \sqrt{2}\varepsilon_{12} \end{bmatrix}, \tag{3.41}$$

[23]*see also*: Slawinski (2015, Chapters 3 and 4)

where tensor $\mathbf{c}$ is denoted by C. $\sqrt{2}$ and 2 ensure that the basis of the stress and strain tensors is the same, namely,

$$\begin{bmatrix}1&0&0\\0&0&0\\0&0&0\end{bmatrix},\begin{bmatrix}0&0&0\\0&1&0\\0&0&0\end{bmatrix},\begin{bmatrix}0&0&0\\0&0&0\\0&0&1\end{bmatrix},\frac{1}{\sqrt{2}}\begin{bmatrix}0&0&0\\0&0&1\\0&1&0\end{bmatrix},\frac{1}{\sqrt{2}}\begin{bmatrix}0&0&1\\0&0&0\\1&0&0\end{bmatrix},\frac{1}{\sqrt{2}}\begin{bmatrix}0&1&0\\1&0&0\\0&0&0\end{bmatrix},$$

which is a convenient property in examining rotations of the elasticity tensor. We refer to matrix (3.41) as the Kelvin notation in view of Thomson (1890, p. 110). This notation is discussed in textbooks, notably, in Chapman (2004, Section 4.4.2). We comment on it in Section 3.2.4.4, below.

3.2.4.3 *Index symmetries*

Symmetries (3.40) are contained in the general definition of a Hookean solid. In other words, any Hookean solid exhibits these symmetries. These symmetries are induced by the index symmetries of the stress and strain tensors and the symmetry of their index pairs, which in turn result from—as does the entire subject of continuum mechanics—the underlying assumptions. The first results from limiting the formulation to the strong form of Newton's third law,[24] the second from considering only infinitesimal deformations,[25] and the third from assuming the existence of the strain-energy function.[26] Relaxing these assumptions is tantamount to reaching beyond the definition of a Hookean solid.

Different index symmetries result in different material-symmetry groups than the ones given by expressions (3.32)–(3.39), which are limited to Hookean solids. Since in this book we choose to remain within these solids, we do not derive such expressions. However, to gain an insight into aspects that are ignored by this choice, let us state, for different index-symmetry classes, the number of linearly independent elasticity parameters required to define the two material-symmetry extremes in a given class, namely, isotropy and general anisotropy of $c_{ijk\ell}$.

If there is no index symmetry, which means that—in general—no exchange among indices is allowed without modifying the values of the corresponding parameters, the general anisotropy requires eighty-one parameters and isotropy requires three,[27] in contrast to twenty-one stated in expression (3.41), above, and two in expression (3.57), below, which correspond to the index symmetries stated in expression (3.40). If there is a com-

[24] *see also*: Slawinski (2015, Section 2.7.3)
[25] *see also*: Slawinski (2015, Section 1.4.2)
[26] *see also*: Slawinski (2015, Section 4.2)
[27] *see also*: Slawinski (2015, Exercise 5.12)

plete index symmetry, which means that the four indices can be exchanged without modifying the value of the corresponding parameters, the general anisotropy requires fifteen parameters and isotropy requires one. Between no index symmetry and a complete index symmetry there is a partial ordering of symmetries, as illustrated in Figure 3.4.

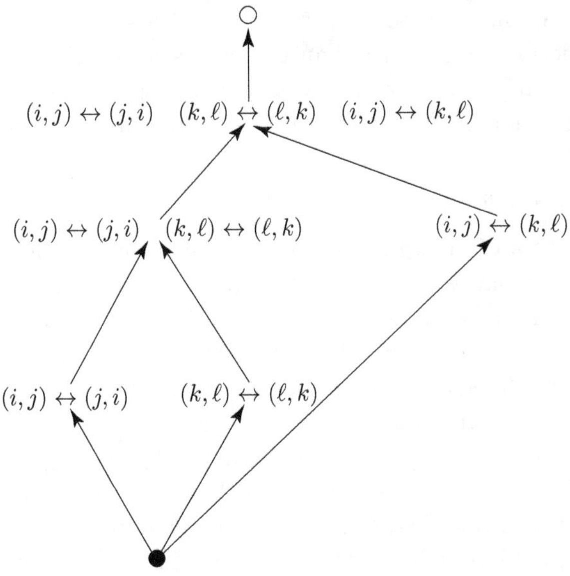

Fig. 3.4: Order relation of selected index-symmetry classes of fourth-rank tensors from no index symmetry to full index symmetry: Arrows indicate subgroups in this partial ordering. Double arrows mean that, given a coordinate system, specific indices can be interchanged without modifying values of the corresponding tensor components.

If either $c_{ijk\ell} = c_{jik\ell}$ or $c_{ijk\ell} = c_{ij\ell k}$, the general anisotropy requires fifty-four parameters and isotropy requires two. These two groups are *isomorphic* to one another. If both $c_{ijk\ell} = c_{jik\ell}$ and $c_{ijk\ell} = c_{ij\ell k}$, the general anisotropy requires thirty-six parameters and isotropy requires two. This is a physically relevant model; it is used in the existing theory of *photoelasticity*. It corresponds to the linear elasticity without the assumption of the strain-energy function,[28] and can be viewed as a natural extension of a Hookean solid.

If only $c_{ijk\ell} = c_{k\ell ij}$, the general anisotropy requires forty-five pa-

[28] *see also*: Slawinski (2015, Section 4.2)

rameters and isotropy requires three. If, as stated in expression (3.40), $c_{ijk\ell} = c_{jik\ell}$, $c_{ijk\ell} = c_{ij\ell k}$ and $c_{ijk\ell} = c_{k\ell ij}$, which corresponds to Hookean solids, the general anisotropy requires twenty-one parameters and isotropy two. Studies of continua exhibiting fewer index symmetries than the ones stated in expression (3.40) would allow us to consider subtleties whose quantitative examination is not possible within behaviours of Hookean solids. However, we do not engage in such studies in this book.[29] Also, there are index-symmetry classes other than the ones selected above and shown in Figure 3.4, such as $c_{ijk\ell} = c_{ikj\ell}$; however, they appear to be less pertinent to a linear relation between two second-rank tensors, σ_{ij} and $\varepsilon_{k\ell}$.

3.2.4.4 *Kelvin notation*

$\sqrt{2}$ and 2 within C in expression (3.41) have an important meaning that, among other instances, becomes apparent in the context of the *Frobenius norm*, which is named after a German mathematician, Ferdinand Georg Frobenius, and is used in Section 4.1, below. That norm is defined as a square root of the sum of the squared entries of C. Examining expression (3.41), we see that, for example, c_{2222}^2 appears only once in that sum; c_{1133}^2 appears twice; c_{2212}^2 appears four times: $2\,(\sqrt{2}\,c_{2212})^2$; c_{1312}^2 appears eight times: $2\,(2\,c_{1312})^2$. To understand these results, we consider expression (3.40), which contains symmetries that reduce eighty-one components of a fourth-rank tensor in $\mathbb{R}^3$ to twenty-one linearly independent components: certain components appear several times. For instance, there is only one occurrence of c_{2222}^2 among the components of $c_{ijk\ell}$. However, in view of symmetries (3.40), c_{1133} appears twice: as c_{1133} and as c_{3311}; c_{2212} appears four times: as c_{2212}, c_{2221}, c_{2122} and c_{1222}; c_{1312} appears eight times: c_{1312} c_{3112}, c_{1321}, c_{1213}, c_{1231}, c_{2113}, c_{3121} and c_{2131}. Thus, $\sqrt{2}$ and 2 account for the number of times that a given component appears in $c_{ijk\ell}$. Hence, the structure of C is as follows. The components on the main diagonal of the upper-left 3×3 submatrix of C appear once in $c_{ijk\ell}$, and the offdiagonal components twice; the components of the upper-right and lower-left submatrices appear four times, as do the components on the main diagonal of the lower-right submatrix; the offdiagonal components of the last submatrix appear eight times.

To examine material symmetries of the elasticity tensor, which are its invariances to rotations, we consider orthogonal transformations in $\mathbb{R}^3$. To apply such rotations to C of equation (3.41), which exists in $\mathbb{R}^6$, we

[29] Interested readers might refer to Yong-Zhong and Del Piero (1991).

must consider transformations in $\mathbb{R}^6$. Transformation $A \in SO(3)$ of $c_{ijk\ell}$ corresponds to a transformation of C given by

$$\tilde{A} = \begin{bmatrix} A_{11}^2 & A_{12}^2 & A_{13}^2 & \sqrt{2}A_{12}A_{13} & \sqrt{2}A_{11}A_{13} & \sqrt{2}A_{11}A_{12} \\ A_{21}^2 & A_{22}^2 & A_{23}^2 & \sqrt{2}A_{22}A_{23} & \sqrt{2}A_{21}A_{23} & \sqrt{2}A_{21}A_{22} \\ A_{31}^2 & A_{32}^2 & A_{33}^2 & \sqrt{2}A_{32}A_{33} & \sqrt{2}A_{31}A_{33} & \sqrt{2}A_{31}A_{32} \\ \sqrt{2}A_{21}A_{31} & \sqrt{2}A_{22}A_{32} & \sqrt{2}A_{23}A_{33} & A_{23}A_{32}+A_{22}A_{33} & A_{23}A_{31}+A_{21}A_{33} & A_{22}A_{31}+A_{21}A_{32} \\ \sqrt{2}A_{11}A_{31} & \sqrt{2}A_{12}A_{32} & \sqrt{2}A_{13}A_{33} & A_{13}A_{32}+A_{12}A_{33} & A_{13}A_{31}+A_{11}A_{33} & A_{12}A_{31}+A_{11}A_{32} \\ \sqrt{2}A_{11}A_{21} & \sqrt{2}A_{12}A_{22} & \sqrt{2}A_{13}A_{23} & A_{13}A_{22}+A_{12}A_{23} & A_{13}A_{21}+A_{11}A_{23} & A_{12}A_{21}+A_{11}A_{22} \end{bmatrix}, \tag{3.42}$$

which is an orthogonal matrix, $\tilde{A} \in SO(6)$.[30,31] The corresponding transformation on the space of elasticity tensors is $C \mapsto \tilde{A}C\tilde{A}^T$. In view of its formulation, $\tilde{A}$ is only a subset of all rotations in $\mathbb{R}^6$. It represents all rotations in $\mathbb{R}^3$, as illustrated by the indices for entries of $\tilde{A}$, which are A_{ij}, with $i,j = 1,2,3$, and correspond to a rotation matrix in $\mathbb{R}^3$.

Having stated the components of a generally anisotropic tensor, which is the 6×6 matrix in expression (3.41), we state the expressions for the remaining seven symmetry classes in the same notation.

For each of these seven classes, the corresponding 6×6 matrix is expressed in its *natural form*, which means that the orientations of the coordinate axes and planes coincide with the orientations of symmetry axes and planes. For instance, the components of a monoclinic tensor, which has a single symmetry plane, are expressed in the coordinate system, whose x_1x_2-plane coincides with the symmetry plane. Such a system is referred to as a *natural coordinate system*. In such a coordinate system, each symmetry class is recognizable by the pattern of its matrix entries, including zeros.

For an arbitrary orientation of coordinate system—except for isotropy, where any system is natural—the easily recognizable pattern of entries, including the appearance of zeros, is lost, making the recognition of symmetry class by inspection impossible. This property—in view of the fact that it is unlikely to know *a priori* the orientation of symmetry planes or axes for the tensor to be used as a model of the subsurface—necessitates other methods to recognize material symmetries of elasticity tensors in seismology.[32]

[30] *see also*: Slawinski (2015, Section 5.2.5)

[31] Readers interested in a mathematical justification for matrix (3.42) might refer to Bóna *et al.* (2008, pp. 126–127).

[32] Readers interested in recognizing material symmetries of tensors given in arbitrary coordinate systems by using eigenproperties of C might refer to Bóna *et al.* (2007a).

3.2.4.5 *Monoclinic tensor*

A monoclinic tensor expressed in a coordinate system, whose x_3-axis is normal to its symmetry plane is

$$C^{\rm mono} = \begin{bmatrix} c_{1111} & c_{1122} & c_{1133} & 0 & 0 & \sqrt{2}c_{1112} \\ c_{1122} & c_{2222} & c_{2233} & 0 & 0 & \sqrt{2}c_{2212} \\ c_{1133} & c_{2233} & c_{3333} & 0 & 0 & \sqrt{2}c_{3312} \\ 0 & 0 & 0 & 2c_{2323} & 2c_{2313} & 0 \\ 0 & 0 & 0 & 2c_{2313} & 2c_{1313} & 0 \\ \sqrt{2}c_{1112} & \sqrt{2}c_{2212} & \sqrt{2}c_{3312} & 0 & 0 & 2c_{1212} \end{bmatrix} . \qquad (3.43)$$

In view of condition (3.30), and as shown in Exercise 3.6, $C^{\rm mono}$ is C that satisfies $C = \left(\tilde{A}^{\rm mono}\right) C \left(\tilde{A}^{\rm mono}\right)^T$, where—in view of group (3.38) and formula (3.42)—transformation $\tilde{A}^{\rm mono}$ corresponds to

$$A^{\rm mono} = \begin{bmatrix} \cos\pi & -\sin\pi & 0 \\ \sin\pi & \cos\pi & 0 \\ 0 & 0 & 1 \end{bmatrix} = \begin{bmatrix} -1 & 0 & 0 \\ 0 & -1 & 0 \\ 0 & 0 & 1 \end{bmatrix} . \qquad (3.44)$$

Condition (3.44) results in $G^{\rm mono} = \{\pm I, \pm M_{\hat{\mathbf{e}}_3}\}$, stated in expression (3.38). Let us examine this statement.

I must be present in any group. $-I$ is a consequence of $\mathbf{c}$ being an even-rank tensor. Since $-I\, M_{\hat{\mathbf{e}}_3} = A^{\rm mono}$, and the product of elements of the group must belong to that group, it follows that a π rotation is tantamount to reflection.

3.2.4.6 *On tensorial properties of monoclinic tensor*

Note. In this section, we use a monoclinic tensor whose symmetry plane coincides with the x_1x_2-plane of the coordinate system to examine properties exhibited by this elasticity tensor. An insight into certain comments herein relies on concepts discussed in Chapter 4, below.

If we rotate the coordinate system about the x_3-axis, the form of the tensor—which is the pattern of the zero and nonzero entries—remains the same, even though, in general, the values of the nonzero coefficients change. These values change in accordance with tensorial properties. The change is such that, among other quantities, the Frobenius norm of $\mathbf{c}$ remains invariant.

The line in Figure 3.5 consists of values of the Euler angles for which the form of tensor (3.43) is invariant. Herein, it corresponds to all rotations

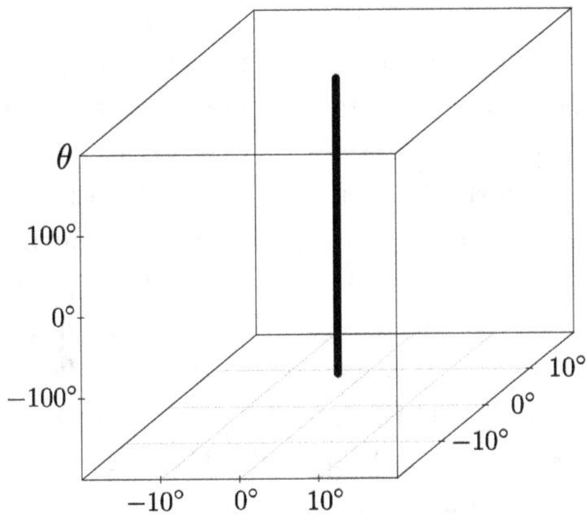

Fig. 3.5: Orientations, in degrees, of a monoclinic tensor under which its form—with its thirteen nonzero independent parameters given in expression (3.43)—and its Frobenius norm remain unchanged. The black line illustrates the fact that the form is invariant under all rotations, θ, about the symmetry-plane normal, but no other rotation. This line includes the four points of Figure 3.6.
Note: This plot represents the space of the Euler angles, not the space of the Cartesian coordinates, in which the tensor is expressed. The angles are between the orientation of the coordinate system of a given generally anisotropic tensor and the orientations of the system of its closest monoclinic counterpart.

about the x_3-axis, and no rotation about the other two axes; thus, the range of values of the angle about the x_3-axis is $[-180°, 180°]$, and only zero about the other axes.

This figure is generated by searching the orientations of the effective monoclinic tensor that corresponds to a generally anisotropic one, as discussed in Section 4.1, below. *A priori*, the sought effective tensor exhibits the form shown in expression (3.43). The search commences with a generally anisotropic tensor—whose components are expressed in an arbitrary coordinate system—and finding its monoclinic counterpart expressed in a coordinate system such that its symmetry plane coincides with the x_1x_2-plane. The Euler angles of Figure 3.5 are the angles between the orientation of the original coordinate system of the generally anisotropic tensor and the orientations of the system of its monoclinic counterpart. The two angles—other than θ—fix the orientation of the x_3-axis for the coordinate system of this counterpart; θ is the rotation about that axis. Figures 3.6 and 3.9 are generated in the same manner, except that the choice of a counterpart of the generally anisotropic tensor is different.

Figure 3.5 illustrates the fact that—given a generally anisotropic tensor—if the orientation of the symmetry plane of the corresponding effective monoclinic tensor is found and is taken in the new coordinate system to be the horizontal plane, as in case of tensor (3.43), the rotation about the symmetry-plane normal can be arbitrary. In other words, the form of a given monoclinic tensor remains invariant under such a rotation. In the context of Section 4.1, below, this property is exhibited by the distance between a generally anisotropic tensor and its monoclinic counterpart, which is the same for all θ, as discussed by Diner *et al.* (2011a). This is tantamount to the norm of the monoclinic counterpart being the same for all θ.

It is important to emphasize that, even though the form of tensor (3.43) remains the same, the values of its entries do not. Consequently, the properties of a Hookean solid defined by this tensor are not invariant under rotations about the x_3-axis. Such an invariance is exhibited by transverse isotropy, discussed in Section 3.2.4.10, below; in that case, not only the form and the norm, but also the values of the entries remain the same.

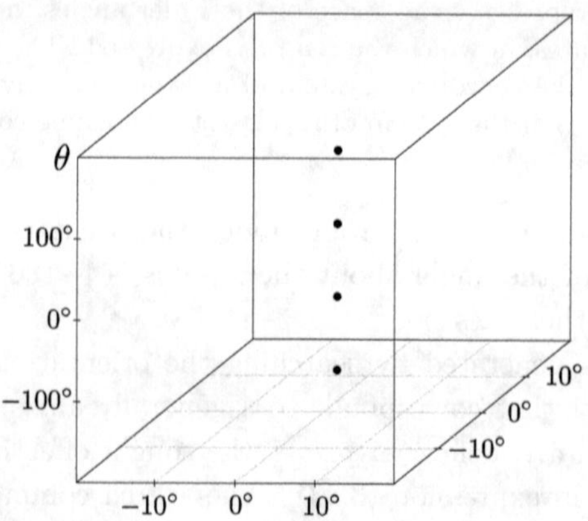

Fig. 3.6: Orientations, in degrees, of a monoclinic tensor under which its form—with twelve nonzero parameters given in expression (3.43) with $c_{2313} = 0$—and its Frobenius norm remain unchanged. The black points illustrate the fact that the form is invariant in only four orientations, θ, of the rotation about the symmetry-plane normal. These points are included in the line of Figure 3.5, above, and correspond to $c_{2313} = 0$ in Figure 3.10, below.
Note: This plot represents the space of the Euler angles, not the space of the Cartesian coordinates, in which the tensor is expressed. The angles are between the orientation of the coordinate system of a given generally anisotropic tensor and the orientations of the system of its closest monoclinic counterpart.

Let us address the issue of the number of the nonzero elasticity parameters for a monoclinic tensor. In a coordinate system, whose orientation is arbitrary with respect to the symmetry plane of the monoclinic tensor, the corresponding expression, in general, has no zeroes; it is an expression of the same tensor, but with respect to a coordinate system, whose axes do not coincide with the symmetry plane. On the other hand, if the x_1x_2-plane coincides with the symmetry plane, a rotation about the x_3-axis, discussed below in Section 3.2.4.15, results—for particular orientations—in matrix (3.43) with $c_{2313} = 0$.[33] Again, it is the same tensor, but expressed with respect to a coordinate system with a particular orientation. The dots in Figure 3.6 illustrate that the form of tensor (3.43) exhibits $c_{2313} = 0$ for four orientations of rotation about the x_3-axis, which correspond to $c_{2313} = 0$ in Figure 3.10.

In a manner similar to Figure 3.5, Figure 3.6 is generated by finding the orientation of the effective monoclinic tensor, discussed in Section 4.1, below. Herein, the only distinction is that the sought effective monoclinic tensor has twelve—not thirteen—nonzero parameters, as is the case in Figure 3.5.

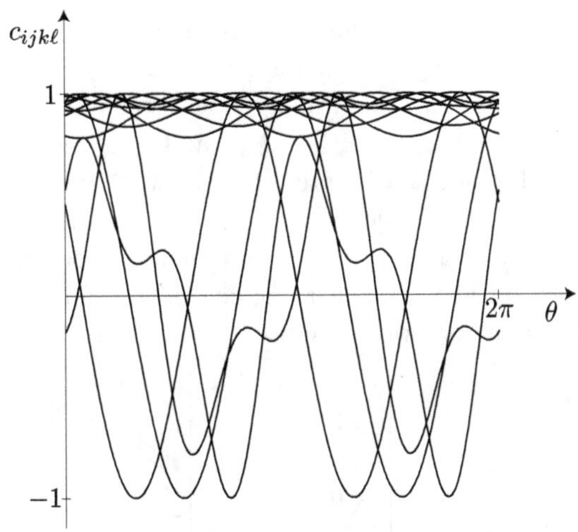

Fig. 3.7: The coordinate coefficients of a monoclinic tensor as a function of rotation about the symmetry-plane normal; note that four coefficients are zero between 0 and 2π. For a convenient display, each coefficient is normalized by its maximum value.

[33] *see also*: Slawinski (2015, Section 5.6.3, expression (5.6.6))

Two related conclusions follow. First, parameter c_{2313} is associated with rotations about the normal to the symmetry plane. Second, since the search for an effective monoclinic tensor—discussed in Section 4.1—can be, in general, performed in terms of two Euler angles, one can perform an *a posteriori* rotation about that normal—whose orientation is fixed by these angles—to obtain $c_{2313} = 0$. The same is true for the tetragonal and trigonal tensors.[34]

To gain further insight into the issue of the number of parameters, let us distinguish between the nonzero parameters and the independent parameters. The monoclinic tensor has thirteen independent parameters, even though—depending on the orientation of the coordinate system, as illustrated in Figure 3.7—it can have as few as twelve nonzero parameters. It can also have as many as twenty-one nonzero parameters, if no symmetry-plane normals coincide with the coordinate axes.[35] Viewed as coordinate coefficients, these parameters change their values in accordance with tensorial transformations in such a manner that the Frobenius norm remains invariant. Also, viewed as a second-rank tensor, expression (3.43) exhibits the eigenvalues and determinant that are invariant under orthogonal transformations, as required.

Since coefficients are continuous functions of orientation and might exhibit both positive and negative values in a 2π-rotation, it follows that their values are zero at certain orientations. Such coefficients are—in order of their first zero crossing—c_{2212}, c_{2313}, c_{3312} and c_{1112}. As stated by expression (3.58) and illustrated in Figure 3.10, below, there is an orientation—for any monoclinic tensor—such that $c_{2313} = 0$. Similar relations exists for $c_{2212} = 0$, $c_{3312} = 0$ and $c_{1112} = 0$. The remaining coordinate coefficients are either nonzero or identically zero, under all transformations allowed by the monoclinic symmetry; we can write

$$\begin{bmatrix} \otimes & \otimes & \otimes & 0 & 0 & \odot \\ \otimes & \otimes & \otimes & 0 & 0 & \odot \\ \otimes & \otimes & \otimes & 0 & 0 & \odot \\ 0 & 0 & 0 & \otimes & \odot & 0 \\ 0 & 0 & 0 & \odot & \otimes & 0 \\ \odot & \odot & \odot & 0 & 0 & \otimes \end{bmatrix}, \tag{3.45}$$

where the nonzero coefficients are marked by $\otimes$ and the identically zero

[34]Readers interested in the number of parameters as a function of a rotation angle might refer to Landau and Lifshitz (1986, Chapter I, § 10) and Helbig (1994, Sections 3.6–3.8).
[35]The values used in this figure are the monoclinic tensor obtained by Danek *et al.* (2013).

coefficients are marked by 0; the others are marked by $\odot$. The entries marked by $\odot$ cannot be all zero at the same orientation, since this would imply an orthotropic tensor, as stated by Theorem 3.1, below.

Theorem 3.1. *If a tensor stated by matrix C in equation (3.41) can be expressed in a natural form of a given symmetry—as given by expressions (3.43), (3.50), (3.51), (3.54), (3.55), (3.56) and (3.57)—it follows that this tensor belongs to that symmetry class.*

Proof. It suffices to consider condition (3.30), which—in the context of matrix (3.41)—is a system of equations given by

$$\begin{aligned} C &= \tilde{A}_1^{\rm sym}\, C\, \tilde{A}_1^{{\rm sym}\,T} \\ &\vdots \\ C &= \tilde{A}_n^{\rm sym}\, C\, \tilde{A}_n^{{\rm sym}\,T}\,, \end{aligned} \tag{3.46}$$

where $\tilde{A}_i^{\rm sym}$ is a symmetry condition expressed by rotation (3.42). C expressed in a natural form of a given symmetry class is tantamount to its satisfying equations (3.46) in a particular orientation of the coordinate system. Since these are tensor equations, if they hold for a particular orientation, they must hold for all orientations. □

Nevertheless, without contradicting Theorem 3.1, we can find a tensor whose entries marked by $\odot$ are zero at almost the same points, as illustrated in Figure 3.8. With an arbitrary orientation, the tensor illustrated in that figure can be expressed as

$$\begin{bmatrix} 5.7036 & 6.1909 & 0.0018 & 0 & 0 & -0.1137 \\ 6.1909 & 8.8250 & 1.8121 & 0 & 0 & 0.0100 \\ 0.0018 & 1.8121 & 7.3910 & 0 & 0 & -0.6379 \\ 0 & 0 & 0 & 9.3531 & -0.0236 & 0 \\ 0 & 0 & 0 & -0.0236 & 4.7056 & 0 \\ -0.1137 & 0.0100 & -0.6379 & 0 & 0 & 8.5526 \end{bmatrix}, \tag{3.47}$$

which has the form of tensor (3.43). It is a positive-definite tensor, as required; its eigenvalues are 13.9719, 9.3533, 8.7759, 7.0389, 4.7055 and 0.6854. Since—as implied by Theorem 3.1 and as can be verified by the method presented by Bóna *et al.* (2007a)—this is not an orthotropic tensor, there exists no orthogonal transformation that could transform tensor (3.47) into an orthotropic tensor.

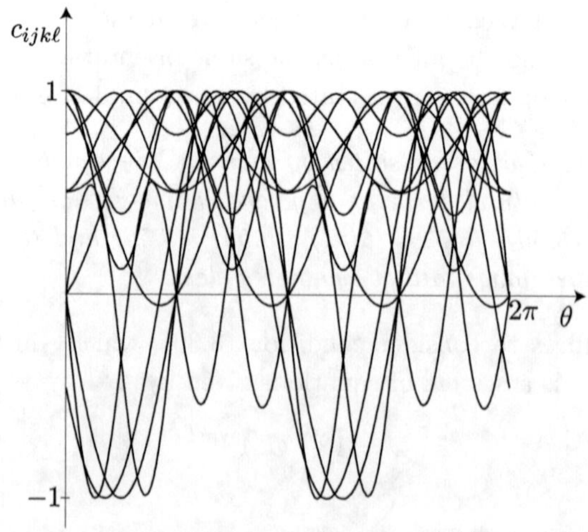

Fig. 3.8: The coordinate coefficients of a monoclinic tensor—different from the one used for Figure 3.7—as a function of rotation about the symmetry-plane normal; note that four coefficients are zero at nearly the same points between 0 and 2π. For a convenient display, each coefficient is normalized by its maximum value.

However, in anticipation of discussions in Section 4.1.3, we state its closest orthotropic counterpart, which—in the Frobenius sense—is

$$\left[\begin{array}{cccccc} 5.7024 & 6.1907 & -0.0093 & 0 & 0 & 0 \\ 6.1907 & 8.8265 & 1.8233 & 0 & 0 & 0 \\ -0.0093 & 1.8233 & 7.3910 & 0 & 0 & 0 \\ 0 & 0 & 0 & 9.3530 & 0 & 0 \\ 0 & 0 & 0 & 0 & 4.7058 & 0 \\ 0 & 0 & 0 & 0 & 0 & 8.5522 \end{array}\right]; \tag{3.48}$$

as expected, it has the form of tensor (3.50), below. Tensor (3.48) is related to tensor (3.47) by the projection stated in expression (4.5), below.

Since, for tensor (3.47), entries marked by $\odot$ in expression (3.45) are zero at almost the same orientation as the orientation of tensor (3.48), we infer several similarities exhibited by both tensors. The eigenvalues of tensor (3.48) are 13.9661, 9.3530, 8.5522, 7.2685, 4.7057 and 0.6854, which are similar to the eigenvalues of tensor (3.47). The similarity of the two tensors is hinted by their Frobenius norms, which are 20.7633, for tensor (3.48), and 20.7816, for tensor (3.47); in other words, the difference of their norms is small. This property, however, can be shared also by tensors that are not similar to each other.

To make a conclusive statement, we need to consider the norm of their difference, which—unlike the comparison of their norms—requires expressing both tensors in the same coordinate system, since one tensor can be subtracted from another only if they are expressed in the same system. Such a subtraction is shown in expressions (4.33) and (4.34), in Section 4.1.3, below, where this issue and the process of achieving such a result is discussed.

According to numerical analysis, the coordinate systems in which tensors (3.47) and (3.48) are expressed are related to one another by

$$\begin{bmatrix} 0.9999 & -0.0126 & -0.0001 \\ 0.0126 & 0.9999 & -0.0001 \\ 0.0001 & 0.0001 & 1 \end{bmatrix}, \tag{3.49}$$

which—in view of numerical computations—we can interpret as a rotation in $\mathbb{R}^3$; for instance, the determinant of matrix (3.49) is 0.9996, not the unity, as required for a rotation. Also, this transformation is close to the identity matrix, which would imply no transformation.

Thus, matrix (3.49) can be viewed as describing a very small rotation about a single axis, from which we conclude that the symmetry plane of tensor (3.47) exhibits the orientation similar to one of the symmetry planes of tensor (3.48) and—as expected in view of Figure 3.5—the orientation of a monoclinic tensor is not a function of the rotation about the symmetry-plane normal. As discussed in Section 5 of Kochetov and Slawinski (2009b), such a relation is also a general property of the monoclinic and orthotropic tensors, if they are the counterparts of the same generally anisotropic tensor; this concept is examined in Section 4.1.3, below.

3.2.4.7 *Orthotropic tensor*

In view of condition (3.30) and group (3.37), an orthotropic tensor—in a system whose coordinate axes are parallel to normals of the symmetry planes—is

$$C^{\mathrm{ortho}} = \begin{bmatrix} c_{1111} & c_{1122} & c_{1133} & 0 & 0 & 0 \\ c_{1122} & c_{2222} & c_{2233} & 0 & 0 & 0 \\ c_{1133} & c_{2233} & c_{3333} & 0 & 0 & 0 \\ 0 & 0 & 0 & 2c_{2323} & 0 & 0 \\ 0 & 0 & 0 & 0 & 2c_{1313} & 0 \\ 0 & 0 & 0 & 0 & 0 & 2c_{1212} \end{bmatrix}. \tag{3.50}$$

In view of group (3.37), matrix (3.50) satisfies a system of three equations, $C = \left(\tilde{A}^{\mathrm{ortho}_i}\right) C \left(\tilde{A}^{\mathrm{ortho}_i}\right)^T$, with $A^{\mathrm{ortho}_1} = A^{\mathrm{mono}}$, given in expression (3.44), and A^{ortho_i}, where $i = 2, 3$, denoting reflections about the x_1x_3-plane and the x_2x_3-plane, respectively. We obtain the same result by using the reflection symmetries about any two orthogonal planes, as discussed in Section 3.2.4.16, below.

3.2.4.8 *Tetragonal tensor*

A tetragonal tensor—in a system whose coordinate axes are parallel to normals of the symmetry planes—is

$$C^{\text{tetra}} = \begin{bmatrix} c_{1111} & c_{1122} & c_{1133} & 0 & 0 & 0 \\ c_{1122} & c_{1111} & c_{1133} & 0 & 0 & 0 \\ c_{1133} & c_{1133} & c_{3333} & 0 & 0 & 0 \\ 0 & 0 & 0 & 2c_{2323} & 0 & 0 \\ 0 & 0 & 0 & 0 & 2c_{2323} & 0 \\ 0 & 0 & 0 & 0 & 0 & 2c_{1212} \end{bmatrix}. \tag{3.51}$$

If the x_1-axis and x_2-axis are contained in the tetragonal-symmetry plane but pointing in an arbitrary direction, we find that $c_{1112} \neq 0$, which results in seven nonzero elasticity parameters. In expression (3.51), the orientation of these axes is quite particular, since—proceeding from an orthotropic tensor—we are *a priori* in a coordinate system, whose axes are parallel to the normals of the three symmetry planes, which results in six nonzero parameters.

3.2.4.9 *On tensorial properties of tetragonal tensor*

Note. In this section, we use a tetragonal tensor—in a system whose coordinate axes are parallel to normals of the symmetry planes—to examine properties exhibited by the elasticity tensor.

To obtain $c_{1112} \neq 0$, which is the case of expression (3.51), it suffices to consider the fourfold invariance under the rotation about the x_3-axis,

$$A_{2\pi/4} \equiv A_{\pi/2} = \begin{bmatrix} 0 & 1 & 0 \\ -1 & 0 & 0 \\ 0 & 0 & 1 \end{bmatrix}. \tag{3.52}$$

We can obtain the required expression by subjecting the stress and strain tensors to $A_{\pi/2}$,

$$\begin{bmatrix} 0 & 1 & 0 \\ -1 & 0 & 0 \\ 0 & 0 & 1 \end{bmatrix} \begin{bmatrix} \sigma_{11} & \sqrt{2}\sigma_{12} & \sqrt{2}\sigma_{13} \\ \sqrt{2}\sigma_{12} & \sigma_{22} & \sqrt{2}\sigma_{23} \\ \sqrt{2}\sigma_{13} & \sqrt{2}\sigma_{23} & \sigma_{33} \end{bmatrix} \begin{bmatrix} 0 & -1 & 0 \\ 1 & 0 & 0 \\ 0 & 0 & 1 \end{bmatrix} = \begin{bmatrix} \sigma_{22} & -\sqrt{2}\sigma_{12} & \sqrt{2}\sigma_{23} \\ -\sqrt{2}\sigma_{12} & \sigma_{11} & -\sqrt{2}\sigma_{13} \\ \sqrt{2}\sigma_{23} & -\sqrt{2}\sigma_{13} & \sigma_{33} \end{bmatrix}. \tag{3.53}$$

The same pattern applies to the strain tensor. Following expression (3.41), we write

$$\sigma_{33} = c_{1133}\varepsilon_{11} + c_{2233}\varepsilon_{22} + c_{3333}\varepsilon_{33} + 2c_{3323}\varepsilon_{23} + 2c_{3313}\varepsilon_{13} + 2c_{3312}\varepsilon_{12}\,.$$

In view of expression (3.53), we write

$$\sigma'_{33} = c_{2233}\varepsilon_{11} + c_{1133}\varepsilon_{22} + c_{3333}\varepsilon_{33} + 2c_{3313}\varepsilon_{23} - 2c_{3323}\varepsilon_{13} - 2c_{3312}\varepsilon_{12}\,.$$

The equality of σ_{33} and σ'_{33} implies $c_{1133} = c_{2233}$ and $c_{3313} = c_{3323} = c_{3312} = 0$. Applying the remaining five conditions for σ_{ij} and σ'_{ij}, we obtain expression (3.59), shown below.

Condition (3.52) is $R_{\frac{\pi}{2},\hat{\mathbf{e}}_3}$ of group (3.35). Let us examine the fact that a single explicit condition suffices to obtain a tetragonal tensor. By intrinsic symmetries, this condition entails $-R_{\frac{\pi}{2},\hat{\mathbf{e}}_3}$ and $\pm R_{-\frac{\pi}{2},\hat{\mathbf{e}}_3}$. $\pm I$, which is a consequence of $\mathbf{c}$ being an even-rank tensor, together with $R_{\frac{\pi}{2},\hat{\mathbf{e}}_3}$ imply $\pm M_{\hat{\mathbf{e}}_3}$. They result in the unchanged form of the zero entries under all rotations about $\hat{\mathbf{e}}_3$; hence, the invariant coordinate-system orientations are the same as the ones shown in Figure 3.5.

The remaining elements, $\pm M_{\hat{\mathbf{e}}_1}, \pm M_{\hat{\mathbf{e}}_2}, \pm M_{(1,1,0)}, \pm M_{(1,-1,0)}$, correspond to particular orientations of the coordinate system for which the planes whose normals are $\hat{\mathbf{e}}_1$ and $\hat{\mathbf{e}}_2$ coincide with the symmetry planes. In such a case, $c_{1112} = 0$. Conditions $\pm M_{\hat{\mathbf{e}}_1}$ and $\pm M_{\hat{\mathbf{e}}_2}$ reduce the invariant coordinate-system rotations to points shown in Figure 3.9; note that $\pm M_{(1,1,0)}$ and $\pm M_{(1,-1,0)}$ are entailed by these conditions. Figure 3.9 is analogous to Figure 3.6, which illustrates a monoclinic counterpart of a generally anisotropic tensor.

Whether or not we insist on $c_{1112} = 0$, we deal with the same tensor denoted by different expressions. It follows that, given a generally anisotropic tensor, it is impossible to find the orientation of its tetragonal counterpart beyond the orientation of its fourfold symmetry axis. However, information about wave propagation—such as velocities as functions of azimuth—provides us with further constraints, if tetragonal symmetry is accurate enough as a description of a generally anisotropic tensor. The same is true for a monoclinic case: its orientation beyond the symmetry-plane normal could be inferred from velocity information within that plane. In particular, we can find the best fit, as a function of azimuth, between the slowness curve in the symmetry plane of a monoclinic or tetragonal tensor and the slowness curve—in the same orientation—of a generally anisotropic tensor. One might start with a single curve, such as the innermost curve, which

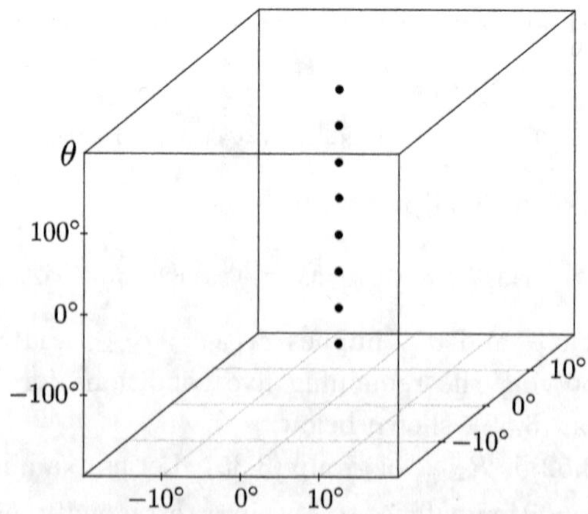

Fig. 3.9: Orientations, in degrees, of a tetragonal tensor under which its form—with only six nonzero parameters given in expression (3.51), where $c_{1112} = 0$—and its Frobenius norm remain unchanged. The black points illustrate the fact that the form is invariant in only eight orientations, $\theta \in [0, 2\pi)$, of the rotation about the four-fold symmetry axis. These points correspond to $c_{1112} = 0$ in Figure 3.11, below.
Note: This plot represents the space of the Euler angles, not the space of the Cartesian coordinates, in which the tensor is expressed. The angles are between the orientation of the coordinate system of a given generally anisotropic tensor and the orientations of the system of its closest tetragonal counterpart.

is convex and corresponds to the fastest wave. One could proceed to the three curves and even consider the fit of the complete slowness surface.[36]

Let us examine the distinction between the aforementioned properties of the monoclinic and tetragonal tensors and analogous properties of a transversely isotropic tensor, which is discussed below. In the last case, neither the form of the zero entries nor values of the nonzero entries change under rotation. In the other cases, the form remains the same, but values of the nonzero entries change. Consequently—even though the tensor is the same, and hence its Frobenius norm, which is the square root of the sum of all entries squared, remains invariant, as shown in Figure 3.12 for a tetragonal tensor—the magnitudes of velocities change with rotation due to the change of values of nonzero entries. For a transversely isotropic tensor, the magnitudes of the velocities are invariant under rotations; hence, the

[36] In evaluating the fit, one could include the errors; in other words, the slowness curves or surfaces would exhibit "error bars".

slowness curves in the plane normal to the rotation-symmetry axis, which correspond to the P waves and the S waves, are circles.

3.2.4.10 *Transversely isotropic tensor*

A transversely isotropic tensor—in a system, whose x_3-axis is parallel to the rotation symmetry axis—is

$$C^{\mathrm{TI}} = \begin{bmatrix} c_{1111} & c_{1122} & c_{1133} & 0 & 0 & 0 \\ c_{1122} & c_{1111} & c_{1133} & 0 & 0 & 0 \\ c_{1133} & c_{1133} & c_{3333} & 0 & 0 & 0 \\ 0 & 0 & 0 & 2c_{2323} & 0 & 0 \\ 0 & 0 & 0 & 0 & 2c_{2323} & 0 \\ 0 & 0 & 0 & 0 & 0 & c_{1111} - c_{1122} \end{bmatrix} . \tag{3.54}$$

The distinction between this expression and expression (3.51) is the dependence among elasticity parameters: $c_{1212} = (c_{1111} - c_{1122})/2$, which—in the Kelvin notation—results in $2((c_{1111} - c_{1122})/2) = c_{1111} - c_{1122}$, as the $(6, 6)$ entry in expression (3.54).

3.2.4.11 *Trigonal tensor*

A trigonal tensor is

$$C^{\mathrm{trig}} = \begin{bmatrix} c_{1111} & c_{1122} & c_{1133} & 0 & \sqrt{2}c_{1113} & 0 \\ c_{1122} & c_{1111} & c_{1133} & 0 & -\sqrt{2}c_{1113} & 0 \\ c_{1133} & c_{1133} & c_{3333} & 0 & 0 & 0 \\ 0 & 0 & 0 & 2c_{2323} & 0 & -2c_{1113} \\ \sqrt{2}c_{1113} & -\sqrt{2}c_{1113} & 0 & 0 & 2c_{2323} & 0 \\ 0 & 0 & 0 & -2c_{1113} & 0 & c_{1111} - c_{1122} \end{bmatrix} . \tag{3.55}$$

If the x_1-axis and the x_2-axis are contained in the trigonal-symmetry plane but pointing in arbitrary directions, we find that $c_{1123} \neq 0$, and we obtain the form given by matrix (3.60), below.[37]

[37]Such a form is derived by Helbig (1994, expression (3.37a)).

3.2.4.12 *Cubic tensor*

A cubic tensor is

$$C^{\text{cubic}} = \begin{bmatrix} c_{1111} & c_{1133} & c_{1133} & 0 & 0 & 0 \\ c_{1133} & c_{1111} & c_{1133} & 0 & 0 & 0 \\ c_{1133} & c_{1133} & c_{1111} & 0 & 0 & 0 \\ 0 & 0 & 0 & 2c_{2323} & 0 & 0 \\ 0 & 0 & 0 & 0 & 2c_{2323} & 0 \\ 0 & 0 & 0 & 0 & 0 & 2c_{2323} \end{bmatrix}. \tag{3.56}$$

3.2.4.13 *Isotropic tensor*

An isotropic tensor is

$$C^{\text{iso}} = \begin{bmatrix} c_{1111} & c_{1111}-2c_{2323} & c_{1111}-2c_{2323} & 0 & 0 & 0 \\ c_{1111}-2c_{2323} & c_{1111} & c_{1111}-2c_{2323} & 0 & 0 & 0 \\ c_{1111}-2c_{2323} & c_{1111}-2c_{2323} & c_{1111} & 0 & 0 & 0 \\ 0 & 0 & 0 & 2c_{2323} & 0 & 0 \\ 0 & 0 & 0 & 0 & 2c_{2323} & 0 \\ 0 & 0 & 0 & 0 & 0 & 2c_{2323} \end{bmatrix}. \tag{3.57}$$

Unlike for the other symmetries, for an isotropic tensor, the arrangement of the zero and nonzero entries and the values of the latter remain the same for all orientations of an orthonormal coordinate system.

3.2.4.14 *On isotropic parameters*

Following Section 3.2.4.13, we are left with only two independent elasticity parameters. In other words, two parameters suffice to define an isotropic Hookean solid.

The use of isotropic Hookean models in many disciplines results in different parametrization of the two linearly independent elasticity parameters. Let us examine three common ones.

In Sections 3.2.4.1–3.2.4.13, we use $c_{ijk\ell}$, in general, to examine Hooke's law stated in expression (3.1). This notation lends itself to an ease of operations within tensor algebra in examining symmetry classes. Consequently, the resulting isotropic parameters presented in Section 3.2.4.13 are c_{1111} and c_{2323}. This notation is also convenient in seismology due to their physical meaning, where speeds of the two waves, P and S, are given by $v_P = \sqrt{c_{1111}/\rho}$ and $v_S = \sqrt{c_{2323}/\rho}$, where ρ is mass density.

In general, the two parameters of an isotropic Hookean solid can be stated as any linear combinations of c_{1111} and c_{2323}. Certain among these combinations exhibit a convenient physical or geometrical meaning.

We can let $\lambda := c_{1111} - 2c_{2323}$ and $\mu := c_{2323}$, which are referred to as the Lamé parameters in honour of the French mathematician Gabriel Lamé. Herein, μ is the measure of the rigidity of the solid, which is its resistance to change in shape, and—if μ tends to zero—λ is the reciprocal of the compressibility.[38] Thus, by invoking the Lamé parameters, an isotropic Hookean solid is expressed in terms of its rigidity and an a quantity akin to compressibility.

We can let $K := c_{1111} - 4c_{2323}/3$ and $\mu := c_{2323}$, which are referred to as the bulk modulus and rigidity modulus, respectively. The former is the ratio of the infinitesimal pressure increase to the resulting relative decrease of the volume. The latter is the ratio of shear stress to the shear strain.[39] Note that—if μ tends to zero—which corresponds to a liquid, $K = \lambda$; both become measures of compressibility.

The reader can suggest other linear combinations of c_{1111} and c_{2323}. It is an easy task, and the number of possibilities is infinite. However, the challenge lies in making them physically meaningful or mathematically convenient, as illustrated in Exercise 3.9.

3.2.4.15 *Relations among elasticity parameters*

Symmetries shown in expressions (3.43)–(3.57) exhibit a different number of elasticity parameters. In general, this number is not conclusive in identifying the symmetry class, since—except for isotropy—a change of orientation of a coordinate system in which an elasticity tensor is expressed results in fewer zero entries in expressions (3.43)–(3.56). Also, in the case of tensor (3.43), we can obtain fewer nonzero entries if we rotate the coordinate system by θ such that[40]

$$\tan(2\theta) = \frac{2c_{2313}}{c_{2323} - c_{1313}}; \tag{3.58}$$

such a rotation results in an expression for which c_{2313} is zero, as illustrated in Figure 3.10.

Remark 3.2. Expression (3.58) is analogous to the linear-algebra expressions for the diagonalization of symmetric matrices. In $\mathbb{R}^2$, the rotation of

[38] *see also*: Slawinski (2015, Section 5.12.4)

[39] Readers interested in a physical interpretation of the relation between K and the Lamé parameters, where $3K = 3\lambda + 2\mu$, might refer to Müller and Weiss (2016, pp. 3-4).

[40] *see also*: Slawinski (2015, Section 5.6.3, expression (5.6.6))

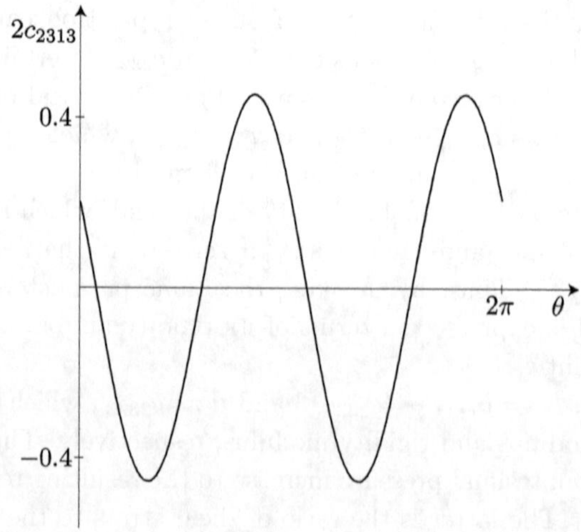

Fig. 3.10: Values of $2c_{2313}$ under rotations of a monoclinic tensor; $c_{2313}=0$, for $\theta=\theta_0+n\pi/2$. At these rotation angles, the tensor can be represented explicitly by matrix (3.43) with c_{2313} set to zero. The tensor used in this figure is obtained from seismic measurements, as described by Danek *et al.* (2013); its units are $\mathrm{km}^2/\mathrm{s}^2$.

a coordinate system by θ such that

$$\tan(2\theta)=\frac{2a_{12}}{a_{22}-a_{11}}$$

results in a diagonal form of a symmetric matrix, which means that, upon rotation, $a'_{12}=a'_{21}=0$.

Herein, in contrast to condition (3.3), $a_{ij}\neq a'_{ij}$. Hence, if we view the symmetric matrix as components of a second-rank tensor in $\mathbb{R}^2$, this tensor is not invariant under the rotation by θ. However—since it is the same tensor but expressed in two different orientations of the coordinate system—it exhibits the same rotational symmetries in either form.

It is a fundamental property of a tensor that its rotation—or, equivalently, a rotation of a coordinate system in which its components are expressed—does not modify the tensor itself, which is a mathematical entity independent of coordinates. Consequently, the symmetry class to which a given tensor belongs cannot be a function of its orientation. To gain insight into this statement, let us consider Table 9 of Nye (1987); in particular, let us examine the tetragonal and trigonal tensors therein. Nye (1987) represents the tetragonal and trigonal classes by two matrices: one has the form of expression (3.51) or (3.55), and the other is

$$
\hat{C}^{\text{tetra}} = \begin{bmatrix}
\hat{c}_{1111} & \hat{c}_{1122} & \hat{c}_{1133} & 0 & 0 & \sqrt{2}\hat{c}_{1112} \\
\hat{c}_{1122} & \hat{c}_{1111} & \hat{c}_{1133} & 0 & 0 & -\sqrt{2}\hat{c}_{1112} \\
\hat{c}_{1133} & \hat{c}_{1133} & \hat{c}_{3333} & 0 & 0 & 0 \\
0 & 0 & 0 & 2\hat{c}_{2323} & 0 & 0 \\
0 & 0 & 0 & 0 & 2\hat{c}_{2323} & 0 \\
\sqrt{2}\hat{c}_{1112} & -\sqrt{2}\hat{c}_{1112} & 0 & 0 & 0 & 2\hat{c}_{1212}
\end{bmatrix} \tag{3.59}
$$

or

$$
\hat{C}^{\text{trig}} = \begin{bmatrix}
\hat{c}_{1111} & \hat{c}_{1122} & \hat{c}_{1133} & \sqrt{2}\hat{c}_{1123} & \sqrt{2}\hat{c}_{1113} & 0 \\
\hat{c}_{1122} & \hat{c}_{1111} & \hat{c}_{1133} & -\sqrt{2}\hat{c}_{1123} & -\sqrt{2}\hat{c}_{1113} & 0 \\
\hat{c}_{1133} & \hat{c}_{1133} & \hat{c}_{3333} & 0 & 0 & 0 \\
\sqrt{2}\hat{c}_{1123} & -\sqrt{2}\hat{c}_{1123} & 0 & 2\hat{c}_{2323} & 0 & -2\hat{c}_{1113} \\
\sqrt{2}\hat{c}_{1113} & -\sqrt{2}\hat{c}_{1113} & 0 & 0 & 2\hat{c}_{2323} & 2\hat{c}_{1123} \\
0 & 0 & 0 & -2\hat{c}_{1113} & 2\hat{c}_{1123} & \hat{c}_{1111} - \hat{c}_{1122}
\end{bmatrix} . \tag{3.60}
$$

In Nye (1987) and several other books, these expressions are viewed—in contrast to expressions (3.51) and (3.55)—as exhibiting the seventh independent parameter: $\hat{c}_{1112}$ and $\hat{c}_{1123}$, respectively. However, as we show below, expressions (3.51) and (3.59) are two coordinate expressions of the same tensor. The same is true for expressions (3.55) and (3.60).

Let us examine expression (3.59), where the apparent independence of $\hat{c}_{1112}$ disappears if we subject a tetragonal tensor to rotation about its fourfold-symmetry axis. Notably, the rotation that results in $\hat{c}_{1112} = 0$ is[41]

$$
\tan(4\theta) = -\frac{4\hat{c}_{1112}}{2\hat{c}_{1212} - \hat{c}_{1111} + \hat{c}_{1122}} .
$$

To show that the same tensor can be expressed with $\hat{c}_{1112} \neq 0$ and with $\hat{c}_{1112} = 0$, let us consider a tetragonal tensor and its Frobenius norm, $\sqrt{\sum c_{ijk\ell}^2}$, which—similarly to the length of a vector—is independent of the orientation of coordinates. In other words, a transformation of coordinates is such that the magnitude of a tensor is unchanged, as is the length of a vector.

In response to rotations in $\mathbb{R}^3$, the values of entries of matrix (3.51) change in such a manner that the sum of $c_{ijk\ell}^2$ remains the same. In other words, the contribution of a given entry changes to ensure that the sum of their squares be the same. For the $\mathbb{R}^2$ rotations about the fourfold-symmetry axis, the matrix form remains as shown in expression (3.59),

[41]Readers interested in a justification of that property might refer to Bóna *et al.* (2004b, Section 3, p. 595).

except the nonzero values change. In particular, as shown in Figure 3.11, $\hat{c}_{1112} = 0$, for $\theta = n\pi/4$. To justify this periodicity, we can represent the fourfold invariance by a square.[42] If the coordinate axes, whose centre is in the middle of this square, bisect its vertices or its sides, we have $\hat{c}_{1112} = 0$, which means that—at these orientations—certain information about the tensor is contained in the value of θ, with $c_{1112} = 0$.[43] In general, the values of $\hat{c}_{1112}$, for all rotations, are small in comparison to other values.

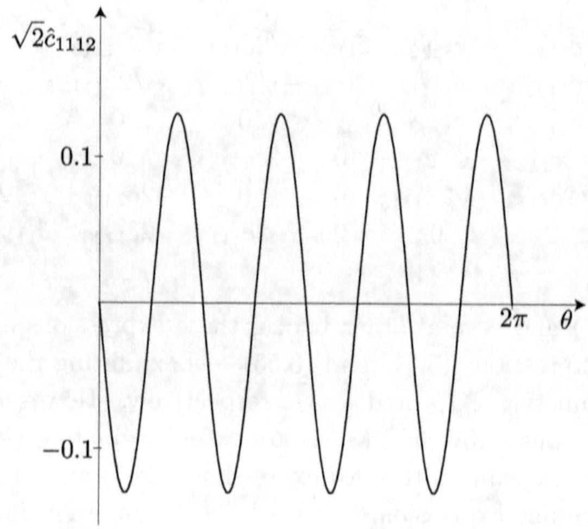

Fig. 3.11: Values of $\sqrt{2}\hat{c}_{1112}$ under rotations of tetragonal tensor. We see that $\hat{c}_{1112} = 0$, for $\theta = n\pi/4$, as expected for tetragonal symmetry. At these rotation angles, the tensor is represented explicitly by matrix (3.51). For other angles, it is represented by matrix (3.59). The tensor used in this figure is obtained from seismic measurements, as described by Danek *et al.* (2013); its units are $\mathrm{km}^2/\mathrm{s}^2$.

Examining Figure 3.12 in the context of the entries of matrix (3.59), we see that the square of the Frobenius norm, which we can view as the sum of contributions of $4\left(\pm\sqrt{2}\hat{c}_{1112}\right)^2 = 8\,\hat{c}_{1112}^2$, together with the sum of the squares of the remaining entries, result in a constant, as expected.[44] These contributions vary with θ in such a manner that their sum is constant, as required for the tensor transformations.

[42] *see also*: Slawinski (2015, Figure 5.9.1)

[43] *see also*: Slawinski (2015, Exercise 5.6)

[44] Values of the components of the tetragonal tensor used for this illustration are stated in Slawinski (2015, Exercise 5.6) and Danek *et al.* (2013).

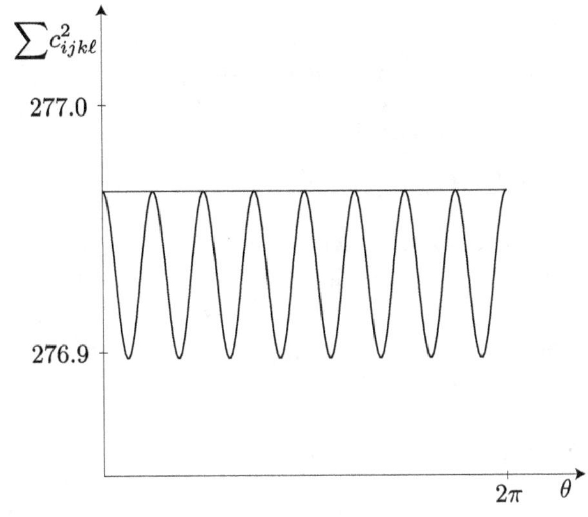

Fig. 3.12: Rotation from 0 to 2π about the fourfold-symmetry axis of a tetragonal tensor. The square of its Frobenius norm is invariant, as required. Contributions of the values of $8\,\hat{c}_{1112}^2$ to this norm—which are below the range of values shown on the vertical axis—and of the sum of squares of the remaining entries are functions of the rotation angle; the angular dependence of $8\,\hat{c}_{1112}^2$ can be inferred from Figure 3.11. These contributions vary in such a manner that their sum is constant; the maximum contributions of the entries shown in this figure correspond to angles for which $\hat{c}_{1112} = 0$, which occurs at $\theta = n\pi/4$. The tensor used in this figure is obtained from seismic measurements, as described by Danek *et al.* (2013); its units are $\mathrm{km}^2/\mathrm{s}^2$.

We can conclude this investigation with a plausibility argument according to which expressions (3.51) and (3.59) represent the same tetragonal tensor. The same Frobenius norm for the 2π rotation leaves us with a strong conjecture of no distinction—other than orientation of coordinates—between expressions (3.51) and (3.59). To prove that this is indeed the case, we could consider eigenproperties of tensors, as discussed by Bóna *et al.* (2007a).

In a similar manner, one could examine expression (3.60) by considering its rotation about the threefold-symmetry axis of a trigonal tensor. The conjecture analogous to the one above follows: expressions (3.55) and (3.60) represent the same trigonal tensor, which can be proven by the method discussed by Bóna *et al.* (2007a). Notably, the rotation of expression (3.60) that results in $\hat{c}_{1123} = 0$ is[45]

$$\tan(3\theta_1) = \frac{\hat{c}_{1123}}{\hat{c}_{1113}};$$

also, the rotation that results in $\hat{c}_{1113} = 0$ is[46]

$$\tan(3\theta_2) = \frac{\hat{c}_{1113}}{\hat{c}_{1123}}.$$

These are statements of dependence among $\hat{c}_{1123}$, $\hat{c}_{1113}$ and θ; as is illustrated in Figure 3.13. Examining this figure, we see that values of $\hat{c}_{1113}$ and $\hat{c}_{1123}$ are zero for $\theta = \theta_0 + n\pi/3$, as expected for trigonal symmetry. Also, since graphs of $\sqrt{2}\hat{c}_{1113}$ and $\sqrt{2}\hat{c}_{1123}$ are $\pi/6$ out of phase with one another, the maximum, 0.104065, or minimum, -0.104065, of one corresponds to the zero of the other. They can be represented by the sine and cosine functions. Hence, the sum of their squares, which is

$$0.104065^2 \sin^2\left(\frac{\theta}{3}\right) + 0.104065^2 \cos^2\left(\frac{\theta}{3}\right) = 0.104065^2 = 0.010829,$$

is constant for $\theta \in [0, 2\pi)$. Thus, we infer that the Frobenius norm of matrix (3.60) is constant regardless of the contribution of $\sqrt{2}\hat{c}_{1113}$ and $\sqrt{2}\hat{c}_{1123}$. In other words, the relation between $\sqrt{2}\hat{c}_{1113}$ and $\sqrt{2}\hat{c}_{1123}$ is tantamount to the effect of rotation about the x_3-axis.

Nye (1987) discusses dependences among parameters for a given orientation by showing, for instance, that certain entries of matrix (3.59)—say, $(1,1)$ and $(2,2)$ or $(1,3)$ and $(2,3)$—are occupied by the same values; it is impossible to relate entries $(1,6)$ and $(2,6)$ to other entries in such a manner, since these relations involve θ, as becomes apparent upon rotation. Thus, we must be careful not to imply intrinsic properties of a tensor from relations that are not rotation-invariant.

3.2.4.16 *Diclinic solids*

In this and the following section, we exemplify cases that illustrate the necessity of distinction between Hookean solids, which are abstract entities, and physical materials for which they can serve as quantitative analogies. In other words, we examine limitations that stem from the definition of these solids.

The concept of symmetry is rooted in the mathematical structure of group theory. Among other properties, the definition of a group requires

[45] *see also*: Slawinski (2015, Section 5.8.2, expression (5.8.6))

[46] Readers interested in a justification of that property might refer to Bóna *et al.* (2004b, Section 3, p. 594).

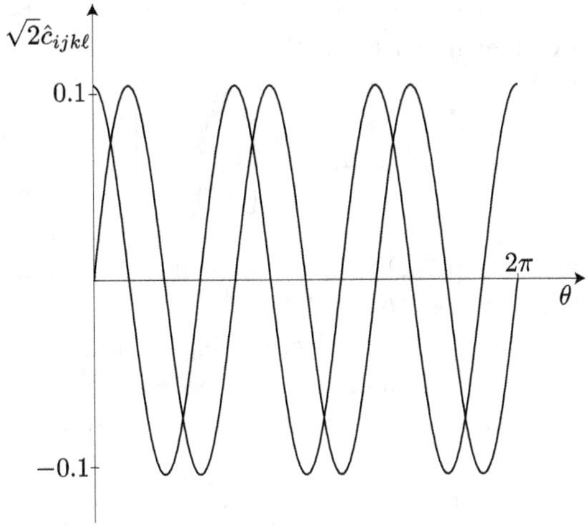

Fig. 3.13: Values of $\sqrt{2}\hat{c}_{1113}$ and $\sqrt{2}\hat{c}_{1123}$ under rotations of a trigonal tensor. At angles for which $\hat{c}_{1123} = 0$, the tensor is represented explicitly by matrix (3.55). For angles for which $\hat{c}_{1113} \neq 0$ and $\hat{c}_{1123} \neq 0$, it is represented by matrix (3.60). The tensor used in this figure is obtained from seismic measurements, as described by Danek *et al.* (2013); its units are $\mathrm{km}^2/\mathrm{s}^2$.

that a product of any of its two elements belongs to the group. In the context of material symmetries, if the reflections in the planes orthogonal to one another belong to a group, so does their product.

For example, consider a Hookean solid that is invariant under the actions of

$$M_{\hat{\mathbf{e}}_1} = \begin{bmatrix} -1 & 0 & 0 \\ 0 & 1 & 0 \\ 0 & 0 & 1 \end{bmatrix}$$

and

$$M_{\hat{\mathbf{e}}_2} = \begin{bmatrix} 1 & 0 & 0 \\ 0 & -1 & 0 \\ 0 & 0 & 1 \end{bmatrix},$$

which are the reflections in the planes whose normals are the x_1-axis and the x_2-axis, respectively. Their product is

$$M_{\mathbf{e}_1} M_{\mathbf{e}_2} = \begin{bmatrix} -1 & 0 & 0 \\ 0 & -1 & 0 \\ 0 & 0 & 1 \end{bmatrix}.$$

Furthermore, any Hookean solid is—as a consequence of its being an even-rank tensor—invariant under the action of $-I$. Thus,

$$-I(M_{\hat{\mathbf{e}}_1}M_{\hat{\mathbf{e}}_2}) = \begin{bmatrix} 1 & 0 & 0 \\ 0 & 1 & 0 \\ 0 & 0 & -1 \end{bmatrix},$$

which is the reflection in the plane orthogonal to the x_3-axis, also belongs to the group that contains $M_{\hat{\mathbf{e}}_1}$ and $M_{\hat{\mathbf{e}}_2}$.

Hence, the symmetry under reflections in two orthogonal planes entails the symmetry under reflection in three orthogonal planes. In notation used in expressions (3.37) and (3.38), we can write concisely $(-M_{\hat{\mathbf{e}}_1})(-M_{\hat{\mathbf{e}}_2}) = -M_{\hat{\mathbf{e}}_3}$.

Thus, there are no Hookean solids with only two orthogonal symmetry planes, since any formulation of the *diclinic*-symmetry class entails the orthotropic class, which exhibits three orthogonal planes. This result is a consequence of the mathematical structure of elasticity theory; this is not a statement about properties of physical materials.

In Earth sciences, this result does not imply that a physical material cannot exhibit only two symmetry planes, such as parallel fractures perpendicular to layering. However, for mathematical modeling, the nonexistence of the diclinic Hookean solids means that there is no intermediate case between the single plane of the monoclinic class and the three planes of the orthotropic one, as illustrated in Figure 3.3.

Let us emphasize that Hookean solids are mathematical entities whose relationship to physical materials is a quantitative analogy that allows us to describe and predict observable phenomena. In our descriptions and predictions, we must be alert to limitations of such analogies to avoid false conclusions about the physical world. Mathematical insights are fruitful as long as we distinguish between the Platonic realm of mathematics and the physical world, and as long as we do not endow analogies with excessive qualities.

3.2.4.17 *Hexagonal solids*

Another limitation of Hookean solids that stems from their definition is the impossibility to distinguish among rotation symmetries greater than fourfold. Herman (1945) brings to our attention an important theorem.[47]

[47] *see also*: Slawinski (2015, Theorem 5.10.1 and Exercise 5.8)

Let us quote it in a manner stated in his paper.

Theorem 3.2. *If the medium has a rotating axis of symmetry $\mathcal{C}_N$ of order N, it is axially isotropic relative to this axis for all the physical properties defined by the tensors of the rank $r = 0, 1, 2, \ldots, N-1$.*

In the context of material symmetry of Hookean solids, which are defined by a fourth-rank tensor, $c_{ijk\ell}$, Theorem 3.2 means that a rotation symmetry greater than fourfold results in isotropic rotations. Hence, there are no pentagonal, hexagonal, or higher discrete-symmetry classes in Figure 3.3. In other words, there are no intermediate cases between the fourfold-rotation symmetry of the tetragonal class and the isotropic rotation about the same axis in transverse isotropy.

Many authors refer to transversely isotropic Hookean solids as hexagonal. Let us examine that issue in view of Theorem 3.2.

Remaining within the realm of tensor analysis and expression (3.1), we cannot distinguish between transversely isotropic and hexagonal continua. Any continuum defined by expression (3.1) and whose rotation symmetry is greater than fourfold is a transversely isotropic Hookean solid. Consequently—if we consider an inverse problem—a transversely isotropic Hookean solid can represent any material whose rotation symmetry is greater than fourfold.

If we choose to associate a transversely isotropic Hookean solid with a material exhibiting a discrete symmetry, we could argue in favour of hexagonal symmetry in the context of *tessellation*, which is fitting geometric shapes to fill a plane surface by congruent shapes that interlock without overlapping or leaving gaps. Since only three regular polygons tessellate in the Euclidean plane: triangles, squares and hexagons, and the former two are associated explicitly with trigonal and tetragonal symmetries, respectively, we might choose to implicitly associate hexagonal polygons with transverse isotropy. In this choice, we tacitly limit our considerations to regular polygons; in other words, we consider only a regular tessellation.

To appreciate consequences of such a limitation, let us consider Dyson (2011) who writes,

> Theorems were proved, establishing the fact that in three-dimensional space discrete symmetry groups could contain only rotations of order three, four, or six. Then in 1984 quasicrystals were discovered, real solid objects growing out of liquid metal

alloys, showing the symmetry of the icosahedral[48] group, which includes five-fold rotations.

There are other tessellations beyond the regular one. For instance, there are eight semiregular tessellations, which—for each case—use several regular polygons with identical arrangements at each vertex. Penrose tiling, named in acknowledgement of contributions by Roger Penrose, uses two polygons and creates a pattern that exhibits the symmetry of both reflection and fivefold rotation. It is selfsimilar, which means that it can be constructed from selfreplicating sets of polygons; in other words, the same patterns occur at all scales and a finite patch of tiling occurs infinitely many times. However, it is nonperiodic, which means that it lacks translational symmetry. Hence, a pentagonal solid composed of Penrose tiling cannot be homogeneous, since homogeneity implies the same properties at different locations along a given direction; by contrast, isotropy implies the same properties at a given location in all directions.

However, arguments based on tessellation, as well as subsequent inferences of symmetry classes, rely on properties that are not contained in the definition of material symmetry. For a Hookean solid, if its rotation symmetry is greater than fourfold, its symmetry is transversely isotropic.

Closing remarks

In this chapter, we introduce the concept of a Hookean solid as a mathematical entity used for analogies of physical materials. We discuss intrinsic limitations that result from the definition of this solid as a fourth-rank tensor with assumed index symmetries.

Anisotropy of materials in seismic measurements is inferred from a mathematical analogy of Hookean solids. A study of anisotropy deals directly with the symmetry of tensors and only indirectly with properties of rocks. Symmetry of a Hookean solid only implies a directional pattern within a material.

There are many situations in the physical world that exemplify a given mathematical analogy. Transverse isotropy might be an analogy for parallel layers in a sedimentary basin or for the preferred orientation of olivine crystals in the Earth's mantle.

[48]The icosahedron is a regular polyhedron with twenty identical equilateral triangular faces. It is one of the five Platonic solids.

Also, we need to be aware of criteria for the choice of a model and the extent to which the model is imposed. For instance, are the data forced into an improper model, if a transversely isotropic solid with the orientation of its rotation-symmetry axis is chosen *a priori*? Many models can be proposed to accommodate the same set of experimental data, particularly if errors are taken into account, as discussed in Chapter 4.

3.3 Exercises

Exercise 3.1. Using the rotation of a vector in $\mathbb{R}^2$, illustrate explicitly that there is no change in the length of the vector.

Solution. Let us consider the rotation of a nonzero vector, $v' = A\,v$, which we write explicitly as

$$\begin{bmatrix} v_1' \\ v_2' \end{bmatrix} = \begin{bmatrix} \cos\theta & -\sin\theta \\ \sin\theta & \cos\theta \end{bmatrix} \begin{bmatrix} v_1 \\ v_2 \end{bmatrix}.$$

The length of the vector before and after deformation is

$$\sqrt{v_1^2 + v_2^2}$$

and

$$\sqrt{(v_1\cos\theta - v_2\sin\theta)^2 + (v_1\sin\theta + v_2\cos\theta)^2} = \sqrt{v_1^2 + v_2^2}\,,$$

respectively. Thus, it remains the same for all θ, as expected.

Exercise 3.2. Using a rotation matrix in $\mathbb{R}^2$, illustrate that a nonzero vector cannot be isotropic.

Solution. Consider a nonzero vector, v, and a rotation matrix, A. The isotropic vector must obey—for all orientations—the following invariance:

$$v = A\,v\,.$$

Rearranging, we write

$$(A - I)\,v = 0\,,$$

which is a linear homogeneous equation for v. The solution of this equation, for $v \neq 0$, requires that

$$\det(A - I) = 0\,.$$

For the rotation matrix in $\mathbb{R}^2$, as shown in Exercise 3.3, $\det(A - I) = 0$ implies that $A = I$. Hence, a nonzero vector cannot be isotropic in $\mathbb{R}^2$. An analogous argument can be presented for $\mathbb{R}^3$. For higher dimensions, the proof requires further subtleties or details (Jeffreys, 1961a, Chapter VII), but—in general—a nonzero vector cannot be isotropic.

Exercise 3.3. Consider a rotation matrix, A, in $\mathbb{R}^2$. Show that the only solution to $\det(A - I) = 0$ is $A = I$, which means that there is no rotation.

Solution. For $\mathbb{R}^2$, we can write

$$A - I = \begin{bmatrix} m_{11} & m_{12} \\ m_{21} & m_{22} \end{bmatrix} =: M .$$

Hence,

$$\det(M) = 0 \Rightarrow m_{11}m_{22} - m_{12}m_{21} = 0 .$$

Also, we can write

$$A = I + M = \begin{bmatrix} m_{11} + 1 & m_{12} \\ m_{21} & m_{22} + 1 \end{bmatrix} = \begin{bmatrix} \cos\theta & -\sin\theta \\ \sin\theta & \cos\theta \end{bmatrix} .$$

Equating the two rightmost matrices, we obtain

$$m_{11} = m_{22} \qquad \text{and} \qquad m_{12} = -m_{21} .$$

Hence,

$$\det(M) = m_{11}^2 + m_{12}^2 = 0 \Rightarrow m_{11} = m_{12} = 0 \Rightarrow m_{21} = m_{22} = 0 .$$

Thus, M must be a zero matrix, which implies that $A = I$ is the only solution to $\det(A - I) = 0$, if A is a rotation matrix.

Exercise 3.4. Using equations (3.4) and (3.5), derive the expression for Mohr's circle, and interpret its meaning. Draw Mohr's circle for the case discussed in Exercise 2.1 on page 34.

Solution. For a symmetric case, where $a_{12} = a_{21}$, equations (3.4) and (3.5) become

$$\hat{a}_{11} = \tfrac{1}{2}\left(a_{11} + a_{22} + (a_{11} - a_{22})\cos(2\theta) - 2a_{12}\sin(2\theta)\right) ,$$

$$\hat{a}_{12} = \tfrac{1}{2}\left(2a_{12}\cos(2\theta) + (a_{11} - a_{22})\sin(2\theta)\right) .$$

They suffice to derive Mohr'scircle, since they contain the three independent components of the symmetric case: a_{11}, a_{12} and a_{22}.

Letting $a_{ij} = \sigma_{ij}$, to represent the stress-tensor components, we rewrite these equations as

$$\hat{\sigma}_{11} - \frac{\sigma_{11} + \sigma_{22}}{2} = \frac{\sigma_{11} - \sigma_{22}}{2} \cos(2\theta) - \sigma_{12} \sin(2\theta) ,$$

$$\hat{\sigma}_{12} = \sigma_{12} \cos(2\theta) + \frac{\sigma_{11} - \sigma_{22}}{2} \sin(2\theta) .$$

Squaring both sides of these equations, adding them and using trigonometric identities, we obtain

$$\left(\hat{\sigma}_{11} - \frac{\sigma_{11} + \sigma_{22}}{2} \right)^2 + \hat{\sigma}_{12}^2 = \left(\frac{\sigma_{11} - \sigma_{22}}{2} \right)^2 + \sigma_{12}^2 ,$$

which is the equation of a circle. The horizontal axis is the normal stress, $\hat{\sigma}_{11}$, and the vertical axis is the shear stress $\hat{\sigma}_{12}$.[49] Thus, the space of Mohr's circle is an abstract entity spanned by the stress-tensor components, not the physical x_1x_2-plane. The circle is centred at the point that corresponds to the average normal stress, $\overline{\sigma}_{ii}$, and to the shear stress of zero, $\sigma_{12} = 0$; hence, its radius is equal to the maximum shear stress.

Mohr's circle illustrates the fact, stated in expression (3.10), that the values of the stress-tensor components depend on the orientation of the plane on which the traction acts. For a given orientation, the state of stress within a medium is given by the coordinates of the opposite points on the circle. One point corresponds to $(\sigma_{11}, \sigma_{12})$ and the other to $(\sigma_{21}, \sigma_{22})$, with $\sigma_{12} = \sigma_{21}$, due to the symmetry of the stress tensor. In the physical x_1x_2-plane the components of the stress tensor at these points are related by $\pi/2$.[50]

Examining Figure 2.3 in Exercise 2.1, we see that the only nonzero component of the stress tensor is σ_{11}; according to our convention (Slawinski, 2015, Section 2.5.1), it is positive, since it represent a tension. The corresponding points of Mohr's circle are denoted by dots in Figure 3.14; they define a diameter of the required circle. Other opposite points correspond to the states of stress with different orientations.

Remark 3.3. The points on the opposite sides of the circle correspond to coordinates that constitute the components of a symmetric second-rank tensor in $\mathbb{R}^2$,

$$\begin{bmatrix} \sigma_{11} & \sigma_{12} \\ \sigma_{12} & \sigma_{22} \end{bmatrix} ;$$

[49] *see also*: Slawinski (2015, Section 2.5.1)
[50] *see also*: Slawinski (2015, Figure 2.5.1)

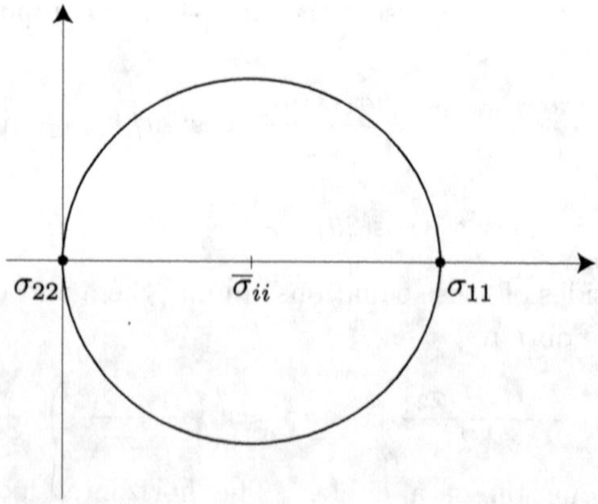

Fig. 3.14: Mohr's circle corresponding to Figure 2.3 on page 35

it is the same tensor expressed in different orientations of the coordinate system; in the original orientation, discussed in Exercise 3.4, we have

$$\begin{bmatrix} \sigma_{11} & 0 \\ 0 & 0 \end{bmatrix}.$$

Exercise 3.5. Prove the following theorem.

Theorem 3.3. *For any antisymmetric tensor, ξ_{hk}, there is a function, $A_{hk}(s)$, that is defined over $\mathbb{R}$ and whose values are in the group of rotations, $SO(3)$, such that*

$$A_{hk}(0) = \delta_{hk}, \qquad A'_{hk}(0) = \xi_{hk}, \qquad h, k = 1, 2, 3,$$

where $'$ denotes the derivative with respect to s.

Solution. Let us proceed with the following proof.

Proof. Consider the initial-value problem given by the set of first-order linear differential equations,

$$A'_{ik} = \sum_{h=1}^{3} A_{ih}\xi_{hk}, \qquad i, k = 1, 2, 3, \tag{3.61}$$

in the unknown functions, $A_{ik}(s)$, with side conditions $A_{ik}(0) = \delta_{ik}$. In view of the existence and uniqueness theorem for solutions of such a problem, we deduce that there is a unique tensor-valued function, $A_{ik}(s)$, such that $A_{ik}(0) = \delta_{ik}$ and

$$A'_{ik}(0) = \sum_{h=1}^{3} A_{ih}(0)\xi_{hk} = \sum_{h=1}^{3} \delta_{ih}\xi_{hk} = \xi_{ik}\,, \qquad i,k = 1,2,3\,.$$

We proceed to show that $A_{ik}(s)$ are the components of a rotation, for each s. We define a tensor-valued function, $Z_{ji}(s)$, as

$$Z_{ji}(s) = \sum_{h=1}^{3} A_{jh}(s)A_{ih}(s)\,, \qquad i,j = 1,2,3\,,$$

and consider its derivative with respect to s,

$$Z'_{ji} = \sum_{h=1}^{3} \left(A'_{jh}A_{ih} + A_{jh}A'_{ih}\right)\,, \qquad i,j = 1,2,3\,,$$

which, in view of expression (3.61), is equivalent to

$$\begin{aligned}
Z'_{ji} &= \sum_{h=1}^{3}\left(\sum_{k=1}^{3} A_{jk}\xi_{kh}A_{ih} + A_{jh}\sum_{k=1}^{3} A_{ik}\xi_{kh}\right) \\
&= \sum_{h=1}^{3}\left(\sum_{k=1}^{3} A_{jk}\xi_{kh}A_{ih} + \sum_{k=1}^{3} A_{jh}\xi_{kh}A_{ik}\right) \\
&= \sum_{h=1}^{3}\sum_{k=1}^{3} A_{jk}\xi_{kh}A_{ih} + \sum_{h=1}^{3}\sum_{k=1}^{3} A_{jh}\xi_{kh}A_{ik} \\
&= \sum_{h=1}^{3}\sum_{k=1}^{3} A_{jk}\xi_{kh}A_{ih} + \sum_{h=1}^{3}\sum_{k=1}^{3} A_{jk}\xi_{hk}A_{ih} \qquad\qquad i,j = 1,2,3\,, \\
&= \sum_{h=1}^{3}\sum_{k=1}^{3} A_{jk}\xi_{kh}A_{ih} - \sum_{h=1}^{3}\sum_{k=1}^{3} A_{jk}\xi_{kh}A_{ih} \\
&= \sum_{h=1}^{3}\sum_{k=1}^{3} \left(A_{jk}\xi_{kh}A_{ih} - A_{jk}\xi_{kh}A_{ih}\right) \\
&= 0\,,
\end{aligned}$$

where we use the fact that the summation indices can be renamed, and that ξ_{hk} is antisymmetric, $\xi_{hk} = -\xi_{kh}$.

We conclude that Z_{ji} is constant and, since

$$Z_{ji}(0) = \sum_{h=1}^{3} A_{jh}(0)A_{ih}(0) = \sum_{h=1}^{3} \delta_{jh}\delta_{ih} = \delta_{ji}\,, \qquad i,j = 1,2,3\,,$$

this implies that, for each s,

$$\sum_{h=1}^{3} A_{jh}(s)A_{ih}(s) = \delta_{ji}\,, \qquad i,j = 1,2,3\,.$$

As a consequence, $A_{jh}(s)$ are the components of an orthogonal tensor and, thus, either $\det[A_{jh}(s)] = 1$ or $\det[A_{jh}(s)] = -1$. Since $\det[A_{jh}(0)] = \det[\delta_{jh}] = 1$, from continuity it follows that, for each s, $\det[A_{jh}(s)] = 1$, and we conclude that the tensor with components $A_{jh}(s)$ is indeed a rotation. □

Exercise 3.6. Using condition (3.46), show explicitly that the monoclinic tensor is given by expression (3.43).

Solution. Using expression (3.42), let us write the element of the monoclinic symmetry group, other than identity, namely, element (3.44), as a 6×6 matrix,

$$A^{\mathrm{mono}} = \begin{bmatrix} -1 & 0 & 0 \\ 0 & -1 & 0 \\ 0 & 0 & 1 \end{bmatrix} \mapsto \begin{bmatrix} 1 & 0 & 0 & 0 & 0 & 0 \\ 0 & 1 & 0 & 0 & 0 & 0 \\ 0 & 0 & 1 & 0 & 0 & 0 \\ 0 & 0 & 0 & -1 & 0 & 0 \\ 0 & 0 & 0 & 0 & -1 & 0 \\ 0 & 0 & 0 & 0 & 0 & 1 \end{bmatrix} = \tilde{A}^{\mathrm{mono}}\,,$$

which is a particular case of transformation (3.42); it is tantamount to the reflection about the x_1x_2-plane. The monoclinic-symmetry condition is

$$C = \left(\tilde{A}^{\mathrm{mono}}\right) C \left(\tilde{A}^{\mathrm{mono}}\right)^T\,, \tag{3.62}$$

where C is the 6×6 matrix in equation (3.41), and T denotes transpose. Hence, computing the right-hand side, we obtain

$$\begin{bmatrix} c_{1111} & c_{1122} & c_{1133} & -\sqrt{2}c_{1123} & -\sqrt{2}c_{1113} & \sqrt{2}c_{1112} \\ c_{1122} & c_{2222} & c_{2233} & -\sqrt{2}c_{2223} & -\sqrt{2}c_{2213} & \sqrt{2}c_{2212} \\ c_{1133} & c_{2233} & c_{3333} & -\sqrt{2}c_{3323} & -\sqrt{2}c_{3313} & \sqrt{2}c_{3312} \\ -\sqrt{2}c_{1123} & -\sqrt{2}c_{2223} & -\sqrt{2}c_{3323} & 2c_{2323} & 2c_{2313} & -2c_{2312} \\ -\sqrt{2}c_{1113} & -\sqrt{2}c_{2213} & -\sqrt{2}c_{3313} & 2c_{2313} & 2c_{1313} & -2c_{1312} \\ \sqrt{2}c_{1112} & \sqrt{2}c_{2212} & \sqrt{2}c_{3312} & -2c_{2312} & -2c_{1312} & 2c_{1212} \end{bmatrix}.$$

The symmetry condition requires the left-hand side to be

$$\begin{bmatrix} c_{1111} & c_{1122} & c_{1133} & 0 & 0 & \sqrt{2}c_{1112} \\ c_{1122} & c_{2222} & c_{2233} & 0 & 0 & \sqrt{2}c_{2212} \\ c_{1133} & c_{2233} & c_{3333} & 0 & 0 & \sqrt{2}c_{3312} \\ 0 & 0 & 0 & 2c_{2323} & 2c_{2313} & 0 \\ 0 & 0 & 0 & 2c_{2313} & 2c_{1313} & 0 \\ \sqrt{2}c_{1112} & \sqrt{2}c_{2212} & \sqrt{2}c_{3312} & 0 & 0 & 2c_{1212} \end{bmatrix} =: C^{\mathrm{mono}}\,, \tag{3.63}$$

which is an elasticity tensor exhibiting a monoclinic symmetry and stated in expression (3.43).

Remark 3.4. The material symmetry of a tensor means that the values of its components remain unchanged under a given operation. Herein, it means that the values of components of C^{mono} remain the same under rotation by π about the x_3-axis.

However, the tensor itself—by the essence of its formulation—maintains its properties regardless of its symmetry or the orientation of the coordinate system in which its components are expressed. Its essential properties, such as norms and eigenvalues, are coordinate-independent, and—to maintain this independence—in general, the values of the components change accordingly.

Exercise 3.7. Write the form of transformation (3.42) under which tensor (3.54) remains invariant.

Solution. Tensor (3.54) is transversely isotropic. Thus, it remains invariant under rotation about a single axis, which, for expression (3.54), is the x_3-axis. In $\mathbb{R}^3$, such a rotation is given by

$$A := \begin{bmatrix} A_{11} & A_{12} & A_{13} \\ A_{21} & A_{22} & A_{23} \\ A_{31} & A_{32} & A_{33} \end{bmatrix} = \begin{bmatrix} \cos\theta & -\sin\theta & 0 \\ \sin\theta & \cos\theta & 0 \\ 0 & 0 & 1 \end{bmatrix}.$$

Thus, in $\mathbb{R}^6$,

$$\tilde{A} = \begin{bmatrix} A_{11}^2 & A_{12}^2 & A_{13}^2 & \sqrt{2}A_{12}A_{13} & \sqrt{2}A_{11}A_{13} & \sqrt{2}A_{11}A_{12} \\ A_{21}^2 & A_{22}^2 & A_{23}^2 & \sqrt{2}A_{22}A_{23} & \sqrt{2}A_{21}A_{23} & \sqrt{2}A_{21}A_{22} \\ A_{31}^2 & A_{32}^2 & A_{33}^2 & \sqrt{2}A_{32}A_{33} & \sqrt{2}A_{31}A_{33} & \sqrt{2}A_{31}A_{32} \\ \sqrt{2}A_{21}A_{31} & \sqrt{2}A_{22}A_{32} & \sqrt{2}A_{23}A_{33} & A_{23}A_{32} + A_{22}A_{33} & A_{23}A_{31} + A_{21}A_{33} & A_{22}A_{31} + A_{21}A_{32} \\ \sqrt{2}A_{11}A_{31} & \sqrt{2}A_{12}A_{32} & \sqrt{2}A_{13}A_{33} & A_{13}A_{32} + A_{12}A_{33} & A_{13}A_{31} + A_{11}A_{33} & A_{12}A_{31} + A_{11}A_{32} \\ \sqrt{2}A_{11}A_{21} & \sqrt{2}A_{12}A_{22} & \sqrt{2}A_{13}A_{23} & A_{13}A_{22} + A_{12}A_{23} & A_{13}A_{21} + A_{11}A_{23} & A_{12}A_{21} + A_{11}A_{22} \end{bmatrix},$$

which, in view of $A_{13} = A_{23} = A_{31} = A_{32} = 0$ and $A_{33} = 1$, becomes

$$\begin{bmatrix} A_{11}^2 & A_{12}^2 & 0 & 0 & 0 & \sqrt{2}A_{11}A_{12} \\ A_{21}^2 & A_{22}^2 & 0 & 0 & 0 & \sqrt{2}A_{21}A_{22} \\ 0 & 0 & 1 & 0 & 0 & 0 \\ 0 & 0 & 0 & A_{22} & A_{21} & 0 \\ 0 & 0 & 0 & A_{12} & A_{11} & 0 \\ \sqrt{2}A_{11}A_{21} & \sqrt{2}A_{12}A_{22} & 0 & 0 & 0 & A_{12}A_{21} + A_{11}A_{22} \end{bmatrix}.$$

Using trigonometric function, we write

$$\begin{bmatrix} \cos^2\theta & \sin^2\theta & 0 & 0 & 0 & -\sqrt{2}\sin\theta\cos\theta \\ \sin^2\theta & \cos^2\theta & 0 & 0 & 0 & \sqrt{2}\sin\theta\cos\theta \\ 0 & 0 & 1 & 0 & 0 & 0 \\ 0 & 0 & 0 & \cos\theta & \sin\theta & 0 \\ 0 & 0 & 0 & -\sin\theta & \cos\theta & 0 \\ \sqrt{2}\sin\theta\cos\theta & -\sqrt{2}\sin\theta\cos\theta & 0 & 0 & 0 & \cos^2\theta-\sin^2\theta \end{bmatrix} =: \tilde{A}^{\mathrm{TI}}\,, \tag{3.64}$$

which is the required form of expression (3.42). The invariance means that $\left(\tilde{A}^{\mathrm{TI}}\right) C^{\mathrm{TI}} \left(\tilde{A}^{\mathrm{TI}}\right)^T = C^{\mathrm{TI}}$, where T denotes transpose.

Remark 3.5. Since, in accordance with Figure 3.3, the monoclinic symmetry is a subgroup of transverse isotropy, it follows that elements of the monoclinic symmetry belong also to transverse isotropy. Thus, in the context of Exercises 3.6 and 3.7,

$$\left(\tilde{A}^{\mathrm{mono}}\right) C^{\mathrm{TI}} \left(\tilde{A}^{\mathrm{mono}}\right)^T = C^{\mathrm{TI}}\,.$$

Remaining in Exercise 3.7, we see that this equality is the consequence of the fact that $\tilde{A}^{\mathrm{mono}}$ is a particular case of $\tilde{A}^{\mathrm{TI}}$, where $\theta = \pi$.

Exercise 3.8. Show that expression (3.64) is an orthonormal transformation that corresponds to rotation.

Solution. For matrix (3.64) to be an orthogonal transformation any two of its rows must be orthogonal to one another; the same property must hold for its columns. To exemplify this property, let us consider the first and last row. Their scalar product is

$$\sqrt{2}\sin\theta\cos^3\theta - \sqrt{2}\sin^3\theta\cos\theta - \sqrt{2}\sin\theta\cos\theta\left(\cos^2\theta - \sin^2\theta\right) = 0,$$

as required.

For normality of this transformation, its rows and columns must be of unit length. Let us exemplify this property for the first column,

$$\sqrt{\cos^4\theta + \sin^4\theta + 2\sin^2\theta\cos^2\theta} = \sqrt{\left(\sin^2\theta + \cos^2\theta\right)^2} = 1\,,$$

as required.

For matrix (3.64) to be a rotation, its determinant must be equal to unity. It is

$$\sin^8\theta + 4\sin^2\theta\cos^6\theta + 6\sin^4\theta\cos^4\theta + 4\sin^6\theta\cos^2\theta + \cos^8\theta = \left(\sin^2\theta + \cos^2\theta\right)^4,$$

which is equal to unity, as required.

Thus—in view of these three properties—expression (3.64) is a rotation matrix in $\mathbb{R}^6$; however, according to its derivation, it is only a subset of all rotations in $\mathbb{R}^6$, since it is limited to their counterparts in $\mathbb{R}^3$. These properties, and the conclusion, remain true for matrix (3.42).

Exercise 3.9. Following Section 3.2.4.14, obtain the expression for the bulk modulus as a function of the P-wave speed, S-wave speed and mass density.

Given, in the SI units, the values of $v_P = 5000$, $v_S = 3000$, $\rho = 2700$ and $v_P = 3000$, $v_S = 1500$, $\rho = 2300$, calculate the corresponding bulk and rigidity moduli for a granite and a sandstone, respectively. Comment on these results.

Solution. According to Section 3.2.4.14, $c_{1111} = \rho v_P^2$ and $c_{2323} = \rho v_S^2$; hence,

$$K = \rho \left(v_P^2 - \frac{4}{3} v_S^2 \right) ,$$

expressed in $\mathrm{N/m^2}$, which are pascals, Pa.

Given the values for a granite, we obtain the bulk modulus of $K = 3.51 \times 10^{10}$, which is 35.1 GPa, and the rigidity modulus of $\mu = \rho v_S^2 = 2.43 \times 10^{10}$, which is 24.3 GPa. Given the values for a sandstone, we obtain the bulk modulus of $K = 1.38 \times 10^{10}$, which is 13.8 GPa, and the rigidity modulus of $\mu = \rho v_S^2 = 0.5175 \times 10^{10}$, which is 5.175 GPa. As expected, a sandstone resists changes in bulk and shape less than a granite does.

Chapter 4

Hookean solid: Effective symmetry and equivalent medium

[...] *an important consideration in deciding whether or not a field is to be regarded as a physically continuous medium rather than a mere mathematical device, lies in its possession of detectable properties other than the one property for which it was introduced. A condition of this kind is often suggested as a criterion of the physical 'reality' of a theoretical entity* [...].

Mary Brenda Hesse (2005)

Preliminary remarks

In this chapter, we consider Hookean models as analogies for information obtained from measurements. Consequently, we examine the choice of a model in the context of both its abstract formulation and the experimental errors of measurements within which the model might serve as an ideal quantitative analogy.

To do so, we invoke the concept of an average and apply it in two different contexts. The first is an average of symmetry groups at a point. Its result is a Hookean solid that exhibits a higher symmetry than the tensor obtained from measurements. It might accommodate cases for which the accuracy of measurements do not justify information required for lower-symmetry models. We refer to results of such averages as effective symmetries.

The second is an average of elasticity parameters over a spatial interval. It might accommodate a case for which the wavelength of a signal used to perform the measurements does not allow for a sufficient resolution to examine individual layers.

We begin this chapter with discussions on effective symmetries and proceed to discuss equivalent media. Even though both concepts can be combined in a single model, their formulations are not related to one another. We conclude this chapter by discussing the commutativity of the two averages.

4.1 Effective symmetries

4.1.1 *On accuracy*

Since a theory is unavoidable to mediate between measurements and interpretations, let us examine—within such a theory—the accuracy of representing a generally anisotropic tensor, inferred from physical measurements, by its more symmetric counterpart.

Prior to that examination, let us acknowledge an abuse of terminology. Below, we consider the components of the elasticity tensor, to which we refer as the elasticity parameters, and their standard deviations. Since the elasticity tensor is not the object of measurements, but a mathematical entity inferred from them, it is nonsensical to discuss whether or not the tensor is accurately measured. Herein, given the tensor components with their standard deviations—and remaining within the realm of mathematics—we discuss the accuracy with which this generally anisotropic tensor is represented by its symmetric counterpart to which we refer as its effective tensor. There are two other aspects of accuracy. First, the generally anisotropic tensor is obtained, through an inverse theory, from seismic measurements. Hence, we could attempt to relate its standard deviations to measurement errors. Secondly, once the effective tensor is chosen, we could examine the accuracy that it provides for modeling physical phenomena. Neither issue is the explicit subject of our discussion. Thus, aware of these subtleties, we use the meaning of accuracy and precision *sensu lato*.

In a manner similar to the statement about the P and S waves propagating in the Earth, which—strictly speaking—refers to a Hookean solid, the statement that a material is transversely isotropic is not to be understood literally. It means that properties of waves in a transversely isotropic Hookean solid allow for sufficiently accurate descriptions of waves in a generally anisotropic continuum, and—subsequently—that waves in such a solid are accurate analogies for propagation of disturbances in the material whose measurements resulted originally in the generally anisotropic tensor. In our subsequent discussion, we tacitly imply the above sequence of statements when we say, as a shorthand, that a given symmetry and orientation of a tensor is an accurate analogy for the material in question.

To examine materials in terms of Hookean solids, which are fourth-rank tensors expressed in terms of twenty-one independent components, we make measurements that allow us to estimate these twenty-one components. It is not an easy task, but the experimental requirements are not discussed in this book. The importance of all twenty-one components is the fact that, as shown by Bóna *et al.* (2007a), they contain the information about the orientation of the coordinate system in which a tensor is expressed. Consequently—having the twenty-one components—it is not necessary to assume this orientation *a priori*, as is the case if we consider *ab initio* a particular symmetry class, such as transverse isotropy with the vertical symmetry axis.

Dewangan and Grechka (2003) obtained the twenty-one components from vertical-seismic-profile measurements in New Mexico. Their estimate is

$$C=\begin{bmatrix} 7.8195 & 3.4495 & 2.5667 & 0.1374 & 0.0558 & 0.1239\\ 3.4495 & 8.1284 & 2.3589 & 0.0812 & 0.0735 & 0.1692\\ 2.5667 & 2.3589 & 7.0908 & -0.0092 & 0.0286 & 0.1655\\ 0.1374 & 0.0812 & -0.0092 & 1.6636 & -0.0787 & 0.1053\\ 0.0558 & 0.0735 & 0.0286 & -0.0787 & 2.0660 & -0.1517\\ 0.1239 & 0.1692 & 0.1655 & 0.1053 & -0.1517 & 2.4270 \end{bmatrix}, \tag{4.1}$$

which are the density-scaled elasticity parameters, whose units are $\mathrm{km}^2/\mathrm{s}^2$.

Once these parameters are estimated, we can proceed to find the symmetry class and orientation of the tensor that—in a certain sense—is the best analogy for the rock subjected to seismic measurements. Such a symmetry allows us to interpret features of the material in question. For instance, a monoclinic symmetry is consistent with obliquely fractured parallel layers, as described by Winterstein (1990).

We must quantify the meaning of the best analogy in the context of the precision and accuracy with which the generally anisotropic tensor is given. To do so, we invoke the concept of proximity, which allows us to say that the closest analogy that is consistent with our material is, for example, a monoclinic Hookean solid with a dipping symmetry plane. In turn, to quantify the concept of proximity, we must define the notion of a distance among fourth-rank tensors. Two distinct norms appear to be of particular interest in seismology: the Frobenius norm and the *operator norm*.[1]

[1]The former norm is discussed in this section. This latter norm is illustrated in Remark 4.4, below. Readers interested in these norms might refer to Danek *et al.* (2018). Also, readers interested in the latter norm might refer to Bos and Slawinski (2015) and Danek and Slawinski (2014).

Once we establish the tensor of a given symmetry that is closest to the twenty-one parameters obtained from the measurements, we can examine whether or not this tensor can be used as an effective counterpart for the generally anisotropic Hookean solid, in the context of the standard deviations of these twenty-one parameters. The higher the symmetry of a given Hookean solid, the further it is from the generally anisotropic Hookean solid; hence, isotropy is the most distant model.

In the case of isotropy, the twenty-one parameters are reduced to two, which are referred to commonly as the Lamé parameters, λ and μ. For instance, Voigt (1910) showed that the closest isotropic solid—in the Frobenius sense—is

$$\lambda := c_{1111}^{\rm iso} - 2\,c_{2323}^{\rm iso} \qquad \text{and} \qquad \mu := c_{2323}^{\rm iso}\,,$$

where

$$c_{1111}^{\rm iso} = \frac{1}{15}\,(3\,(c_{1111}+c_{2222}+c_{3333}) + 2\,(c_{1122}+c_{1133}+c_{2233}) \qquad (4.2)$$
$$+\,4\,(c_{1212}+c_{1313}+c_{2323}))$$

and

$$c_{2323}^{\rm iso} = \frac{1}{15}\,(c_{1111}+c_{2222}+c_{3333} - (c_{1122}+c_{1133}+c_{2233}) \qquad (4.3)$$
$$+\,3\,(c_{1212}+c_{1313}+c_{2323}))\,.$$

Herein, $c_{ijk\ell}$ are the components of the generally anisotropic Hookean solid; $c_{1111}^{\rm iso}$ and $c_{2323}^{\rm iso}$ are the components of its closest isotropic counterpart.

As illustrated in Exercise 4.2, these expressions can be derived in a straightforward manner by minimizing the Frobenius-norm distance between general anisotropy and isotropy; furthermore, due to isotropy, they are rotation-invariant, which is not the case for other symmetries, as discussed in Section 4.1.3, below.

From the fact that the higher the symmetry of a given Hookean solid, the further is its tensor from the generally anisotropic Hookean solid, it follows that the closest counterpart is necessarily the least symmetric tensor. However, the effective tensor is not the closest one, but the most symmetric one, which means the furthest tensor that is acceptable within the precision with which the twenty-one parameters are given.

This is an epistemological, as opposed to ontological, statement of *Occam's razor*, named in honour of William of Occam, which is a principle of parsimony and succinctness of a hypothesis, *lex parsimoniæ*. Herein, we do not claim Nature's propensity for simplicity, we just recognize our limitations in analyzing her complexities. Let us examine that issue.

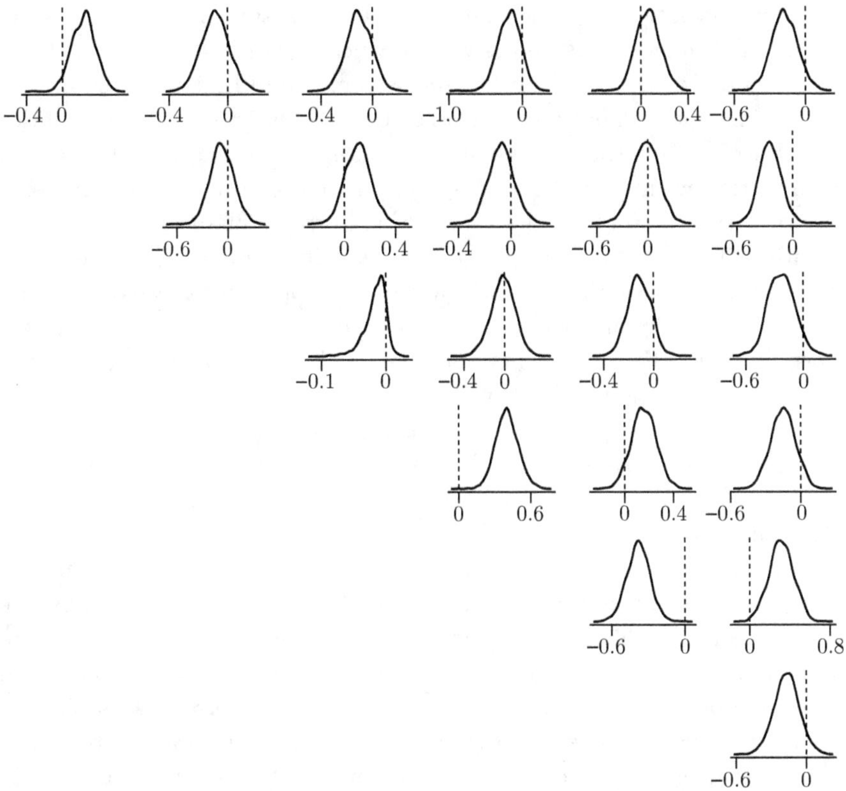

Fig. 4.1: Differences between the elasticity parameters of realizations of the generally anisotropic Hookean solid shown in expression (4.1), subject to random errors, and the corresponding parameters of the effective transversely isotropic tensors.

The standard deviations of estimate (4.1), in the study of Dewangan and Grechka (2003), are

$$\pm\begin{bmatrix} 0.1656 & 0.1122 & 0.1216 & 0.1176 & 0.0774 & 0.0741 \\ 0.1122 & 0.1862 & 0.1551 & 0.0797 & 0.1137 & 0.0832 \\ 0.1216 & 0.1551 & 0.1439 & 0.0856 & 0.0662 & 0.1010 \\ 0.1176 & 0.0797 & 0.0856 & 0.0714 & 0.0496 & 0.0542 \\ 0.0774 & 0.1137 & 0.0662 & 0.0496 & 0.0626 & 0.0621 \\ 0.0741 & 0.0832 & 0.1010 & 0.0542 & 0.0621 & 0.0802 \end{bmatrix}; \tag{4.4}$$

these values are not included in Dewangan and Grechka (2003), but obtained from Vladimir Grechka (email comm., 2007). Considering these standard deviations, we view the parameters of the effective tensors not as specific values but as ranges within which lie the best-fit values. These ranges are obtained with the Monte-Carlo method (Tarantola, 2006).

To find the symmetry class of the effective tensor—not to be confused with a tensor from the closest symmetry class, which is necessarily the least symmetric—we require the most symmetric tensor such that the difference between that tensor and the generally anisotropic one is contained sufficiently within the error ranges of the elasticity parameters of the latter. For instance, considering the transversely isotropic tensor, we obtain—for each of the twenty-one parameters—the empirical distribution of the differences between the components of the generally anisotropic and symmetric tensors. Examining the distributions in Figure 4.1, we might conclude that the transversely isotropic Hookean solid is not a sufficiently good analogy for the material examined by Dewangan and Grechka (2003). Were it a good analogy, most zero differences would be close to the centres of the distributions.

Figure 4.1 illustrates the differences between the elasticity parameters of realizations of the generally anisotropic Hookean solid obtained by Dewangan and Grechka (2003), and shown in expression (4.1)—subject to random errors whose standard deviations are stated in matrix (4.4)—and the corresponding parameters of the effective transversely isotropic tensors. Differences between respective elasticity parameters are displayed as the upper-diagonal part of a matrix given by $C - C_{\rm eff}^{\rm TI}$, where C corresponds to tensor (4.1), subject to random errors, and $C_{\rm eff}^{\rm TI}$ follows from the formulation shown in Section 4.1.2.5 and the approach discussed in Section 4.1.3, below. As discussed in Section 4.1.3, $C_{\rm eff}^{\rm TI}$ results from finding its optimal orientation. However, to obtain the difference illustrated in Figure 4.1, all tensor components are expressed in the coordinate system of tensor (4.1) and of matrix (4.4), which is not a tensor and hence, cannot be subject to rotation.

To illustrate quantitatively the appropriateness of a given analogy, let us consider the highest symmetry: isotropy, for which the effective tensor is independent of orientation, and—unlike for other symmetry classes—for all orientations, all $c_{ijk\ell}^{\rm iso} = 0$, except $c_{1111}^{\rm iso} = c_{2222}^{\rm iso} = c_{3333}^{\rm iso}$, $c_{1122}^{\rm iso} = c_{1133}^{\rm iso} = c_{2233}^{\rm iso}$ and $c_{1212}^{\rm iso} = c_{1313}^{\rm iso} = c_{2323}^{\rm iso}$, as shown in tensor (3.57) on page 82. Using the twenty-one parameters obtained by Dewangan and Grechka (2003), we can find the closest isotropic tensor by invoking the Voigt (1910) formulæ, stated in expressions (4.2) and (4.3).

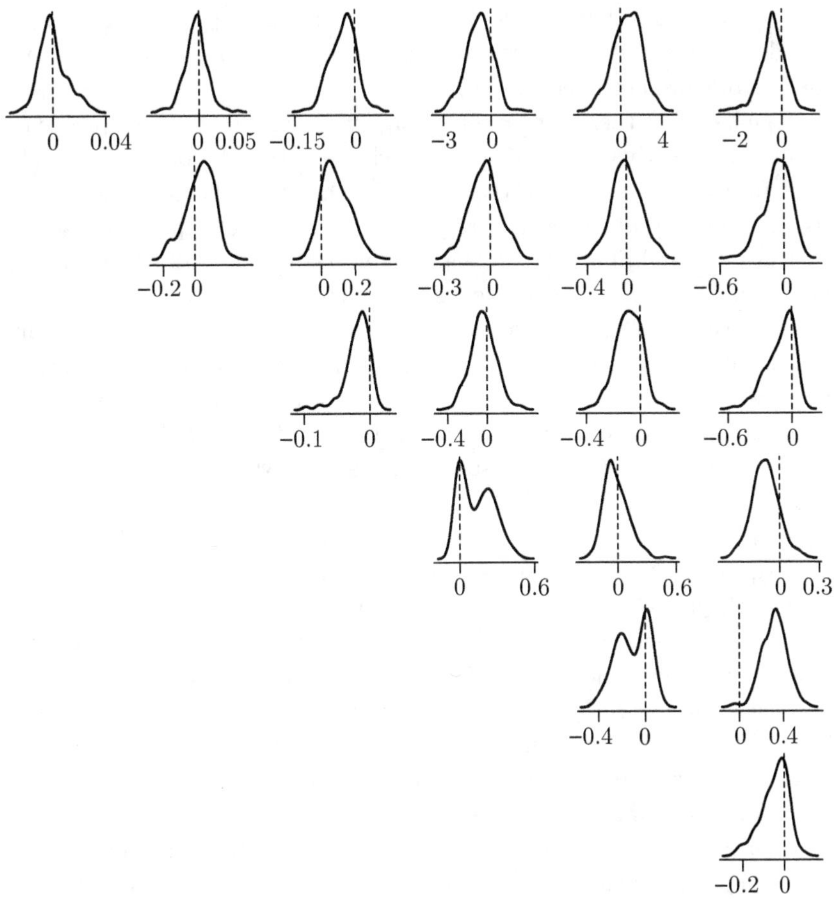

Fig. 4.2: Differences between the elasticity parameters of realizations of the generally anisotropic Hookean solid shown in expression (4.1), subject to random errors, and the corresponding parameters of the effective orthotropic tensors.

The question, however, remains whether or not isotropy is a good enough analogy for the physical material under consideration. To shed light on this issue, let us consider the entries that are zero for an isotropic tensor, shown in expression (3.57). For instance, considering expressions (4.1) and (4.4), we see therein that the entry at the first row and the fourth column is 0.1374 ± 0.1176, which means that zero required by isotropy is more than one standard deviation away. It follows, from properties of the Gaussian distribution, that probability of the required zero is less than 30%. For the entry at the second row and the sixth column, in expressions (4.1) and (4.4), which is 0.1692 ± 0.0832, the zero required by isotropy is more than two standard deviations away; its probability is less than 5%. Thus, conditions for isotropy are not likely to be satisfied.

The fact that there is no orientation dependence renders the problem of finding the effective isotropic tensor much easier than for other classes. The need to involve rotations in the search for the other effective tensors leads to a nonlinear problem, which requires numerical techniques, as discussed by Kochetov and Slawinski (2009a,b,c) and Danek *et al.* (2013, 2015). The requirement for such techniques is the reason why, a century ago, Voigt (1910) formulated the closest tensor for isotropy only; he was aware of limitations of computational methods at his disposal. Also, in the isotropic case, we can consider analytically the concept of normal distributions and standard deviations. For other cases, we need to study the error propagation numerically—by such approaches as the Monte-Carlo method—to obtain empirical distributions.

It is important to ensure that our inferences are justifiable by accuracy of the generally anisotropic tensor. In general, the higher the symmetry, the simpler the model; an isotropic Hookean solid contains less information than any solid of lesser symmetry. To invoke a less symmetric analogy, we need more information, such as a sufficient resolution to distinguish among the elasticity parameters. Consequently, proposing an isotropic model might be a statement about intrinsic properties of the material in question or it might be a statement about the limited accuracy or precision of data.

Also, it is important to ensure that we take from the data as much information as possible. To do so we might have to invoke advanced mathematical methods to deal with nonlinearity and the absence of analytical expressions.

In the experiment of Dewangan and Grechka (2003), the accuracy is sufficiently high to consider an analogy not only with two values—which would correspond to isotropy, and are discussed on page 109—but with twelve values: nine independent elasticity parameters, which is an orthotropic Hookean solid, and three independent parameters stating the orientations of its symmetry planes with respect to the coordinate system in which this tensor is expressed.[2] Since most zero differences shown in Figure 4.2 are close to the centres of the distributions, we might conclude that the orthotropic Hookean solid is an acceptable analogy for the material examined by Dewangan and Grechka (2003).

[2]Readers interested in this effective orthotropic tensor resulting from the experiment of Dewangan and Grechka (2003) might refer to Kochetov and Slawinski (2009b).

Figure 4.2 illustrates the differences between the elasticity parameters of realizations of tensor (4.1), subject to random errors, and the corresponding parameters of the effective orthotropic tensors. $C - C_{\rm eff}^{\rm ortho}$, where C corresponds to tensor (4.1) and $C_{\rm eff}^{\rm ortho}$ follows from the formulation shown in Section 4.1.2.3 and the approach discussed in Section 4.1.3, below.

Comparing Figures 4.1 and 4.2, we see that a seven-parameter model of transverse isotropy is insufficient to accommodate—with required accuracy—information available from expressions (4.1), subject to errors (4.4). We need, at least, twelve degrees of freedom provided by an orthotropic tensor.

4.1.2 *Fixed orientation of coordinate system*

4.1.2.1 *Introduction*

Since Hookean solids are Platonic analogies for physical materials, we cannot say that a physical material belongs to a particular symmetry class of Hookean solids. However, we can study a macroscopic behaviour of such a material by considering symmetry classes. In seismology, these behaviours are, for instance, speed of propagation, and magnitude, of mechanical disturbances within the Earth.

The most accurate Hookean model is generally anisotropic; its description requires twenty-one elasticity parameters. However, following the Occam-razor approach, we wish to use the simplest model that is sufficient to represent information obtained from measurements in the context of experimental errors. The effect of these errors might limit the resolution necessary to justify a twenty-one-parameter description. This is an epistemological view of Occam's razor not an ontological one; in other words, our view stems from limitations of methodology not from a metaphysical belief in simplicity of nature. Hence, the question arises, of relating a generally anisotropic elasticity tensor, $\mathbf{c}$, to its counterpart, $\mathbf{c}_{\rm eff}^{\rm sym}$, which belongs to a particular symmetry.

To do so, we let $\mathbf{c}_{\rm eff}^{\rm sym}$ be the orthogonal projection of $\mathbf{c}$, in the sense of the Frobenius norm, on the linear space containing all tensors of that symmetry, which are $\mathbf{c}^{\rm sym}$. This projection is the average given by

$$\mathbf{c}_{\rm eff}^{\rm sym} := \int\limits_{G^{\rm sym}} (g \circ \mathbf{c})\, \mathrm{d}\mu(g)\,, \tag{4.5}$$

where the integration is over the symmetry group, $G^{\rm sym}$, whose elements are g, with respect to the invariant measure, μ, normalized so that

$\mu(G^{\rm sym}) = 1$, as described by Denos Gazis, Iradj Tadjbakhsh and Richard Toupin, in Gazis *et al.* (1963). Integral (4.5) reduces to a finite sum for the classes whose symmetry groups are finite, which are all classes except isotropy and transverse isotropy. As shown by Gazis *et al.* (1963, Theorem 2.2), projection (4.5) ensures that a positive-definite tensor is projected to another positive-definite tensor, as required by Hookean solids.

Our choice of Frobenius norm, which is also referred to as Euclidean norm, is determined by its simplicity and mathematical convenience. The invariance of Frobenius norm under orthogonal transformations leads to elegant formulæ for projections of a generally anisotropic elasticity tensor onto nontrivial symmetry classes with fixed orientations. Indeed, this property allows us to use expression (4.5). Another norm with this invariance property is the so-called log-Euclidean norm, discussed by Norris (2006) and defined as the Frobenius norm of the logarithm of elasticity tensor regarded as a positive-definite matrix in the form of the 6×6 matrix in expression (3.41). This norm has an additional advantage of having the same value for both the stiffness and compliance tensors. Methods based on the Euclidean norm can be adapted to deal with the log-Euclidean norm as illustrated by Kochetov and Slawinski (2009c).

Also, let us comment on the operator norm, which—for Hookean solids, whose components are stated in a symmetric 6×6 matrix, C—is given by the largest eigenvalue of C. Similarly to the Frobenius and log-Euclidean norms, the operator norm exhibits the invariance property. Furthermore, following Hooke's law, stated in expression (3.41), the operator norm of C is induced by the Frobenius norm of the stress and strain tensors. For the Frobenius and log-Euclidean norms of C, no such relation is known. The operator norm is discussed in Remark 4.4, on page 159.[3]

4.1.2.2 *Monoclinic tensor*

Invoking the symmetry groups in discussing effective tensors, we consider $SO(3)$. The symmetry group of the monoclinic tensor, which is stated in expression (3.38), can be written explicitly—in terms of $SO(3)$—as

$$G^{\rm mono} = \left\{ \begin{bmatrix} 1 & 0 & 0 \\ 0 & 1 & 0 \\ 0 & 0 & 1 \end{bmatrix}, \begin{bmatrix} -1 & 0 & 0 \\ 0 & -1 & 0 \\ 0 & 0 & 1 \end{bmatrix} \right\} =: \{A_\ell^{\rm mono}\}, \ell \in \{1,2\};$$

[3]Readers interested in the operator norm in the context of effective tensors might refer to Bos and Slawinski (2015), Danek and Slawinski (2014) and Danek *et al.* (2018).

in other words, the symmetry group is composed of the identity, and the rotation by π about the x_3-axis, which is tantamount to the reflection about the x_1x_2-plane: $(-I)(A_2^{\rm mono})$, in other words, the reflection about the plane whose normal is $\hat{\mathbf{e}}_3$, as stated in expression (3.38). Using matrix (3.42), we get the corresponding two 6×6 transformation matrices, and the resulting projection is

$$C_{\rm eff}^{\rm mono}=\frac{1}{2}\sum_{\ell=1}^{2}\left(\tilde{A}_\ell^{\rm mono}\right)C\left(\tilde{A}_\ell^{\rm mono}\right)^T,$$

which is the average of the transformations that constitute the symmetry group; 2 in the denominator is the number of elements in the group. As shown in Exercise 4.1, we obtain $c_{ijk\ell}^{\rm mono}=c_{ijk\ell}$, for the entries that correspond to nonzeros of matrix (3.43), and zeros for the ones corresponding to zero; namely,

$$C_{\rm eff}^{\rm mono}=\begin{bmatrix} c_{1111}^{\rm mono} & c_{1122}^{\rm mono} & c_{1133}^{\rm mono} & 0 & 0 & \sqrt{2}c_{1112}^{\rm mono}\\ c_{1122}^{\rm mono} & c_{2222}^{\rm mono} & c_{2233}^{\rm mono} & 0 & 0 & \sqrt{2}c_{2212}^{\rm mono}\\ c_{1133}^{\rm mono} & c_{2233}^{\rm mono} & c_{3333}^{\rm mono} & 0 & 0 & \sqrt{2}c_{3312}^{\rm mono}\\ 0 & 0 & 0 & 2c_{2323}^{\rm mono} & 2c_{2313}^{\rm mono} & 0\\ 0 & 0 & 0 & 2c_{2313}^{\rm mono} & 2c_{1313}^{\rm mono} & 0\\ \sqrt{2}c_{1112}^{\rm mono} & \sqrt{2}c_{2212}^{\rm mono} & \sqrt{2}c_{3312}^{\rm mono} & 0 & 0 & 2c_{1212}^{\rm mono} \end{bmatrix}, \tag{4.6}$$

which has the form of matrix (3.43) with nonzero entries of the matrix in expression (3.41). As discussed by Kochetov and Slawinski (2009b) and illustrated in Exercises 3.6 and 4.1, such a simple relation between $C_{\rm eff}^{\rm sym}$ and C is true also for orthotropic symmetry, but not for the other symmetry classes.

4.1.2.3 *Orthotropic tensor*

To relate $C_{\rm eff}^{\rm ortho}$ to C, we use expression (4.5) and the symmetry group of the orthotropic tensors, which is stated in expression (3.37), and can be written explicitly as

$$G^{\rm ortho}=\left\{\begin{bmatrix}1&0&0\\0&1&0\\0&0&1\end{bmatrix},\begin{bmatrix}-1&0&0\\0&-1&0\\0&0&1\end{bmatrix},\begin{bmatrix}1&0&0\\0&-1&0\\0&0&-1\end{bmatrix},\begin{bmatrix}-1&0&0\\0&1&0\\0&0&-1\end{bmatrix}\right\}=:\left\{A_\ell^{\rm ortho}\right\}, \tag{4.7}$$

where $\ell\in\{1,\ldots,4\}$. The symmetry group is composed of the identity and the rotations by π about the three Cartesian axes; there are two more rotations than in $G^{\rm mono}$. Using matrix (3.42), we get the corresponding four transformation matrices. The resulting projection (4.5) is

$$C^{\rm ortho}_{\rm eff} = \frac{1}{4}\sum_{\ell=1}^{4}\left(\tilde{A}^{\rm ortho}_{\ell}\right) C \left(\tilde{A}^{\rm ortho}_{\ell}\right)^{T}.$$

Herein, 4 in the denominator is the number of elements in the group. As in the case of the monoclinic symmetry, we obtain $c^{\rm ortho}_{ijk\ell} = c_{ijk\ell}$, for the entries that correspond to nonzeros of matrix (3.50), and zeros otherwise, as discussed by Kochetov and Slawinski (2009b).

4.1.2.4 *Tetragonal tensor*

The symmetry group of the tetragonal tensor, which is stated in expression (3.35), can be written explicitly as

$$G^{\rm tetra} = \left\{ \begin{bmatrix} 1&0&0\\0&1&0\\0&0&1 \end{bmatrix}, \begin{bmatrix} -1&0&0\\0&-1&0\\0&0&1 \end{bmatrix}, \begin{bmatrix} 1&0&0\\0&-1&0\\0&0&-1 \end{bmatrix}, \begin{bmatrix} -1&0&0\\0&1&0\\0&0&-1 \end{bmatrix}, \right. \tag{4.8}$$
$$\left. \begin{bmatrix} 0&1&0\\-1&0&0\\0&0&1 \end{bmatrix}, \begin{bmatrix} 0&-1&0\\1&0&0\\0&0&1 \end{bmatrix}, \begin{bmatrix} 0&-1&0\\-1&0&0\\0&0&-1 \end{bmatrix}, \begin{bmatrix} 0&1&0\\1&0&0\\0&0&-1 \end{bmatrix} \right\},$$

which we denote by $\{A^{\rm tetra}_{\ell}\}$, where $\ell \in \{1,\dots,8\}$. The upper row of $G^{\rm tetra}$ is the same as $G^{\rm ortho}$, since $G^{\rm ortho} \subset G^{\rm tetra}$. Since the transformation matrices in group (4.9) belong to $SO(3)$, their determinants are equal to positive unity. The meaning of the elements in this group, in terms of rotations, is discussed by Bóna *et al.* (2004b) below their expression (2.5). In view of transformation (3.42) and projection (4.5), we write

$$C^{\rm tetra}_{\rm eff} = \frac{1}{8}\sum_{\ell=1}^{8}\left(\tilde{A}^{\rm tetra}_{\ell}\right) C \left(\tilde{A}^{\rm tetra}_{\ell}\right)^{T},$$

where 8 in the denominator is the number of elements in the group; hence, we obtain the entries of the effective tetragonal tensor, which, as expected, result in the form of matrix (3.51); they are

$$c^{\rm tetra}_{1111} = \frac{1}{2}\left(c_{1111} + c_{2222}\right), \tag{4.9}$$

$$c^{\rm tetra}_{1122} = c_{1122}, \tag{4.10}$$

$$c^{\rm tetra}_{1133} = \frac{1}{2}\left(c_{1133} + c_{2233}\right), \tag{4.11}$$

$$c^{\rm tetra}_{3333} = c_{3333}, \tag{4.12}$$

$$c^{\rm tetra}_{2323} = \frac{1}{2}\left(c_{2323} + c_{1313}\right), \tag{4.13}$$

$$c^{\rm tetra}_{1212} = c_{1212}, \tag{4.14}$$

as shown by Diner *et al.* (2011b). As required by Theorem 4.5 of Bóna *et al.* (2007a), $C_{\rm eff}^{\rm tetra}$ has five distinct eigenvalues.

4.1.2.5 *Transversely isotropic tensor*

The projection for $\mathbf{c}^{\rm TI}$ is tensor (3.54), whose effective components, in view of group (3.34), are

$$c_{1111}^{\rm TI} = \frac{1}{8}\left(3c_{1111} + 3c_{2222} + 2c_{1122} + 4c_{1212}\right), \tag{4.15}$$

$$c_{1122}^{\rm TI} = \frac{1}{8}\left(c_{1111} + c_{2222} + 6c_{1122} - 4c_{1212}\right), \tag{4.16}$$

$$c_{1133}^{\rm TI} = \frac{1}{2}\left(c_{1133} + c_{2233}\right), \tag{4.17}$$

$$c_{3333}^{\rm TI} = c_{3333}, \tag{4.18}$$

$$c_{2323}^{\rm TI} = \frac{1}{2}\left(c_{2323} + c_{1313}\right), \tag{4.19}$$

as shown by Moakher and Norris (2006) and Bucataru and Slawinski (2009a).

4.1.2.6 *Trigonal tensor*

The symmetry group of the trigonal tensor, which is stated in expression (3.36), can be written explicitly as

$$G^{\rm trig} = \left\{ \begin{bmatrix} 1 & 0 & 0 \\ 0 & 1 & 0 \\ 0 & 0 & 1 \end{bmatrix}, \begin{bmatrix} -1 & 0 & 0 \\ 0 & 1 & 0 \\ 0 & 0 & -1 \end{bmatrix}, \begin{bmatrix} \frac{1}{2} & -\frac{\sqrt{3}}{2} & 0 \\ -\frac{\sqrt{3}}{2} & -\frac{1}{2} & 0 \\ 0 & 0 & -1 \end{bmatrix}, \right. \tag{4.20}$$

$$\left. \begin{bmatrix} \frac{1}{2} & \frac{\sqrt{3}}{2} & 0 \\ \frac{\sqrt{3}}{2} & -\frac{1}{2} & 0 \\ 0 & 0 & -1 \end{bmatrix}, \begin{bmatrix} -\frac{1}{2} & \frac{\sqrt{3}}{2} & 0 \\ -\frac{\sqrt{3}}{2} & -\frac{1}{2} & 0 \\ 0 & 0 & 1 \end{bmatrix}, \begin{bmatrix} -\frac{1}{2} & -\frac{\sqrt{3}}{2} & 0 \\ \frac{\sqrt{3}}{2} & -\frac{1}{2} & 0 \\ 0 & 0 & 1 \end{bmatrix} \right\},$$

which we denote by $\{A_\ell^{\rm trig}\}$, where $\ell \in \{1, \ldots, 6\}$.

Let us comment on elements of group (4.20). The first one is the invariance of the tensor under the identity transformation. The following three elements refer to invariances under reflection about three planes containing the x_3-axis and whose unit normals are oriented at $2\pi/3$ with respect to each other. The first one among them is tantamount to the invariance under reflection about the plane whose normal is $\hat{\mathbf{e}}_2:\ (-I)(A_2^{\rm trig})$, namely, $M_{\hat{\mathbf{e}}_2}=M_{(0,1,0)}$. The following two elements are tantamount to reflections about the plane whose normals are also contained in the x_1x_2-plane and whose x_1 components are $\cos(\pi/3)$ and $\cos(2\pi/3)$, and whose x_2 components are $\sin(\pi/3)$ and $\sin(2\pi/3)$, namely, $M_{(\cos\frac{\pi}{3},\,\sin\frac{\pi}{3},\,0)}$ and $M_{(\cos\frac{2\pi}{3},\,\sin\frac{2\pi}{3},\,0)}$. The remaining two elements are tantamount to invariances under clockwise and counterclockwise rotations about the x_3-axis by $2\pi/3$, namely, $R_{\frac{2\pi}{3},\hat{\mathbf{e}}_3}$ and $R_{-\frac{2\pi}{3},\hat{\mathbf{e}}_3}$.

As in group (4.9), $\det A_\ell^{\rm trig}=+1$, and the meaning of the elements in group (4.20) is discussed by Bóna *et al.* (2004b) below their expression (2.5). Using transformation (3.42), we write the projection as

$$C_{\rm eff}^{\rm trig}=\frac{1}{6}\sum_{\ell=1}^{6}\left(\tilde{A}_\ell^{\rm trig}\right)C\left(\tilde{A}_\ell^{\rm trig}\right)^T,$$

where 6 in the denominator is the number of elements in the group and which, as expected, results in the form of matrix (3.55) with

$$c_{1111}^{\rm trig}=\frac{1}{8}\left(3c_{1111}+2c_{1122}+3c_{2222}+4c_{1212}\right),\tag{4.21}$$

$$c_{1122}^{\rm trig}=\frac{1}{8}\left(c_{1111}+6c_{1122}+c_{2222}-4c_{1212}\right),\tag{4.22}$$

$$c_{1133}^{\rm trig}=\frac{1}{2}\left(c_{1133}+6c_{2233}\right),\tag{4.23}$$

$$c_{1113}^{\rm trig}=\frac{1}{3\sqrt{2}}\left(c_{1113}+c_{2213}-2c_{2312}\right),\tag{4.24}$$

$$c_{2323}^{\rm trig}=\frac{1}{2}\left(c_{2323}+c_{1313}\right),\tag{4.25}$$

$$c_{2312}^{\rm trig}=\frac{1}{4}\left(-c_{1113}+c_{2213}+2c_{2312}\right),\tag{4.26}$$

as shown by Diner *et al.* (2011b). As required by Theorem 4.4 of Bóna *et al.* (2007a), two of the six eigenvalues of $C_{\rm eff}^{\rm trig}$ are distinct and the remaining ones are doubled.

4.1.2.7 *Cubic tensor*

Projection for $\mathbf{c}^{\mathrm{cubic}}$ results in matrix (3.56), whose entries are

$$c_{1111}^{\mathrm{cubic}} = \frac{1}{3}\left(c_{1111} + c_{2222} + c_{3333}\right) , \tag{4.27}$$

$$c_{1122}^{\mathrm{cubic}} = \frac{1}{3}\left(c_{1122} + c_{1133} + c_{2233}\right) , \tag{4.28}$$

$$c_{2323}^{\mathrm{cubic}} = \frac{1}{3}\left(c_{1212} + c_{1313} + c_{2323}\right) , \tag{4.29}$$

as shown by Moakher and Norris (2006) and Bucataru and Slawinski (2009a).

4.1.2.8 *Isotropic tensor*

The entries of the effective isotropic tensor that result in matrix (3.57) are

$$\begin{aligned} c_{1111}^{\mathrm{iso}} =& \frac{1}{15}\left(3\left(c_{1111} + c_{2222} + c_{3333}\right) + 2\left(c_{1122} + c_{1133} + c_{2233}\right)\right. \\ &\left. + 4\left(c_{1212} + c_{1313} + c_{2323}\right)\right) \end{aligned} \tag{4.30}$$

and

$$\begin{aligned} c_{2323}^{\mathrm{iso}} =& \frac{1}{15}\left(c_{1111} + c_{2222} + c_{3333} - \left(c_{1122} + c_{1133} + c_{2233}\right)\right. \\ &\left. + 3\left(c_{1212} + c_{1313} + c_{2323}\right)\right) , \end{aligned} \tag{4.31}$$

as shown by Voigt (1910).

As stated by expression (3.57), c_{1111}^{iso} and c_{2323}^{iso} are the two independent elasticity parameters of an isotropic Hookean solid. However, in view of tensor $\mathbf{c}$, expressions (4.30) and (4.31) are not independent; they are related by parameters of $\mathbf{c}$. This relation can be expressed as

$$c_{1111}^{\mathrm{iso}} = c_{2323}^{\mathrm{iso}} \frac{3(c_{1111} + c_{2222} + c_{3333}) + 2(c_{1122} + c_{1133} + c_{2233}) + 4(c_{1212} + c_{1313} + c_{2323})}{c_{1111} + c_{2222} + c_{3333} - (c_{1122} + c_{1133} + c_{2233}) + 3(c_{1212} + c_{1313} + c_{2323})} .$$

To clarify these statements, which—at first sight—appear contradictory, and which are pertinent to effective tensors of all symmetry classes, let us understand that, in general, $\mathbf{c} \mapsto \mathbf{c}_{\mathrm{eff}}^{\mathrm{sym}}$ is an averaging process; hence, it is a many-to-one map. For instance, in expressions (4.30) and (4.31), nine elements of the original tensor, $\mathbf{c}$, are mapped onto two elements of the effective tensor, $\mathbf{c}_{\mathrm{eff}}^{\mathrm{iso}}$. Thus, parameters of an effective tensor are combinations of parameters of the original tensor, but—if we ignore their origins—within each $\mathbf{c}_{\rm eff}^{\rm sym}$, they are independent of each other.

Unlike for the other symmetries, the values of expressions (4.30) and (4.31) remain the same regardless of the orientations of the coordinate systems in which the generally anisotropic tensor—whose components are stated by the square matrix in expression (3.41)—and the effective isotropic tensor are expressed. The expressions for the effective tensors of the monoclinic, orthotropic, tetragonal, transversely isotropic, trigonal and cubic symmetries are valid only in the coordinate system of tensor (3.41). However, we can proceed to find effective tensors in optimal orientations of coordinate systems by considering the concept of distance in the space of elasticity tensors.

4.1.3 *Optimal orientation of coordinate system*

[4]Expressions (4.30) and (4.31) describe the isotropic proxy that is "closest" to a generally anisotropic tensor. Similarly, other effective tensors are representatives of a given symmetry class that are "closest" to a generally anisotropic tensor. To consider the concept of closeness between two tensors, we must formulate the notion of distance in the space of such tensors.

Let us begin by stating the Frobenius product, which—similarly to the inner product of vectors—is the sum of the componentwise multiplications of tensors. Thus, the Frobenius product of two tensors, C and C', whose components are stated as in the square matrix in expression (3.41), are $C_{11}C'_{11}+C_{12}C'_{12}+\ldots+C_{66}C'_{66}$.

The Frobenius norm of C, which we denote by $\|C\|$, is the square root of the sum of the squares of its components. As stated on page 112, this quantity is invariant under orthogonal transformations. In other words, the value of the Frobenius norm remains the same under rotations of the coordinate system.

If the components of both C and C' are expressed with respect to the same coordinate system, the Frobenius distance between them is the square root of the sum of $(C_{mn}-C'_{mn})^2$ for $m,n=1,\ldots,6$; concisely, we can write the distance as $\|C-C'\|$. If the components are not expressed in the same system, this expression, in general, does not result in the value of distance; note that—for differently oriented coordinate systems—the distance

[4]In this section, we discuss the subject examined in several publications (Bucataru and Slawinski, 2009a; Danek *et al.*, 2013, 2015; Kochetov and Slawinski, 2009a,b,c). Therein, the interested reader might find further information on this subject.

from a tensor to itself is not zero, which contradicts the definition of distance. However, since the Frobenius product is invariant under orthogonal transformations, to find the distance, we can minimize the above expressions under all orientations of the coordinate system, $\min\|C(\tilde{A}) - C'\|$, where $\tilde{A} \in SO(3)$ is the rotation matrix shown in expression (3.42). Finding the minimum is tantamount to expressing both C and C' in the same coordinate system. Notably, if they happen to represent the same tensor, the minimum is zero; otherwise, it is a positive number. To illustrate the former statement let us examine a natural expression for a material symmetry in the context of an effective tensor of that symmetry. For instance, let us consider transverse isotropy, whose expression is given in matrix (3.54), which we write as

$$C^{\mathrm{TI}} = \begin{bmatrix} c_{1111} & c_{1122} & c_{1133} & 0 & 0 & 0 \\ c_{1122} & c_{1111} & c_{1133} & 0 & 0 & 0 \\ c_{1133} & c_{1133} & c_{3333} & 0 & 0 & 0 \\ 0 & 0 & 0 & 2c_{2323} & 0 & 0 \\ 0 & 0 & 0 & 0 & 2c_{2323} & 0 \\ 0 & 0 & 0 & 0 & 0 & 2\,\dfrac{c_{1111} - c_{1122}}{2} \end{bmatrix}, \tag{4.32}$$

to emphasize the presence of 2 in the $(6,6)$ entry. Inserting these entries into expressions (4.15)–(4.19), in a manner prescribed by matrix (3.41), we see that $c^{\mathrm{TI}}_{ijk\ell} = c_{ijk\ell}$, for all i,j,k,ℓ. For example,

$$\begin{aligned} c^{\mathrm{TI}}_{1111} &= \frac{1}{8}\left(3c_{1111} + 3c_{2222} + 2c_{1122} + 4c_{1212}\right) \\ &= \frac{1}{8}\left(3c_{1111} + 3c_{1111} + 2c_{1122} + 4\frac{c_{1111} - c_{1122}}{2}\right) = c_{1111}\,. \end{aligned}$$

This implies that $\|C^{\mathrm{TI}}_{\mathrm{eff}} - C^{\mathrm{TI}}\| = 0$, as expected.

A minimization requires that C' can be expressed in terms of C. The relation that is of interest to us is C' being an effective tensor of C, which results in expressions stated in Section 4.1.2. In other words, C is a generally anisotropic tensor and $C' \equiv C^{\mathrm{sym}}_{\mathrm{eff}}$ is its analogue that exhibits a chosen symmetry.

As stated on page 111, the closeness viewed as the orthogonal projection is tantamount to averaging stated in expression (4.5). Since, according to the definition of orthogonal projection, tensors $\mathbf{c} - \mathbf{c}^{\mathrm{sym}}_{\mathrm{eff}}$ and $\mathbf{c}^{\mathrm{sym}}_{\mathrm{eff}}$ are normal to one another, we write the squared distance between $\mathbf{c}$ and $\mathbf{c}^{\mathrm{sym}}_{\mathrm{eff}}$ as

$$d^2_{\mathrm{sym}} = \|\,\mathbf{c} - \mathbf{c}^{\mathrm{sym}}_{\mathrm{eff}}\|^2 = \|\,\mathbf{c}\|^2 - \|\mathbf{c}^{\mathrm{sym}}_{\mathrm{eff}}\|^2\,, \tag{4.33}$$

which, using the Kelvin notation, we can express as

$$d_{\rm sym}^2 = \|C\|^2 - \|C_{\rm eff}^{\rm sym}\|^2 \,. \tag{4.34}$$

In all cases—except for isotropy—the values of expressions (4.33) and (4.34) depend on the orientation of the coordinate system in which the tensor exhibiting the chosen symmetry is expressed with respect to the original tensor, since—in general—each orientation results in a distinct candidate for the effective tensor. If the pair of $\|\mathbf{c}\|$ and $\|\mathbf{c}_{\rm eff}^{\rm sym}\|$ or the pair of $\|C\|$ and $\|C^{\rm sym}\|$ is expressed in the same coordinate system, the value obtained from either expression is the value of distance. Otherwise, to find the effective tensor, we must minimize either expression under all orientations, which we can do since the Frobenius product is invariant under rotations. Finding the minimum as a function of orientation is a computationally challenging task due to the nonlinearity of $\|C(\tilde{A}) - C_{\rm eff}^{\rm sym}\|$, which—in view of the form of $\tilde{A}$—is a high-order polynomial in several variables. Let us discuss this minimization process.

It follows from projection (4.5) that the effective tensor relative to the orientation of $G^{\rm sym}$ that differs from the original orientation by rotation $A \in SO(3)$ is given by $\tilde{A}\,{\rm pr}_{\rm sym}(\tilde{A}^T\, C\, \tilde{A})\, \tilde{A}^T$, where ${\rm pr}_{\rm sym}$ stands for a projection onto a symmetry and where we perform the minimization of $d_{\rm sym}$ under all $A \in SO(3)$. To carry out computations for finding effective tensors, we parametrize $SO(3)$ by quaternions, which were introduced by William Rowan Hamilton in the first half of the nineteenth century, and are described in the context of rotations in such textbooks as Stillwell (2008).[5] The use of quaternions—not Euler angles—results in algebraic, not trigonometric, polynomials to be minimized. Such a form is advantageous for computational purposes.

We write a quaternion as

$$q = a + b\,\mathbf{i} + c\,\mathbf{j} + d\,\mathbf{k}\,,$$

where $\mathbf{i}^2 = \mathbf{j}^2 = \mathbf{k}^2 = \mathbf{i}\,\mathbf{j}\,\mathbf{k} = -1$. We can view the quaternion as a scalar-vector pair, $(a, [b, c, d]) =: (a, \mathbf{q})$, with a being referred to as the *real part* and $\mathbf{q}$ as the *imaginary part*. The *norm* of a quaternion is given by $\|q\|^2 = q\,\bar{q} = \bar{q}\,q = a^2 + \mathbf{q}\cdot\mathbf{q} = a^2 + b^2 + c^2 + d^2$, where $\bar{q} = (a, -\mathbf{q})$ is the conjugate of q. Unit quaternions, $\|q\| = 1$, give rise to rotations in the space of purely imaginary quaternions, p, as follows: $p \mapsto q\,p\,\bar{q}$. It is easy to verify that this mapping indeed sends purely imaginary quaternions to purely

[5]Readers interested in the formulation of quaternions as a logical consequence of extending the multiplication of the complex numbers to higher dimensions might refer to Leng (2011); therein, we see that such an extension is not possible for $a + b\mathbf{i} + c\mathbf{j}$.

imaginary quaternions and also preserves the norm. The corresponding orthogonal 3×3 matrix is given by

$$A(q) = \begin{bmatrix} a^2+b^2-c^2-d^2 & -2ad+2bc & 2ac+2bd \\ 2ad+2bc & a^2-b^2+c^2-d^2 & -2ab+2cd \\ -2ac+2bd & 2ab+2cd & a^2-b^2-c^2+d^2 \end{bmatrix}; \quad (4.35)$$

this is the rotation by angle $\theta = 2\arccos a$ about the axis whose components are $[b, c, d]$ (relative to basis $\{\mathbf{i}, \mathbf{j}, \mathbf{k}\}$). The mapping $q \mapsto A(q)$ is not one-to-one, since $\pm q$ correspond to the same rotation: $(-q)\,p\,(-\bar{q}) = q\,p\,\bar{q}$. Following expression (3.42), $\tilde{A}(q)$ gives rise to an orthogonal transformation of the space of elasticity tensors, $C \mapsto \tilde{A}(q)^T\, C\, \tilde{A}(q)$. Substituting $C(q) = \tilde{A}(q)^T\, C\, \tilde{A}(q)$ for C in expression (4.34) and using the $O(3)$ invariance of the Frobenius norm, we obtain the following optimization problem:

$$\|C(q)\|^2 - \|C_{\rm eff}^{\rm sym}\|^2 \to \min\,. \quad (4.36)$$

The solutions of this problem are the values of q that result in the absolute minimum of the distance. The corresponding $C(q)$ is the effective tensor expressed in the coordinate system obtained from the original one by rotation. To find an absolute minimum of an expression that exhibits many local minima, Danek *et al.* (2013, 2015) solve problem (4.36) by a global optimization method called particle-swarm optimization (PSO).

We note that finding the closest tensor of a given symmetry is tantamount to finding a tensor that exhibits the Frobenius norm that is closest to the original tensor. This norm can be also associated with energy.

4.2 Equivalent media

4.2.1 *Introduction*

[6]To infer physical properties from observations, seismologists are implicitly or explicitly engaged in inverse problems. Their inferences rely on inductive processes, even though the theory of quantitative seismology is formulated, from its axioms or principles, by a deductive process. For instance, the P and S waves are deduced from the principle of the balance of linear momentum in the context of the definition of a Hookean solid. The usefulness of these waves—which reside in the mathematical realm—is a consequence of their providing satisfactory analogies for empirical data, as established in the nineteenth century by Oldham.

[6]Section 4.2 benefitted from research collaborations with Len Bos, Colin Brisco, David Dalton and Theodore Stanoev.

It is important to distinguish between properties of deductive and inductive processes. As Peirce (1998, p. 443) states laconically in his 1908 paper, deduction and induction

> render the indefinite definite: deduction explicates, induction evaluates: that is all.

George Edward Backus (1962) uses deduction within the realm of Hookean solids to explain how the layering results in a *seismic anisotropy*, which is the anisotropy inferred from seismic measurements. Data that appear to be consistent with such an anisotropy might allow us to evaluate the layered composition. Given the axioms, the former is certain: thin layers result in anisotropic behaviour of long waves; the latter is an interpretation: the observed anisotropic behaviour is a consequence of thin layers.

Let us consider several cases. An accurate enough analogy of a seismic behaviour with a model of a transversely isotropic Hookean solid is consistent with parallel layers. An accurate enough analogy of a seismic behaviour with an orthotropic solid is consistent with parallel layers and a perpendicular set of parallel fractures. An accurate enough analogy of a seismic behaviour with a monoclinic solid is consistent with parallel layers and an oblique set of parallel fractures. However, none of these cases is the only possible interpretation, and the choice of a preferred interpretation might rely on information beyond seismic anisotropy, and even beyond geophysics.

Let us emphasize that the analogy between the anisotropy and layers or the anisotropy and fractures relies on the assumption that the thicknesses of parallel layers or the separations between parallel fractures are much smaller than the wavelength used for their examination. Heuristically—as well as formally, in the context of the Backus (1962) formulation—we can view the passage of such a wave as a process of averaging the properties of discrete objects over a wavelength. As a consequence of averaging, the anisotropy might serve as an accurate-enough analogy within the realm of continuum mechanics. For continuum mechanics, anisotropy is a property of a continuum, and as such—by the definition of a continuum—it is scale independent: a continuum exhibits the same macroscopic and microscopic properties. Thus, for instance, the same continuum with its anisotropy might serve as an analogy for the behaviour of waves both in a crystal and in a sedimentary basin.

In the case of the Backus (1962) average, however, we deal with a scale-dependent phenomenon examined in terms of continuum mechanics, and not with an intrinsic property of a continuum. Hence, we might refer to the resulting effect as a seismic anisotropy.

Furthermore, achieving the most accurate analogy is not our purpose either. If we wish to gain an insight into directional patterns in the subsurface, we must consider symmetric Hookean solids, at the expense of accuracy, by comparison to the accuracy achieved with a generally anisotropic tensor. Thus, we need criteria other than the fit itself; for instance, a criterion that examines the increase in accuracy as a function of the number of parameters. The transversely isotropic model has seven parameters, which are five elasticity parameters to describe mechanical properties and two Euler angles to describe orientation of the symmetry axis; the orthotropic model has twelve parameters; the monoclinic model has fifteen parameters. To answer quantitatively whether or not the increase in the number of parameters is justified by the improvement of the fit, we might consider information criteria, such as Bayesian analysis.

Also, let us mention that even perfect knowledge of the wavefront symmetries is insufficient to infer the material symmetry.[7] According to *Curie's principle* (Curie, 1894), the symmetries of the causes must be found in their effects, but the converse is not true; that is, the effects can be more symmetric than the causes. In other words, Hookean solids can be less symmetric than wavefronts within them. However, Bóna *et al.* (2007b) show that we can determine the symmetries of the solid if we combine the information about the symmetries of wavefronts and about symmetries of the wave polarizations, which are their displacement vectors.

Be that as it may, continua are abstract media whose properties are, in a certain sense, analogous to granular materials. Remaining within the realm of continuum mechanics, we can represent an inhomogeneous continuum by a homogeneous one as a result of averaging its properties, as shown by Backus (1962).

In particular, we can view layered isotropic solids—from the viewpoint of mechanical properties—as a single transversely isotropic medium. In other words, we can combine mechanical properties of an isotropic inhomogeneous solid to obtain an anisotropic homogeneous solid. As described by Backus (1962), a disturbance propagating through a series of horizontal isotropic layers whose thickness is smaller than the wavelength behaves as if it were a homogeneous transversely isotropic solid.

In the abstract of the paper by Backus (1962), whose title is "Long-wave

[7]Readers interested in a mathematical discussion of the material versus wavefront symmetries might refer to Bóna *et al.* (2007b).

elastic anisotropy produced by horizontal layering", we read the following.

> A horizontally layered inhomogeneous medium, isotropic or transversely isotropic, is considered, whose properties are constant or nearly so when averaged over some vertical height ℓ'. For waves longer than ℓ' the medium is shown to behave like a homogeneous, or nearly homogeneous, transversely isotropic medium whose density is the average density and whose elastic coefficients are algebraic combinations of averages of algebraic combinations of the elastic coefficients of the original medium. The nearly homogeneous medium is said to be 'long-wave equivalent' to the original medium.

Unlike the effective symmetries, discussed in Section 4.1, the media discussed by Backus (1962) and examined herein are not related to distance in the space of elasticity tensors. The two approaches are based on distinct averages. Expression (4.5), formulated by Gazis *et al.* (1963), is based on the average over a symmetry group at a given point of a continuum; hence it refers to anisotropy only. The Backus (1962) average is based on the properties over an interval, ℓ'; hence, it relates the concepts of anisotropy and inhomogeneity: variation of properties with direction at a given position and with position in a given direction. Thus, the former is intrinsic to Hookean solids, regardless of wave propagation and wavelength therein. The latter, which involves the wavelength, is an examination of properties of a Hookean solid in terms of a seismic experiment. As mentioned in "Summary of Conclusions" of Backus (1962),

> The results which follow are true for any ℓ', but are useful only if ℓ' is large enough so that the properties of the medium are significantly smoothed by averaging over a vertical distance ℓ'.

We refer to the approach proposed by Gazis *et al.* (1963) as the effective solid and to the one of Backus (1962) as *equivalent*. This nomenclature is a matter of convention. Herein, the meaning of "equivalent" is, of course, *sensu lato*, since an average is not equivalent to its ingredients; it carries only partial information about them.

4.2.2 *Equivalence parameters for isotropic layers*

4.2.2.1 *Formulæ*

As shown by Backus (1962), the behaviour of waves propagating through parallel isotropic layers—with a wavelength much greater than thicknesses of individual layers—is akin to the propagation through a homogeneous transversely isotropic medium. If each of the aforementioned layers is described by the elasticity parameters, c_{1111} and c_{2323}, the corresponding parameters of the transversely isotropic medium are

$$c^{\mathrm{TI}}_{1111} = \overline{\left(\frac{c_{1111}-2c_{2323}}{c_{1111}}\right)}^{2}\,\overline{\left(\frac{1}{c_{1111}}\right)}^{-1} + \overline{\left(\frac{4(c_{1111}-c_{2323})c_{2323}}{c_{1111}}\right)}\,, \tag{4.37}$$

$$c^{\mathrm{TI}}_{1122} = \overline{\left(\frac{c_{1111}-2c_{2323}}{c_{1111}}\right)}^{2}\,\overline{\left(\frac{1}{c_{1111}}\right)}^{-1} + \overline{\left(\frac{2(c_{1111}-2c_{2323})c_{2323}}{c_{1111}}\right)}\,, \tag{4.38}$$

$$c^{\mathrm{TI}}_{1133} = \overline{\left(\frac{c_{1111}-2c_{2323}}{c_{1111}}\right)}\,\overline{\left(\frac{1}{c_{1111}}\right)}^{-1}\,, \tag{4.39}$$

$$c^{\mathrm{TI}}_{1212} = \overline{c_{2323}}\,, \tag{4.40}$$

$$c^{\mathrm{TI}}_{2323} = \overline{\left(\frac{1}{c_{2323}}\right)}^{-1}\,, \tag{4.41}$$

$$c^{\mathrm{TI}}_{3333} = \overline{\left(\frac{1}{c_{1111}}\right)}^{-1}\,, \tag{4.42}$$

which are the *Backus parameters* for isotropic layers. Herein, the bar indicates an average, which—according to the formulation of Backus (1962)—is given by expression (4.44), below.

The simplest case is an arithmetic average. For instance, if the layers are assumed to be of equal thickness, expression (4.40), becomes $c^{\mathrm{TI}}_{1212} = \frac{1}{n}\sum_{i=1}^{n}(c_{2323})_i$, where n is the number of layers. Such an average is illustrated in Exercise 4.9. Since the layer thickness is the weight of the value of c_{2323} for a given layer, if the layers are of different thicknesses,

$$c^{\mathrm{TI}}_{1212} = \frac{\sum\limits_{i=1}^{n} h_i\,(c_{2323})_i}{\sum\limits_{i=1}^{n} h_i}\,, \tag{4.43}$$

where h_i is the thickness of the ith layer.

Independently of the layer thickness, the average can be also weighted by the averaging operator, denoted by w in expression (4.44), below. Such an average is illustrated in Exercise 4.11.

In general, weighting the values of $c_{ijk\ell}$ by the thickness of the corresponding layer has an implicit assumption that waves travel a distance in each layer that is proportional to its thickness. This condition is satisfied for propagation normal to interfaces. Otherwise, the accuracy of the Backus (1962) average is diminished, particularly, if the speed of propagation varies significantly between layers and hence—according to Fermat's principle—so does the proportion of distance travelled in each layer, which is a function of the source-receiver offset.[8]

In expressions (4.37)–(4.42), we include, for convenience, both expressions (4.38) and (4.40), even though—as shown in expression (4.32), in a homogeneous transversely isotropic medium, they are related by expression (4.37), and this relation holds for an equivalent medium—$\overline{c^{\mathrm{TI}}_{1212}} = (\overline{c^{\mathrm{TI}}_{1111}} - \overline{c^{\mathrm{TI}}_{1122}})/2$. To see this, it suffices to subtract expression (4.38) from expression (4.37) and divide by 2, which results in expression (4.40). In doing so, we use the fact that the first summands of expressions (4.37) and (4.38) cancel one another, and we use the linearity of the averaging operation to combine the second summands.

Expressions (4.37)–(4.42)—being averages of the parameter values for individual layers—are the elasticity parameters of a medium representing these layers. Properties and applications of these averages, which are the Backus parameters, are illustrated in Exercises 4.7–4.15. Herein, these parameters correspond to expressions (13) of Backus (1962), where they are stated in terms of the averaged Lamé parameters. As shown by Backus (1962) and as discussed in Section 4.2.2.2, below, expressions (4.37)–(4.42) can be extended to the case of each layer exhibiting transverse isotropy. Also, as illustrated in Exercise 4.6, they can be extended to the case of orthotropic layers. Furthermore, they can be generalized for layers exhibiting a general anisotropy.[9]

For anisotropic layers, the strength of anisotropy of an equivalent medium is a function of both the strength of anisotropy of individual layers and the inhomogeneity of their stack. For isotropic layers, the strength of anisotropy of the resulting transversely isotropic medium is solely a function of inhomogeneity of that stack, as illustrated in Remark 4.10, on page 177. In other words, in such a case, the strength of anisotropy is a measure of inhomogeneity.

---

[8]Readers interested in the effect of waves propagating obliquely to interfaces on the accuracy of the Backus (1962) average for traveltime estimates might refer to Dalton and Slawinski (2016).

[9]Readers interested in a general-anisotropy formulation might refer to Bos *et al.* (2017b).

While referring, for the conciseness of nomenclature, to the *Backus parameters* and the *Backus average*, it is important to remember that they stem from the work of several researchers. Among them, as cited by Backus (1962), there are Rudzki (1911, 2003), Riznichenko (1949), Thomson (1950), Haskell (1953), Postma (1955), Helbig (1958) and Anderson (1961).[10]

4.2.2.2 *Justification*

Prior to proceeding to discussion of Exercises 4.7–4.15, let us justify expressions (4.37)–(4.42) in the context of an averaging process associated with a wave propagation. This justification requires certain mathematical subtlety.

Physically, the averaging is a consequence of a disturbance whose wavelength is much longer than inhomogeneities of the material. In essence, it is an averaging akin to the one that allows fundamentally the tensorial quantities of continuum mechanics to represent discrete physical objects. Herein, we consider a wavelength that is much longer than layer thicknesses. Hence, mechanical properties of these layers have an effect on the behaviour of the wave as an ensemble, not as individual layers. Formally, the Backus (1962) average requires $\ell'/\lambda \ll 1$, where λ is the wavelength and ℓ' is the distance over which mechanical properties of the medium change significantly.

Let us begin by considering a load applied to a horizontal plane of a medium in static equilibrium that is composed of thin horizontal layers. We set the Cartesian coordinate system in such a manner that the x_3-axis is vertical. Hence, the stress-tensor components of the applied load are σ_{13}, σ_{23} and σ_{33}.[11]

In view of the static equilibrium, these stress-tensor components are constant throughout the medium. The remaining components, σ_{11}, σ_{12}, σ_{22}, can vary significantly along the x_3-axis due to distinct properties of different layers. Furthermore, the components of the displacement vector, u_1, u_2, u_3, are continuous.[12] In view of the lateral homogeneity of the medium, $\partial u_i/\partial x_1$ and $\partial u_i/\partial x_2$ are smoothly varying functions of x_3, but $\partial u_i/\partial x_3$ might not be, due to the vertical inhomogeneity of layers.

To represent a given isotropic layer, let us invoke definition (3.14) to

[10]For a detailed summary of the work preceding Backus (1962), readers might refer therein to Section 2.

[11]*see also*: Slawinski (2015, Figure 2.5.1)

[12]*see also*: Slawinski (2015, Section 10.2.1)

write expression (3.41) with the isotropic tensor stated in matrix (3.57) as a system of equations,

$$\sigma_{11} = c_{1111}\frac{\partial u_1}{\partial x_1} + (c_{1111} - 2c_{2323})\frac{\partial u_2}{\partial x_2} + (c_{1111} - 2c_{2323})\frac{\partial u_3}{\partial x_3},$$

$$\sigma_{22} = (c_{1111} - 2c_{2323})\frac{\partial u_1}{\partial x_1} + c_{1111}\frac{\partial u_2}{\partial x_2} + (c_{1111} - 2c_{2323})\frac{\partial u_3}{\partial x_3},$$

$$\sigma_{33} = (c_{1111} - 2c_{2323})\frac{\partial u_1}{\partial x_1} + (c_{1111} - 2c_{2323})\frac{\partial u_2}{\partial x_2} + c_{1111}\frac{\partial u_3}{\partial x_3},$$

$$\sigma_{23} = c_{2323}\frac{\partial u_2}{\partial x_3} + c_{2323}\frac{\partial u_3}{\partial x_2},$$

$$\sigma_{31} = c_{2323}\frac{\partial u_3}{\partial x_1} + c_{2323}\frac{\partial u_1}{\partial x_3},$$

$$\sigma_{12} = c_{2323}\frac{\partial u_1}{\partial x_2} + c_{2323}\frac{\partial u_2}{\partial x_1}.$$

Let us write these equations in such a manner that the stress-tensor components and the partial derivatives that might vary significantly along the x_3-axis are brought to one side. On the other side, there remain the elasticity parameters, whose values can also change abruptly along the x_3-axis, but they are multiplied by the components that are constant or by the derivatives that vary smoothly. In accordance with this requirement, we write the last four equations as

$$\frac{\partial}{\partial x_3}u_3 = \left(\frac{1}{c_{1111}}\right)\sigma_{33} - \left(\frac{c_{1111} - 2c_{2323}}{c_{1111}}\right)\frac{\partial}{\partial x_1}u_1 - \left(\frac{c_{1111} - 2c_{2323}}{c_{1111}}\right)\frac{\partial}{\partial x_2}u_2,$$

$$\frac{\partial}{\partial x_3}u_2 = \left(\frac{1}{c_{2323}}\right)\sigma_{23} - \frac{\partial}{\partial x_2}u_3,$$

$$\frac{\partial}{\partial x_3}u_1 = \left(\frac{1}{c_{2323}}\right)\sigma_{31} - \frac{\partial}{\partial x_1}u_3,$$

$$\sigma_{12} = c_{2323}\frac{\partial}{\partial x_2}u_1 + c_{2323}\frac{\partial}{\partial x_1}u_2.$$

These equations are written in a manner that allows us to obtain the averages. The left-hand side consists of the stress-tensor components and of the partial derivatives whose values vary significantly along the x_3-axis. Their variations are due solely to abrupt changes of the elasticity parameters, whose averages along this axis we seek.

We cannot include the first two equations of the system on page 128 in the averaging process, since the product in the last term of either equation is a product of two functions varying rapidly along the x_3-axis. As stated by Backus (1962), the average of a product of two functions can be approximated by the product of their averages, if at least one of these functions is nearly constant over the averaging window. Notably, this is the only approximation used explicitly in this formulation.

As indicated in Section 4.2.2.1, arithmetic averages are straightforward to calculate. However, since, in general, we can consider moving averages with a weight function, calculations could be more involved than would be suggested at the first sight by expressions (4.37)–(4.42). The average, as defined by Backus (1962) is

$$\overline{f}(x_3) = \int_{-\infty}^{\infty} w(\xi - x_3) f(\xi)\,\mathrm{d}\xi\,, \tag{4.44}$$

where the weight, $w(x_3)$, allows us the use of many functions, since the conditions imposed on it are not restrictive; w is required to be a continuous nonnegative function tending to zero at infinities and exhibiting the following properties:

$$\int_{-\infty}^{\infty} w(x_3)\,\mathrm{d}x_3 = 1\,, \tag{4.45}$$

$$\int_{-\infty}^{\infty} x_3\, w(x_3)\,\mathrm{d}x_3 = 0 \qquad \text{and} \qquad \int_{-\infty}^{\infty} x_3^2\, w(x_3)\,\mathrm{d}x_3 = \ell'^2\,.$$

Definition (4.44) is the average of function $f(x_3)$ over "width" ℓ', where $w(x_3)$ plays the role akin to Dirac's delta centred at 0. Integral (4.45) and the two integrals below it are, respectively, the zeroth moment, the first moment and the second moment of $w(x_3)$. Since $w(x_3) \geqslant 0$ and its integral over $\mathbb{R}$, which is the zeroth moment, is equal to unity—as required for a probability measure—$w(x_3)$ is a distribution-density function with prescribed values of the first and second moments. The first moment is the expected value, the square root of the second moment is the standard deviation and the difference between the second and first moments is the variance. Thus, w has the expected value of 0, which is its mean, with the standard deviation of ℓ', which explains the term "width", and the variance of ℓ'^2.

According to expression (4.44), the average, $\overline{f}(x_3)$, is defined as the convolution of f and $w(-x_3)$; hence, it can be viewed as a moving average. Note that the mirror image of $w(x_3)$, namely: $w(-x_3)$, has the same properties as $w(x_3)$. Being a moving average, $\overline{f}$ is a continuous function of x_3, as illustrated in Exercise 4.10. Exercise 4.12 illustrates also a result of the convolution.

The Fourier transform, $\widehat{w}(k)$, evaluated at $k=0$, is equal to one; its derivative at $k=0$ is zero and its second derivative at $k=0$ is $-(\ell')^2$. If we approximate $\widehat{w}(k)$ in a neighbourhood of $k=0$ with a parabola, it results in $1-(1/2)(\ell')^2k^2$, which drops off to zero at $k=\pm\sqrt{2}/\ell'$. Since the Fourier transform of a convolution is the product of Fourier transforms, the effect of taking convolution of f with w is to carve out, in the Fourier domain, the spatial frequencies from about 0 up to approximately $\sqrt{2}/\ell'$; hence, the interpretation in terms of wavelength.

In obtaining the averages, we use the fact that the average of the derivative is the derivative of the average,

$$\overline{\frac{\partial u_i}{\partial x_j}} = \frac{\partial}{\partial x_j}\overline{u_i}\,, \qquad i,j=1,2,3\,, \tag{4.46}$$

as stated by Backus (1962); as shown in Exercise 4.4, this property is obvious for $j=1,2$, and results from integration by parts of integral (4.44), for $j=3$.

Proceeding with the first of the four equations on page 128, we write the averages of both sides as

$$\overline{\frac{\partial}{\partial x_3}u_3} = \overline{\left(\frac{1}{c_{1111}}\right)\sigma_{33} - \left(\frac{c_{1111}-2c_{2323}}{c_{1111}}\right)\frac{\partial}{\partial x_1}u_1 - \left(\frac{c_{1111}-2c_{2323}}{c_{1111}}\right)\frac{\partial}{\partial x_2}u_2}\,.$$

Since, by linearity of the averaging operator, which is defined by integral (4.44), the average of a sum is the sum of the averages, we write

$$\overline{\frac{\partial}{\partial x_3}u_3} = \overline{\left(\frac{1}{c_{1111}}\right)\sigma_{33}} - \overline{\left(\frac{c_{1111}-2c_{2323}}{c_{1111}}\right)\frac{\partial}{\partial x_1}u_1} - \overline{\left(\frac{c_{1111}-2c_{2323}}{c_{1111}}\right)\frac{\partial}{\partial x_2}u_2}\,.$$

Approximating the average of a product by the product of their averages, given that one of the factors is nearly constant, we write

$$\overline{\frac{\partial}{\partial x_3}u_3} = \overline{\left(\frac{1}{c_{1111}}\right)}\,\overline{\sigma_{33}} - \overline{\left(\frac{c_{1111}-2c_{2323}}{c_{1111}}\right)}\,\overline{\frac{\partial}{\partial x_1}u_1} - \overline{\left(\frac{c_{1111}-2c_{2323}}{c_{1111}}\right)}\,\overline{\frac{\partial}{\partial x_2}u_2}\,.$$

Finally, using property (4.46), we write

$$\frac{\partial}{\partial x_3}\overline{u_3} = \overline{\left(\frac{1}{c_{1111}}\right)}\,\overline{\sigma_{33}} - \overline{\left(\frac{c_{1111}-2c_{2323}}{c_{1111}}\right)}\frac{\partial}{\partial x_1}\overline{u_1} - \overline{\left(\frac{c_{1111}-2c_{2323}}{c_{1111}}\right)}\frac{\partial}{\partial x_2}\overline{u_2}\,.$$

Also, for the remaining three equations, we obtain

$$\frac{\partial}{\partial x_3}\overline{u_2} = \overline{\left(\frac{1}{c_{2323}}\right)}\overline{\sigma_{23}} - \frac{\partial}{\partial x_2}\overline{u_3}\,,$$

$$\frac{\partial}{\partial x_3}\overline{u_1} = \overline{\left(\frac{1}{c_{2323}}\right)}\overline{\sigma_{31}} - \frac{\partial}{\partial x_1}\overline{u_3}\,,$$

$$\overline{\sigma_{12}} = \overline{c_{2323}}\frac{\partial}{\partial x_2}\overline{u_1} + \overline{c_{2323}}\frac{\partial}{\partial x_1}\overline{u_2}\,.$$

Let us express these four equations in the standard form of the stress-strain equations by reorganizing them in such a manner that the stress-tensor components alone appear on the left-hand side, and by using the definition of the strain tensor, $\varepsilon_{k\ell} := (\partial u_k/\partial x_\ell + \partial u_\ell/\partial x_k)/2$, stated in expression (3.14), to combine the expressions for the partial derivatives, where we use the fact that the sum of averages is the average of the sum. In that manner, the first among the above four equations can be written as

$$\overline{\sigma_{33}} = \overline{\left(\frac{c_{1111} - 2c_{2323}}{c_{1111}}\right)}\,\overline{\left(\frac{1}{c_{1111}}\right)}^{-1}\overline{\varepsilon_{11}} + \overline{\left(\frac{c_{1111} - 2c_{2323}}{c_{1111}}\right)}\,\overline{\left(\frac{1}{c_{1111}}\right)}^{-1}\overline{\varepsilon_{22}} + \overline{\left(\frac{1}{c_{1111}}\right)}^{-1}\overline{\varepsilon_{33}}\,; \tag{4.47}$$

comparing it with the third row in equation (3.41), we see that

$$\langle c_{1133}\rangle = \langle c_{2233}\rangle = \overline{\left(\frac{c_{1111} - 2c_{2323}}{c_{1111}}\right)}\,\overline{\left(\frac{1}{c_{1111}}\right)}^{-1}\,, \tag{4.48}$$

which is the equality between elasticity parameters exhibited by the symmetry classes discussed in Sections 3.2.4.8–3.2.4.13, and that

$$\langle c_{3333}\rangle = \overline{\left(\frac{1}{c_{1111}}\right)}^{-1}\,. \tag{4.49}$$

Herein, $\langle\,\rangle$ denotes the equivalent elasticity parameter. The other parameters of the third row, by comparison with equation (3.41), are zero, which is exhibited by the symmetry classes discussed in Sections 3.2.4.7–3.2.4.13. A progressive restriction of symmetry classes to which the equivalent medium could belong is facilitated by an iterative examination of Figure 3.3.

The remaining three equations are

$$\overline{\sigma_{23}} = \overline{\left(\frac{1}{c_{2323}}\right)}^{-1}\overline{\left(\frac{\partial u_2}{\partial x_3} + \frac{\partial u_3}{\partial x_2}\right)} = 2\overline{\left(\frac{1}{c_{2323}}\right)}^{-1}\overline{\varepsilon_{23}}\,,$$

$$\overline{\sigma_{13}} = \overline{\left(\frac{1}{c_{2323}}\right)}^{-1} \overline{\left(\frac{\partial u_1}{\partial x_3} + \frac{\partial u_3}{\partial x_1}\right)} = 2\overline{\left(\frac{1}{c_{2323}}\right)}^{-1} \overline{\varepsilon_{13}}\,,$$

$$\overline{\sigma_{12}} = \overline{c_{2323}} \overline{\left(\frac{\partial u_1}{\partial x_2} + \frac{\partial u_2}{\partial x_1}\right)} = 2\,\overline{c_{2323}}\,\overline{\varepsilon_{12}}\,.$$

Examining—in view of the fourth row of equation (3.41)—the first, among the three equations, we see that

$$\langle c_{2323}\rangle = \overline{\left(\frac{1}{c_{2323}}\right)}^{-1},$$

and that the other parameters of that row are zero, which is exhibited by the symmetry classes discussed in Sections 3.2.4.7–3.2.4.13, except for the class in Section 3.2.4.11. Examining Figure 3.3, we see that the classes to which the equivalent medium could belong exhibit at least the tetragonal symmetry.

From the second equation, we see that

$$\langle c_{1313}\rangle = \overline{\left(\frac{1}{c_{2323}}\right)}^{-1}.$$

Comparing these two expressions, we deduce that $\langle c_{1313}\rangle = \langle c_{2323}\rangle$. In other words, $\overline{\sigma_{13}}$ and $\overline{\sigma_{23}}$ are related to their corresponding strain-tensor components by the same elasticity parameter, which is a common property for several symmetry classes, as seen in Sections 3.2.4.8–3.2.4.13. From the third equation, and in view of the last row of equation (3.41), we see that

$$\langle c_{1212}\rangle = \overline{c_{2323}}\,,$$

where $\langle c_{1212}\rangle$ is the equivalent elasticity parameter relating $\overline{\sigma_{12}}$ and $\overline{\varepsilon_{12}}$. The fact that $\langle c_{1212}\rangle \neq \langle c_{1313}\rangle = \langle c_{2323}\rangle$ means, in view of discussions in Sections 3.2.4.13 and 3.2.4.12, respectively, that the equivalent medium is neither isotropic nor cubic. The remaining choice is between transverse isotropy and tetragonal symmetry.

Obtaining the expressions for the remaining two stress-tensor components requires solving the first three equations of the system on page 128, which do not contain averaged quantities, such that rapidly varying stress-tensor and strain-tensor components are on the left-hand side only, then perform the averaging, and inserting, for $\overline{\sigma_{33}}$, equation (4.47). Again, this is necessary since both $c_{1111} - 2c_{2323}$ and ε_{33} are rapidly varying.

Expressing the partial derivatives of the first three equations of the unaveraged system on page 128 as the strain-tensor components, we write

$$\sigma_{11} = c_{1111}\,\varepsilon_{11} + (c_{1111} - 2c_{2323})\,\varepsilon_{22} + (c_{1111} - 2c_{2323})\,\varepsilon_{33}\,, \qquad (4.50)$$

$$\sigma_{22} = (c_{1111} - 2c_{2323})\,\varepsilon_{11} + c_{1111}\,\varepsilon_{22} + (c_{1111} - 2c_{2323})\,\varepsilon_{33}\,, \tag{4.51}$$

$$\varepsilon_{33} = \left(\frac{1}{c_{1111}}\right)\sigma_{33} - \left(\frac{c_{1111} - 2c_{2323}}{c_{1111}}\right)\varepsilon_{11} - \left(\frac{c_{1111} - 2c_{2323}}{c_{1111}}\right)\varepsilon_{22}\,. \tag{4.52}$$

We insert expression (4.52) into equation (4.50), so that stress and strain components on the right hand side are all slowly varying, and perform the averages to obtain

$$\overline{\sigma_{11}} = \overline{\left(\frac{c_{1111} - 2c_{2323}}{c_{1111}}\right)}\,\overline{\sigma_{33}} + \overline{\left(\frac{4(c_{1111} - c_{2323})c_{2323}}{c_{1111}}\right)}\,\overline{\varepsilon_{11}}$$
$$+ \overline{\left(\frac{2(c_{1111} - 2c_{2323})c_{2323}}{c_{1111}}\right)}\,\overline{\varepsilon_{22}}\,.$$

Inserting expression (4.47), upon algebraic manipulations, we obtain

$$\overline{\sigma_{11}} = \left(\overline{\left(\frac{c_{1111} - 2c_{2323}}{c_{1111}}\right)}^{2}\,\overline{\left(\frac{1}{c_{1111}}\right)}^{-1} + \overline{\left(\frac{4(c_{1111} - c_{2323})c_{2323}}{c_{1111}}\right)}\right)\overline{\varepsilon_{11}}$$
$$+ \left(\overline{\left(\frac{c_{1111} - 2c_{2323}}{c_{1111}}\right)}^{2}\,\overline{\left(\frac{1}{c_{1111}}\right)}^{-1} + \overline{\left(\frac{2(c_{1111} - 2c_{2323})c_{2323}}{c_{1111}}\right)}\right)\overline{\varepsilon_{22}}$$
$$+ \overline{\left(\frac{c_{1111} - 2c_{2323}}{c_{1111}}\right)}\,\overline{\left(\frac{1}{c_{1111}}\right)}^{-1}\,\overline{\varepsilon_{33}}\,.$$

In an analogous manner, for equation (4.51), we obtain

$$\overline{\sigma_{22}} = \left(\overline{\left(\frac{c_{1111} - 2c_{2323}}{c_{1111}}\right)}^{2}\,\overline{\left(\frac{1}{c_{1111}}\right)}^{-1} + \overline{\left(\frac{2(c_{1111} - 2c_{2323})c_{2323}}{c_{1111}}\right)}\right)\overline{\varepsilon_{11}}$$
$$+ \left(\overline{\left(\frac{c_{1111} - 2c_{2323}}{c_{1111}}\right)}^{2}\,\overline{\left(\frac{1}{c_{1111}}\right)}^{-1} + \overline{\left(\frac{4(c_{1111} - c_{2323})c_{2323}}{c_{1111}}\right)}\right)\overline{\varepsilon_{22}}$$
$$+ \overline{\left(\frac{c_{1111} - 2c_{2323}}{c_{1111}}\right)}\,\overline{\left(\frac{1}{c_{1111}}\right)}^{-1}\,\overline{\varepsilon_{33}}\,.$$

Comparing these two results with the first and second row of equation (3.41), respectively, we see that

$$\langle c_{1111}\rangle = \langle c_{2222}\rangle = \overline{\left(\frac{c_{1111} - 2c_{2323}}{c_{1111}}\right)}^{2}\,\overline{\left(\frac{1}{c_{1111}}\right)}^{-1} + \overline{\left(\frac{4(c_{1111} - c_{2323})c_{2323}}{c_{1111}}\right)},$$

which, as expected, is the equality between elasticity parameters exhibited by the symmetry classes discussed in Sections 3.2.4.8–3.2.4.13, and that

$$\langle c_{1122}\rangle = \overline{\left(\frac{c_{1111} - 2c_{2323}}{c_{1111}}\right)}^{2}\,\overline{\left(\frac{1}{c_{1111}}\right)}^{-1} + \overline{\left(\frac{2(c_{1111} - 2c_{2323})c_{2323}}{c_{1111}}\right)}.$$

A subsequent examination allows us to recognize, as discussed on page 126, that $\langle c_{1212}\rangle = (\langle c_{1111}\rangle - \langle c_{1122}\rangle)/2$, and, hence, we conclude that there are only five linearly independent elasticity parameters, whose relations are stated in matrix (3.54); this is a defining property of transverse isotropy.[13]

Following this conclusion, we use the form of the corresponding elasticity tensor, given in matrix (3.54), to write

$$C^{\overline{\mathrm{TI}}} = \begin{bmatrix} c_{1111}^{\overline{\mathrm{TI}}} & c_{1122}^{\overline{\mathrm{TI}}} & c_{1133}^{\overline{\mathrm{TI}}} & 0 & 0 & 0 \\ c_{1122}^{\overline{\mathrm{TI}}} & c_{1111}^{\overline{\mathrm{TI}}} & c_{1133}^{\overline{\mathrm{TI}}} & 0 & 0 & 0 \\ c_{1133}^{\overline{\mathrm{TI}}} & c_{1133}^{\overline{\mathrm{TI}}} & c_{3333}^{\overline{\mathrm{TI}}} & 0 & 0 & 0 \\ 0 & 0 & 0 & 2c_{2323}^{\overline{\mathrm{TI}}} & 0 & 0 \\ 0 & 0 & 0 & 0 & 2c_{2323}^{\overline{\mathrm{TI}}} & 0 \\ 0 & 0 & 0 & 0 & 0 & 2c_{1212}^{\overline{\mathrm{TI}}} \end{bmatrix},$$

where we denote $\langle c_{ijk\ell}\rangle$ by $c_{ijk\ell}^{\overline{\mathrm{TI}}}$. As shown, $c_{ijk\ell}^{\overline{\mathrm{TI}}}$ correspond to expressions (4.37)–(4.42).

Having justified expressions (4.37)–(4.42) for a static case, let us comment on their validity for the wave propagation, which is a dynamic process. Considering a farfield case, which amounts to examining the wave propagation far from its source, σ_{13}, σ_{23} and σ_{33} vary only slightly and u_1, u_2 and u_3 remain continuous. Consequently, we can—with a satisfactory accuracy—apply the averaging results obtained for the case of a static equilibrium.

There remains another aspect to complete the justification for expressing a series of layers as an equivalent medium. We must examine whether or not—given layer-parameters that satisfy the stability conditions[14]—the resulting average also satisfies these conditions. In other words, we must examine if the averaging process, defined by integral (4.44), preserves the positive definiteness of the elasticity tensor.

As shown by Gazis *et al.* (1963, Theorem 2.2), in the case of effective media, discussed in Section 4.1, a Frobenius-norm counterpart of a positive-definite tensor is positive-definite.[15] A similar relation is true for equivalent media. An average of layers whose tensors are positive-definite is positive-definite, and, as stated by Backus (1962, p. 4433), for transversely isotropic layers resulting in a transversely isotropic medium, "[t]he proof is straightforward and will be omitted". Herein, we illustrate a general argument, which we present as a proposition and its proof.

[13]Readers interested in identification of material symmetries might refer to Bóna *et al.* (2007a).

[14]*see also*: Slawinski (2015, Section 4.3)

[15]This is not necessarily so for other norms (Bos and Slawinski, 2015).

Proposition 4.1. *In the context of the Backus (1962) average, the stability of layers is inherited by the stability of an equivalent medium.*

Proof. The stability of layers means that their deformation requires work. Mathematically, it means that, for each layer,

$$W = \frac{1}{2}\sigma \cdot \varepsilon > 0\,,$$

where W stands for work, and σ and ε denote the stress and strain tensors, respectively, which are expressed as columns in equation (3.41): $\sigma = C\,\varepsilon$. As an aside, we can say that, herein, $W > 0$ is equivalent to the positive definiteness of C, for each layer.[16]

Performing the average of W over all layers and using—in the scalar product—the fact that the average of a sum is the sum of averages, we write

$$\overline{W} = \frac{1}{2}\overline{\sigma \cdot \varepsilon} > 0\,,$$

which refers to the equivalent medium. Thus, $W > 0 \implies \overline{W} > 0$.

Let us proceed to show that, this implication—in turn—entails the stability of the equivalent medium, which is tantamount to the positive definiteness of $\langle C \rangle$.

Following Backus (1962, equation (3))—if one of two functions is nearly constant—we can approximate the average of their product by the product of their averages,

$$\overline{W} = \frac{1}{2}\overline{\sigma} \cdot \overline{\varepsilon} > 0\,; \tag{4.53}$$

an analytical formulation of this property, in the context of the Backus (1962) average, is given by Proposition 4.4, in Exercise 4.5. A numerical and empirical examination of this approximation could be based on that proposition.

Herein, we use that property and the fact that, as stated by Backus (1962, p. 4429) and quoted with the present notation,

> stresses, σ_{13}, σ_{23}, and σ_{33} will be constant, and the values of $\partial u_i/\partial x_1$, and $\partial u_i/\partial x_2$ will be smoothly varying functions of x_3 on which are superposed very small wiggles, while σ_{11}, σ_{12}, σ_{22}, $\partial u_1/\partial x_3$, $\partial u_2/\partial x_3$, and $\partial u_3/\partial x_3$ will vary widely and rapidly with x_3,

[16] *see also*: Slawinski (2015, Section 4.3)

together with the fact that each product in expression (4.53) is such that one function is smoothly varying and the other varies rapidly.

For instance, for

$$\sigma_{22}\,\varepsilon_{22} = \sigma_{22}\,\frac{\partial u_2}{\partial x_2}\,,$$

even though the first factor varies rapidly, the second factor does not; hence, $\overline{\sigma_{22}\varepsilon_{22}} \approx \overline{\sigma_{22}}\;\overline{\varepsilon_{22}}$. Analogously, for

$$\sigma_{13}\,\varepsilon_{13} = \sigma_{13}\left(\frac{1}{2}\left(\frac{\partial u_1}{\partial x_3} + \frac{\partial u_3}{\partial x_1}\right)\right),$$

even though the factor in parentheses varies rapidly, the other factor is constant; hence, $\overline{\sigma_{13}\varepsilon_{13}} \approx \overline{\sigma_{13}}\;\overline{\varepsilon_{13}}$. Due to the lateral homogeneity, such a pattern exists for all six products within the inner product, $\overline{\sigma \cdot \varepsilon}$.

By definition of Hooke's law, $\overline{\sigma} = \langle C\rangle\,\overline{\varepsilon}$, expression (4.53) can be written as

$$\frac{1}{2}\left(\langle C\rangle\,\overline{\varepsilon}\right)\cdot\overline{\varepsilon} > 0\,,$$

for all $\overline{\varepsilon} \neq 0$, which means that $\langle C\rangle$ is positive-definite, and which—in view of this derivation—proves that the equivalent medium inherits the stability of individual layers. □

Both Gazis *et al.* (1963, Theorem 2.2) and Proposition 4.1 relate tensors to their respective averages. However, there arises a reverse question. Can any tensor be a result of such averages? In other words, is integral (4.44) a surjective map from the domain, consisting of layers, to the codomain, consisting of all media whose material symmetry is the symmetry of the equivalent medium. In the case discussed herein, this question can be phrased in the following manner. Can any transversely isotropic medium be a result of the Backus (1962) average? For isotropic layers, Backus (1962, Section 8) shows that not every transversely isotropic medium could arise as their long-wavelength equivalent.

> [T]here are many stable [transversely isotropic] media which are not long-wave equivalent to any layered, stable, isotropic medium. If the elastic coefficients of an apparently transversely isotropic medium observed by long waves in the field fail to satisfy [required] inequalities, then it is certain that some intrinsic anisotropy is present; no layered isotropic medium can reproduce the observations.

Aspects of this reverse question are addressed, for effective media, by Diner *et al.* (2011a,b), and, for equivalent media, by Backus (1962), by Berryman (1997), by Schoenberg and Muir (1989) and by Helbig (1998, 2000). However, there remain further issues of that, and other, questions to be addressed.

According to Gazis *et al.* (1963, Theorem 2.2) and Proposition 4.1, in general, the respective averages preserve the positive definiteness. It is true for the case discussed in this section, remains true for the case in Section 4.2.3, below, and for Exercise 4.6, but, there remains a question about Proposition 4.1. Could our approximation of the average of a product by the product of averages be avoided in completing a proof for stability of equivalent media?

If such a proof is not possible, then this approximation becomes a requirement for the existence of a stable equivalent medium as a mathematical entity. However, one could argue that a proof without invoking this approximation might be neither possible nor pertinent for the formulation of equivalent media that relies on that approximation. Furthermore, the validity of any approximation in constructing a model—as an analogy to perform a quantitative analysis of empirical data—is justified *a posteriori* by the accuracy of resulting predictions or retrodictions, which is a standard approach for justifying approximations within continuum-mechanics, as a hypotheticodeductive theory. This approach allows us, within its constraints, to propose mechanical explanations. Herein, the interpretation of observed transverse isotropy as a possible seismic effect due to a series of thin parallel isotropic layers is an explanation consistent with hypotheses.

4.2.2.3 *Interpretation*

As illustrated in Exercises 4.7 and 4.8, if all isotropic layers have the same properties, their equivalent medium is isotropic, since the same properties for all isotropic layers are tantamount to a single isotropic layer. Thus, as expected, the formulation of Backus (1962) is reduced to isotropy. Also in Exercises 4.7 and 4.8, we illustrate that—along the rotation-symmetry axis—the behaviour of the qP, qSV and SH waves is the same as their behaviour in an isotropic medium; this is true for all transversely isotropic media, whose key property is the rotation-symmetry axis, and which is the reason for the name due to the invariance of properties under rotations about that axis.

As another note on terminology, let us comment on the fact that formally—for a convenience of description with respect to a coordinate system and in spite of the impossibility of determining the specific orientation of displacement in isotropic media[17]—S waves whose displacement vectors are horizontal are denoted by SH, and S waves whose displacement vectors are vertical by SV. Such a terminology is *a priori* coordinate-dependent and is commonly associated with the vertical direction assumed to be perpendicular to the orientation of interfaces; it does not originate in an intrinsic property of these waves.

Given a hypothetical model of isotropic layers whose properties are stated in Table 4.1 of Exercise 4.9, expressions (4.37)–(4.42) are used to obtain the Backus (1962) parameters. These parameters are used in several remaining exercises. In Exercise 4.15, we reduce further the Backus (1962) model to the isotropic case. To do so, we invoke the Voigt (1910) results stated in expressions (4.30) and (4.31). The effectiveness of such a reduction relies on weak anisotropy of the transversely isotropic medium that results from the averaging process. A convenient measure of the strength of anisotropy are the Thomsen (1986) parameters, discussed in Exercise 4.15.

Traveltimes through the layered model and through its Backus (1962) and Voigt (1910) averages are presented in Table 4.2; the three models exhibit similar traveltimes. One might argue that a discrepancy between traveltimes for the Backus (1962) model and its Voigt (1910) reduction would be indicative of the averaged medium being quite anisotropic and, hence, the reduction being inappropriate. However, a discrepancy between traveltimes for the layered model and its Backus (1962) average might be a consequence of distinct wave responses to mechanical properties. In other words, the traveltime discrepancy between the infinitesimal-wavelength and long-wavelength cases might illustrate different responses to individual layer and bulk material properties. This is not to say that wave velocities are explicit functions of frequency, which—in Hookean solids—they are not.

4.2.2.4 *Possible issues with Backus averaging*

Let us examine the case for which the Backus (1962) average gives unreliable results. Backus (1962) writes

> The only approximation that we make in the present paper is the following: if $f(x_3)$ is nearly constant when x_3 changes by no more than ℓ', while $g(x_3)$ may vary by a large fraction over this distance, then, approximately, $\overline{fg} \approx \overline{f}\,\overline{g}$.

[17] *see also*: Slawinski (2015, Remark 9.4.1)

To discuss the reliability of this approximation, let us consider the relative error defined by

$$\frac{\overline{fg} - \overline{f}\,\overline{g}}{\overline{fg}} \times 100\% \,. \tag{4.54}$$

We see that if $\overline{g} = 0$, this error becomes 100%; hence, the case of $\overline{g} = 0$ is of concern. Such a case is illustrated in Figure 4.3, where g corresponds to

$$g(x) = \left(\frac{c_{1111} - 2c_{2323}}{c_{1111}}\right) , \tag{4.55}$$

which appears in Section 4.2.2.2.

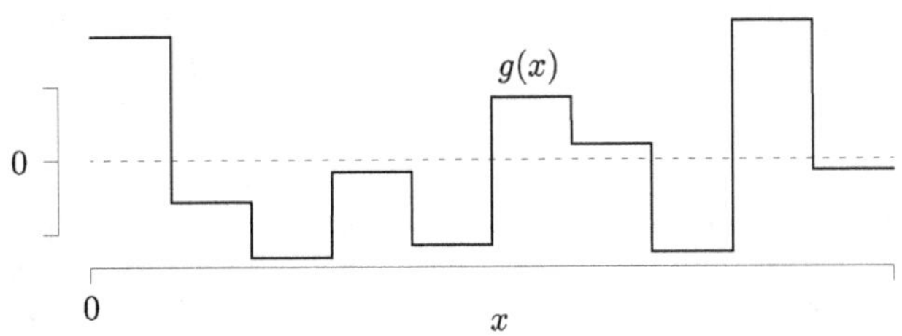

Fig. 4.3: An example of function g for which the product approximation in the Backus (1962) average over x exhibits the relative error of 100%, due to $\overline{g} = 0$.

As discussed by Bos *et al.* (2017c), only the case of $\overline{g} \approx 0$, where g is the quickly varying function that represents properties of Hookean layers, raises concerns with respect to error (4.54). Furthermore, for sedimentary layers—which is a common scenario for the application of the Backus average— $\overline{g} \approx 0$ is unlikely to occur, since it would require the value of expression (4.55) to exhibit both positive and negative values within the region considered by the averaging process, which in Figure 4.3 is denoted by x.

However, Hookean models of the Earth—notably the *Preliminary Reference Earth Model* (PREM), developed by Dziewoński and Anderson (1981)—exhibit only positive values of expression (4.55). Thus, the problematic issue of the approximation, $\overline{fg} \approx \overline{f}\,\overline{g}$, is not likely to appear in seismology. It might, however, appear in other aspects of material sciences, where the Hookean model approaches the minimum value of the allowable range of Poisson's ratio, which is $(-1, \frac{1}{2})$,[18] and corresponds to $3\,c_{1111} = 4\,c_{2323}$. The negative values of expression (4.55) are contained

[18] *see also*: Slawinski (2015, Exercise 5.18)

within $(-1,0)$, which is an uncommon range for naturally occurring materials.

Examining expression (4.55), Bos *et al.* (2017c) write as follows.

> Stability conditions of Hookean solids, which are expressed as the positive definiteness of the elasticity tensor, require that both c_{1111} and c_{2323} be positive. Thus, if $c_{1111} > 2\,c_{2323}$, for all layers, g is positive for all x. If, in any layer, $\frac{4}{3}\,c_{2323} < c_{1111} < 2\,c_{2323}$, g is negative in that layer. The lower limit is also required by the stability conditions.
>
> The range of the elasticity parameters resulting in negative values of g appears to be less common in modelling natural materials. It corresponds to Hookean solids exhibiting high rigidity. Expressed in terms of α and β, which are the P-wave and S-wave speeds, respectively, the negative values occur if and only if
>
> $$2\beta/\sqrt{3} < \alpha < \sqrt{2}\beta\,.$$
>
> The lower limit is the closest allowable case of the two speeds. The upper limit is still below the case of the so-called Poisson's solid, whose $\alpha = \sqrt{3}\,\beta$; for such a solid, the Poisson ratio is $1/4$, and the two Lamé parameters are equal to one another.
>
> Poisson's solid is representative of common sedimentary rocks. Thus, the change of sign for the term composed of elasticity parameters, although it might occur, appears to be limited to values that are not common for seismic measurements in sedimentary basins. Therein, the values of the quickly varying function are expected to remain positive.

Herein, x corresponds to x_3 in expression (4.44).

The case of $\overline{g} \approx 0$ might also occur for anisotropic layers. For such cases, there are more expressions analogous to expression (4.55), as exemplified for orthotropic layers in Exercise 4.6. However, the stability conditions for anisotropic solids form a set of complicated inequalities that tend to prevent changes of sign of these expressions, which would lead to $\overline{g} = 0$. Hence, $\overline{fg} \approx \overline{f}\,\overline{g}$ remains a reasonable approximation for anisotropic layers.

4.2.3 *Equivalence parameters for TI layers*

In a manner similar to the one presented in Section 4.2.2, by invoking definition (3.14) to write expression (3.41) with the transversely isotropic tensor stated in matrix (3.54) and using expression (4.46) for averaging, we can obtain formulæ analogous to expressions (4.37)–(4.42), but for a medium composed of parallel transversely isotropic layers whose rotation-symmetry axes are collinear. Again, as shown by Backus (1962, expression (9)), the result of such an averaging is a transversely isotropic medium defined by

$$c_{1111}^{\overline{\mathrm{TI}}} = \overline{\left(c_{1111} - \frac{c_{1133}^2}{c_{3333}}\right)} + \overline{\left(\frac{c_{1133}}{c_{3333}}\right)}^2 \overline{\left(\frac{1}{c_{3333}}\right)}^{-1} , \tag{4.56}$$

$$c_{1122}^{\overline{\mathrm{TI}}} = \overline{\left(c_{1122} - \frac{c_{1133}^2}{c_{3333}}\right)} + \overline{\left(\frac{c_{1133}}{c_{3333}}\right)}^2 \overline{\left(\frac{1}{c_{3333}}\right)}^{-1} , \tag{4.57}$$

$$c_{1133}^{\overline{\mathrm{TI}}} = \overline{\left(\frac{c_{1133}}{c_{3333}}\right)}\ \overline{\left(\frac{1}{c_{3333}}\right)}^{-1} , \tag{4.58}$$

$$c_{1212}^{\overline{\mathrm{TI}}} = \overline{c_{1212}} , \tag{4.59}$$

$$c_{2323}^{\overline{\mathrm{TI}}} = \overline{\left(\frac{1}{c_{2323}}\right)}^{-1} , \tag{4.60}$$

$$c_{3333}^{\overline{\mathrm{TI}}} = \overline{\left(\frac{1}{c_{3333}}\right)}^{-1} , \tag{4.61}$$

which are the Backus parameters for a series of transversely isotropic layers. For isotropy, where $c_{1212} = c_{2323}$, $c_{1122} = c_{1133} = c_{1111} - 2c_{2323}$ and $c_{3333} = c_{1111}$, these formulæ reduce to expressions (4.37)–(4.42), as expected.

At first sight this reduction might not be obvious due to several relations among the elasticity parameters of an isotropic tensor; for instance, Lamé parameters exemplify one of several parameterizations. Thus, let us reduce expression (4.56) to its counterpart resulting from isotropic layers,

$$\begin{aligned}
&\overline{\left(c_{1111} - \frac{c_{1133}^2}{c_{3333}}\right)} + \overline{\left(\frac{c_{1133}}{c_{3333}}\right)}^2 \overline{\left(\frac{1}{c_{3333}}\right)}^{-1} \\
&\quad = \overline{\left(c_{1111} - \frac{(c_{1111} - 2c_{2323})^2}{c_{1111}}\right)} + \overline{\left(\frac{c_{1111} - 2c_{2323}}{c_{1111}}\right)}^2 \overline{\left(\frac{1}{c_{1111}}\right)}^{-1} \\
&\quad = \overline{\left(\frac{4(c_{1111} - c_{2323})c_{2323}}{c_{1111}}\right)} + \overline{\left(\frac{c_{1111} - 2c_{2323}}{c_{1111}}\right)}^2 \overline{\left(\frac{1}{c_{1111}}\right)}^{-1} ,
\end{aligned}$$

which is expression (4.37), as required.

Note that both expressions (4.37) and (4.56) are denoted by $c_{1111}^{\overline{\mathrm{TI}}}$, since they both represent an equivalent medium of the transversely isotropic symmetry. Also, as shown herein, the former is a special case of the latter. However, even though they share the same material symmetry, they are distinct media; the former is a result of isotropic layers and the latter of transversely isotropic layers; their origins are distinguishable by different parameters on the right-hand side of the corresponding expressions in systems (4.37)–(4.42) and (4.56)–(4.61).

Following the presented pattern of averaging, one can formulate equivalent-media expressions for lower-symmetry layers, including general anisotropy.[19] The constraining assumptions, with an allowance for minor departures, are the static equilibrium, which results in the constancy of σ_{13}, σ_{23} and σ_{33}, and lateral homogeneity, which results in $\partial u_i/\partial x_1$ and $\partial u_i/\partial x_2$, where $i = 1, 2, 3$, varying smoothly along the x_3-axis. For instance, the Backus parameters of orthotropic layers are illustrated in Exercise 4.6.[20]

A consideration for an empirical examination of general anisotropy would be to ensure that the elasticity parameters of all layers be expressed with respect to the same orientation of the coordinate system. Otherwise, we could not perform the averaging.

An insight could be gained by considering various anisotropic and isotropic layers, which might allow us to quantify the fact that—in spite of the presence of strongly anisotropic individual layers—the overall anisotropy inferred from seismic measurements is commonly weak. For that reason, a formulation of seismic theory under the assumption of weak anisotropy, or even isotropy, has often resulted in sufficiently accurate interpretations.

4.3 Relation between effective and equivalent formulations

4.3.1 *Introduction*

[21] The distinction between anisotropy and inhomogeneity is a property of continuum mechanics. Physically, the two concepts are interrelated as a consequence of the scale of investigation, such as the microscopic and macroscopic considerations. As discussed in this chapter, both concepts

[19] Readers interested in a general-anisotropy formulation might refer to Bos *et al.* (2017b).

[20] Backus parameters for orthotropic layers are also given by Tiwary (2007, equation (5.1)) and Shermergor (1977, equation (2.4)).

[21] Section 4.3 is based on Bos *et al.* (2017a).

can be subject to averaging whose limits are isotropy and homogeneity, respectively. In Section 4.1, we discuss an average in the context of rotation, and in Section 4.2 of translation; the former is intrinsically related to anisotropy, and the latter to inhomogeneity. Let us discuss the commutativity of these two averaging operations.

The fundamental distinction between the two averages is their domain of operation. The Gazis *et al.* (1963) average is an average over symmetry groups at a point and the Backus (1962) average is a spatial average over a distance. These averages can be used separately or together, and the latter raises the question of commutativity, which we address in this section.

The interplay between anisotropy and inhomogeneity is important in modelling signal traveltimes. The same traveltime can be obtained by considering anisotropy, inhomogeneity or their combination. However—if the purpose of modelling is to infer properties of the medium, not only to account for the measured traveltimes—we need to examine this interplay in the context of symmetry increase, homogenization and their commutativity.

The commutator of two operators is defined as

$$[A, B] := AB - BA\,, \tag{4.62}$$

which means that it is zero if A and B commute; more generally, the size of the commutator gives an indication of how close they are to commuting. In our examination, we consider two types of averages that may be perturbed by parameter h. Thus, we may consider the commutator $[A, B] =: F(h)$ to be a function of h. If, for no perturbation, which means that $h = 0$, the two averages commute, we have $F(0) = 0$. For $h \neq 0$, we expect,

$$[A, B] = F(0) + F'(0)\,h + \frac{F''(0)}{2!}h^2 + \cdots = F'(0)\,h + \frac{F''(0)}{2!}h^2 + \cdots$$

to be of order h. In this section, we show that—in certain cases pertinent to seismology—we have $[A, B] = O(h^2)$, which means that the commutator is much smaller than expected; in other words, we have near commutativity.

We begin this section by formulating analytically the commutativity diagrams between the two averages. We proceed from generally anisotropic layers to a monoclinic medium, from monoclinic layers to an orthotropic medium, and from orthotropic layers to a tetragonal medium. Also, we discuss transversely isotropic layers, which—depending on the order of operations—result in a transversely isotropic or isotropic medium.

4.3.2 *Analytical formulation*

4.3.2.1 *From generally anisotropic layers to monoclinic medium*

Let us consider a stack of generally anisotropic layers to obtain a monoclinic medium. To examine the commutativity between the Backus (1962) and Gazis *et al.* (1963) averages, let us study the following diagram,

$$\begin{array}{ccc} \text{aniso} & \xrightarrow{\text{B}} & \text{aniso} \\ \text{G}\big\downarrow & & \big\downarrow\text{G} \\ \text{mono} & \xrightarrow[\text{B}]{} & \text{mono} \end{array} \tag{4.63}$$

together with Theorem 4.1 and its corollary.

Diagram 4.63 illustrates two ways of obtaining a monoclinic medium from generally anisotropic layers; B and G denote the Backus (1962) and Gazis *et al.* (1963) averages, respectively. In the first step of the clockwise path, we use the Backus (1962) average to obtain a generally anisotropic medium, followed by the Gazis *et al.* (1963) average to obtain a monoclinic medium. In the first step of the counterclockwise path, we use the Gazis *et al.* (1963) average to obtain monoclinic layers, followed by the Backus (1962) average to obtain a monoclinic medium. The purpose is to examine whether or not the resulting monoclinic media are the same, regardless of the path, and if not, to estimate the discrepancy between them.

Theorem 4.1. *In general, the Backus (1962) and Gazis* et al. *(1963) averages do not commute.*

Proof. This is a consequence of the specific case, stated by the following proposition.

Proposition 4.2. *For the generally anisotropic and monoclinic symmetries, the Backus (1962) and Gazis* et al. *(1963) averages do not commute.*

To proceed, we invoke the following lemma.

Lemma 4.1. *For the effective monoclinic symmetry, the result of the Gazis* et al. *(1963) average is tantamount to replacing each* $c_{ijk\ell}$*, in a generally anisotropic tensor, by its corresponding* $c_{ijk\ell}$ *of the monoclinic tensor, expressed in the natural coordinate system, including replacements of the anisotropic-tensor components by the zeros of the corresponding monoclinic components.*

Proof. For discrete symmetries, which includes the monoclinic symmetry, we can write integral (4.5) as a sum,

$$C_{\text{eff}}^{\text{sym}} = \frac{1}{n}\sum_{i=1}^{n} \tilde{A}_i^{\text{sym}}\, C\, \tilde{A}_i^{\text{sym}\,T}\,, \tag{4.64}$$

where n is the number of elements in the symmetry group. To write the elements of the monoclinic symmetry group as 6×6 matrices, we must consider its two orthogonal transformations in $\mathbb{R}^3$. In general, transformation $A \in SO(3)$ of $c_{ijk\ell}$ corresponds to transformation of C given by matrix (3.42), which is an orthogonal transformation, $\tilde{A} \in SO(6)$.

Following steps whose details are shown in the solution to Exercise 4.1, we obtain

$$C_{\text{eff}}^{\text{mono}} = \begin{bmatrix} c_{1111} & c_{1122} & c_{1133} & 0 & 0 & \sqrt{2}c_{1112} \\ c_{1122} & c_{2222} & c_{2233} & 0 & 0 & \sqrt{2}c_{2212} \\ c_{1133} & c_{2233} & c_{3333} & 0 & 0 & \sqrt{2}c_{3312} \\ 0 & 0 & 0 & 2c_{2323} & 2c_{2313} & 0 \\ 0 & 0 & 0 & 2c_{2313} & 2c_{1313} & 0 \\ \sqrt{2}c_{1112} & \sqrt{2}c_{2212} & \sqrt{2}c_{3312} & 0 & 0 & 2c_{1212} \end{bmatrix},$$

which, as required, are the components of the monoclinic tensor expressed in its natural coordinate system, given in matrix (3.43). In other words, $C_{\text{eff}}^{\text{sym}} = C^{\text{mono}}$, which is Corollary 4.1, below. □

Corollary 4.1. *For the effective monoclinic symmetry, given a generally anisotropic tensor,* C,

$$C_{\text{eff}}^{\text{mono}} = C^{\text{mono}}; \tag{4.65}$$

where $C_{\text{eff}}^{\text{mono}}$ *is the Gazis* et al. *(1963) average of* C, *and* C^{mono} *is the monoclinic tensor whose nonzero entries are the same as for* C.

Let us first examine the first step of the counterclockwise path of Diagram 4.63. According to Corollary 4.1, the effective monoclinic tensor is obtained simply by setting to zero—in the generally anisotropic tensor—the components that are zero for a monoclinic tensor. Then, the second step of the counterclockwise path of Diagram 4.63 is performed as follows. Applying the Backus (1962) average, we obtain (Bos *et al.*, 2017b)

$$\langle c_{3333}\rangle = \overline{\left(\frac{1}{c_{3333}}\right)}^{\,-1}, \qquad \langle c_{2323}\rangle = \frac{\overline{\left(\frac{c_{2323}}{D}\right)}}{2D_2},$$

$$\langle c_{1313}\rangle = \frac{\overline{\left(\frac{c_{1313}}{D}\right)}}{2D_2}\,,\qquad \langle c_{2313}\rangle = \frac{\overline{\left(\frac{c_{2313}}{D}\right)}}{2D_2}\,,$$

where

$$D := 2\left(c_{2323}c_{1313} - c_{2313}^2\right)$$

and

$$D_2 := \overline{\left(\frac{c_{1313}}{D}\right)}\;\overline{\left(\frac{c_{2323}}{D}\right)} - \overline{\left(\frac{c_{2313}}{D}\right)}^2\,.$$

We also obtain

$$\langle c_{1133}\rangle = \overline{\left(\frac{1}{c_{3333}}\right)}^{-1}\overline{\left(\frac{c_{1133}}{c_{3333}}\right)}\,,\quad \langle c_{2233}\rangle = \overline{\left(\frac{1}{c_{3333}}\right)}^{-1}\overline{\left(\frac{c_{2233}}{c_{3333}}\right)}\,,$$

$$\langle c_{3312}\rangle = \overline{\left(\frac{1}{c_{3333}}\right)}^{-1}\overline{\left(\frac{c_{3312}}{c_{3333}}\right)}\,,$$

$$\langle c_{1111}\rangle = \overline{c_{1111}} - \overline{\left(\frac{c_{1133}^2}{c_{3333}}\right)} + \overline{\left(\frac{1}{c_{3333}}\right)}^{-1}\overline{\left(\frac{c_{1133}}{c_{3333}}\right)}^2\,,$$

$$\langle c_{1122}\rangle = \overline{c_{1122}} - \overline{\left(\frac{c_{1133}\,c_{2233}}{c_{3333}}\right)} + \overline{\left(\frac{1}{c_{3333}}\right)}^{-1}\overline{\left(\frac{c_{1133}}{c_{3333}}\right)}\;\overline{\left(\frac{c_{2233}}{c_{3333}}\right)}\,,$$

$$\langle c_{2222}\rangle = \overline{c_{2222}} - \overline{\left(\frac{c_{2233}^2}{c_{3333}}\right)} + \overline{\left(\frac{1}{c_{3333}}\right)}^{-1}\overline{\left(\frac{c_{2233}}{c_{3333}}\right)}^2\,,$$

$$\langle c_{1212}\rangle = \overline{c_{1212}} - \overline{\left(\frac{c_{3312}^2}{c_{3333}}\right)} + \overline{\left(\frac{1}{c_{3333}}\right)}^{-1}\overline{\left(\frac{c_{3312}}{c_{3333}}\right)}^2\,,$$

$$\langle c_{1112}\rangle = \overline{c_{1112}} - \overline{\left(\frac{c_{3312}\,c_{1133}}{c_{3333}}\right)} + \overline{\left(\frac{1}{c_{3333}}\right)}^{-1}\overline{\left(\frac{c_{1133}}{c_{3333}}\right)}\;\overline{\left(\frac{c_{3312}}{c_{3333}}\right)}\,,$$

$$\langle c_{2212}\rangle = \overline{c_{2212}} - \overline{\left(\frac{c_{3312}\,c_{2233}}{c_{3333}}\right)} + \overline{\left(\frac{1}{c_{3333}}\right)}^{-1}\overline{\left(\frac{c_{2233}}{c_{3333}}\right)}\;\overline{\left(\frac{c_{3312}}{c_{3333}}\right)}\,,$$

where angle brackets denote the equivalent-medium elasticity parameters. The other equivalent-medium elasticity parameters are zero.

Following the clockwise path of Diagram 4.63, the upper branch is derived in matrix form by Bos *et al.* (2017b). Then, according to Corollary 4.1, the result of the right-hand branch is derived by setting entries in the generally anisotropic tensor that are zero for a monoclinic tensor to zero. The nonzero entries, which appear too complicated to display explicitly, are—in general—not the same as the result of the counterclockwise path. Hence, for generally anisotropic and monoclinic symmetries, the Backus (1962) and Gazis *et al.* (1963) averages do not commute, which completes the proof of Theorem 4.1. □

4.3.2.2 *From monoclinic layers to orthotropic medium*

Theorem 4.1 remains valid for layers exhibiting higher material symmetries. However, for such symmetries, simpler expressions of the corresponding elasticity tensors allow us to examine special cases that result in commutativity. Let us consider the following case of Theorem 4.1.

Proposition 4.3. *For the monoclinic and orthotropic symmetries, the Backus (1962) and Gazis* et al. *(1963) averages do not commute.*

Proof. To prove this proposition, let us consider—in a manner similar to Diagram 4.63—the following diagram,

$$\begin{array}{ccc} \text{mono} & \xrightarrow{\text{B}} & \text{mono} \\ {\scriptstyle\text{G}}\big\downarrow & & \big\downarrow{\scriptstyle\text{G}} \\ \text{ortho} & \xrightarrow[\text{B}]{} & \text{ortho} \end{array} \tag{4.66}$$

and the following lemma.

Lemma 4.2. *For the effective orthotropic symmetry, the result of the Gazis* et al. *(1963) average is tantamount to replacing each $c_{ijk\ell}$, in a generally anisotropic—or monoclinic—tensor, by its corresponding $c_{ijk\ell}$ of an orthotropic tensor, expressed in the natural coordinate system, including the replacements by the corresponding zeros.*

Proof. The group of orthotropic symmetry contains two orthogonal transformations beyond $\tilde{A}_1^{\rm ortho} = \tilde{A}_1^{\rm mono}$, $\tilde{A}_2^{\rm ortho} = \tilde{A}_2^{\rm mono}$, which also belong to the monoclinic group. Following matrix (3.42), we write these two transformations as

$$A_3^{\rm ortho} = \begin{bmatrix} -1 & 0 & 0 \\ 0 & 1 & 0 \\ 0 & 0 & -1 \end{bmatrix} \mapsto \begin{bmatrix} 1 & 0 & 0 & 0 & 0 & 0 \\ 0 & 1 & 0 & 0 & 0 & 0 \\ 0 & 0 & 1 & 0 & 0 & 0 \\ 0 & 0 & 0 & -1 & 0 & 0 \\ 0 & 0 & 0 & 0 & 1 & 0 \\ 0 & 0 & 0 & 0 & 0 & -1 \end{bmatrix} = \tilde{A}_3^{\rm ortho}$$

and

$$A_4^{\rm ortho} = \begin{bmatrix} 1 & 0 & 0 \\ 0 & -1 & 0 \\ 0 & 0 & -1 \end{bmatrix} \mapsto \begin{bmatrix} 1 & 0 & 0 & 0 & 0 & 0 \\ 0 & 1 & 0 & 0 & 0 & 0 \\ 0 & 0 & 1 & 0 & 0 & 0 \\ 0 & 0 & 0 & 1 & 0 & 0 \\ 0 & 0 & 0 & 0 & -1 & 0 \\ 0 & 0 & 0 & 0 & 0 & -1 \end{bmatrix} = \tilde{A}_4^{\rm ortho}.$$

According to expression (4.64),

$$C_{\text{eff}}^{\text{ortho}} = \frac{1}{4}\left[\left(\tilde{A}_1^{\text{ortho}}\right) C \left(\tilde{A}_1^{\text{ortho}}\right)^T + \left(\tilde{A}_2^{\text{ortho}}\right) C \left(\tilde{A}_2^{\text{ortho}}\right)^T + \left(\tilde{A}_3^{\text{ortho}}\right) C \left(\tilde{A}_3^{\text{ortho}}\right)^T + \left(\tilde{A}_4^{\text{ortho}}\right) C \left(\tilde{A}_4^{\text{ortho}}\right)^T\right],$$

which results in

$$C_{\text{eff}}^{\text{ortho}} = \begin{bmatrix} c_{1111} & c_{1122} & c_{1133} & 0 & 0 & 0 \\ c_{1122} & c_{2222} & c_{2233} & 0 & 0 & 0 \\ c_{1133} & c_{2233} & c_{3333} & 0 & 0 & 0 \\ 0 & 0 & 0 & 2c_{2323} & 0 & 0 \\ 0 & 0 & 0 & 0 & 2c_{1313} & 0 \\ 0 & 0 & 0 & 0 & 0 & 2c_{1212} \end{bmatrix}, \tag{4.67}$$

which are the components of an orthotropic tensor in its natural coordinate system, stated in matrix (3.50). In other words, $C_{\text{eff}}^{\text{ortho}} = C^{\text{ortho}}$, which is Corollary 4.2, below. □

Corollary 4.2. *For the effective orthotropic symmetry, given a generally anisotropic or monoclinic tensor,* C*,*

$$C_{\text{eff}}^{\text{ortho}} = C^{\text{ortho}}, \tag{4.68}$$

where $C_{\text{eff}}^{\text{ortho}}$ *is the Gazis* et al. *(1963) average of* C*, and* C^{ortho} *is an orthotropic tensor whose nonzero entries are the same as for* C*.*

Let us consider a monoclinic tensor and proceed counterclockwise along the first branch of Diagram 4.66. For the first step, we use the fact that the monoclinic symmetry is a special case of general anisotropy, and invoke Corollary 4.2 to conclude that $C_{\text{eff}}^{\text{ortho}} = C^{\text{ortho}}$, which is equivalent to setting c_{1112}, c_{2212}, c_{3312} and c_{2313} to zero, in each monoclinic tensor, which herein represents a layer. Following the second counterclockwise step, we obtain the nine parameters of an orthotropic medium; among them,

$$c_{1212}^{\circlearrowleft} = \overline{c_{1212}}, \quad c_{1313}^{\circlearrowleft} = \overline{\left(\frac{1}{c_{1313}}\right)}^{-1}, \quad c_{2323}^{\circlearrowleft} = \overline{\left(\frac{1}{c_{2323}}\right)}^{-1}. \tag{4.69}$$

Following the clockwise path, we obtain the nine parameters of an orthotropic medium; among them,

$$c_{1212}^{\circlearrowright} = \overline{c_{1212}} - \overline{\left(\frac{c_{3312}^2}{c_{3333}}\right)} + \overline{\left(\frac{1}{c_{3333}}\right)}^{-1}\overline{\left(\frac{c_{3312}}{c_{3333}}\right)}^2,$$

$$c^{\circlearrowleft}_{1313} = \frac{\overline{\left(\frac{c_{1313}}{D}\right)}}{2D_2}\,, \qquad c^{\circlearrowleft}_{2323} = \frac{\overline{\left(\frac{c_{2323}}{D}\right)}}{2D_2}\,,$$

which are distinct from their counterparts in expression (4.69). Thus, the Backus (1962) average and Gazis *et al.* (1963) average do not commute, which completes the proof of Proposition 4.3. □

Remark 4.1. Since the other entries are the same for both the counterclockwise and clockwise paths of Diagram 4.66, we can say that the Backus (1962) average and Gazis *et al.* (1963) average commute if $c_{2313} = c_{3312} = 0$, which is a special case of monoclinic symmetry.

In view of Remark 4.1, let us consider the case of weak anisotropy by letting c_{2313} and c_{3312}, which are zero for isotropy, be small. To study the commutativity of the two averages, we write commutator (4.62) as

$$\mathscr{C} = [\mathrm{B}, \mathrm{G}] = \mathrm{BG} - \mathrm{GB}\,,$$

where BG and GB are the clockwise and counterclockwise paths, respectively.

Since—if $c_{2313} = c_{3312} = 0$—the commutator is zero, it is expected that, in the neighbourhood, we have near commutativity. If both c_{2313} and c_{3312} are of order ϵ, then $\mathscr{C}$ is expected to be of the same order. However, a stronger statement is true. For c_{2313} and c_{3312} of order ϵ, $\mathscr{C}$ is of order ϵ^2, thus indicating a stronger near commutativity than expected *a priori*. This follows from the following calculation.

Consider $\mathscr{C} = [\mathscr{C}_1, \mathscr{C}_2, \mathscr{C}_3]$, where

$$\mathscr{C}_1 = c^{\circlearrowleft}_{2323} - c^{\circlearrowright}_{2323} = \frac{\overline{\left(\frac{c_{2323}}{D}\right)}}{2D_2} - \overline{\left(\frac{1}{c_{2323}}\right)}^{-1}\,,$$

$$\mathscr{C}_2 = c^{\circlearrowleft}_{1313} - c^{\circlearrowright}_{1313} = \frac{\overline{\left(\frac{c_{1313}}{D}\right)}}{2D_2} - \overline{\left(\frac{1}{c_{1313}}\right)}^{-1}$$

and

$$\mathscr{C}_3 = c^{\circlearrowleft}_{1212} - c^{\circlearrowright}_{1212} = \overline{\left(\frac{1}{c_{3333}}\right)}^{-1}\overline{\left(\frac{c_{3312}}{c_{3333}}\right)}^{2} - \overline{\left(\frac{c^2_{3312}}{c_{3333}}\right)}\,.$$

The starting parameters are

$$x = c^i_{3333}\,, c^i_{2323}\,, c^i_{1313}\,, c^i_{2313}\,, c^i_{3312}\,, \quad i = 1, \ldots, n\,,$$

and we have commutativity if

$$c^i_{2313} = c^i_{3312} = 0\,, \quad i = 1, \ldots, n\,,$$

which we denote by $x = a$, such that $\mathscr{C}(a) = [\,0\,]$.

For the arithmetic average and assuming that n layers have the same thickness, we have

$$\overline{f} = \frac{1}{n}\sum_{i=1}^{n} f_i\,,$$

where f is a piecewise-constant function and f_i is its constant value in the i-th layer. Also, we let the $3 \times 5n$ Jacobi matrix be

$$\mathscr{C}'(x) = \left[\frac{\partial \mathscr{C}}{\partial x}\right].$$

If we evaluate this matrix (Bos *et al.*, 2017a, Appendix B), we find that $\mathscr{C}'(a) = [\,0\,]$. If we expand $\mathscr{C}(x)$ in a Taylor series,

$$\mathscr{C}(x) = \mathscr{C}(a) + \mathscr{C}'(a)(x - a) + \cdots = \mathscr{C}'(a)(x - a) + \cdots\,,$$

we see that, near $x = a$, $\|\mathscr{C}(x)\| = O\left(\|x - a\|^2\right)$, so there is a near commutativity in a neighbourhood of $x = a$.

4.3.2.3 *From orthotropic layers to tetragonal medium*

In a manner analogous to Diagram 4.66, proceeding from the upper-left-hand corner, which is an orthotropic tensor, by the counterclockwise path to the lower-right-hand corner, which is a tetragonal tensor,

$$\begin{array}{ccc} \text{ortho} & \xrightarrow{\ \text{B}\ } & \text{ortho} \\ {\scriptstyle \text{G}}\big\downarrow & & \big\downarrow{\scriptstyle \text{G}} \\ \text{tetra} & \xrightarrow[\text{B}]{} & \text{tetra} \end{array} \tag{4.70}$$

we obtain

$$c^{\circlearrowleft}_{1111} = \overline{\frac{c_{1111} + c_{2222}}{2} - \frac{\left(\dfrac{c_{1111} + c_{2222}}{2}\right)^2}{c_{3333}}} + \overline{\left(\frac{c_{1111} + c_{2222}}{2c_{3333}}\right)}^{2}\,\overline{\left(\frac{1}{c_{3333}}\right)}^{-1}.$$

Following the clockwise path, we obtain

$$c^{\circlearrowright}_{1111} = \overline{\frac{c_{1111}+c_{2222}}{2} - \frac{c_{1133}^2+c_{2233}^2}{2c_{3333}}} + \frac{1}{2}\left[\overline{\left(\frac{c_{1133}}{c_{3333}}\right)}^2 + \overline{\left(\frac{c_{2233}}{c_{3333}}\right)}^2\right]\overline{\left(\frac{1}{c_{3333}}\right)}^{-1} .$$

These results are not equal to one another, unless $c_{1133} = c_{2233}$, which is a special case of orthotropic symmetry. The same is true for $c^{\circlearrowleft}_{1122}$ and $c^{\circlearrowright}_{1122}$. Also, c_{2323} must equal c_{1313} for $c^{\circlearrowright}_{2323} = c^{\circlearrowleft}_{2323}$. The other entries are the same for both paths. Thus, the Backus (1962) average and Gazis *et al.* (1963) average do commute for $c_{1133} = c_{2233}$ and $c_{2323} = c_{1313}$, which is a special case of orthotropic symmetry, but they do not commute in general.

Similarly to our discussion in Section 4.3.2.2, we examine the commutator. Herein, the commutator is $\mathscr{C} = [\mathscr{C}_1, \mathscr{C}_2, \mathscr{C}_3]$, where

$$\mathscr{C}_1 = c^{\circlearrowright}_{1111} - c^{\circlearrowleft}_{1111} , \quad \mathscr{C}_2 = c^{\circlearrowright}_{1122} - c^{\circlearrowleft}_{1122} , \quad \mathscr{C}_3 = c^{\circlearrowright}_{2323} - c^{\circlearrowleft}_{2323} .$$

The starting parameters that in this commutator are

$$x = c^i_{1133} , c^i_{2233} , c^i_{3333} , c^i_{2323} , c^i_{1313} , \quad i = 1, \ldots, n ,$$

and we have commutativity if

$$c^i_{1133} = c^i_{2233} \quad \text{and} \quad c^i_{2323} = c^i_{1313} , \quad i = 1, \ldots, n ,$$

which we denote by $x = a$, such that $\mathscr{C}(a) = [0]$. Again, as in Section 4.3.2.2, we let the $3 \times 5n$ Jacobi matrix be

$$\mathscr{C}'(x) = \left[\frac{\partial \mathscr{C}}{\partial x}\right] ,$$

to find that $\mathscr{C}'(a) = [\,0\,]$, which entails $\|\mathscr{C}(x)\| = O\left(\|x-a\|^2\right)$.

4.3.2.4 *Transversely isotropic layers*

Lack of commutativity between the two averages can be also exemplified by the case of transversely isotropic layers. The Backus (1962) average results in a transversely isotropic medium, whose Gazis *et al.* (1963) average—in accordance with Figure 3.3—is isotropic. On the other hand, reversing the sequence, the Gazis *et al.* (1963) average results in isotropic layers, whose Backus (1962) average is transversely isotropic. Thus, not only the elasticity parameters, but even the resulting material-symmetry classes differ.

4.3.2.5 *Combining rotational averages*

Let us also examine the process of combining the Gazis *et al.* (1963) averages, which is tantamount to combining Diagrams (4.66) and (4.70),

$$\begin{array}{ccc} \text{mono} & \xrightarrow{\text{B}} & \text{mono} \\ \text{G}\big\downarrow & & \big\downarrow\text{G} \\ \text{ortho} & \xrightarrow[\text{B}]{} & \text{ortho} \\ \text{G}\big\downarrow & & \big\downarrow\text{G} \\ \text{tetra} & \xrightarrow[\text{B}]{} & \text{tetra} \end{array} \tag{4.71}$$

In accordance with Theorem 4.1, in general, there is no commutativity. However, the outcomes are the same as for the corresponding steps in Sections 4.3.2.2 and 4.3.2.3. In general, for the Gazis *et al.* (1963) average, proceeding directly, aniso $\xrightarrow{\text{G}}$ iso, is tantamount to proceeding along arrows in Figure 3.3, aniso $\xrightarrow{\text{G}} \cdots \xrightarrow{\text{G}}$ iso. No such combining of the Backus (1962) averages is possible, since, for each step, layers become a homogeneous medium.

4.3.3 *Discussion*

In general, the Backus (1962) and Gazis *et al.* (1963) averages do not commute. There are, however, particular cases of given material symmetries for which the averaging processes commute, as discussed in Sections 4.3.2.2 and 4.3.2.3. More importantly—under weak anisotropy—which is a common case of seismological studies, the averages nearly commute for all cases that can be examined by a commutator; this is not the case of transverse isotropy, discussed in Section 4.3.2.4. Furthermore, and perhaps surprisingly, a perturbation of the elasticity parameters about a point of weak anisotropy results in the commutator of the two types of averaging being of the order of the square of this perturbation. In other words, the order of a sequence of these two averages is nearly indistinguishable.

Closing remarks

In this chapter, we examine the effective and equivalent tensors, which are discussed in Sections 4.1 and 4.2, respectively. The former is an average of symmetry groups at a point; the latter is an average of properties over an interval. The former, given by integral (4.5), stems from expression (3.23), which is the definition of material symmetry and, in accordance with this definition, is limited to directional properties at a point. The latter is defined by integral (4.44), which invokes a concept of spatial averaging and provides further insights into mathematical analogies of physical phenomena. In particular, it allows us to distinguish between behaviours resulting from local properties, which are associated with a point, and bulk properties, which are associated with a finite interval or a volume.

In Section 4.3, we discuss relations between anisotropy and inhomogeneity in the context of commutativity of the two averages discussed in Sections 4.1 and 4.2. Both averages can be used to examine seismic phenomena, and—in the case of weak anisotropy, which is an empirically adequate model in many seismological studies—the two averages nearly commute. In general, they do not commute; hence, the resulting medium depends on the sequence of operations.

From a foundational viewpoint, these averages could be considered, in the context of emergence, which refers to properties that arise from other entities and yet are novel with respect to, and might not be reducible to, these entities.

In view of Chapter 3, we bring to the reader's attention the distinction between a symmetric tensor, discussed therein, and an effective tensor exhibiting a particular symmetry, discussed herein, in Section 4.1. The formula for the former is a statement of invariance based on condition (3.30), and the formula for the latter is expression (4.5), which is a statement of averaging.

As stated in expression (3.46), we can write the condition of invariance as a system of equations:

$$\begin{aligned} C &= (\tilde{A}_1^{\rm sym})\, C\, (\tilde{A}_1^{\rm sym})^T \\ &\ \vdots \\ C &= (\tilde{A}_n^{\rm sym})\, C\, (\tilde{A}_n^{\rm sym})^T\,, \end{aligned} \tag{4.72}$$

where n is the number of elements in the symmetry group and $\tilde{A}_i^{\rm sym}$ are their expressions according to matrix (3.42); C is stated according to expression (3.41). C satisfying this system is $C^{\rm sym}$.

For discrete symmetries, we can write integral (4.5) as a sum shown in expression (4.64),

$$C_{\rm eff}^{\rm sym} = \frac{1}{n}\left((\tilde{A}_1^{\rm sym})\, C\, (\tilde{A}_1^{\rm sym})^T + \ldots + (\tilde{A}_n^{\rm sym})\, C\, (\tilde{A}_n^{\rm sym})^T\right)\,.$$

As discussed in Section 4.1.3, if the tensor to be averaged, C, is generally anisotropic, its orientation is implicitly contained within its twenty-one components; hence—upon searching for the closest tensor under all orientations—not only the averaged elasticity parameters, but also the corresponding optimal orientation of the symmetry axes or planes is obtained.

For the remainder of the book, Hookean solids are the key mathematical analogy of physical materials. Their simplicity and elegance allow us to achieve many accurate descriptions and insightful conclusions about those materials, in spite of obvious limitations such as stress being independent of the temporal rate of strain or the incapacity of considering explicitly hexagonal symmetries.

Accommodating such limitations by postulating constitutive equations exhibiting nonlinearity or containing higher-rank tensors might increase the complexity of a theory without increasing sufficiently the agreement between observations and model results. Since we rely on an *a posteriori* justification, we must ensure a sufficient increase of accuracy to increase the complexity of a model.

It might be interesting to mention that the approaches of both Backus (1962) and Gazis *et al.* (1963), which we discuss in this chapter, were formulated at a similar time. The former was motivated by seismological considerations and follows the work of such researchers as Rudzki (1911, 2003), Postma (1955) and Helbig (1958). The latter is related to an increase of interest in the mathematical foundations of continuum mechanics and, in particular, to the work of Clifford Truesdell; insightful descriptions of continuum mechanics can be found in Truesdell (1966), Bunge (1967), Noll (1974b) and Truesdell (1977). This historical comment might suggest the interrelation of research interests between quantitative seismology and continuum mechanics.

In conclusion, the awareness of the necessity for a theory to mediate between measurements and interpretations—and hence the unavoidable presence of abstract concepts, such as symmetry of tensors, for analysis of physical properties—is crucial for seismology, as, in general, is the awareness of pitfalls in using any theory to explain, interpret or predict physical phenomena.

4.4 Exercises

Exercise 4.1. Using formula (4.64) show explicitly that the effective monoclinic tensor is given by expression (4.6).

Solution. Let us write the elements of the monoclinic symmetry group as 6×6 matrices; they are

$$A_1^{\text{mono}} = \begin{bmatrix} 1 & 0 & 0 \\ 0 & 1 & 0 \\ 0 & 0 & 1 \end{bmatrix} \mapsto \begin{bmatrix} 1 & 0 & 0 & 0 & 0 & 0 \\ 0 & 1 & 0 & 0 & 0 & 0 \\ 0 & 0 & 1 & 0 & 0 & 0 \\ 0 & 0 & 0 & 1 & 0 & 0 \\ 0 & 0 & 0 & 0 & 1 & 0 \\ 0 & 0 & 0 & 0 & 0 & 1 \end{bmatrix} = \tilde{A}_1^{\text{mono}}$$

$$A_2^{\text{mono}} = \begin{bmatrix} -1 & 0 & 0 \\ 0 & -1 & 0 \\ 0 & 0 & 1 \end{bmatrix} \mapsto \begin{bmatrix} 1 & 0 & 0 & 0 & 0 & 0 \\ 0 & 1 & 0 & 0 & 0 & 0 \\ 0 & 0 & 1 & 0 & 0 & 0 \\ 0 & 0 & 0 & -1 & 0 & 0 \\ 0 & 0 & 0 & 0 & -1 & 0 \\ 0 & 0 & 0 & 0 & 0 & 1 \end{bmatrix} = \tilde{A}_2^{\text{mono}} .$$

For the monoclinic case, expression (4.64) can be stated explicitly as

$$C_{\text{eff}}^{\text{mono}} = \frac{\left(\tilde{A}_1^{\text{mono}}\right) C \left(\tilde{A}_1^{\text{mono}}\right)^T + \left(\tilde{A}_2^{\text{mono}}\right) C \left(\tilde{A}_2^{\text{mono}}\right)^T}{2}, \tag{4.73}$$

where C is the 6×6 matrix in equation (3.41), and T denotes transpose. Performing matrix operations, we obtain

$$C_{\text{eff}}^{\text{mono}} = \begin{bmatrix} c_{1111} & c_{1122} & c_{1133} & 0 & 0 & \sqrt{2}c_{1112} \\ c_{1122} & c_{2222} & c_{2233} & 0 & 0 & \sqrt{2}c_{2212} \\ c_{1133} & c_{2233} & c_{3333} & 0 & 0 & \sqrt{2}c_{3312} \\ 0 & 0 & 0 & 2c_{2323} & 2c_{2313} & 0 \\ 0 & 0 & 0 & 2c_{2313} & 2c_{1313} & 0 \\ \sqrt{2}c_{1112} & \sqrt{2}c_{2212} & \sqrt{2}c_{3312} & 0 & 0 & 2c_{1212} \end{bmatrix}, \tag{4.74}$$

which is expression (4.6), as required. Note that—unlike in the conditions for monoclinic symmetry, which is invariance under rotation (3.44)—the identity must be included in the averaging process for effective media, even though it does not affect C. In this particular case, matrices (3.63) and (4.74) are equal to one another, which is not a general property. A similar pattern is exhibited by $\tilde{A}_\ell^{\text{ortho}}$, $\ell = 2, 3, 4$; where 1 and -1 permute in the lower three diagonal entries, which—as in the monoclinic case—results in $c_{ijk\ell}^{\text{ortho}} = c_{ijk\ell}$, for the nonzero entries of the orthotropic tensor expressed in natural coordinates.[22]

[22]Readers interested in explicit orthotropic formulation might refer to Bos *et al.* (2017a, Appendix A.2).

Remark 4.2. Formulæ (3.62) and (4.73) exemplify, using the case of monoclinic symmetry, the distinction between the material and effective symmetries. The former results from imposing a condition on C and the latter from averaging it.

Exercise 4.2. Using matrices (3.41) and (3.57), obtain the expression for the Frobenius-norm distance between the generally anisotropic and isotropic tensors. Using that expression, show that—for the Frobenius norm—the elasticity parameters of the closest isotropic tensor are given by expressions (4.2) and (4.3), namely,

$$c_{1111}^{\rm iso} = \frac{1}{15}\left(3\left(c_{1111}+c_{2222}+c_{3333}\right)+2\left(c_{1122}+c_{1133}+c_{2233}\right)\right.$$
$$\left.+4\left(c_{1212}+c_{1313}+c_{2323}\right)\right) \tag{4.75}$$

and

$$c_{2323}^{\rm iso} = \frac{1}{15}\left(c_{1111}+c_{2222}+c_{3333}-\left(c_{1122}+c_{1133}+c_{2233}\right)\right.$$
$$\left.+3\left(c_{1212}+c_{1313}+c_{2323}\right)\right), \tag{4.76}$$

where the parameters on the right-hand sides correspond to a generally anisotropic tensor.

Solution. The Frobenius norm is the square root of the sum of the squared entries. Hence, in view of expressions (3.41) and (3.57), the Frobenius-norm distance between the generally anisotropic and isotropic tensors is the square root of

$$\Delta := \left(c_{1111}-c_{1111}^{\rm iso}\right)^2+\ldots+\left(2\,c_{1212}-2\,c_{2323}^{\rm iso}\right)^2,$$

where superscript $^{\rm iso}$ distinguishes between isotropy and general anisotropy. To minimize the distance function with respect to $c_{1111}^{\rm iso}$ and $c_{2323}^{\rm iso}$, we consider all $c_{ijk\ell}$ without the superscript as constants and both $c_{ijk\ell}^{\rm iso}$ as variables. This is tantamount to searching for an isotropic counterpart of a given generally anisotropic tensor.

Since minimizing the square root is equivalent to minimizing its radicand, the elasticity parameters of the closest isotropic tensor are the solutions of

$$\frac{\partial\Delta}{\partial c_{1111}^{\rm iso}} = -\,2\left(c_{1111}+c_{2222}+c_{3333}+2\left(c_{1122}+c_{1133}+c_{2233}\right)\right.$$
$$\left.-\,9\,c_{1111}^{\rm iso}+12\,c_{2323}^{\rm iso}\right)=0$$

and

$$\frac{\partial\Delta}{\partial c^{\rm iso}_{2323}} = 8\left(c_{1122} + c_{1133} + c_{2233} - (c_{1212} + c_{1313} + c_{2323})\right.$$
$$\left. - 3\,c^{\rm iso}_{1111} + 9\,c^{\rm iso}_{2323}\right) = 0\,.$$

This is a system of two linear equations for two unknowns, $c^{\rm iso}_{1111}$ and $c^{\rm iso}_{2323}$, whose solution is the required expressions.

To ensure that this solution corresponds to the global minimum, we perform the second-derivatives test (e.g., Stewart (1995, Section 12.7)), by calculating

$$\det\begin{bmatrix} \dfrac{\partial^2\Delta}{\partial(c^{\rm iso}_{1111})^2} & \dfrac{\partial^2\Delta}{\partial c^{\rm iso}_{1111}\partial c^{\rm iso}_{2323}} \\ \dfrac{\partial^2\Delta}{\partial c^{\rm iso}_{2323}\partial c^{\rm iso}_{1111}} & \dfrac{\partial^2\Delta}{\partial(c^{\rm iso}_{2323})^2} \end{bmatrix}$$
$$= \frac{\partial^2\Delta}{\partial(c^{\rm iso}_{1111})^2}\frac{\partial^2\Delta}{\partial(c^{\rm iso}_{2323})^2} - \left(\frac{\partial^2\Delta}{\partial c^{\rm iso}_{1111}\partial c^{\rm iso}_{2323}}\right)\left(\frac{\partial^2\Delta}{\partial c^{\rm iso}_{2323}\partial c^{\rm iso}_{1111}}\right),$$

which is the determinant of the Hessian matrix of Δ, and

$$\frac{\partial^2\Delta}{\partial(c^{\rm iso}_{1111})^2}\,,$$

which is its first entry.

Since their values—which are 720 and 18, respectively—are both greater than zero, for all $c_{ijk\ell}$, it follows, according to this test, that Δ has only one local minimum and no local maxima; hence, this minimum is also global. This confirms that the obtained expressions correspond to the shortest distance between general anisotropy and isotropy.

Since 720 and 18 are independent of $c_{ijk\ell}$, it follows that the same values are obtained for all elasticity tensors. Hence, for all elasticity tensors, there is only one extremum, which is the global minimum.

The same effective-tensor expressions could be obtained following the approach illustrated in Exercise 4.1.

Remark 4.3. There are two common Frobenius norms, which can be denoted by $\|\cdot\|_{F36}$ and $\|\cdot\|_{F21}$, depending on the number of entries included in the norm. In Exercise 4.2, we use $\|\cdot\|_{F36}$, which contains all entries. $\|\cdot\|_{F21}$ considers only twenty-one entries that correspond to all linearly independent parameters. We can view $\|\cdot\|_{F36}$ and $\|\cdot\|_{F21}$ as the same norm with distinct weights. As discussed by Danek *et al.* (2015, Section 3), the latter norm possesses certain properties required for statistical analysis. Following the approach of Exercise 4.2, we find that the effective isotropic tensor corresponding to $\|\cdot\|_{F21}$ is

$$c_{1111}^{\mathrm{iso}_{F21}} = \frac{1}{9}\left(2\left(c_{1111}+c_{2222}+c_{3333}+c_{1212}+c_{1313}+c_{2323}\right)\right.$$
$$\left.+\,c_{1122}+c_{1133}+c_{2233}\right) \tag{4.77}$$

and

$$c_{2323}^{\mathrm{iso}_{F21}} = \frac{1}{18}\left(c_{1111}+c_{2222}+c_{3333}+4\left(c_{1212}+c_{1313}+c_{2323}\right)\right.$$
$$\left.-\left(c_{1122}+c_{1133}+c_{2233}\right)\right). \tag{4.78}$$

Herein, the values for the second-derivative test are 432 and 12.

The approach of Exercise 4.2 can be used to obtain the effective-tensor expressions for all symmetry classes discussed in Section 4.1.2, including the case considered in Exercise 4.1. In each case, the number of linear equations is equal to the number of unknowns, which is the number of the independent elasticity parameters for a given class.

However—except for isotropy, discussed in Exercise 4.2 and in this remark—Δ depends not only on these parameters but also on the orientations of the coordinate systems in which the components of a generally anisotropic tensor and of a symmetric tensor are expressed. In such a case, Δ exhibits several local minima, and—due to its nonlinearity—the search for its global minimum is not soluble analytically. As discussed in Section 4.1.3, numerical methods must be employed to obtain an effective tensor that is not isotropic.[23]

Exercise 4.3. Consider the elasticity parameters of a rutile (Auld, 1990, *modified from* Vol. I, Table A.5), which is a mineral composed primarily of titanium dioxide,

$$C = 10^{10}\begin{bmatrix} 26.60 & 17.33 & 13.62 & 0 & 0 & 0 \\ 17.33 & 26.60 & 13.62 & 0 & 0 & 0 \\ 13.62 & 13.62 & 46.99 & 0 & 0 & 0 \\ 0 & 0 & 0 & 2(12.39) & 0 & 0 \\ 0 & 0 & 0 & 0 & 2(12.39) & 0 \\ 0 & 0 & 0 & 0 & 0 & 2(14.86) \end{bmatrix}\frac{\mathrm{N}}{\mathrm{m}^2}. \tag{4.79}$$

Obtain its effective isotropic tensors in the sense of $\|\cdot\|_{F36}$ and $\|\cdot\|_{F21}$. Comment on the results.

[23]Readers interested in a numerical approach might refer to Danek *et al.* (2015).

Solution. Following formulæ (4.75) and (4.76), we obtain $c_{1111}^{\mathrm{iso}_{F36}} = 36.55$ and $c_{2323}^{\mathrm{iso}_{F36}} = 11.64$, respectively. Following formulæ (4.77) and (4.78), we obtain $c_{1111}^{\mathrm{iso}_{F21}} = 36.03$ and $c_{2323}^{\mathrm{iso}_{F21}} = 11.90$, respectively; each value is multiplied by 10^{10}.

The two isotropic counterparts are similar to one another, even though they are obtained with different norms. The difference of norms can be exemplified by the fact that $\|C\|_{F36} = 84.13$ and $\|C\|_{F21} = 80.04$, both values multiplied by 10^{10}. Herein, however, doubling the weight of the three offdiagonal elements has little effect on the final outcome, which are the two isotropic counterparts of C.

Remark 4.4. Another norm used in such a context is the operator norm, mentioned on page 112. The effective isotropic counterpart, according to that norm, is given by the values of c_{1111} and c_{2323} that minimize the largest eigenvalue of $C - C^{\mathrm{iso}}$, which are expressions (4.79) and (3.57), respectively.

There are no analytical expressions to obtain these values. For tensor (4.79), a numerical search results in $c_{1111}^{\mathrm{iso}_\lambda} = 37.04$ and $c_{2323}^{\mathrm{iso}_\lambda} = 10.10$, both values multiplied by 10^{10}.

The operator norm of tensor (4.79) is $\|C\|_\lambda = 64.78$, multiplied by 10^{10}. This value, however, cannot be compared to either $\|C\|_{F36}$ or $\|C\|_{F21}$. The operator norm, which is the largest eigenvalue of tensor (4.79), does not exhibit standard units; the units of both $\|C\|_{F36}$ and $\|C\|_{F21}$ are $\mathrm{N/m^2}$. Yet, the values of the resulting elasticity parameters—which are $\mathrm{N/m^2}$, for all norms—are similar to the ones obtained with the Frobenius norms.

This similarity might result from the properties imposed on the effective tensor, regardless of which among the three standard norms is to be applied. Each counterpart of the tensor that exhibits tetragonal symmetry is to be isotropic. To visualize this requirement in terms of corresponding wavefronts, we can view it as fitting—using each norm—a sphere that is, in accordance with a given norm, the best representation of a surface that exhibits a tetragonal symmetry.

Exercise 4.4. Invoking integral (4.44),

$$\overline{f}(x_3) = \int_{-\infty}^{\infty} w(\xi - x_3) f(\xi)\,\mathrm{d}\xi\,, \tag{4.80}$$

justify expression (4.46),

$$\overline{\frac{\partial u_i}{\partial x_j}} = \frac{\partial}{\partial x_j}\overline{u_i}\,, \qquad i,j = 1,2,3\,. \tag{4.81}$$

Solution. Since each u_i, with $i = 1,2,3$, is a scalar-valued function of x_1, x_2, x_3, let us refer generically to any of them as $f(x_1, x_2, x_3)$. In view of expression (4.80), the derivatives with respect to x_1 and x_2 can be written as

$$\frac{\partial \overline{f}}{\partial x_i} = \frac{\partial}{\partial x_i} \int_{-\infty}^{\infty} w(\xi - x_3) f(x_1, x_2, \xi)\, \mathrm{d}\xi$$

$$= \int_{-\infty}^{\infty} w(\xi - x_3) \frac{\partial f(x_1, x_2, \xi)}{\partial x_i}\, \mathrm{d}\xi =: \overline{\frac{\partial f}{\partial x_i}}, \qquad i = 1, 2,$$

where the last equality is the statement of definition (4.80), as required. In other words,

$$\frac{\partial\, \overline{u_i}}{\partial x_j} = \overline{\frac{\partial u_i}{\partial x_j}}, \qquad i = 1, 2, 3, \quad j = 1, 2.$$

For the derivatives with respect to x_3, we need to verify that

$$\frac{\partial}{\partial x_3} \int_{-\infty}^{\infty} w(\xi - x_3) f(x_1, x_2, \xi)\, \mathrm{d}\xi = \int_{-\infty}^{\infty} w(\xi - x_3) \frac{\partial f(x_1, x_2, \xi)}{\partial \xi}\, \mathrm{d}\xi. \quad (4.82)$$

Applying the integration by parts, we write the right-hand side as

$$w(\xi - x_3) f(x_1, x_2, \xi)|_{-\infty}^{\infty} - \int_{-\infty}^{\infty} w'(\xi - x_3)\, f(x_1, x_2, \xi)\, \mathrm{d}\xi,$$

where w is a function of a single variable. Since

$$\lim_{x_3 \to \pm\infty} w(x_3) = 0,$$

the product of w and f vanishes at $\pm\infty$, and we are left with

$$-\int_{-\infty}^{\infty} w'(\xi - x_3)\, f(x_1, x_2, \xi)\, \mathrm{d}\xi.$$

Let us consider the left-hand side of expression (4.82). Since only w is a function of x_3, we can interchange the operations of integration and differentiation to write

$$-\int_{-\infty}^{\infty} w'(\xi - x_3)\, f(x_1, x_2, \xi)\, \mathrm{d}\xi;$$

the negative sign arises from the chain rule,

$$\frac{\partial w(\xi - x_3)}{\partial x_3} = w'(\xi - x_3)\frac{\partial(\xi - x_3)}{\partial x_3} = -w'(\xi - x_3)\,.$$

Thus, both sides of expression (4.82) are equal to one another, as required. In other words,

$$\frac{\partial \overline{u_i}}{\partial x_j} = \overline{\frac{\partial u_i}{\partial x_j}}\,, \qquad i = 1, 2, 3, \quad j = 3\,,$$

which completes the justification of expression (4.81).

Exercise 4.5. Given the following proposition,[24] argue for its applicability even in the case in which one function is nearly constant but the other varies rapidly.

Proposition 4.4. *For functions $f(x)$ and $g(x)$, where both*

$$\|f'\|_\infty := \sup_{-\infty<x<\infty} |f'(x)| \qquad \text{and} \qquad \|g'\|_\infty := \sup_{-\infty<x<\infty} |g'(x)|$$

are finite, it follows that

$$|\overline{fg} - \overline{f}\,\overline{g}| \leqslant 2\,(\ell')^2\,\|f'\|_\infty\,\|g'\|_\infty\,, \tag{4.83}$$

where the bar above a function denotes its average defined by integral (4.44), and ℓ' is tantamount to the extent over which this average is taken.

Hence—according to Proposition 4.4—if f and g are nearly constant, which means that $\|f'\|_\infty$ and $\|g'\|_\infty$ are small, then $\overline{fg} \approx \overline{f}\,\overline{g}$.

Solution. Proposition 4.4 means that the difference between the average of the product, $\overline{f\,g}$, and the product of averages, $\overline{f}\,\overline{g}$, has the upper bound given by the right-hand side of equation (4.83), provided the effect of ℓ' can be considered as finite. If f or g is constant, then $f' = 0$ or $g' = 0$, and $\overline{f\,g} = \overline{f}\,\overline{g}$; otherwise, $\overline{f\,g} \approx \overline{f}\,\overline{g}$, with different accuracies of approximation, depending on properties of f and g over interval ℓ'. The smaller the interval of averaging, the lesser the variability of functions and hence the better the approximation. In the limit—w in integral (4.44) is Dirac's delta and hence ℓ' is a point—$\overline{f\,g} = \overline{f}\,\overline{g}$; in such a case, however, the essence of averaging is lost.

Since the error estimate involves the product of the norms of the derivatives, it follows that if, over extent ℓ', one of them is sufficiently small and the other is not exceedingly large, their product can remain small enough for $\overline{fg} \approx \overline{f}\,\overline{g}$ to be applicable.

[24]Readers interested in a proof of that proposition might refer to Bos *et al.* (2017b).

Exercise 4.6. Following the formulation presented in Section 4.2.2.2 and using matrix (3.50) to represent each orthotropic layer, obtain the expressions for the equivalent-medium elasticity parameters. Assume that—within each layer—one of the symmetry planes is parallel to interfaces, and another symmetry plane is parallel to the corresponding planes in all other layers.

Solution. Since—for each layer—one of the planes is parallel to interfaces, we can express the components of the tensor in its natural coordinates. Furthermore, since another symmetry plane is parallel for all layers, we can consider that the elasticity parameters for all layers are expressed with respect to the same coordinate system, and hence can be averaged. Following the approach presented on page 128, we write the last four stress-strain equations as

$$\frac{\partial}{\partial x_3}u_3=\left(\frac{1}{c_{3333}}\right)\sigma_{33}-\left(\frac{c_{1133}}{c_{3333}}\right)\frac{\partial}{\partial x_1}u_1-\left(\frac{c_{2233}}{c_{3333}}\right)\frac{\partial}{\partial x_2}u_2\,,$$

$$\frac{\partial}{\partial x_3}u_2=\left(\frac{1}{c_{2323}}\right)\sigma_{23}-\frac{\partial}{\partial x_2}u_3\,,$$

$$\frac{\partial}{\partial x_3}u_1=\left(\frac{1}{c_{1313}}\right)\sigma_{13}-\frac{\partial}{\partial x_1}u_3\,,$$

$$\sigma_{12}=c_{1212}\frac{\partial}{\partial x_2}u_1+c_{1212}\frac{\partial}{\partial x_1}u_2\,.$$

Performing the averaging and writing the results in terms of the strain-tensor components, we have

$$\overline{\sigma_{33}}=\overline{\left(\frac{1}{c_{3333}}\right)}^{-1}\overline{\left(\frac{c_{1133}}{c_{3333}}\right)}\,\overline{\varepsilon_{11}}+\overline{\left(\frac{1}{c_{3333}}\right)}^{-1}\overline{\left(\frac{c_{2233}}{c_{3333}}\right)}\,\overline{\varepsilon_{22}}+\overline{\left(\frac{1}{c_{3333}}\right)}^{-1}\overline{\varepsilon_{33}}\,,\tag{4.84}$$

$$\overline{\sigma_{23}}=2\,\overline{\left(\frac{1}{c_{2323}}\right)}^{-1}\overline{\varepsilon_{23}}\,,$$

$$\overline{\sigma_{13}}=2\,\overline{\left(\frac{1}{c_{1313}}\right)}^{-1}\overline{\varepsilon_{13}}\,,$$

$$\overline{\sigma_{12}}=2\,\overline{c_{1212}}\;\overline{\varepsilon_{12}}\,.$$

In view of the last four equations of system (3.41), we infer six elasticity parameters of the equivalent medium. They are $\langle c_{1133}\rangle =$

$\overline{(1/c_{3333})}^{\,-1}\,\overline{(c_{1133}/c_{3333})}$, $\langle c_{2233}\rangle = \overline{(1/c_{3333})}^{\,-1}\,\overline{(c_{2233}/c_{3333})}$, $\langle c_{3333}\rangle = \overline{(1/c_{3333})}^{\,-1}$, $\langle c_{2323}\rangle = \overline{(1/c_{2323})}^{\,-1}$, $\langle c_{1313}\rangle = \overline{(1/c_{1313})}^{\,-1}$ and $\langle c_{1212}\rangle = \overline{c_{1212}}$.

To proceed, we consider the first three stress-strain equations, in the context of matrix (3.50); thus,

$$\sigma_{11} = c_{1111}\,\varepsilon_{11} + c_{1122}\,\varepsilon_{22} + c_{1133}\,\varepsilon_{33}\,,$$

$$\sigma_{22} = c_{1122}\,\varepsilon_{11} + c_{2222}\,\varepsilon_{22} + c_{2233}\,\varepsilon_{33}\,,$$

$$\varepsilon_{33} = \left(\frac{1}{c_{3333}}\right)\sigma_{33} - \left(\frac{c_{1133}}{c_{3333}}\right)\varepsilon_{11} - \left(\frac{c_{2233}}{c_{3333}}\right)\varepsilon_{22}\,.$$

Inserting the third expression into the first one and performing the averaging, we get

$$\overline{\sigma_{11}} = \overline{\left(c_{1111} - \frac{c_{1133}^2}{c_{3333}}\right)}\,\overline{\varepsilon_{11}} + \overline{\left(c_{1122} - \frac{c_{1133}\,c_{2233}}{c_{3333}}\right)}\,\overline{\varepsilon_{22}} + \overline{\left(\frac{c_{1133}}{c_{3333}}\right)}\,\overline{\sigma_{33}}\,.$$

Substituting expression (4.84) for $\overline{\sigma_{33}}$, we obtain

$$\begin{aligned}\overline{\sigma_{11}} = {} & \left(\overline{\left(c_{1111} - \frac{c_{1133}^2}{c_{3333}}\right)} + \overline{\left(\frac{c_{1133}}{c_{3333}}\right)}^{2}\,\overline{\left(\frac{1}{c_{3333}}\right)}^{-1}\right)\overline{\varepsilon_{11}} \\ & + \left(\overline{\left(c_{1122} - \frac{c_{1133}\,c_{2233}}{c_{3333}}\right)} + \overline{\left(\frac{c_{1133}}{c_{3333}}\right)}\,\overline{\left(\frac{c_{2233}}{c_{3333}}\right)}\,\overline{\left(\frac{1}{c_{3333}}\right)}^{-1}\right)\overline{\varepsilon_{22}} \\ & + \overline{\left(\frac{c_{1133}}{c_{3333}}\right)}\,\overline{\left(\frac{1}{c_{3333}}\right)}^{-1}\,\overline{\varepsilon_{33}}\,.\end{aligned}$$

Comparing this result to the first equation of system (3.41), we infer another two elasticity parameters of the equivalent medium, $\langle c_{1111}\rangle$ and $\langle c_{1122}\rangle$, while $\langle c_{1133}\rangle$ is already known from expression (4.84). The other parameters of the first row, by comparison with equation (3.41), are zero, which is exhibited by the symmetry classes discussed in Sections 3.2.4.7–3.2.4.13, except for the class in Section 3.2.4.11.

Again, inserting the third expression—but into the second one—performing the averaging and using expression (4.84), we get

$$\begin{aligned}\overline{\sigma_{22}} = {} & \left(\overline{\left(c_{1122} - \frac{c_{1133}\,c_{2233}}{c_{3333}}\right)} + \overline{\left(\frac{c_{1133}}{c_{3333}}\right)}\,\overline{\left(\frac{c_{1133}}{c_{3333}}\right)}\,\overline{\left(\frac{1}{c_{3333}}\right)}^{-1}\right)\overline{\varepsilon_{11}} \\ & + \left(\overline{\left(c_{2222} - \frac{c_{2233}^2}{c_{3333}}\right)} + \overline{\left(\frac{c_{2233}}{c_{3333}}\right)}^{2}\,\overline{\left(\frac{1}{c_{3333}}\right)}^{-1}\right)\overline{\varepsilon_{22}} \\ & + \overline{\left(\frac{c_{2233}}{c_{3333}}\right)}\,\overline{\left(\frac{1}{c_{3333}}\right)}^{-1}\,\overline{\varepsilon_{33}}\,,\end{aligned}$$

from which, by comparison with the second equation of system (3.41), we infer $\langle c_{2222}\rangle$. Since there are nine independent elasticity parameters, we conclude that the equivalent medium is orthotropic.[25] Thus, similarly to the case of transversely isotropic layers, discussed in Section 4.2.3, the equivalent medium exhibits the same symmetry as do individual layers. Herein, the equivalent-medium elasticity parameters are

$$\overline{c_{1111}^{\rm ortho}} = \overline{\left(c_{1111} - \frac{c_{1133}^2}{c_{3333}}\right)} + \overline{\left(\frac{c_{1133}}{c_{3333}}\right)}^2 \overline{\left(\frac{1}{c_{3333}}\right)}^{-1},$$

$$\overline{c_{1122}^{\rm ortho}} = \overline{\left(c_{1122} - \frac{c_{1133}\,c_{2233}}{c_{3333}}\right)} + \overline{\left(\frac{c_{1133}}{c_{3333}}\right)}\,\overline{\left(\frac{c_{2233}}{c_{3333}}\right)}\,\overline{\left(\frac{1}{c_{3333}}\right)}^{-1},$$

$$\overline{c_{1133}^{\rm ortho}} = \overline{\left(\frac{c_{1133}}{c_{3333}}\right)}\,\overline{\left(\frac{1}{c_{3333}}\right)}^{-1},$$

$$\overline{c_{1313}^{\rm ortho}} = \overline{\left(\frac{1}{c_{1313}}\right)}^{-1},$$

$$\overline{c_{1212}^{\rm ortho}} = \overline{c_{1212}}\,,$$

$$\overline{c_{2222}^{\rm ortho}} = \overline{\left(c_{2222} - \frac{c_{2233}^2}{c_{3333}}\right)} + \overline{\left(\frac{c_{2233}}{c_{3333}}\right)}^2 \overline{\left(\frac{1}{c_{3333}}\right)}^{-1},$$

$$\overline{c_{2233}^{\rm ortho}} = \overline{\left(\frac{c_{2233}}{c_{3333}}\right)}\,\overline{\left(\frac{1}{c_{3333}}\right)}^{-1},$$

$$\overline{c_{2323}^{\rm ortho}} = \overline{\left(\frac{1}{c_{2323}}\right)}^{-1},$$

$$\overline{c_{3333}^{\rm ortho}} = \overline{\left(\frac{1}{c_{3333}}\right)}^{-1},$$

which are the Backus parameters for layers exhibiting the orthotropic symmetry. As expected, by introducing appropriate relations among parameters, these expressions reduce to the case derived for the transversely isotropic layers, discussed in Section 4.2.3, and—with further relations—to the case derived for isotropic layers stated by parameters (4.37)–(4.42).

[25]Readers interested in identification of material symmetries might refer to Bóna *et al.* (2007a).

The requirement of parallel symmetry planes among layers imposes a limitation on applying this approach, which is not shared by the case of isotropic or transversely isotropic layers. However, this approach could be used to quantify the seismic behaviour of parallel layers with common vertical fractures. Also an insight could be gained by considering orthotropic layers interspersed with transversely isotropic and isotropic layers, which might be a reasonable representation of geological situations. Either approach could be followed by an effective-symmetry consideration, discussed in Section 4.1.

Perhaps a fitting remark to conclude this exercise, which is also a comment that is pertinent to both Chapter 3 and to the entire book, is a consideration of a model as a mathematical formulation, with its hypotheses and assumptions, to be examined in the context of the available accuracy of measurements for which that model is to serve as a quantitative analogy.

Exercise 4.7. Consider a stack of isotropic layers, where the S-wave velocity for each layer is $\sqrt{c_{2323}}$; herein, c_{2323} is the density-scaled elasticity parameter for a given layer.

Using the expression for the wavefront velocity of the SH waves, which in the equivalent transversely isotropic medium is (Slawinski, 2015, Section 9.2.3, expressions (9.2.18) and (9.2.22))

$$v_{SH}^{\overline{\mathrm{TI}}}(\vartheta) = \sqrt{c_{1212}^{\overline{\mathrm{TI}}} \sin^2 \vartheta + c_{2323}^{\overline{\mathrm{TI}}} \cos^2 \vartheta}\,, \tag{4.85}$$

where ϑ is the wavefront angle, show that if the values of c_{2323} are equal to each other for all layers, the behaviour of the SH waves in the resulting medium reduces to an isotropic one.

Also, for the c_{2323} parameters not equal to each other, comment on $v_{SH}^{\overline{\mathrm{TI}}}(\vartheta)$, for the propagation along the rotation-symmetry axis, which corresponds to $\vartheta = 0$.

Solution. If the values of c_{2323} are equal to each other for all layers, then—according to expressions (4.40) and (4.41)—we have

$$c_{1212}^{\overline{\mathrm{TI}}} = c_{2323}^{\overline{\mathrm{TI}}} = c_{2323}\,.$$

Setting $c_{1212}^{\overline{\mathrm{TI}}} = c_{2323}^{\overline{\mathrm{TI}}} = c_{2323}$ in expression (4.85), we obtain

$$v_{SH}^{\overline{\mathrm{TI}}} = \sqrt{c_{2323}}\,, \tag{4.86}$$

which—having no angular dependence—implies an isotropic behaviour of the SH waves, as required.

Along the symmetry axis, expression (4.85) is reduced to

$$v_{SH}^{\overline{\mathrm{TI}}}(\vartheta) = \sqrt{c_{2323}^{\overline{\mathrm{TI}}}}\,,$$

which, in view of expression (4.41), we write as

$$v_{SH}^{\overline{\mathrm{TI}}}(\vartheta) = \frac{1}{\sqrt{\overline{\left(\frac{1}{c_{2323}}\right)}^{-1}}}\,;$$

this average is the S-wave velocity along the rotation-symmetry axis.

Remark 4.5. The SH ray velocity in a transversely isotropic medium is (Slawinski, 2015, Exercise 8.5, expression (8.7.13)),

$$V(\theta) = \sqrt{\frac{c_{2323}^{\overline{\mathrm{TI}}}\left(\tan^2\theta + 1\right)}{\frac{c_{2323}^{\overline{\mathrm{TI}}}}{c_{1212}^{\overline{\mathrm{TI}}}}\tan^2\theta + 1}}\,, \tag{4.87}$$

where θ is the ray angle. The explicit and closed form of expression (4.87) is possible, since the SH waves in transversely isotropic media exhibit an elliptical velocity dependence, as shown by expression (4.85). Also, for the SH waves in transversely isotropic media (Slawinski, 2015, Exercise 9.10, expression (9.4.8)),

$$\tan\theta = \frac{c_{1212}^{\overline{\mathrm{TI}}}}{c_{2323}^{\overline{\mathrm{TI}}}}\tan\vartheta\,. \tag{4.88}$$

The ray-velocity dependences of the qP and qSV waves on direction, even in transversely isotropic media, is more complicated and does not allow for closed-form expressions, as illustrated in Exercise 4.13.

Exercise 4.8. Consider a stack of isotropic layers, where c_{1111} and c_{2323} stand for the density-scaled elasticity parameters for each layer.

Using the expression for the wavefront velocity of the qP and qSV waves, which in the equivalent transversely isotropic medium is (Slawinski, 2015, Section 9.2.3, expressions (9.2.19)–(9.19.23))

$$v(\vartheta) = \sqrt{\frac{\left(c_{3333}^{\overline{\mathrm{TI}}} - c_{1111}^{\overline{\mathrm{TI}}}\right)\cos^2\vartheta + c_{1111}^{\overline{\mathrm{TI}}} + c_{2323}^{\overline{\mathrm{TI}}} \pm \sqrt{\Delta}}{2}}\,, \tag{4.89}$$

where

$$\Delta := \left(\left(c_{1111}^{\overline{\mathrm{TI}}} - c_{2323}^{\overline{\mathrm{TI}}}\right)\sin^2\vartheta - \left(c_{3333}^{\overline{\mathrm{TI}}} - c_{2323}^{\overline{\mathrm{TI}}}\right)\cos^2\vartheta\right)^2 + 4\left(c_{2323}^{\overline{\mathrm{TI}}} + c_{1133}^{\overline{\mathrm{TI}}}\right)^2\sin^2\vartheta\cos^2\vartheta\,,$$

and ϑ is the wavefront angle, show that if the values of c_{1111} are equal to each other for all layers and the values of c_{2323} are equal to each other for all layers, the equivalent behaviour of the qP and qSV waves reduces to an isotropic one. Expression (4.89) corresponds to the qP waves for $+\sqrt{\Delta}$ and to the qSV waves for $-\sqrt{\Delta}$.

Solution. If the values of c_{1111} and c_{2323} remain the same for all layers, expressions (4.37), (4.39), (4.38) and (4.42) become

$$c_{1111}^{\overline{\mathrm{TI}}} = \left(\frac{c_{1111} - 2c_{2323}}{c_{1111}}\right)^2 c_{1111} + 4\frac{(c_{1111} - c_{2323})c_{2323}}{c_{1111}} = c_{1111}\,,$$

$$c_{1133}^{\overline{\mathrm{TI}}} = c_{1111} - 2c_{2323}\,,$$

$$c_{1122}^{\overline{\mathrm{TI}}} = \left(\frac{c_{1111} - 2c_{2323}}{c_{1111}}\right)^2 c_{1111} + 2\frac{(c_{1111} - 2c_{2323})c_{2323}}{c_{1111}} = c_{1111} - 2c_{2323}$$

and

$$c_{3333}^{\overline{\mathrm{TI}}} = c_{1111}\,;$$

also, according to Exercise 4.7, expression (4.41) is $c_{2323}^{\overline{\mathrm{TI}}} = c_{2323}$. Thus, expression (4.89) becomes

$$v = \sqrt{\frac{c_{1111} + c_{2323} \pm \sqrt{\Delta}}{2}}\,,$$

with

$$\Delta = (c_{1111} - c_{2323})^2\,;$$

hence, $v_{qP} = \sqrt{c_{1111}}$ and $v_{qS} = \sqrt{c_{2323}}$, which, as required, are tantamount to the isotropic cases for the P-wave and S-wave velocities.

Remark 4.6. Averages (4.41) and (4.42), respectively, imply that—along the rotation-symmetry axis, where $\vartheta = 0$—the ray and wavefront velocities are

$$V_{qSV}^2(0) = V_{SH}^2(0) = v_{qSV}^2(0) = v_{SH}^2(0) = \overline{\left(\frac{1}{v_S^2}\right)}^{-1}$$

and

$$V_{qP}^2(0) = v_{qP}^2(0) = \overline{\left(\frac{1}{v_P^2}\right)}^{-1}\,,$$

where v_S and v_P stand for values in the individual layers. Along this axis, there is no distinction between the SH and qSV waves, since—for both waves—the polarization is horizontal and contained within the plane that is normal to the rotation-symmetry axis, which is isotropic; along this axis, qSV becomes SH.

Layer	c_{1111}	c_{2323}	v_P	v_S
1	10.56	2.02	3.25	1.42
2	20.52	4.45	4.53	2.11
3	31.14	2.89	5.58	1.70
4	14.82	2.62	3.85	1.62
5	32.15	2.92	5.67	1.71
6	16.00	2.56	4.00	1.60
7	16.40	6.35	4.05	2.52
8	18.06	4.33	4.25	2.08
9	31.47	8.01	5.61	2.83
10	17.31	3.76	4.16	1.94

Table 4.1: Density-scaled elasticity parameters, whose units are $10^6\,\mathrm{m^2s^{-2}}$, for a stack of isotropic layers, and the corresponding P-wave and S-wave velocities in $\mathrm{km\,s^{-1}}$.

Exercise 4.9. Given the density-scaled elasticity parameters stated in Table 4.1, use expressions (4.37)–(4.42) to obtain the corresponding Backus (1962) parameters. Use the arithmetic average.

Solution. Following expressions (4.37)–(4.42) and facilitating calculations with the aid of a computer, we obtain the Backus (1962) parameters for the arithmetic average: $\overline{c^{\mathrm{TI}}_{1111}} = 18.84$, $\overline{c^{\mathrm{TI}}_{1133}} = 10.96$, $\overline{c^{\mathrm{TI}}_{1212}} = 3.99$, $\overline{c^{\mathrm{TI}}_{2323}} = 3.38$ and $\overline{c^{\mathrm{TI}}_{3333}} = 18.43$; these values are multiplied by 10^6, and their units are $\mathrm{m^2s^{-2}}$.

Exercise 4.10. Given parameters stated in Table 4.1 and expressions (4.37)–(4.42), illustrate the moving average. For that purpose, let the size of the averaging window be equal to a width of six layers and each step be equal to the thickness of a layer. Plot the result.

Solution. The first position of the averaging window extends from the surface to the interface with the seventh layer. We use the elasticity parameters in Table 4.1 for the layers 1 to 6, and—by expressions (4.37)–(4.42)—we obtain $\overline{c^{\mathrm{TI}}_{1111}} = 18.30$, $\overline{c^{\mathrm{TI}}_{3333}} = 17.83$, $\overline{c^{\mathrm{TI}}_{1133}} = 12.31$, $\overline{c^{\mathrm{TI}}_{1212}} = 2.91$, $\overline{c^{\mathrm{TI}}_{2323}} = 2.75$; these values are multiplied by 10^6, and their units are $\mathrm{m^2s^{-2}}$. We associate these values with the centre of the averaging window, which, herein, is the interface between the third layer and the fourth layer. We repeat this process at four other, equally spaced, locations. The result is shown in Figure 4.4.

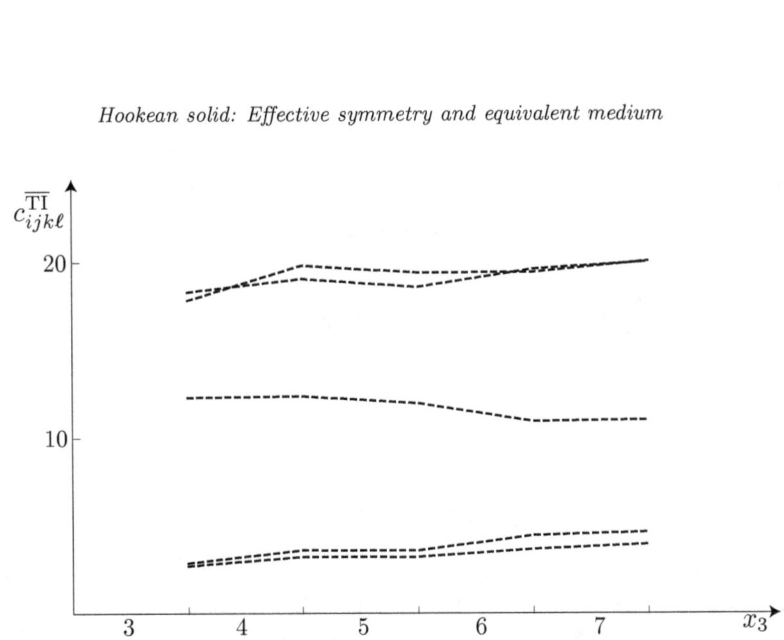

Fig. 4.4: Plot of the elasticity parameters for the Backus (1962) moving average, in the solution of Exercise 4.10: The horizontal axis corresponds to depth in terms of layers, with ticks representing interfaces between them; the vertical axis corresponds to the Backus parameters, with values to be scaled by 10^6. The dashed lines correspond to the values of the elasticity parameters in equivalent media, as obtained using the averaging window centred at each interface from the interface between the third and fourth layer to the interface between the seventh and eighth layer. These lines are dashed to indicate that they represent only the straight-line interpolations between the five computed values that are positioned on these interfaces, which is not the case of Figure 4.6, below, where the values are given continuously along the x_3-axis.

Examining this figure, we can conclude that—in view of the similarity of values for $c^{\overline{\mathrm{TI}}}_{1111}$ and $c^{\overline{\mathrm{TI}}}_{3333}$ as well as for $c^{\overline{\mathrm{TI}}}_{1212}$ and $c^{\overline{\mathrm{TI}}}_{2323}$, which are equal to one another for isotropy—herein, the equivalent medium is only weakly anisotropic. This conclusion is supported by results of Exercise 4.15.

Exercise 4.11. Given the density-scaled elasticity parameters stated in Table 4.1 and adjusting expressions (4.37)–(4.42), obtain the parameters of the equivalent medium using the standard normal distribution as the weight of the averaging operator. Let the span of one standard deviation, on either side of the mean, be equal to the thickness of ten layers.

Solution. Since—for the standard normal distribution—68.27% of the area under the curve lies within one standard deviation on either side of

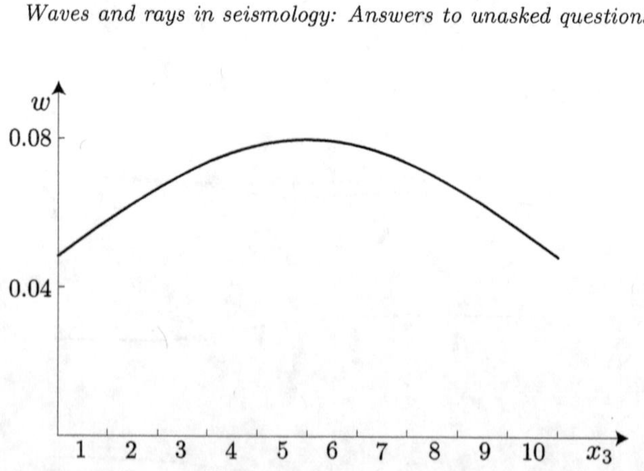

Fig. 4.5: A plot of expression (4.90) for the solution of Exercise 4.11: A standard normal distribution as the weight function, w. The horizontal axis corresponds to depth in terms of layers; the ticks represent interfaces between layers. The value of the integral of this curve—covering the ten layers—is 0.6827; $\int_{-\infty}^{\infty} w(x_3)\,\mathrm{d}x_3 = 1$, as required by expression (4.45).

the mean, we let the effect of each layer on the average be the ratio of the value, w_i, and 0.6827, where the former is the value of the distribution curve in the middle of a given layer. For instance, following expression (4.43), we write

$$\overline{c_{1212}^{\mathrm{TI}}} = \frac{\sum\limits_{i=1}^{10} w_i\,(c_{2323})_i}{0.6827}\,;$$

the expressions for other parameters exhibit similar patterns with weights for quantities to be averaged. To find the values of w_i for each layer, we write the standard normal distribution as

$$w = \frac{1}{5\sqrt{2\pi}} \exp\left(-\frac{(x-5)^2}{2(5)^2}\right), \tag{4.90}$$

where both the mean, which appears in the numerator, and the standard deviation, which appears in the denominators, is 5. Using this expression, which is illustrated in Figure 4.5, we find the weight values in the centre of each layer. Dividing each value by 0.6827, we get $\hat{w}_1 = 0.0780$, $\hat{w}_2 = 0.0915$, $\hat{w}_3 = 0.1031$, $\hat{w}_4 = 0.1117$, $\hat{w}_5 = 0.1163$, $\hat{w}_6 = 0.1163$, $\hat{w}_7 = 0.1117$, $\hat{w}_8 = 0.1031$, $\hat{w}_9 = 0.0915$ and $\hat{w}_{10} = 0.0780$.

To include these weights in expressions (4.37)–(4.42), we replace each average by its weighted analogue. To do so, we view—for each layer—the weights and the quantities to be averaged as components of vectors. Let us exemplify it with expression (4.41),

$$c_{2323}^{\overline{\mathrm{TI}}} = \overline{\left(\frac{1}{c_{2323}}\right)}^{-1} .$$

Using the weights, $\hat{w}_i$, and the values from Table 4.1, we write their scalar product as

$$c_{2323}^{\overline{\mathrm{TI}}} = \left([0.0780, \ldots, 0.0780] \cdot \left[\frac{1}{2.02}, \ldots, \frac{1}{3.76}\right]^T\right)^{-1} = 3.3750 ,$$

where T denotes the transpose. The remaining values are $c_{3333}^{\overline{\mathrm{TI}}} = 18.6879$, $c_{1111}^{\overline{\mathrm{TI}}} = 18.9545$, $c_{1122}^{\overline{\mathrm{TI}}} = 12.8219$, $c_{1133}^{\overline{\mathrm{TI}}} = 11.1807$, $c_{1212}^{\overline{\mathrm{TI}}} = 3.9746$; these values are multiplied by 10^6, and their units are $\mathrm{m^2 s^{-2}}$. They are analogous to the values obtained in Exercise 4.9, where all ten layers are considered with equal weights, which—for $c_{2323}^{\overline{\mathrm{TI}}}$—is tantamount to

$$c_{2323}^{\overline{\mathrm{TI}}} = \left([0.1, \ldots, 0.1] \cdot \left[\frac{1}{2.02}, \ldots, \frac{1}{3.76}\right]^T\right)^{-1} = 3.3786 .$$

By analogy to Figure 4.5, in Exercise 4.9, w could be illustrated by a boxcar, whose length is 10, height is $1/10$, and hence, $\int_{-\infty}^{\infty} w(x_3)\,\mathrm{d}x_3 = 1$, as required by expression (4.45); in that exercise—since a boxcar is compactly supported: $w = 0$ outside of a compact set—this integral is tantamount to the integral over ten layers.

Exercise 4.12. Consider the c_{2323} values in Table 4.1; also, on either side of this ten-layer sequence, let there be two halfspaces, whose $c_{2323} = 3.99 \cdot 10^6$, which is the arithmetic average of the layer values. Following expression (4.40) and definition (4.44), plot the moving average for $c_{1212}^{\overline{\mathrm{TI}}}$. In that definition, let the weight function, w, be the standard normal distribution, stated in expression (4.90), except with the mean being equal to zero, and f be the stepwise function representing the c_{2323} values. Use a computer to calculate the convolution of w and f. Discuss the results.

Solution. Expression (4.44) is a definition of the convolution. In functional analysis, convolution is an operation on two functions that results in a third function. Geometrically, that result is the changing amount of the overlap between these functions due to a translation of one of them.

Commonly, this result is interpreted as a modification of one of the original functions. If the arguments are temporal variables, the convolution is interpreted as filtering, with the resulting function being the filtered version of the original one; in such a case, the other function is the filter.

If the arguments are spatial variables, the convolution is interpreted as a moving average, with the resulting function being the averaged version of the original one, and the other being the averaging operator. Both interpretations appear in seismology, in particular, in signal analysis and in modelling of materials, respectively. In the former case, the modified function is a filtered signal. In the latter, it is an averaged counterpart of the material properties.

Expression (4.44) is the convolution of f and $w(-x_3)$. Since the mirror image of $w(x_3)$, which is $w(-x_3)$, has the same properties as $w(x_3)$, and x_3 is a spatial variable, that expression can be viewed as an average of f due to the moving averaging operator, w. Let us examine the convolution illustrated in Figure 4.6, which is obtained with the aid of a computer.

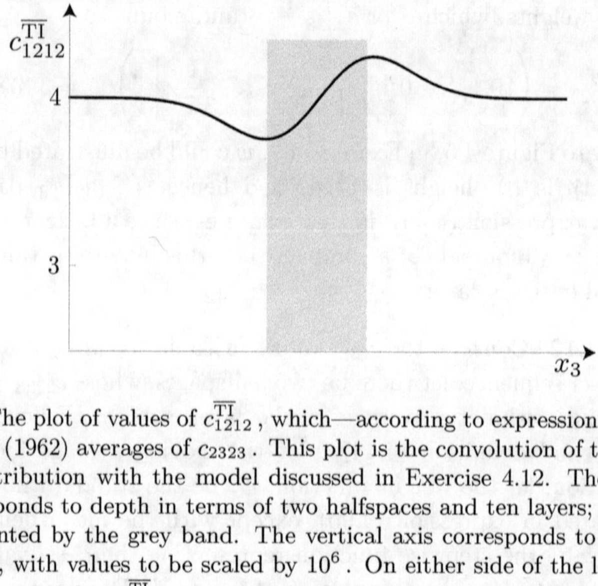

Fig. 4.6: The plot of values of $c_{1212}^{\overline{\mathrm{TI}}}$, which—according to expression (4.40)—are the Backus (1962) averages of c_{2323}. This plot is the convolution of the standard normal distribution with the model discussed in Exercise 4.12. The horizontal axis corresponds to depth in terms of two halfspaces and ten layers; these layers are represented by the grey band. The vertical axis corresponds to the Backus parameters, with values to be scaled by 10^6. On either side of the layering, the average values tend to $c_{1212}^{\overline{\mathrm{TI}}} = c_{2323} = 3.99$, which is the value for the halfspaces.

Since, according to expression (4.40), $c_{1212}^{\overline{\mathrm{TI}}}$ is only a function of c_{2323i}, where i is the number of a given layer, let us examine, in the context of Figure 4.6, the values of c_{2323} for each layer in Table 4.1.

The flat segments of the curve correspond to the two halfspaces, since, as expected, the average for a homogeneous medium is equal to the value exhibited by that medium. The maximum of the $c_{1212}^{\overline{\mathrm{TI}}}$ curve is due to the high values of c_{2323} within individual layers, as exhibited, in particular, by the seventh and the ninth layer.

From these observations, we can infer—with limitations due to resolution of the averaging operator—certain information about the parameters within layers. Also, the average provides us with information about the long-wavelength trends, along the x_3-axis, as indicated by the flat segments corresponding to the homogeneous halfspaces.

Perhaps most importantly, however, this average is a good analogy for the scale of resolution achievable with seismic data. It illustrates the fact that—as an effect of the seismic-wave propagation, which is akin to the averaging process—even abrupt changes, marked by interfaces between thin layers, appear smooth on a seismic record.

Remark 4.7. The larger the window, the fewer the details in the resulting averaging. To illustrate it, let us consider the model discussed in Exercise 4.12. In a manner presented in that exercise, in Figure 4.7, we plot the moving average for $\overline{c_{1212}^{\mathrm{TI}}}$, using the standard deviations of 1, 2.5, 5 and 10. As expected, the larger the standard deviation, the lesser the resolution; for instance, the value of 10 results in the presence of layers—which are represented by the grey band—being nearly undetectable. On the other hand, for the value of 1, we can infer certain properties of these layers.

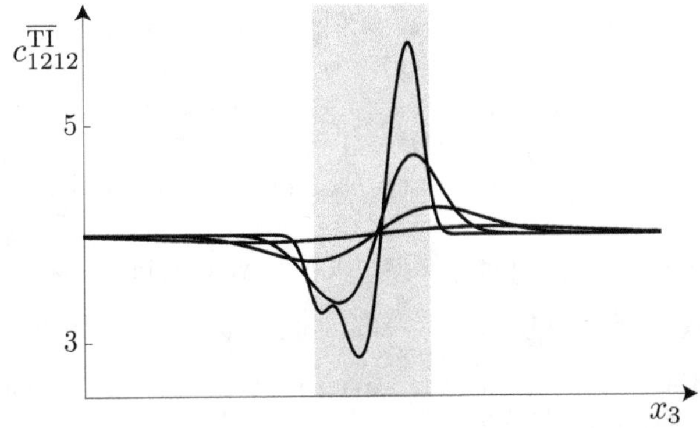

Fig. 4.7: The plot of values of $\overline{c_{1212}^{\mathrm{TI}}}$. These are the convolutions of the standard normal distribution with the model discussed in Exercise 4.12; the standard deviations of this distribution are 1, 2.5, 5 and 10. The horizontal and vertical axes are the same as in Figure 4.6.

Exercise 4.13. Using the values of the Backus (1962) parameters obtained in Exercise 4.9 and the relation between the wavefront and ray angles in transverse isotropy,

$$\tan\theta = \frac{\tan\vartheta + \frac{1}{v}\left(\frac{\partial v}{\partial\vartheta}\right)}{1 - \frac{\tan\vartheta}{v}\left(\frac{\partial v}{\partial\vartheta}\right)}, \tag{4.91}$$

with v given by expression (4.89), obtain—for the qP and qSV waves—the values of their wavefront angles, ϑ, that correspond to $\theta = \pi/4$; use a computer to obtain these results, since expression (4.91) is not soluble explicitly for ϑ.

For these values of the wavefront angles, use expression (4.89) to obtain the magnitudes of the corresponding wavefront velocities for the qP and qSV waves.

Subsequently, use the relation between the magnitudes of the wavefront and ray velocities (Slawinski, 2015, Section 8.4.2, expression (8.4.9)),

$$V(\vartheta) = \sqrt{(v(\vartheta))^2 + \left(\frac{\partial v(\vartheta)}{\partial\vartheta}\right)^2}, \tag{4.92}$$

with v given by expression (4.89), to obtain the magnitude of the qP and qSV ray velocities that corresponds to $\theta = \pi/4$.

Solution. We invoke expression (4.91), together with expression (4.89). Using the values of the Backus (1962) parameters from Exercise 4.9 and setting $\theta = \pi/4$, in expression (4.91), we use a computer to obtain $\vartheta_{qP} = 0.775$, for the qP waves, and $\vartheta_{qSV} = 0.789$, for the qSV waves; the former value corresponds to the case of $+\sqrt{\Delta}$ and the latter to $-\sqrt{\Delta}$, in expression (4.89).

Using these results in expression (4.89), we obtain

$$v_{qP} = 4.2631\,\mathrm{km\,s^{-1}} \qquad \text{and} \qquad v_{qSV} = 1.9588\,\mathrm{km\,s^{-1}},$$

which are the magnitudes of the wavefront velocities. Subsequently, using expression (4.92), we obtain

$$V_{qP}(\pi/4) = 4.2633\,\mathrm{km\,s^{-1}} \qquad \text{and} \qquad V_{qSV}(\pi/4) = 1.9588\,\mathrm{km\,s^{-1}},$$

which are the magnitudes of ray velocities.

The similarity of values between the corresponding ray and wavefront quantities, indicated by $\theta = \pi/4 \approx 0.785$, $\vartheta_{qP} = 0.775$ and $\vartheta_{qSV} = 0.789$, is a consequence of weak anisotropy, illustrated in Exercise 4.10, above, and discussed in Exercise 4.15, below. Also, in the case of the qSV waves, the similarity of the wavefront-velocity and ray-velocity magnitudes is a result of a property of the qSV wavefronts, which—in transversely isotropic media—exhibit a reflection symmetry about $\pi/4$.[26]

Exercise 4.14. Using expressions (4.85), (4.87), (4.88) and the values obtained in Exercise 4.9, find the SH wavefront and ray velocities that correspond to $\theta = \pi/4$.

Solution. Following expression (4.88) with $\theta = \pi/4$, we obtain $\vartheta = 0.703$; the corresponding wavefront velocity, given by expression (4.85), is $v_{SH}(\vartheta) = 1.907\,\mathrm{km\,s^{-1}}$, and the corresponding ray velocity, given by expression (4.87), is

$$V_{SH}\left(\frac{\pi}{4}\right) = \sqrt{\frac{2c_{2323}^{\overline{\mathrm{TI}}}}{\dfrac{c_{2323}^{\overline{\mathrm{TI}}}}{c_{1212}^{\overline{\mathrm{TI}}}} + 1}} = 1.913\,\mathrm{km\,s^{-1}}.$$

Remark 4.8. The difference between $V_{qSV}(\pi/4) = 1.959$, obtained in Exercise 4.13, and $V_{SH}(\pi/4) = 1.913$—albeit small due to weak anisotropy discussed in Exercise 4.15, below—is an example of the *shear-wave splitting*: a term originating from the perspective of isotropy, where both shear waves coincide.

Remark 4.9. Using the relation between the magnitudes of the velocities and angles (Slawinski, 2015, Section 8.4.4, expression (8.4.14)),

$$V(\theta) = \frac{v(\vartheta)}{\cos(\theta - \vartheta)},$$

one could verify the consistency of values obtained in Exercises 4.13 and 4.14, where

$$V\left(\frac{\pi}{4}\right) = \frac{v(\vartheta)}{\cos\left(\dfrac{\pi}{4} - \vartheta\right)},$$

for each wave.

Exercise 4.15. Use the Backus (1962) and Voigt (1910) averages to obtain the isotropic Hookean solid. The former approximates a series of parallel isotropic layers by a single transversely isotropic medium; the latter approximates a transversely isotropic medium by its isotropic counterpart.

[26] For an illustration of the qSV wavefront-slowness curves, readers might refer to Slawinski (1996, Appendix 2).

Thus, using the values of the Backus (1962) parameters from Exercise 4.9, obtain the values of $c_{1111}^{\overline{\rm iso}}$ and $c_{2323}^{\overline{\rm iso}}$, and of the corresponding P-wave and S-wave velocities.

Comment on the reduction of the Backus (1962) model to the Voigt (1910) model, by invoking the Thomsen (1986) parameters,

$$\gamma := \frac{c_{1212}^{\overline{\rm TI}} - c_{2323}^{\overline{\rm TI}}}{2c_{2323}^{\overline{\rm TI}}}\,, \tag{4.93}$$

$$\delta := \frac{\left(c_{1133}^{\overline{\rm TI}} + c_{2323}^{\overline{\rm TI}}\right)^2 - \left(c_{3333}^{\overline{\rm TI}} - c_{2323}^{\overline{\rm TI}}\right)^2}{2c_{3333}^{\overline{\rm TI}}\left(c_{3333}^{\overline{\rm TI}} - c_{2323}^{\overline{\rm TI}}\right)}\,, \tag{4.94}$$

$$\epsilon := \frac{c_{1111}^{\overline{\rm TI}} - c_{3333}^{\overline{\rm TI}}}{2c_{3333}^{\overline{\rm TI}}}\,, \tag{4.95}$$

which become zero in the case of isotropy, and by examining traveltimes in Table 4.2.

Solution. Consider expressions (4.30) and (4.31), in the context of transverse isotropy; we obtain

$$c_{1111}^{\overline{\rm iso}} = \frac{1}{15}\left(8c_{1111}^{\overline{\rm TI}} + 4c_{1133}^{\overline{\rm TI}} + 8c_{2323}^{\overline{\rm TI}} + 3c_{3333}^{\overline{\rm TI}}\right) \tag{4.96}$$

and

$$c_{2323}^{\overline{\rm iso}} = \frac{1}{15}\left(c_{1111}^{\overline{\rm TI}} - 2c_{1133}^{\overline{\rm TI}} + 5c_{1212}^{\overline{\rm TI}} + 6c_{2323}^{\overline{\rm TI}} + c_{3333}^{\overline{\rm TI}}\right)\,; \tag{4.97}$$

the required expressions for $c_{ijk\ell}^{\overline{\rm TI}}$ are in expressions (4.37)–(4.42). For the values of the Backus (1962) parameters obtained in Exercise 4.9, we have $c_{1111}^{\overline{\rm iso}} = 18.46\times10^6$ and $c_{2323}^{\overline{\rm iso}} = 3.71\times10^6$, and the corresponding P-wave and S-wave speeds—which are the square roots of the parameters—are $v_P = 4.30\,{\rm km\,s}^{-1}$ and $v_S = 1.93\,{\rm km\,s}^{-1}$.

The values of the Thomsen (1986) parameters are $\gamma = 0.09$, $\delta = -0.04$ and $\epsilon = 0.01$. Since these values are close to zero, we conclude that—for the layer parameters in Table 4.1—the resulting Backus (1962) model is only weakly anisotropic.

To comment on the Voigt (1910) reduction, let us compare traveltimes in Table 4.2. The $\overline{qP}$-wave traveltimes are similar to the $\overline{P}$-wave traveltimes, which is consistent with very small values of δ and ϵ: the Thomsen (1986) parameters relating velocities of these waves. The $\overline{SH}$-wave and $\overline{S}$-wave

Receiver	P	S	$\overline{qP}$	$\overline{qSV}$	$\overline{SH}$	$\overline{P}$	$\overline{S}$
1	229.4	534.0	232.9	544.0	544.0	232.7	519.4
2	230.5	536.6	234.1	545.8	546.4	233.9	522.1
3	233.8	544.2	237.8	551.2	553.3	237.4	529.8
4	239.2	556.6	243.9	560.0	564.5	243.0	542.4
5	246.2	573.4	252.0	572.4	580.0	250.7	559.6
6	255.6	594.1	262.0	588.3	599.0	260.3	581.0
7	266.2	618.2	273.5	607.7	621.6	271.5	606.0
8	278.0	645.1	286.5	630.6	647.3	284.2	634.3
9	291.0	674.4	300.6	657.3	675.7	298.2	665.2
10	304.9	705.5	315.8	687.8	706.6	313.2	699.2
11	319.6	738.0	331.8	722.2	739.5	329.3	734.9

Table 4.2: Traveltimes, in milliseconds, through a stack of isotropic layers whose properties are given in Table 4.1 and whose thicknesses are 100 metres. The source is directly opposite the first receiver, and other receivers are spaced in 100-metre intervals. The P-wave and S-wave traveltimes correspond to the layered medium and are obtained by invoking Fermat's principle. The remaining traveltimes, indicated by bars, refer to the averaged homogeneous medium; they are the ratios of distance and, respectively, V_{qP} , V_{qSV} ,V_{SH} , for the Backus (1962) model, and v_P , v_S , for the Voigt (1910) model. Except for the cases of V_{SH} , v_P and v_S , obtaining the traveltimes relies on numerical computations to find the Fermat traveltimes or to get $V_{qP}(\theta)$ and $V_{qSV}(\theta)$.[27]

traveltimes are quite distinct from one another, which is consistent with a larger value of γ, which directly relates velocities of these waves. The $\overline{qSV}$-wave traveltimes are also quite distinct from the $\overline{S}$-wave traveltimes; they are related by δ and ϵ, but the presence of two types of secondary waves in anisotropy renders the relation more subtle than in the case of the primary waves that results in similarity between the $\overline{qP}$-wave and $\overline{P}$-wave traveltimes.

Remark 4.10. [28]Thomsen (1986) parameters can be used to investigate the effects of inhomogeneity between layers on anisotropy of a homogeneous medium resulting from the Backus (1962) average. In general, as the inhomogeneity within a stack of layers increases, so does the anisotropy of the medium.

[27]Values in Table 4.2 are obtained using the code of Brisco (2014).
[28]Remark 4.10 is based on Adamus *et al.* (2018).

To exemplify this increase, let us consider a stack of identical isotropic layers. To introduce inhomogeneity, we multiply the two elasticity parameters of every second layer by a. Considering two adjacent layers, c_{1111}, c_{2323} and $a\,c_{1111}$, $a\,c_{2323}$, we use expressions (4.37)–(4.42) to obtain the parameters of a transversely isotropic medium, $c^{\overline{\mathrm{TI}}}_{ijk\ell}$, which we use in expressions (4.93)–(4.95) to obtain

$$\gamma = \frac{(a-1)^2}{8\,a}\,, \tag{4.98}$$

$$\delta = 0\,,$$

$$\epsilon = \frac{(a-1)^2\,(c_{1111}-c_{2323})\,c_{2323}}{2\,a\,c_{1111}^2}\,. \tag{4.99}$$

In contrast to parameters (4.93) and (4.95), their counterparts (4.98) and (4.99) can be only nonnegative. Herein, $\delta = 0$ is a consequence of alternating layers whose both parameters are scaled by the same value of a; it is not a general property for alternating isotropic layers in the context of the Backus (1962) average.

If $a = 1$, which means that all layers are the same, then also $\gamma = \epsilon = 0$; hence, in such a case, the averaged medium is isotropic, as expected. If $a \to 0$ or $a \to \infty$, which is tantamount to increasing inhomogeneity between layers, then γ and ϵ tend to infinity; in such a case, the averaged medium is extremely anisotropic.

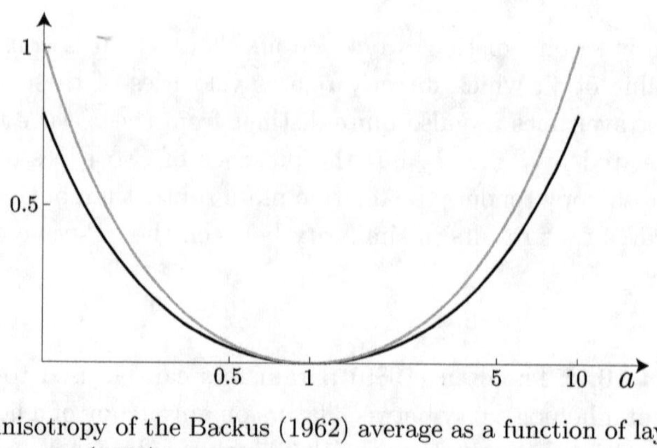

Fig. 4.8: Anisotropy of the Backus (1962) average as a function of layer inhomogeneity: Thomsen (1986) parameters, γ and ϵ, plotted as grey and black lines, respectively, against logarithmic values of $a \in (10^{-1}, 10^{1})$.

To illustrate the relationship between inhomogeneity and anisotropy, let us consider a numerical example. We use $c_{1111} = 12.15$ and $c_{2323} = 3.24$, which are density-scaled elasticity parameters that correspond to sandstone. Their SI units are $\mathrm{km}^2/\mathrm{s}^2$, and their square roots are P-wave and S-wave speeds, respectively. Figure 4.8 illustrates a monotonic increase in anisotropy of the averaged medium with an increase of inhomogeneity between layers. At $a = 1$, which means that all layers are the same, $\gamma = \epsilon = 0$. As a tends to zero or to infinity, γ and ϵ tend to infinity. For $a \in (10^{-1}, 10^{0})$, the values of the elasticity parameters of the alternating layer are progressively diminished by up to one order of magnitude; for $a \in (10^{0}, 10^{1})$, they are progressively increased by up to one order.

For the SH and qP waves, respectively, γ and ϵ are measures of difference between propagation speeds along, and perpendicular to, the layers,

$$\frac{v_{\parallel}^2 - v_{\perp}^2}{2v_{\perp}^2}\,.$$

Parameter δ, whose definition does not have such a geometrical interpretation, remains equal to zero. If, however, the elasticity parameters of the alternate layers are $a\,c_{1111}$ and $\sqrt{a}\,c_{2323}$, δ asymptotically approaches a finite value, as a tends to infinity; γ and ϵ still tend to infinity and, as such, they are symptomatic of inhomogeneity among layers.

As illustrated in Figure 4.8, for a stack of isotropic layers, the strength of anisotropy of the resulting transversely isotropic medium is solely a function of inhomogeneity of that stack. In other words, herein, the strength of anisotropy is a measure of inhomogeneity.

A rather slow increase of values of γ and ϵ as functions of a supports the adequacy of weakly anisotropic models in many quantitative studies in seismology. Herein, according to the Backus (1962) average, even moderately inhomogeneous alternating layers result only in a weakly anisotropic medium.

Let us conclude this remark by discussing the case of inhomogeneous layers that—following the Backus (1962) average—result in an isotropic homogeneous medium. To do so, we let the inhomogeneity among layers be expressed only by differences in c_{1111}; c_{2323} remains constant. Following expression (4.37), (4.39)–(4.42), and using a symbolic-calculation software, we obtain

$$c_{1111}^{\mathrm{TI}} = \overline{\left(\frac{1}{c_{1111}}\right)}^{-1},$$

$$c_{1133}^{\mathrm{TI}} = \overline{\left(\frac{1}{c_{1111}}\right)}^{-1} - 2c_{2323},$$

$$c_{1212}^{\overline{\mathrm{TI}}} = c_{2323}\,,$$

$$c_{2323}^{\overline{\mathrm{TI}}} = c_{2323}\,,$$

$$c_{3333}^{\overline{\mathrm{TI}}} = \overline{\left(\frac{1}{c_{1111}}\right)}^{-1}\,,$$

respectively. Since $c_{1111}^{\overline{\mathrm{TI}}} = c_{3333}^{\overline{\mathrm{TI}}}$, $c_{1212}^{\overline{\mathrm{TI}}} = c_{2323}^{\overline{\mathrm{TI}}}$ and $c_{1133}^{\overline{\mathrm{TI}}} = c_{1111}^{\overline{\mathrm{TI}}} - 2c_{2323}^{\overline{\mathrm{TI}}}$, the medium is isotropic; consequently, its three Thomsen (1986) parameters, γ, δ, ϵ, are zero.

This result indicates that the anisotropy of the Backus (1962) average is the consequence of the rigidity of the layers, but not of their compressibility.[29] A physical counterpart of such a mathematical model might be a porous rock of constant rigidity, whose compressibility varies depending on the amount of liquid within its pores. Following such a physical interpretation, and according to the Backus (1962) average, the level of saturation has no effect on the isotropy of the medium, even though it has an effect on the value of $c_{1111}^{\overline{\mathrm{TI}}}$ and mass density, whose values, in turn, determine the P-wave propagation speed.

[29] *see also*: Slawinski (2015, Remark 5.12.4)

Chapter 5

Body waves

It is true that the mechanics of systems of finite number of mass points has been on a sufficiently rigorous basis since Newton. Many textbooks on theoretical mechanics dismiss continuous bodies with the remark that they can be regarded as the limiting case of a particle system with an increasing number of particles. They cannot. [...] *It is known that continuous matter is really made up of elementary particles. The basic laws governing the elementary particles are those of quantum mechanics. The science that provides the link between these basic laws and the laws describing the behaviour of gross matter is statistical mechanics.*[1]

Walter Noll (1974a)

Preliminary remarks

In this chapter, we discuss the description of dynamical processes within the realm of continuum mechanics. We examine these processes in the context of Hookean solids.

We begin this chapter with a formulation of the equation of motion within Hookean solids, the elastodynamic equation, to discuss the resulting body waves.

We proceed with formulations of wave equations whose derivations do not originate in the elastodynamic equation. Such formulations allow us to examine aspects of inhomogeneity and friction, which do not appear in the classic derivation of the wave equation.[2]

To conclude this chapter, we examine several methods of solution for the wave equation, and comment on relations between physical phenomena

[1] Readers interested in "the mechanical view" of thermodynamics might refer to Berdichevsky (2009, Chapter 2).

[2] *see also*: Slawinski (2015, Chapter 6)

and their analogies in the mathematical realm. In examining solutions, we restrict our discussion to isotropic homogeneous solids, except in the final section, where we comment on approximations of solutions for elastodynamic equations of anisotropic inhomogeneous Hookean solids.

5.1 Wave equations

5.1.1 *Assumptions and formulation*

To relate the field equation that describes, in general, motion within continua to the particular type of materials of interest, which herein are the Hookean solids, we consider Cauchy's equations of motion (2.16), which stem from the hypothesis of the balance of linear momentum, in the context of Hooke's law (3.1); we write[3,4]

$$\rho\left(\mathbf{x}\right)\frac{\partial^2 u_i\left(\mathbf{x},t\right)}{\partial t^2}=\sum_{j=1}^{3}\sum_{k=1}^{3}\sum_{\ell=1}^{3}\frac{\partial c_{ijk\ell}\left(\mathbf{x}\right)}{\partial x_j}\frac{\partial u_k\left(\mathbf{x},t\right)}{\partial x_\ell}+\sum_{j=1}^{3}\sum_{k=1}^{3}\sum_{\ell=1}^{3}c_{ijk\ell}\left(\mathbf{x}\right)\frac{\partial^2 u_k\left(\mathbf{x},t\right)}{\partial x_j\partial x_\ell}, \tag{5.1}$$

where $i = 1,2,3$. In equation (5.1), we ignore body forces, as discussed in Section 8.1, below, which is a consequence of the fact that—for sufficiently high frequencies, which are pertinent to seismic waves—the dominant effect in the balance of linear momentum is due to the contact forces. Thus, in the equation of motion for seismic waves, the effects of elasticity, expressed as contact forces of traction, dominate the effects of gravitation, expressed as body forces. Hence, strictly speaking, expression (5.1) is called the *elastodynamic equation* due to the force on the left-hand side being balanced—on the right-hand side—by the elasticity effects only.

In this chapter, we focus our attention on waves propagating in unbounded media, which we refer to as *body waves*. These media are unbounded within their spatial domains, which is $\mathbb{R}^3$ in both equation (5.1) and Section 5.1.2, below, and $\mathbb{R}^1$ in Sections 5.1.3 and 5.1.4. In Chapter 6, we discuss waves in bounded media; they are *surface waves*, *guided waves* and *interface waves*.

[3]Readers interested in a formulation of equation (5.1) might refer to Bóna *et al.* (2010, Section 2).

[4]*see also*: Slawinski (2015, Section 7.1) and Bóna and Slawinski (2015, Appendix B.3, equation (B.24))

5.1.2 *Particular case: Isotropy and homogeneity*

Apart from the main subject of this section, which is a formulation of the P-wave and S-wave equations based on the derivation by Rochester (2010), this section is an illustration of the fact that quantitative seismology is a hypotheticodeductive formulation, whose theory is rooted in mathematics, with its required standards of theorems and proofs.

We show that equation (5.1)—restricted to the isotropic form of its elasticity tensor—entails two modes of propagation: the P and S waves. The necessary and sufficient conditions for their existence are stated in terms of vector-calculus expressions.

The properties of these waves, such as their propagation speed, which are quantities existing only within the realm of mathematics, can be used as analogies for physical properties of seismic disturbances, which are obtained from experiments. Also, from the necessary and sufficient conditions, which are only mathematical entities, we can infer information about the physical behaviour of these disturbances.

Let us consider a particular case of equation (5.1)—namely, the equation of motion in an isotropic homogeneous Hookean solid—which is used commonly in seismic studies. It is common to write this equation as[5]

$$\rho \frac{\partial^2 \mathbf{u}}{\partial t^2} = (\lambda + 2\mu)\, \nabla\varphi - \mu\, \nabla \times \boldsymbol{\Psi}\,, \tag{5.2}$$

where $\lambda := c_{1111} - 2c_{2323}$ and $\mu = c_{2323}$ are the two independent elasticity parameters of expression (3.57), herein expressed as the Lamé parameters, φ is the dilatation and $\boldsymbol{\Psi}$ is the rotation vector. Dilatation is defined as a scalar given by

$$\varphi := \frac{\partial u_1}{\partial x_1} + \frac{\partial u_2}{\partial x_2} + \frac{\partial u_3}{\partial x_3} =: \nabla \cdot \mathbf{u} \tag{5.3}$$

and the rotation vector is given by

$$\boldsymbol{\Psi} := \nabla \times \mathbf{u}\,; \tag{5.4}$$

similarly to other steps in the formulation of the elastodynamic equation, these definitions rely on linearization and the assumption of continuous and small deformations.[6] In view of geometrical interpretations of effects of these differential operators, $\nabla\cdot$ and $\nabla\times$, if $\varphi = 0$, the motion is called *equivoluminal* and if $\boldsymbol{\Psi} = \mathbf{0}$, it is called *irrotational.*

[5] *see also*: Slawinski (2015, Section 6.1.1, equation (6.1.9))
[6] *see also*: Slawinski (2015, Sections 1.4 and 1.5)

To formulate the wave equations for the P and S waves, and to show that the irrotational and equivoluminal displacements are, respectively, the necessary and sufficient conditions for the existence of these waves, we use definitions (5.3) and (5.4) to write equation (5.2) as

$$\rho\frac{\partial^2\mathbf{u}}{\partial t^2} = c_{1111}\nabla\left(\nabla\cdot\mathbf{u}\right) - c_{2323}\nabla\times\left(\nabla\times\mathbf{u}\right), \tag{5.5}$$

where—following Section 3.2.4.14—we express the Lamé parameters in terms of the elasticity-tensor components; c_{1111} and c_{2323} are the two linearly independent components of the isotropic elasticity tensor, whose components are shown in matrix (3.57). Writing in terms of $\mathbf{u}$, c_{1111} and c_{2323}, we see that equation (5.2) is a particular case of equation (5.1): its isotropic case.

Below, we follow the formulation of Landau and Lifshitz (1986, pp. 87-89) and Rochester (2010).[7]

If the displacement is irrotational, $\nabla\times\mathbf{u} = \mathbf{0}$, equation (5.5) is reduced to

$$\rho\frac{\partial^2\mathbf{u}}{\partial t^2} = c_{1111}\nabla\left(\nabla\cdot\mathbf{u}\right). \tag{5.6}$$

If we invoke the vector identity given by

$$\nabla^2\mathbf{u} = \nabla\left(\nabla\cdot\mathbf{u}\right) - \nabla\times\left(\nabla\times\mathbf{u}\right), \tag{5.7}$$

which—in view of $\mathbf{u}$ being a vector field—can be regarded as a definition of the *vector Laplacian*, a differential operator named in acknowledgement of contributions by Pierre-Simon de Laplace,[8] and which, in this case, is reduced to

$$\nabla^2\mathbf{u} = \nabla\left(\nabla\cdot\mathbf{u}\right), \tag{5.8}$$

equation (5.6) becomes

$$\nabla^2\mathbf{u} = \frac{1}{\frac{c_{1111}}{\rho}}\frac{\partial^2\mathbf{u}}{\partial t^2}, \tag{5.9}$$

which is the wave equation for P waves, with $\sqrt{c_{1111}/\rho}$ expressing their speed of propagation, as can be interpreted by examining the solution of that equation.[9] Thus, given equation (5.5), condition $\nabla\times\mathbf{u} = 0$ is sufficient for the displacement vector to result in the P-wave equation.

[7] *see also*: Slawinski (2015, Sections 6.1.2 and 6.1.3)

[8] Readers interested in the life and times of de Laplace might refer to McGrayne (2011, Chapter 2).

[9] *see also*: Slawinski (2015, Sections 6.5.1 and 6.5.4)

As an aside, let us address properties of the vector Laplacian. Only in Cartesian coordinates, its operation on $\mathbf{u}$ is the operation on each of its components alone, $(\nabla^2\mathbf{u}(\mathbf{x}))_i = \nabla^2 u_i(\mathbf{x})$.[10,11] In other coordinates, all components are involved; for instance, in spherical coordinates,

$$\left(\nabla^2\mathbf{u}(r,\alpha,\beta)\right)_r = \nabla^2 u_r - \frac{2}{r^2}\left(u_r + \cot\alpha\, u_\alpha + \frac{\partial u_\alpha}{\partial\alpha} + \csc\alpha\frac{\partial u_\beta}{\partial\beta}\right),$$

with the first term on the right-hand side given in expression (5.38), below; similar relations hold for $(\nabla^2\mathbf{u})_\alpha$ and $(\nabla^2\mathbf{u})_\beta$.

Also, it might be interesting to gain an insight into the geometrical meaning of the Laplacian. As shown in Exercises 5.1 and 5.2, the standard expression for the Laplacian of a scalar function at a given point is proportional to the difference between an average value of that function on the surface of a sphere, whose centre is that point, and the value at that point. If there is no difference, the Laplacian is zero. To illustrate it, we can write the Laplacian for a one-variable function in terms of the definition of the second derivative,

$$\left.\frac{\mathrm{d}^2 f(x)}{\mathrm{d}x^2}\right|_{x_0} = \lim_{h\to 0}\frac{\dfrac{f(x_0+h)-f(x_0)}{h} - \dfrac{f(x_0)-f(x_0-h)}{h}}{h}$$
$$= \lim_{h\to 0}\left(\frac{2}{h^2}\left(\frac{f(x_0-h)+f(x_0+h)}{2} - f(x_0)\right)\right),$$

which is equivalent to the Laplacian in one dimension, $\partial^2 f/\partial x^2$. Herein, the first term in the inner parentheses is an average of f over the interval centred at x_0 ; the second term is the value of that function at x_0 .

In general—and as shown in Exercise 5.1, for three spatial dimensions—if the value of a function at a point is the same as the average over surrounding points, the Laplacian at that point is zero. Otherwise, it is not. In view of a definition of the Laplacian as the divergence of the gradient, we see that a sufficient condition for $\nabla\cdot\nabla f \equiv \nabla^2 f = 0$ is $\nabla f = 0$, for which—in turn—a sufficient condition is the constancy of f in the neighbourhood of x_0 .

In the context of the wave equation, the Laplacian is a measure of a difference in displacement between a given point and its surroundings. It might be viewed as a measure of displacement of that point from an equilibrium position.

[10] *see also*: Slawinski (2015, Section 6.1.1, identity (6.1.7))

[11] Readers interested in the vector-calculus *pitfalls* might refer to Feynman *et al.* (1964, Vol. II, Section 2-8).

Let us return to equation (5.5); if the displacement is equivoluminal, $\nabla \cdot \mathbf{u} = 0$, this equation becomes

$$\rho \frac{\partial^2 \mathbf{u}}{\partial t^2} = -c_{2323} \nabla \times (\nabla \times \mathbf{u}) \,. \tag{5.10}$$

In this case, the aforementioned identity is reduced to

$$\nabla^2 \mathbf{u} = -\nabla \times (\nabla \times \mathbf{u}) \,, \tag{5.11}$$

and equation (5.6) becomes

$$\nabla^2 \mathbf{u} = \frac{1}{\frac{c_{2323}}{\rho}} \frac{\partial^2 \mathbf{u}}{\partial t^2} \,, \tag{5.12}$$

which is the wave equation for S waves with $\sqrt{c_{2323}/\rho}$ representing their propagation speed. Thus, condition $\nabla \cdot \mathbf{u} = 0$ is sufficient for the displacement vector to result in the S-wave equation.

In summary, requiring $\nabla \times \mathbf{u} = \mathbf{0}$ in equation (5.5) is a sufficient condition to obtain the P-wave equation, and requiring $\nabla \cdot \mathbf{u} = 0$ is a sufficient condition to obtain the S-wave equation.

The decomposition of equation (5.5) into equations (5.9) and (5.12) is an illustration of Helmholtz's decomposition theorem, named in recognition of work by a German physicist Hermann Ludwig Ferdinand von Helmholtz, which states that an arbitrary vector field, $\mathbf{u}(\mathbf{x})$ can be expressed as the sum of its longitudinal component, $\mathbf{u}_{\parallel}$, and its transverse component, $\mathbf{u}_{\perp}$, where $\nabla \times \mathbf{u}_{\parallel} = \mathbf{0}$ and $\nabla \cdot \mathbf{u}_{\perp} = 0$; in other words, any vector field can be decomposed into its curl-free part and its divergence-free part. In view of this decomposition, the P and S waves are also called the *longitudinal waves* and *transverse waves*.[12]

In general, the gradient of a scalar field, $\phi(\mathbf{x})$, is longitudinal, since $\nabla \times (\nabla \phi) = \mathbf{0}$, and the curl of a vector field, $\mathbf{A}(\mathbf{x})$, is transverse, since $\nabla \cdot (\nabla \times \mathbf{A}) = 0$. Conversely, if $\nabla \times \mathbf{u} = \mathbf{0}$ then $\mathbf{u} = \nabla \phi$, and if $\nabla \cdot \mathbf{u} = 0$ then $\mathbf{u} = \nabla \times \mathbf{A}$. Thus, we can let $\mathbf{u}_{\parallel}$ be $\nabla \phi$ and $\mathbf{u}_{\perp}$ be $\nabla \times \mathbf{A}$, where ϕ and $\mathbf{A}$ are called the *scalar* and *vector potentials*.[13]

It can be shown that the P-wave equation, which results from equation (5.5) by setting $\nabla \times \mathbf{u} = \mathbf{0}$, refers to propagation of $\nabla \cdot \mathbf{u}$,[14] which—in view of a geometrical interpretation of the effects of this operator—can be interpreted as propagation of the change in volume. The S-wave equation,

[12] *see also*: Slawinski (2015, Remark 9.4.1)
[13] *see also*: Bóna and Slawinski (2015, Appendix B.4.2)
[14] *see also*: Bóna and Slawinski (2015, Appendix B.4.3)

which results from equation (5.5) by setting $\nabla \cdot \mathbf{u} = 0$, refers to the propagation of $\nabla \times \mathbf{u}$, which can be interpreted as propagation of the change in shape.

Analogies between mathematical and physical properties allow us to infer information about materials. For instance, given proper experimental apparatus with sources and receivers, the absence of detected disturbances akin to the S waves implies the impossibility of propagation of such disturbances, which—in turn—suggests the lack of resistance of the material to the change in shape, as exemplified by liquids.

Let us proceed to show that $\nabla \times \mathbf{u} = \mathbf{0}$ and $\nabla \cdot \mathbf{u} = 0$ are, respectively, the necessary conditions to obtain the P-wave equation and the S-wave equation. To assert the necessity, we have to show that equation (5.9) implies the former condition and equation (5.12) implies the latter.

Let us take the divergence of equation (5.5), which—in view of $\nabla \cdot (\nabla \times \mathbf{u}) = 0$, and using identity (5.8)—we write as

$$\rho \frac{\partial^2}{\partial t^2} \nabla \cdot \mathbf{u} = c_{1111} \nabla^2 (\nabla \cdot \mathbf{u}) \,. \tag{5.13}$$

Using the linearity of differential operators, we rewrite it as

$$\left(\nabla^2 - \frac{1}{\frac{c_{1111}}{\rho}} \frac{\partial^2}{\partial t^2} \right) \nabla \cdot \mathbf{u} = 0 \,. \tag{5.14}$$

In anticipation of the argument for the necessary condition, we view the solutions of equation (5.14) as either $\nabla \cdot \mathbf{u} = 0$ or the result of the differential operator in parentheses acting on $\nabla \cdot \mathbf{u} \neq 0$ being zero; we do not view $\nabla \cdot \mathbf{u} = 0$ as the special case of a general solution. Denoting $\nabla \cdot \mathbf{u}$ by φ and the operator by $\Box_P$, we express these alternatives as

$$\varphi = 0 \qquad \text{or} \qquad \varphi \neq 0 \quad \text{and} \quad \Box_P \, \varphi = 0 \,; \tag{5.15}$$

this wave-equation operator is called the *d'Alembertian* in honour of Jean-Baptiste le Rond d'Alembert. The first choice is valid for an arbitrary differential operator but the second requires the solubility for $\Box_P$, which is the P-wave d'Alembertian.

Taking the curl of equation (5.5), which, due to $\nabla \times \nabla (\nabla \cdot \mathbf{u}) = \mathbf{0}$, becomes

$$\rho \frac{\partial^2}{\partial t^2} \nabla \times \mathbf{u} = -c_{2323} \nabla \times (\nabla \times \nabla \times \mathbf{u}) \,, \tag{5.16}$$

using identity (5.11) and the linearity of differential operators, we write

$$\left(\nabla^2 - \frac{1}{\frac{c_{2323}}{\rho}}\frac{\partial^2}{\partial t^2}\right)\nabla\times\mathbf{u} = \mathbf{0}\,. \tag{5.17}$$

We consider that either $\nabla\times\mathbf{u}=\mathbf{0}$ or the result of operating on $\nabla\times\mathbf{u}$ by the term in parentheses is zero. Denoting $\nabla\times\mathbf{u}$ by Ψ and the operator by $\Box_S$, we express these alternatives as

$$\Psi = \mathbf{0} \qquad \text{or} \qquad \Psi\neq\mathbf{0} \;\text{ and }\; \Box_S\Psi = \mathbf{0}\,. \tag{5.18}$$

Again, the first choice is valid for an arbitrary differential operator but the second requires the solubility for the S-wave d'Alembertian, $\Box_S$.

To complete an argument that $\nabla\cdot\mathbf{u}=0$ is a necessary condition for S waves, we invoke equation (5.12), which is the S-wave equation. Taking the divergence of this equation and using the linearity of differential operators together with the equality of mixed partial derivatives,[15] we obtain $\Box_S\,\varphi=0$. However, if φ is to be a general solution for $\Box_S\,\varphi=0$, it cannot be a general solution for $\Box_P\,\varphi=0$, since these operators exhibit different elasticity parameters in front of the temporal derivative, and, as shown in Exercise 5.3, the stability conditions for isotropic media—which are tantamount to the positive definiteness of matrix (3.57)—require that both c_{1111} and c_{2323} be positive and $3c_{1111}>4c_{2323}$; equality of c_{1111} and c_{2323} would imply equality of the propagation speed for both waves, which—as a consequence of the stability condition—is not allowed.

[15]This equality is also known as Clairaut's theorem since it was formulated by Alexis Claude Clairaut, who was a prominent French mathematician, astronomer and geophysicist. His work exemplifies a relation between geophysics and mathematics. In 1736, together with de Maupertuis—who appears in Chapter 7, in the context of the calculus of variations—he took part in the expedition to Lapland to estimate a degree of the meridian arc. Subsequently, in 1743, he published *Théorie de la figure de la terre*, where he presented another Clairaut's theorem; this theorem connects the gravity at points on the surface of a rotating ellipsoid with the compression and the centrifugal force at the equator. Notably, this hydrostatic model of the shape of the Earth was based on the work of Maclaurin, a Scottish mathematician known for his power series. Maclaurin also showed that a mass of homogeneous fluid set in rotation about a line through its centre of mass would—under the mutual attraction of its particles—take the form of an ellipsoid. Under the assumption that the Earth was composed of concentric ellipsoidal shells of uniform density, Clairaut's theorem could be applied to it, and allowed the ellipticity of the Earth to be calculated from surface measurements of gravity. Furthermore, in 1849, Stokes, an Irish mathematician and physicist, whose theorem of vector calculus—as discussed by Bóna and Slawinski (2015, Appendix A)—is essential for our continuum-mechanics formulations, showed that Clairaut's result is true regardless of the interior constitution and density of the Earth, provided the surface was a spheroid of equilibrium of small ellipticity.

Let us illustrate—for a single spatial dimension—the issue of distinct solutions. As shown in Section 5.2.3, below, the general solutions are

$$f\left(x+\sqrt{\frac{c_{2323}}{\rho}}\,t\right)+g\left(x-\sqrt{\frac{c_{2323}}{\rho}}\,t\right),$$

for the $\Box_S$ operator, and

$$f\left(x+\sqrt{\frac{c_{1111}}{\rho}}\,t\right)+g\left(x-\sqrt{\frac{c_{1111}}{\rho}}\,t\right),$$

for the $\Box_P$ operator; f and g are arbitrary functions.

Thus, since $\Box_S\,\varphi = 0$ is not consistent with the second of alternatives (5.15), we must choose the first alternative, $\varphi = 0$, which—being valid for an arbitrary differential operator—is consistent with equation (5.12). This means that equation (5.12) implies $\nabla\cdot\mathbf{u} = 0$, and hence it is a necessary condition.

To complete an argument that $\nabla\times\mathbf{u} = \mathbf{0}$ is a necessary condition for P waves, we invoke equation (5.9), which is the P-wave equation. Taking the curl of equation (5.9), we obtain $\Box_P\Psi = \mathbf{0}$. This result disagrees with the second of alternatives (5.18); thus, we must choose the first alternative, which is consistent with equation (5.9). This means that equation (5.9) implies $\nabla\times\mathbf{u} = \mathbf{0}$, and hence it is a necessary condition.

Based on the necessary conditions, we can state the following. The presence of P waves implies an irrotational deformation, and the presence of S waves implies an equivoluminal deformation. In view of the geometrical and physical meaning of the Laplacian, together with meanings of the curl and divergence, we can interpret these waves as propagations of disequilibrium in the form of either an irrotational or equivoluminal deformation.

The P and S waves can be adequate mathematical analogies to study physical disturbances within the Earth. As mentioned in Section 1.1, Oldham was the first to interpret disturbances recorded by a seismograph in the context of the P and S waves. Perhaps, as stated in the title by Wigner (1960), the accurate analogy between Helmholtz's decomposition theorem, with its P and S waves, and the observed physical disturbances follows from *the unreasonable effectiveness of mathematics in the natural sciences.*

However, it is important to remember that these waves exist only within the realm of mathematics and the inferences of physical properties rely on analogies, not on deductive arguments. Also, since Hookean solids are mathematical entities, there is no causal relation between their behaviour and the behaviour of physical materials. Hence, their properties might be inappropriate or insufficient as analogies for physical phenomena.

5.1.3 *Particular case: Inhomogeneous string*

As shown in Section 5.1.2, proceeding from the elastodynamic equation in three spatial dimensions, stated in expression (5.1), we obtain the wave equations by assuming homogeneity and isotropy of the continuum.[16] Its generic form can be written as

$$\nabla^2 \mathbf{u}(\mathbf{x},t) = \frac{1}{v^2}\frac{\partial^2 \mathbf{u}(\mathbf{x},t)}{\partial t^2}\,,$$

where $\mathbf{u}$ is the displacement vector, which is a function of position, $\mathbf{x}$, and time, t. To achieve the above form, the speed of propagation, v, must be constant; if it is not, the derivation from expression (5.1) has to proceed in a different manner.[17]

Nevertheless, it is tempting to hypothesize—for an inhomogeneous continuum—the wave equation given by

$$\nabla^2 \mathbf{u}(\mathbf{x},t) = \frac{1}{v^2(\mathbf{x})}\frac{\partial^2 \mathbf{u}(\mathbf{x},t)}{\partial t^2}\,,$$

whose form is such that the propagation speed, v, changes with position, $\mathbf{x}$, which is symptomatic of the inhomogeneity of the Hookean solid. In general, such an equation is an *a posteriori* heuristic proposition, not the result of an axiomatic approach, since it cannot be derived from the balance of linear momentum and the constitutive equation of an inhomogeneous Hookean solid.

However, in a single spatial dimension, we can formulate the equation of motion for an inhomogeneous string that has the form akin to the wave equation, namely,

$$\frac{\partial^2 u(x,t)}{\partial x^2} = \frac{1}{v^2(x)}\frac{\partial^2 u(x,t)}{\partial t^2}\,.$$

Let us consider a string that is subject to displacements that are perpendicular to its length, which we assume to coincide with the x-axis. As illustrated in Figure 5.1, the length element of this string is $\sqrt{(\Delta x)^2 + (\Delta u)^2}$, where u is the vertical displacement; the other displacement components are zero. The mass of this element is $\rho(x)\sqrt{(\Delta x)^2 + (\Delta u)^2}$, where $\rho(x)$ is the mass density. We write the vertical component of Newton's second law as

$$\begin{aligned}\rho(x)\sqrt{(\Delta x)^2 + (\Delta u)^2}\,\frac{\partial^2 u(x,t)}{\partial t^2} \qquad\qquad (5.19)\\ = \mathbf{T}(x+\Delta x,t)\sin(\theta(x+\Delta x,t)) - \mathbf{T}(x,t)\sin(\theta(x,t)) + f(x,t)\Delta x\,,\end{aligned}$$

[16] *see also*: Slawinski (2015, Chapter 6)
[17] *see also*: Slawinski (2015, Chapter 7)

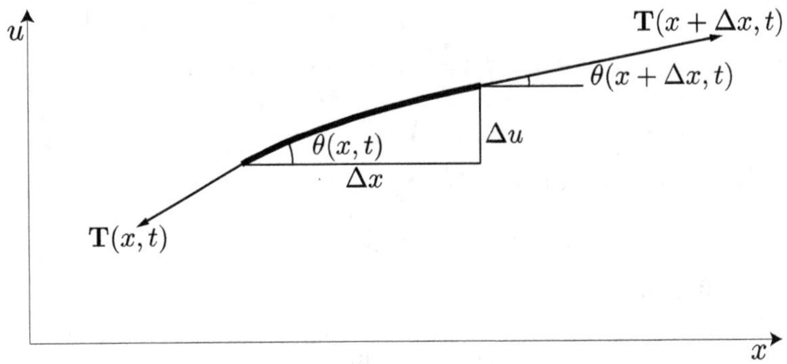

Fig. 5.1: Segment of a string subject to vertical displacement. $\mathbf{T}$ is tangential to the string.

where the first two terms on the right-hand side represent the net vertical component of the force due to the tension, $\mathbf{T}$, on the element, and the third term represents that component for external forces. This is a scalar, not a vector equation, since it refers to the single component, which in the examined problem is the only one resulting in a nonzero net force.

Dividing the right-hand side of equation (5.19) by Δx, letting it tend to zero and invoking the definition of derivative, namely,

$$\lim_{\Delta x\to 0}\frac{\mathbf{T}(x+\Delta x,t)\sin(\theta(x+\Delta x,t))-\mathbf{T}(x,t)\sin(\theta(x,t))}{\Delta x}$$
$$=:\frac{\partial}{\partial x}\left(\mathbf{T}(x,t)\sin(\theta(x,t))\right),$$

we get

$$\rho(x)\sqrt{1+\left(\frac{\partial u(x,t)}{\partial x}\right)^2}\,\frac{\partial^2 u(x,t)}{\partial t^2}=\frac{\partial}{\partial x}\left(\mathbf{T}(x,t)\sin(\theta(x,t))\right)+f(x,t)\,. \tag{5.20}$$

Using the product and chain rules, we write

$$\rho(x)\sqrt{1+\left(\frac{\partial u(x,t)}{\partial x}\right)^2}\,\frac{\partial^2 u(x,t)}{\partial t^2} \tag{5.21}$$
$$=\frac{\partial \mathbf{T}(x,t)}{\partial x}\sin(\theta(x,t))+\mathbf{T}(x,t)\cos(\theta(x,t))\frac{\partial\theta(x,t)}{\partial x}+f(x,t)\,.$$

To eliminate trigonometric functions, we consider

$$\sin(\theta(x,t))=\frac{\dfrac{\partial u(x,t)}{\partial x}}{\sqrt{1+\left(\dfrac{\partial u(x,t)}{\partial x}\right)^2}}$$

and

$$\cos(\theta(x,t)) = \frac{1}{\sqrt{1+\left(\frac{\partial u(x,t)}{\partial x}\right)^2}},$$

where, as illustrated in Figure 5.1,

$$\tan(\theta(x,t)) = \lim_{\Delta x\to 0}\frac{\Delta u}{\Delta x} = \frac{\partial u(x,t)}{\partial x},$$

which implies that

$$\theta(x,t) = \arctan\frac{\partial u(x,t)}{\partial x}$$

and hence

$$\frac{\partial\,\theta(x,t)}{\partial x} = \frac{\frac{\partial^2 u(x,t)}{\partial x^2}}{1+\left(\frac{\partial u(x,t)}{\partial x}\right)^2}.$$

Thus—under the assumption of infinitesimal displacements, which allows us to approximate all radicands by unity—we write expression (5.21) as

$$\rho(x)\frac{\partial^2 u(x,t)}{\partial t^2} = \frac{\partial T(x,t)}{\partial x}\frac{\partial u(x,t)}{\partial x} + T(x,t)\frac{\partial^2 u(x,t)}{\partial x^2} + f(x,t),$$

which is a nonlinear partial differential equation, where T stands for the vertical component of $\mathbf{T}$. If we assume that f is known, the two unknowns are u and T. Considering infinitesimal displacements, we ignore the product of two derivatives to write

$$\rho(x)\frac{\partial^2 u(x,t)}{\partial t^2} = T(x,t)\frac{\partial^2 u(x,t)}{\partial x^2} + f(x,t).$$

If we assume also T to be known, we have a linear partial differential equation for u. If ρ and T are constant and $f = 0$, we have the wave equation, where constancy of ρ is tantamount to the string being homogeneous. Otherwise—if ρ and T are not constant, but $f = 0$—we can write

$$\frac{\partial^2 u(x,t)}{\partial t^2} = \frac{1}{v^2(x,t)}\frac{\partial^2 u(x,t)}{\partial x^2},$$

where $v^2(x,t) = T(x,t)/\rho(x)$.

This derivation relies on several simplifications that could be avoided by returning to equation (5.21) and—as a consequence of the assumption of vertical motion only—introducing another equation for $\mathbf{T}$. Since the magnitude of the net horizontal force is zero,

$$\mathbf{T}(x+\Delta x,t)\cos(\theta(x+\Delta x,t)) - \mathbf{T}(x,t)\cos(\theta(x,t)) = 0.$$

Dividing by Δx, letting it tend to zero and invoking definition of derivative, we get

$$\frac{\partial}{\partial x}\left(\mathbf{T}(x,t)\cos(\theta(x,t))\right)=0\,,$$

which, together with equation (5.21), forms a system of two equations for two unknowns.

5.1.4 *Particular case: String with friction*

Similarly to the discussion in Section 5.1.3, let us consider another equation whose purpose is to study wave propagation and whose form is akin to the wave equation, but whose formulation is not entailed explicitly by Cauchy's equations of motion and Hooke's law.

[18]As illustrated by Morse and Feshbach (1953), the equation for waves on a string immersed in a viscous medium can be written as

$$T\frac{\partial^2 u}{\partial x^2}=\rho\frac{\partial^2 u}{\partial t^2}+R\frac{\partial u}{\partial t}\,,$$

where u denotes the displacement, ρ denotes mass density per unit length of string, R is the frictional resistance of the surrounding medium, and T is the tension of the string; the speed of propagation is $\sqrt{T/\rho}$. As expected, if $R=0$, we obtain the standard wave equation in one spatial dimension. Herein, we assume that R is nonzero, but it is a constant. In particular, we ignore its dependence on frequency; this dependence implies that R opposes the motion of the string in proportion to its speed, which is in agreement with common experience. A justification for this restricting assumption requires that viscosity be sufficiently large.

In the limit, if R is very large, the viscous effects dominate the inertial effects: the wave equation becomes the *diffusion equation*,

$$T\frac{\partial^2 u}{\partial x^2}=R\frac{\partial u}{\partial t}\,.$$

In other words, a hyperbolic partial differential equation becomes parabolic. In such a case, there is no finite propagation speed, which is one of the key properties of hyperbolic equations.

---

[18]Readers interested in the formulation and the solution of the wave equation for an elastic string with friction might refer to Morse and Feshbach (1953, Part. I, p. 137, Part. II, pp. 1335–1339 and pp. 1341–1343).

5.2 Solutions of wave equation

5.2.1 *Introduction*

In this section, to gain an insight into the general solution of the wave equation, we examine five methods for solving it, within the context of isotropic, homogeneous and unbounded Hookean solids. For each method, we consider explicitly the wave equation in a single spatial dimension. The purpose of this examination—beyond gaining familiarity with technicalities of each method—is an implicit comparison of intrinsic limitations associated with each approach. A consequence of such a comparison is the distinction between a physical problem of interest and mathematical entities used as analogies in its investigation, which is a consequence of no causal relation between the world of physics and the realm of mathematics.

5.2.2 *Product solution*

Let us consider

$$\frac{\partial^2 u(x,t)}{\partial x^2} - \frac{1}{v^2}\frac{\partial^2 u(x,t)}{\partial t^2} = 0\,, \tag{5.22}$$

which is the wave equation in its generic form, without explicit reference to P or S waves; u stands for the displacement and x and t denote the position and time, respectively

Let us assume that the solution can be expressed as $u(x,t) = X(x)T(t)$, which is a product of two single-variable functions, X and T, whose independent variables are x and t, respectively. Inserting this expression into equation (5.22), we obtain

$$\frac{v^2}{X}\frac{\mathrm{d}^2 X}{\mathrm{d}x^2} = \frac{1}{T}\frac{\mathrm{d}^2 T}{\mathrm{d}t^2}\,,$$

which means that the product solution leads to the *separation of variables*, in which algebra allows us to rewrite equation (5.22) in such a way that either variable occurs on a different side of the equation. Since the left-hand and right-hand sides are independent of one another, which means that they do not share a variable, their equality implies that they are equal to the same constant,

$$\frac{v^2}{X}\frac{\mathrm{d}^2 X}{\mathrm{d}x^2} = -\omega^2 \qquad \text{and} \qquad \frac{1}{T}\frac{\mathrm{d}^2 T}{\mathrm{d}t^2} = -\omega^2\,.$$

In view of the theory of ordinary differential equations, we set this *separation constant* to be negative to ensure an oscillatory solution. We can write the resulting equations as

$$\frac{\mathrm{d}^2 X}{\mathrm{d}x^2} + \kappa^2 X = 0 \tag{5.23}$$

and

$$\frac{\mathrm{d}^2 T}{\mathrm{d}t^2} + \omega^2 T = 0\,, \tag{5.24}$$

where $\kappa := \omega/v$ is the spatial frequency, also known as the *wave number*: a number of wavelengths in a unit of distance. Invoking standard solutions of ordinary differential equations,

$$X(x) = A\sin(\kappa x) + B\cos(\kappa x)$$

and

$$T(t) = C\sin(\omega t) + D\cos(\omega t)\,,$$

we write a solution of equation (5.22) as their product,

$$u(x,t) = \big(A\sin(\kappa x) + B\cos(\kappa x)\big)\big(C\sin(\omega t) + D\cos(\omega t)\big)\,,$$

which we can rewrite as

$$u(x,t) = \cdots + AD\sin(\kappa x)\cos(\omega t) + \dots\,.$$

Using trigonometric identities, we write

$$\begin{aligned} u(x,t) &= \cdots + \frac{AD}{2}\big(\sin(\kappa x + \omega t) + \sin(\kappa x - \omega t)\big) + \dots \\ &= \cdots + \frac{AD}{2}\Big(\sin\big(\kappa(x+vt)\big) + \sin\big(\kappa(x-vt)\big)\Big) + \dots\,; \end{aligned} \tag{5.25}$$

this summand describes harmonic waves with amplitudes $AD/2$ propagating in opposite directions with velocity $v = \omega/\kappa$, along the x-axis.

The assumption of a two-variable function, $u(x,t)$, being represented by a product of two single-variable functions, $X(x)$ and $T(t)$, imposes restrictions. It necessitates that the problem at hand be composed of linear homogeneous equations with linear homogeneous side conditions. The side conditions given by the initial conditions do not lend themselves to a standard method of separation of variables, except its special case known as the method of characteristics.[19]

The method of separation of variables can be generalized to functions of several variables by separating one variable at a time with introduction of a new separation constant each time. In general, this method belongs to the *Sturm-Liouville theory*, which deals with second-order partial differential equations. In that context, equations (5.23) and (5.24) constitute an *eigenvalue problem*, with κ and ω being the corresponding eigenvalues. Commonly, a *Sturm-Liouville problem* would also contain side conditions.

[19] *see also*: Bóna and Slawinski (2015)

5.2.3 d'Alembert solution

5.2.3.1 d'Alembert's approach

The solution discussed herein, which historically—dating to the eighteenth century—is the first solution of the wave equation, is limited to a single spatial dimension. To examine this solution, let us return to equation (5.22) and let $y = x + vt$ and $z = x - vt$ to write this equation as[20]

$$\frac{\partial^2 u(y,z)}{\partial y \partial z} = 0 \,. \tag{5.26}$$

This approach is called the *d'Alembert method*, in honour of Jean-Baptiste le Rond d'Alembert.[21]. The solution of equation (5.26) is $f(y) + g(z)$,[22] which—in view of the substitutions—we write as

$$u(x,t) = f(x + vt) + g(x - vt) \,. \tag{5.27}$$

Solution (5.27) can be viewed as a more general form of solution (5.25), which, recalling that a linear combination of any summand of the solution is also a solution, we write as

$$u(x,t) = \sin\left(\kappa(x + vt)\right) + \sin\left(\kappa(x - vt)\right) .$$

Since solution (5.25) results from solving ordinary differential equations with constant coefficients, it follows that its u must be a smooth function. This is not the case of solution (5.27), where u can exhibit discontinuities or even be a distribution. This distinction is an important difference between ordinary and partial differential equations.

Solution (5.27), which is the general solution of equation (5.22), was obtained by d'Alembert in 1746. Functions f, g and $\exp(\,\cdot\,)$ are single variable functions. However, solution (5.25) results from a generic separation into the variables present in equation (5.22): x and t. Solution (5.27) results from replacing x and t by the variables whose form is explicitly pertinent to equation (5.22); these variables, $x + vt$ and $x - vt$, are the characteristics of that equation, as discussed below.

The method of characteristics can be viewed as a coordinate transformation that allows us to solve a partial differential equation.[23] This method lends itself to initial-value problems; it is of special importance for hyperbolic partial differential equations, which are the only equations with

[20] *see also*: Slawinski (2015, Lemma 6.5.1)

[21] Readers interested in the times of d'Alembert, and his importance during that period, might refer to McGrayne (2011, Chapter 2).

[22] *see also*: Slawinski (2015, Lemma 6.5.2)

[23] *see also*: Bóna and Slawinski (2015, Exercise 2.1)

real characteristics and whose wellposedness requires the initial—not the boundary—conditions.[24,25]

Even though solution (5.27) is more general, it does not introduce the concept of the wave number or the angular frequency. At times, a general formulation cannot provide certain details.

5.2.3.2 *Euler's approach*

There is another approach to obtain d'Alembert's solution, which is credited to Euler's work of 1748.[26] Let us write equation (5.22) as

$$\left(\frac{\partial^2}{\partial x^2}-\frac{1}{v^2}\frac{\partial^2}{\partial t^2}\right)u(x,t)=0\,. \tag{5.28}$$

We can restate this equation as

$$\left(\frac{\partial}{\partial x}+\frac{1}{v}\frac{\partial}{\partial t}\right)\left(\frac{\partial}{\partial x}-\frac{1}{v}\frac{\partial}{\partial t}\right)u=0\,.$$

Denoting $\partial u/\partial x$ by u' and $\partial u/\partial t$ by $v\dot{u}$, we write

$$\left(\frac{\partial}{\partial x}+\frac{1}{v}\frac{\partial}{\partial t}\right)(u'-\dot{u})=0$$

and, equivalently,

$$\left(\frac{\partial}{\partial x}-\frac{1}{v}\frac{\partial}{\partial t}\right)(u'+\dot{u})=0\,,$$

which means that $u'-\dot{u}$ and $u'+\dot{u}$ are constants, respectively, along $x-vt=C_1$ and $x+vt=C_2$, where C_1 and C_2 are arbitrary constants; these lines are the characteristic curves. Hence, this approach to solving the wave equation is a paradigmatic example of the method of characteristics; for instance, we encounter it again in a similar form in equation (8.13).

Rearranging terms, we write

$$\dot{u}(x,t_0)=\frac{1}{2}\Big(\dot{u}(x,t_0)+u'(x,t_0)\Big)+\frac{1}{2}\Big(\dot{u}(x,t_0)-u'(x,t_0)\Big)\,,$$

whose right-hand side is equivalent to

$$\frac{1}{2}\Big(u'(x+vt_0,0)-u'(x-vt_0,0)\Big)+\frac{1}{2}\Big(\dot{u}(x+vt_0,0)+\dot{u}(x-vt_0,0)\Big)\,, \tag{5.29}$$

[24] *see also*: Bóna and Slawinski (2015, Chapter 2)

[25] Readers interested in wave propagation and hyperbolic partial differential equations might refer to Hadamard (1903, 1932).

[26] For further explanations, readers might refer to Stone and Goldbart (2009, Section 6.3.1).

expressed along the initial characteristics. To obtain the solution, $u(x,t)$, we integrate expression (5.29). Letting $x + vt_0 =: y$ and hence $\mathrm{d}x = \mathrm{d}y$, we integrate the first term in parentheses between $t = 0$ and $t = t_0$,

$$\int_{x}^{x+vt_0} u'(y,0)\,\mathrm{d}y = u(x+vt_0,0) - u(x,0)\,. \tag{5.30}$$

Letting $x - vt_0 =: z$ and hence $\mathrm{d}x = \mathrm{d}z$, we integrate the second term in parentheses between $t = 0$ and $t = t_0$,

$$\int_{x}^{x-vt_0} u'(z,0)\,\mathrm{d}z = u(x-vt_0,0) - u(x,0)\,. \tag{5.31}$$

In equations (5.30) and (5.31), u' denotes the derivative with respect to the integration variable. Adding both equations—upon reversing the integration limits of the latter—and using the fact that $\mathrm{d}z = \mathrm{d}y$, we write

$$\int_{x-vt_0}^{x+vt_0} u'(y,0)\,\mathrm{d}y = u(x+vt_0,0) - u(x-vt_0,0)\,,$$

Imposing an initial condition, $u(x,0) = \gamma(x)$, we write the result of the integration of the first parenthesis in expression (5.29) as

$$\frac{1}{2}\Big(\gamma(x+vt_0) - \gamma(x-vt_0)\Big)\,. \tag{5.32}$$

Again, letting $x + vt_0 =: y$ and hence $v\,\mathrm{d}t_0 = \mathrm{d}y$, we write the integral of the first term in the second parentheses as

$$\frac{1}{v}\int_{x}^{x+vt_0} \dot{u}(y,0)\,\mathrm{d}y\,. \tag{5.33}$$

Letting $x - vt_0 =: z$ and hence $-v\,\mathrm{d}t_0 = \mathrm{d}z$, we write the integral of the second term in the second parentheses as

$$-\frac{1}{v}\int_{x}^{x-vt_0} \dot{u}(z,0)\,\mathrm{d}z\,. \tag{5.34}$$

In equations (5.33) and (5.34), $\dot{u}$ denotes the derivative with respect to the integration variable. Adding both equations—upon reversing the integration limits of the latter—and using the fact that $\mathrm{d}z = -\mathrm{d}y$, we write

$$\frac{1}{v}\int_{x-vt_0}^{x+vt_0} \dot{u}(y,0)\,\mathrm{d}y\,. \tag{5.35}$$

Imposing an initial condition, $\dot{u}(x,0) = \eta(x)$, where the $\dot{}$ donates the temporal derivative, we write the result of the integration of the second parenthesis in expression (5.29) as

$$\frac{1}{2v}\int\limits_{x-vt_0}^{x+vt_0} \eta(y)\,\mathrm{d}y\,. \tag{5.36}$$

Formally, we could proceed to integrate expression (5.35) and obtain u. However, since equation (5.28) is a second-order equation, we need—to obtain a particular solution—to impose constraints on both u and its derivative.

Combining expression (5.32) and (5.36), we write the solution of equation (5.28), subject to conditions $u|_{t=0} = \gamma$ and $\dot{u}|_{t=0} = \eta$, which is referred to as the *Cauchy problem*, as

$$u(x,t_0) = \frac{1}{2}\left(\gamma(x+vt_0) - \gamma(x-vt_0) + \frac{1}{v}\int\limits_{x-vt_0}^{x+vt_0} \eta(y)\,\mathrm{d}y\right), \tag{5.37}$$

where y is the integration variable. This extension of solution (5.27), due to Euler, includes the effect along x due to forces at $t_0 = 0$, which is expressed by the integral.

5.2.3.3 *Spherical-symmetry approach*

As stated in Section 5.2.3.1, the d'Alembert solution is limited to a single spatial dimension. Herein, we discuss a modification whose result can be interpreted in three spatial dimensions, even though the mathematical dependence remains a function of a single spatial variable. This variable is a radius from the origin, which entails the spherical symmetry of a solution.

If the solution is to possess a spherical symmetry, such a symmetry must be imposed explicitly on the initial conditions within a Cauchy problem. This means that, as shown in Exercise 5.5, the initial conditions must be spherically symmetric. The wave equation is spherically symmetric by its formulation, which is limited to isotropic continua; in other words, the wave equation does not exhibit any directional dependence. This is not the case for an elastodynamic equation, given in expression (5.1), which is an analogue of the wave equation for anisotropic Hookean solids.

Furthermore, for the modification of d'Alembert's method, which is formulated for a single spatial dimension, the displacement, $u(x,t)$, is a scalar-valued function. Hence, both its Laplacian and its second time derivative, which appear in the wave equation, are scalar quantities; this is not so for the case discussed in the context of expression (5.7), which is the vector Laplacian. Thus, herein, the wave-equation quantities are rotationally invariant, which is tantamount to being spherically symmetric.

Physically, the spherical symmetry of the initial conditions implies a constant-amplitude spherical wavefront with the same displacement speed at each point. Mathematically, in spherical coordinates, we write $u(r,\alpha,\beta,t)|_{t_0} = \gamma(r)$ and $\partial u(r,\alpha,\beta,t)/\partial t|_{t_0} = \eta(r)$. In other words, both γ and η are only functions of r, which is the distance from the origin; they are not functions of direction.

The spherical-coordinate Laplacian, acting on a scalar-valued function, is

$$\nabla^2 := \frac{1}{r^2}\frac{\partial}{\partial r}\left(r^2\frac{\partial}{\partial r}\right) + \frac{1}{r^2\sin\alpha}\frac{\partial}{\partial\alpha}\left(\sin\alpha\frac{\partial}{\partial\alpha}\right) + \frac{1}{r^2\sin^2\alpha}\frac{\partial^2}{\partial\beta^2}\,. \tag{5.38}$$

If a given function is spherically symmetric, $F(r,\alpha,\beta) = F(r)$, the operations of the second term and the third term result in zeros; thus—for such functions—we write this operator as[27]

$$\nabla^2 F = \frac{1}{r^2}\frac{\partial}{\partial r}\left(r^2\frac{\partial F}{\partial r}\right). \tag{5.39}$$

Using the product rule, we write expression (5.39) as

$$\nabla^2 F = \frac{\partial^2 F}{\partial r^2} + \frac{2}{r}\frac{\partial F}{\partial r}\,, \tag{5.40}$$

which we rewrite as

$$\nabla^2 F = \frac{1}{r}\frac{\partial}{\partial r}\left(F + r\frac{\partial F}{\partial r}\right) = \frac{1}{r}\frac{\partial^2 (rF)}{\partial r^2}\,.$$

The corresponding wave equation can be written as

$$\nabla^2 F - \frac{1}{v^2}\frac{\partial^2 F}{\partial t^2} = \frac{1}{r}\frac{\partial^2 (rF)}{\partial r^2} - \frac{1}{v^2}\frac{\partial^2 F}{\partial t^2} = 0\,.$$

Since r is independent of t, we write

$$\frac{\partial^2 (rF)}{\partial r^2} - \frac{1}{v^2}\frac{\partial^2 (rF)}{\partial t^2} = 0\,,$$

which is analogous to equation (5.22) with $rF \equiv u$; hence, we invoke solution (5.27) to write the general solution as[28]

$$u(r,t) = \frac{1}{r}f(r+vt) + \frac{1}{r}g(r-vt)\,, \tag{5.41}$$

[27] *see also*: Slawinski (2015, equation (6.12.15))
[28] *see also*: Slawinski (2015, Exercise 6.4)

where the first term on the right-hand side represents a spherically symmetric wave that propagates towards the origin and the second one away from it;[29] r is the distance from the origin, it is the magnitude of the radius. In this manner, the discussed extension remains formally a problem of a single spatial dimension.

Since the equation in question, which is a scalar wave equation, corresponds to pressure waves, f and g denote pressure. Examining solution (5.41), we see that, as expected, the pressure of the wave propagating towards the origin increases and the pressure of the wave propagating away decreases. The pressure on a spherical wavefront changes with the distance from the origin as r^{-1}, not as r^{-2}; this issue is addressed in Exercise 5.6 and Remark 5.1.

5.2.4 *Fourier-transform solution*

[30]Taking the Fourier transform of equation (5.22),

$$\frac{\partial^2 u(x,t)}{\partial x^2} - \frac{1}{v^2}\frac{\partial^2 u(x,t)}{\partial t^2} = 0\,, \tag{5.42}$$

with respect to x, and using the property of the transform of the second derivative, $\widehat{\partial^2 u(x,t)/\partial x^2} = -\kappa^2 \hat{u}(\kappa,t)$, we write

$$\frac{\mathrm{d}^2 \hat{u}(\kappa,t)}{\mathrm{d}t^2} + (\kappa v)^2 \hat{u}(\kappa,t) = 0\,, \tag{5.43}$$

where κ is the wave number; notably, herein—appearing as the transform variable paired with the space variable, x—κ exhibits its property of the spatial frequency. Such an interpretation is expected in view of the more familiar case: the Fourier-transform counterpart of the time variable, t, is ω, which is the temporal frequency.

Equation (5.43) is an ordinary differential equation, whose general solution is

$$\hat{u}(\kappa,t) = A(\kappa)\exp(\iota\kappa v t) + B(\kappa)\exp(-\iota\kappa v t)\,. \tag{5.44}$$

To determine A and B, we set the side conditions, $\hat{u}(\kappa,0) = \hat{\gamma}(\kappa)$ and $\mathrm{d}\hat{u}(\kappa,0)/\mathrm{d}t = \hat{\eta}(\kappa)$. Inserting these expressions into solution (5.44) and its derivative, with $t = 0$, we get two linear equations

$$A(\kappa) + B(\kappa) = \hat{\gamma}(\kappa)\,,$$
$$\iota v\kappa A(\kappa) - \iota v\kappa B(\kappa) = \hat{\eta}(\kappa)\,,$$

[29] *see also*: Slawinski (2015, expression (6.5.12))

[30] For further explanations, readers might refer to Bos (2003, Section 1.2).

whose solutions are

$$A = \frac{1}{2}\left(\hat{\gamma} + \frac{1}{\iota v \kappa}\hat{\eta}\right) \quad \text{and} \quad B = \frac{1}{2}\left(\hat{\gamma} - \frac{1}{\iota v \kappa}\hat{\eta}\right).$$

Inserting expressions for A and B into solution (5.44), we write

$$\hat{u}(\kappa, t) = \hat{\gamma}(\kappa)\frac{\exp(\iota\kappa v t) + \exp(-\iota\kappa v t)}{2} + \frac{\hat{\eta}(\kappa)}{\kappa v}\frac{\exp(\iota\kappa v t) - \exp(-\iota\kappa v t)}{2\iota}. \tag{5.45}$$

The solution of equation (5.42) is the inverse Fourier transform of expression (5.45), with κ and x being the transformation variables. Let us write the first term of this expression as

$$\hat{\gamma}(\kappa)\frac{\exp(\iota\kappa v t)}{2} + \hat{\gamma}(\kappa)\frac{\exp(-\iota\kappa v t)}{2}. \tag{5.46}$$

To obtain its inverse, we define the translation operator, $(T_{\pm vt}\gamma)(x) := \gamma(x \pm vt)$, which means that γ is translated along x to the left or right by vt. Its Fourier transform is $\widehat{T_{\pm vt}\gamma}(\kappa) = \hat{\gamma}(\kappa)\exp(\mp\iota\kappa v t)$, which appears in expression (5.46). Hence, the sought inverse is $\gamma(x \mp vt)$; thus, we write the inverse of the first term in expression (5.46) as

$$\frac{1}{2}(\gamma(x + vt) + \gamma(x - vt)). \tag{5.47}$$

In view of definitions of trigonometric functions, we write the second term of expression (5.45) as

$$\hat{\eta}(\kappa)\frac{\sin(\kappa v t)}{\kappa v}. \tag{5.48}$$

This is the product of two transforms, $\hat{\eta}\hat{f}$. Since, in general, $\hat{\eta}\hat{f} = \widehat{\eta * f}$, where $*$ denotes the convolution, we write

$$\hat{\eta}(\kappa)\,\hat{f}(\kappa) \equiv \hat{\eta}(\kappa)\frac{\sin(\kappa v t)}{\kappa v} = \widehat{(\eta * f)}(\kappa).$$

Hence, the sought inverse is $(\eta * f)(x)$, which—following the definition of convolution—we write as

$$(\eta * f)(x) := \int\limits_{-\infty}^{\infty} \eta(x - y) f(y)\, \mathrm{d}y,$$

where y is the integration variable. For a given t, the inverse of $\sin(\kappa v t)/(\kappa v)$ is $f = 1/(2v)$, for $x \in (-vt, vt)$ and zero, otherwise. Thus, we write

$$\frac{1}{2v}\int\limits_{-vt}^{vt} \eta(x - y)\, \mathrm{d}y.$$

Letting $z = x - y$, with $\mathrm{d}y = -\mathrm{d}z$, and interchanging the integration limits, we write the inverse of expression (5.48) as

$$\frac{1}{2v}\int\limits_{x-vt}^{x+vt}\eta(z)\,\mathrm{d}z\,.$$

Combining this result with expression (5.47), we write

$$u(x,t) = \frac{1}{2}\left(\gamma(x+vt)+\gamma(x-vt)+\frac{1}{v}\int\limits_{x-vt}^{x+vt}\eta(z)\,\mathrm{d}z\right)\,, \tag{5.49}$$

which is equivalent to expression (5.37), as expected. Unlike the approach that results in solution (5.37), the Fourier-transform approach can be used for wave equations beyond the single spatial dimension.

Herein, however, γ and η are restricted by the integrability properties required by the Fourier transform. No such restrictions are imposed on solution (5.37), even though, as maintained by d'Alembert in his discussions with Euler, one could argue that γ and η should be twice differentiable, since they are solutions of a second-order differential equation. Referring to these eighteenth-century discussions, we know that this requirement is not necessary, if we consider distributional solutions, formulated in the twentieth century. Also, as remarked by Arnold (2004, p. 69),

> [a] singularity in the data for the Cauchy problem for the wave equation propagates along the characteristics,[31] so that the solutions of the wave equation cannot be smoother than the initial data.

To complete this section, let us state, and comment briefly on, the Fourier-transform solution of the wave equation in three spatial dimensions,

$$\nabla^2 u(\mathbf{x},t) - \frac{1}{v^2}\frac{\partial^2 u(\mathbf{x},t)}{\partial t^2} = 0\,,$$

whose side conditions are

$$u(\mathbf{x},0) = \gamma(\mathbf{x}) \qquad \text{and} \qquad \frac{\partial u}{\partial t}(\mathbf{x},0) = \eta(\mathbf{x})\,,$$

[31]Readers interested in the importance of characteristics for propagation of singularities might refer to Bóna and Slawinski (2015), who state in the closing remarks to Chapter 4 that

> [i]f a solution has discontinuities, they happen along the characteristics only.

with $\mathbf{x} \in \mathbb{R}^3$, and whose solution is[32]

$$u(\mathbf{x}, t) = \frac{1}{4\pi v^2 t} \iint\limits_{S(\mathbf{x}, vt)} \eta(\mathbf{y}) \, \mathrm{d}\sigma + \frac{\mathrm{d}}{\mathrm{d}t} \left(\frac{1}{4\pi v^2 t} \iint\limits_{S(\mathbf{x}, vt)} \gamma(\mathbf{y}) \, \mathrm{d}\sigma \right), \tag{5.50}$$

where $\mathbf{y}$ is the variable on the sphere, $S(\mathbf{x}, vt)$, and σ is the spherical measure. This expression is known as the *Kirchhoff formula* in honour of the nineteenth-century German physicist Gustav Kirchhoff. As expected, it bears a resemblance to expressions (5.37) and (5.49).

Let us examine expression (5.50).[33] The limits of integration are a sphere, $S(\mathbf{x}, vt)$, centred at $\mathbf{x}$, and with radius vt. The term in parentheses is the Fourier transform of the *spherical mean* distribution, where $\mathrm{d}\sigma$ is the element of the surface area of that sphere.

Denoting

$$\mathcal{O}(\cdot) := \frac{1}{4\pi v^2 t} \iint\limits_{S(\mathbf{x}, vt)} (\cdot) \, \mathrm{d}\sigma$$

we can write concisely expression (5.50) as

$$u(\mathbf{x}, t) = \mathcal{O}(\eta) + \frac{\mathrm{d}}{\mathrm{d}t} \mathcal{O}(\gamma),$$

where $\mathcal{O}$ is called the *source operator.*[34] This operator corresponds to the solution of the wave equation whose side conditions are $\gamma = 0$ and $\eta = \delta$, which is an impulse response, discussed in Section 5.2.5, below.

5.2.5 *Green's-function solution*

An insightful method of solving the wave equation stems from the knowledge of the effect of an impulse acting on a system: the *impulse-response functions.* This method is known as the *method of singularities*, and the impulse-response functions are commonly referred to as *Green's functions*, in honour of George Green, whose famous mill is shown in Figure 5.2.[35] This term was introduced by Carl Neumann in 1861; Green (1828) used them in the context of electromagnetism and called them the *potential functions.*

[32] Readers interested in details of obtaining this solution might refer to Bos (2003, Section 1.3).
[33] *see also*: Slawinski (2015, Sections 6.6.1–6.6.2)
[34] For an insightful explanation of this term, readers might refer to Bos (2003, pp. 27–28).
[35] Readers interested in the life of George Green might refer to Cannell (2001).

photo: Elena Patarini

Fig. 5.2: George Green's mill in Sneinton near Nottingham.

Definition 5.1. Green's function, G, is a solution of any linear partial differential equation with constant coefficients such that

$$\mathfrak{L}\, G(x) = \delta(x) \tag{5.51}$$

where $\mathfrak{L}$ is a differential operator, δ is Dirac's delta and x stands for the variables, which—in a physical context—can represent either space or time.[36]

Solution G is referred to as the *fundamental solution* or as the *elementary solution*, since, if we consider $\mathfrak{L}\, u(x) = F(x)$—and we know its Green's function—the solution of that equation is given by the following theorem.

[36] *see also*: Bóna and Slawinski (2015, Appendix F)

Theorem 5.1. *If Green's function for equation (5.51) is known, then the solution of*

$$\mathcal{L}\, u(x) = F(x)$$

is

$$u(x) = (G * F)(x) := \int_{\mathbb{R}} G(\xi)\, F(x - \xi)\, \mathrm{d}\xi\,,$$

where $*$ is the convolution and ξ stands for the integration variables.

This theorem has an insightful corollary, which can be interpreted as a superposition of the elementary solutions.

Corollary 5.1. *Following Definition 5.1 and Theorem 5.1, and considering the commutativity of the differential and integral operators, we can write*

$$\mathcal{L}\, u(x) = \int_{\mathbb{R}} \mathcal{L}\, G(x - \xi)\, F(\xi)\, \mathrm{d}\xi = \int_{\mathbb{R}} \delta(x - \xi)\, F(\xi)\, \mathrm{d}\xi = F(x)\,.$$

There is an inconvenience in application of Definition 5.1, Theorem 5.1 and Corollary 5.1 due to the lack of general expressions for Green's functions; each equation has its Green's function. Nor is there a generic method to find these functions. In mathematical physics, Green's functions appear in the potential, diffusion and propagation equations, and—for each equation—they have a different form depending on the spatial dimensions.

However, there is an insightful piece of information about physics carried by different forms, and, in particular, about the intrinsic dependence of physical phenomena on dimensions. As stated by Arfken *et al.* (2013), it

> illustrates the fact that there is a real difference between flatland (2-D) physics and actual (3-D) physics, even when the latter is applied to problems with translational symmetry in one direction.

Barton (1989) agrees and emphasizes that, rather than attempting to hide this issue by a particular formulation,

> it is truer to the physics to simply recognize that there are deep differences between waves in spaces of different dimensionalities.

Let us illustrate the application of Green's function by considering equation (5.28), which corresponds to one spatial dimension,

$$\left(\frac{\partial^2}{\partial x^2} - \frac{1}{v^2}\frac{\partial^2}{\partial t^2}\right) u(x,t) =: \Box\, u(x,t) = F(x,t)\,, \tag{5.52}$$

where x and t represent space and time, respectively, and $\square$ is the d'Alembertian. Unlike equation (5.28), equation (5.52) contains F, which can be interpreted as the external force acting at $t \geqslant 0$. According to Definition 5.1,

$$\square G(x,t) = \delta(x - x_0)\,\delta(t - t_0)\,, \tag{5.53}$$

where x_0 is the impulse location and t_0 is the time of its occurrence; in view of properties of δ, we could write the right-hand side of equation (5.53) equivalently as $\delta\big((x,t) - (x_0,t_0)\big) \equiv \delta(x - x_0, t - t_0)$.

Green's function for equation (5.53) can be written in terms of the Heaviside step as[37]

$$G(x,t) = \frac{v}{2}H(t-t_0)\Big(H\left(\|x - x_0\| + v(t - t_0)\right) - H\left(\|x - x_0\| - v(t-t_0)\right)\Big)\,, \tag{5.54}$$

where we include explicitly the causality as $H(t-t_0)$. There is no effect of the impulse until the arrival of its signal, which—at point x—happens at $t = t_0 + \|x - x_0\|/v$, where v is the propagation speed. At that time, the amplitude jumps from zero to a finite value, which is rather intuitive; however, the persistence of the effect of the impulse after its passage might not be. Consistently with his statement cited on page 206, Barton (1989) expresses a view that

> one could say, speaking very loosely, that the lower-dimensional spaces are so cramped that the flux from the unit source cannot diverge fast enough [...] to vanish even at infinity.

In the context of expression (5.54), the infinity refers to both space and time, since the impulse propagates with a constant amplitude for all x and, at a given point, its effect remains constant for all $t \geqslant t_0 + \|x - x_0\|/v$.

Following Theorem 5.1 and using expression (5.54), we write

$$\begin{aligned}(G * F)(x,t) &\equiv \iint\limits_{\mathbb{R}^2} G(x - x_0, t - t_0)F(x_0,t_0)\,\mathrm{d}x_0\,\mathrm{d}t_0 \\ &= \frac{v}{2}\iint\limits_{\mathbb{R}^2} H(t-t_0)\Big(H\big((x-x_0) + v(t-t_0)\big) \\ &\qquad - H\big((x - x_0) - v(t-t_0)\big)\Big)F(x_0,t_0)\,\mathrm{d}x_0\,\mathrm{d}t_0\,;\end{aligned}$$

[37] Bóna and Slawinski (2015, Appendix F.2) derive this function by descending from Green's function for the wave equation in three spatial dimensions using the translational symmetry mentioned in the quote on page 206. Readers interested in the formulation of that function by integrating equation (5.53) with respect to time might refer to Stone and Goldbart (2009, Section 6.3.3). Also, see Exercise 5.7, below.

herein, $\mathbb{R}^2$ is the domain of integration and x_0 and t_0 are the integration variables. Invoking the properties of the Heaviside step, we write

$$(G * F)(x,t) = \frac{v}{2} \int_{-\infty}^{t} \int_{x-v(t-t_0)}^{x+v(t-t_0)} F(x_0, t_0)\, \mathrm{d}x_0\, \mathrm{d}t_0\,. \tag{5.55}$$

If we exclude from expression (5.55) the temporal integral and set $t_0 = 0$, the remaining integral is the integral in d'Alembert's solution, stated in expression (5.37), which represents the response along x to forces at $t = 0$.[38] In general, the double integral, which is tantamount to Corollary 5.1, contains the superposition of the responses of the continuum at all positions from the initial moment to the present one. Thus, expression (5.55) is an extension of d'Alembert's solution that includes forces acting after the initial instant. Furthermore, unlike the approach that results in solution (5.37), the Green's-function approach can be used for the wave equations beyond the single spatial dimension.

5.3 On approximations

No relation between the physical world and mathematical realm can be viewed as exact, since there is no deductive or causal relation between them. Hence, any examination of exactness of solutions—or of equations themselves—considered in Sections 5.2.2–5.2.5 must remain within the realm of mathematics. Examinations of relations between physics and mathematics rely on accuracy of analogies, which can be inferred from comparing theories with experiments.

In Sections 5.2.2–5.2.5, we present solutions to the wave equation that results from restricting Cauchy's equation of motion to isotropic and homogeneous Hookean solids. To complement Section 5.2, let us comment briefly on equations whose derivation does not impose restrictions of isotropy and homogeneity on Hookean solids.

If the purpose of a theory built on Cauchy's equation of motion within the general Hooke's law—to which we refer as the wave theory—is to examine solutions of the elastodynamic equations, we must consider their approximations since no exact solution appears to be possible. A common approach to find solutions invokes the method of characteristics for which solutions of the eikonal equation are paths of propagation of the side condi-

[38] *see also*: Slawinski (2015, Chapter 6, expression (6.5.11))

tions and the transport equation results in the evolution of their amplitudes. Such an approach is referred to as the ray theory, and can be viewed as an approximation to the wave theory.

To solve the eikonal and transport equations, we rely on the method of stationary phase, for which we assume a high frequency of the signal.[39] Consequently—in the context of solutions of the elastodynamic equations—we view the ray theory as a high-frequency approximation of the wave theory. However, rays themselves are not approximations; they are real entities within any hyperbolic partial differential equation; they are trajectories of propagation of its side conditions.[40]

Conceptually, as stated by Polkinghorne (1986),

> geometrical optics [...] is an excellent and well-understood approximation to the wave theory, provided the wavelength of the light is small compared to the physical dimensions of the system under consideration. The rays of geometrical optics are not unlike the particle trajectories of classical mechanics.

Hence, both subjects invoke the same mathematical tools, as exemplified by the $JWKB$ method. As stated by Dyson (2007),

> This passage to the classical theory is precisely analogous to the passage from wave-optics to geometrical optics when the wave-length of light is allowed to tend to zero. The WKB approximation is gotten by taking $\hbar$ small but not quite zero.

Both approaches obey the *correspondence principle*, since the more general theory reproduces the results of the more restricted one, within the realm of the latter.

For historical reasons and in agreement with Freeman Dyson (pers. comm., 2015), who, as stated in his quote on page xxvii, personally knew Sir Harold Jeffreys, we include J in the acronym. Moreover, even though it is common to write $WKBJ$, as exemplified by Morse and Feshbach (1953), we follow the style of Dahlen and Tromp (1998): J is placed at the beginning of the acronym to recognize the priority of Jeffreys's work.

WKB stands for Gregor Wentzel, Hendrik Anthony Kramers and Louis Marcel Brillouin, who developed this method in 1926 to work on the Schrödinger equation. However, Jeffreys developed this method in 1923

[39] *see also*: Bóna and Slawinski (2015, Section 5.3.3)

[40] Readers interested in the distinction between the ray theory—as a high-frequency approximation of the wave theory—and rays therein—as intrinsic entities of Cauchy's equation of motion might refer to Bos and Slawinski (2010).

to approximate solutions to linear second-order differential equations (Jeffreys, 1925), which implicitly include the Schrödinger equation. Methods for solving the elastodynamic equation in the context of ray theory, including the $JWKB$ method, rely on the asymptotic series derived from this equation and on the hyperbolicity of that equation.

However, the asymptotic behaviour of that series for the case of an anisotropic inhomogeneous Hookean solid remains a conjecture. This conjecture is based on proofs for simple cases of these solids and, for other cases, on the numerical studies of this series.

The definition of hyperbolicity,[41] which allows us to interpret characteristics and bicharacteristics of the elastodynamic equations as wavefronts and rays, respectively, applies to the scalar wave equations, where we assume both isotropy and homogeneity of a Hookean solid, not the general case of the elastodynamic equation, which is a system of equations whose solution is a vector-value function. We conjecture the hyperbolicity of the elastodynamic equation based on examinations of its second-order derivatives and on numerical studies of the behaviour of its solutions.

While mathematical efforts to prove these conjectures continue, so do many applications of methods relying on these conjectures. Proceeding in such a manner is a common occurrence in mathematical physics and in applied sciences, in general. The full convergence of Fourier series,[42] formulated at the beginning of the nineteenth century, was proven in the second half of the twentieth century. Until then, its applicability relied on proofs for particular cases and on results that were deemed satisfactory. Similarly, Dirac's delta had been used for several decades prior to its rigorous formulation within the theory of distributions, in the middle of the twentieth century.[43] Aspects of theoretical understanding of solutions of the Navier-Stokes equation remain an open problem, while these solutions are commonly used in fluid mechanics. Examples abound.

Closing remarks

In the search for more accurate mathematical analogies of physical phenomena, we could consider—within Cauchy's equations of motion—a variety of modifications, extensions and hybrids of Hookean solids. Presently, such ap-

[41] *see also*: Bóna and Slawinski (2015, Section 2.2)
[42] *see also*: Bóna and Slawinski (2015, Appendix D.2)
[43] *see also*: Bóna and Slawinski (2015, Appendix E.1)

proaches are feasible due to availability of computational techniques, since examinations of models involving complicated constitutive equations commonly require numerical techniques. In particular, such techniques might be necessary for solutions of equations of motion, whose analytical solutions are limited to simple cases. Even for Hookean solids, if we consider anisotropic inhomogeneous media, analytical solutions appear to be impossible.

Among solids, other than Hookean solids, that allow us to examine seismic phenomena, are the Cosserat media, named for Eugène and François Cosserat.[44] These media allow for the propagation of more types of waves than do Hookean solids by accounting for rotation of elements within continua. Examination of physical phenomena within these media can be a fruitful platform for studies that correspond to high-accuracy measurements whose results are affected directly—not only in the form of a macroscopic average—by the granular properties of materials. Notably, an examination of the relation between wave phenomena in Hookean solids and in the Cosserat media might allow for a quantitative correspondence between seismic and laboratory measurements. The choice of a particular medium as a mathematical analogy for physical phenomena depends, in part, on its justification by the accuracy of measurements and, in part, on theoretical and pragmatic issues arising in the resulting formulation and computations.

5.4 Exercises

Exercise 5.1. Show that the Laplacian of a scalar-valued function, $f(\mathbf{x})$, can be expressed in terms of a difference between the average value of f in the neighbourhood of a point, $\mathbf{x}_0$, and the value of f at that point.

Solution. The average of f over the volume of a sphere is

$$\overline{f}(\mathbf{x}) = \frac{1}{V}\iiint\limits_V f(\mathbf{x})\,\mathrm{d}V\,, \tag{5.56}$$

[44]Readers interested in the classic paper of the Cosserat brothers might refer to Cosserat and Cosserat (1909), and many subsequent works inspired by it.

where $V = 4\pi R^3/3$, and R is the radius. Expanding f into a Taylor series about $\mathbf{x}_0$, we write

$$f(\mathbf{x}) = f(\mathbf{x}_0) + \sum_{i=1}^{3}\left(x_i \left.\frac{\partial f}{\partial x_i}\right|_{\mathbf{x}_0}\right) + \frac{1}{2}\sum_{i=1}^{3}\sum_{j=1}^{3}\left(x_i x_j \left.\frac{\partial^2 f}{\partial x_i \partial x_j}\right|_{\mathbf{x}_0}\right) + \dots \quad .$$

Integrating, we write

$$\iiint\limits_V f(\mathbf{x})\,\mathrm{d}V = V f(\mathbf{x}_0) + \sum_{i=1}^{3}\left(\left.\frac{\partial f}{\partial x_i}\right|_{\mathbf{x}_0} \iiint\limits_V x_i\,\mathrm{d}V\right) \tag{5.57}$$

$$+ \frac{1}{2}\sum_{i=1}^{3}\sum_{j=1}^{3}\left(\left.\frac{\partial^2 f}{\partial x_i \partial x_j}\right|_{\mathbf{x}_0} \iiint\limits_V x_i x_j\,\mathrm{d}V\right) + \dots \,,$$

where we use the fact that f and its derivatives, evaluated at $\mathbf{x}_0$, are constants.

By spherical symmetry, $\iiint_V x_i\,\mathrm{d}V = 0$, as discussed in Exercise 5.2, and

$$\delta_{ij}\iiint\limits_V x_i x_j\,\mathrm{d}V = \iiint\limits_V x_i x_i\,\mathrm{d}V = \frac{1}{3}\iiint\limits_V r^2\,\mathrm{d}V \tag{5.58}$$

$$= \frac{1}{3}\int\limits_0^R r^2\left(4\pi r^2\right)\mathrm{d}r = \frac{V R^2}{5}\,, \qquad i,j = 1,2,3\,.$$

where δ_{ij} is the Kronecker delta and $r^2 = \sum_{i=1}^{3} x_i x_i$ represents the squared radius of any point on a sphere with centre at $\mathbf{x}_0$. The second equality in expression (5.58) requires a change of coordinates; this step is discussed also in Exercise 5.2.

Proceeding with these results, we write equation (5.57) as

$$\iiint\limits_V f(\mathbf{x})\,\mathrm{d}V = V f(\mathbf{x}_0) + \frac{V R^2}{10}\sum_{i=1}^{3}\left.\frac{\partial^2 f}{\partial x_i^2}\right|_{\mathbf{x}_0} + \dots \quad . \tag{5.59}$$

Ignoring higher order terms, we write the average stated in expression (5.56) as

$$\overline{f}(\mathbf{x}) = \frac{1}{V}\iiint\limits_V f(\mathbf{x})\,\mathrm{d}V \approx f(\mathbf{x}_0) + \frac{R^2}{10}\left.\nabla^2 f(\mathbf{x})\right|_{\mathbf{x}_0}\,,$$

where we recognize that the sum in expression (5.59) is the Laplacian evaluated at $\mathbf{x}_0$. Thus,

$$\left.\nabla^2 f(\mathbf{x})\right|_{\mathbf{x}_0} \approx \frac{10}{R^2}\left(\overline{f}(\mathbf{x}) - f(\mathbf{x}_0)\right)\,.$$

Examining this expression, we see that the Laplacian is a measure of the departure from the spatial average.

Exercise 5.2. To explain steps in Exercise 5.1, show that, by spherical symmetry, $\iiint_V x_i \,\mathrm{d}V = 0$, for $i = 1, 2, 3$. Also, justify the second equality in expression (5.58), namely,

$$\iiint\limits_V x_i x_i \,\mathrm{d}V \equiv \iiint\limits_V x_i^2 \,\mathrm{d}V = \frac{1}{3} \iiint\limits_V r^2 \,\mathrm{d}V\,, \qquad i = 1, 2, 3\,.$$

Solution. To show that $\iiint_V x_i \,\mathrm{d}V = 0$, let us express $\mathbf{x}$ in spherical coordinates and consider $x_3 = r\cos\alpha$. Using $\mathrm{d}V = r^2 \sin\alpha \,\mathrm{d}r \,\mathrm{d}\alpha \,\mathrm{d}\beta$, we write

$$\iiint\limits_V x_3 \,\mathrm{d}V = \int\limits_0^{2\pi} \int\limits_0^R \int\limits_0^\pi \sin\alpha \cos\alpha \,\mathrm{d}\alpha \, r^3 \,\mathrm{d}r \,\mathrm{d}\beta\,. \tag{5.60}$$

Since the innermost integral is zero, so is the triple integral. The same result would be obtained for $i = 1$ or $i = 2$, with appropriate expressions for x_1 and x_2, in spherical coordinates.

To justify the second equality in expression (5.58), again we consider $x_3 = r\cos\alpha$ to write

$$\iiint\limits_V x_3^2 \,\mathrm{d}V = \int\limits_0^{2\pi} \int\limits_0^R \int\limits_0^\pi \cos^2\alpha \sin\alpha \,\mathrm{d}\alpha \, r^4 \,\mathrm{d}r \,\mathrm{d}\beta\,.$$

Since the integral with respect to α is $2/3$, we write

$$\iiint\limits_V x_3^2 \,\mathrm{d}V = \frac{1}{3} \int\limits_0^{2\pi} \int\limits_0^R 2\, r^4 \,\mathrm{d}r \,\mathrm{d}\beta = \frac{1}{3} \iiint\limits_V r^2 \,\mathrm{d}V\,,$$

where $V = 4\pi R^3/3$, which is the required result. In accordance with the spherical symmetry, the same result would be obtained for $i = 1$ or $i = 2$, with appropriate expressions for x_1 and x_2, in spherical coordinates. Thus—with the corresponding expressions for $\mathrm{d}V$, in Cartesian and spherical coordinates—we can write

$$\iiint\limits_V \left(x_1^2 + x_2^2 + x_3^2\right) \,\mathrm{d}V = \iiint\limits_V r^2 \,\mathrm{d}V\,,$$

as expected.

Exercise 5.3. Using expression (3.57),

$$C^{\text{iso}} = \begin{bmatrix} c_{1111} & c_{1111} - 2c_{2323} & c_{1111} - 2c_{2323} & 0 & 0 & 0 \\ c_{1111} - 2c_{2323} & c_{1111} & c_{1111} - 2c_{2323} & 0 & 0 & 0 \\ c_{1111} - 2c_{2323} & c_{1111} - 2c_{2323} & c_{1111} & 0 & 0 & 0 \\ 0 & 0 & 0 & 2c_{2323} & 0 & 0 \\ 0 & 0 & 0 & 0 & 2c_{2323} & 0 \\ 0 & 0 & 0 & 0 & 0 & 2c_{2323} \end{bmatrix}, \tag{5.61}$$

find the stability conditions for isotropic solids. Express these conditions in terms of the ratio of the P-wave and S-wave speeds.

Solution. In general, the stability conditions imply the positive definiteness of the elasticity tensor,[45] which, for isotropic solids, means the positive definiteness of C^{iso}. Mathematically, positive definiteness is equivalent to the positivity of all eigenvalues, which herein are $\lambda_1 = 3\,c_{1111} - 4\,c_{2323}$ and $\lambda_2 = \ldots = \lambda_6 = 2\,c_{2323}$.[46] Thus, the stability conditions are $c_{2323} > 0$ and $c_{1111} > 4\,c_{2323}/3$.

Since $c_{ijk\ell}$ are the density-scaled elasticity parameters, it follows that the P-wave and S-wave speeds within isotropic solids are $v_P = \sqrt{c_{1111}}$ and $v_S = \sqrt{c_{2323}}$, respectively. Thus, the stability conditions are tantamount to $v_P/v_S > 2/\sqrt{3}$, which means that P waves are at least 15% faster than S waves.

Exercise 5.4. Consider the empirical relation of Gardner *et al.* (1974), which—in the *SI* units—is

$$\rho = 0.31\,\sqrt[4]{v_P}\,, \tag{5.62}$$

where ρ stands for mass density and v_P for the P-wave propagation speed. Also, consider the propagation speed in equation (5.9),

$$v_P = \sqrt{\frac{c_{1111}}{\rho}}\,. \tag{5.63}$$

[45] *see also*: Slawinski (2015, Section 4.3)

[46] *see also*: Bóna *et al.* (2007a, Section 4.1)

Note therein a different parametrization of the elasticity parameters in expression (9), which results in its being positive-definite, even if $c_{12} = 0$. However, the case of $c_{12} = 0$ is excluded from the theorem that characterizes isotropy (Bóna *et al.*, 2007a, Theorem 4.1). According to our notation, $c_{12} \equiv c_{1122}$.

Expression (5.61) becomes positive-semidefinite if $c_{2323} = 0$, and it can be interpreted as corresponding to a fluid. If $c_{1111} = 0$, however, expression (5.61) is not even positive-semidefinite, and we do not have its physical interpretation.

Explain the apparent paradox between the direct proportionality and the inverse proportionality in formulæ $v_P \propto \rho^4$ and $v_P \propto 1/\sqrt{\rho}$, stated in expressions (5.62) and (5.63), respectively.

Solution. Let us write relation (5.62), as

$$v_P = \left(\frac{1}{0.31}\right)^4 \rho^4 = 108.28\,\rho^4\,.$$

Substituting this result for the left-hand side of expression (5.63), we write

$$108.28\,\rho^4 = \sqrt{\frac{c_{1111}}{\rho}}\,.$$

Solving for c_{1111}, we obtain

$$c_{1111} = 11724.8\,\rho^9\,.$$

Using this empirical relation, we infer that the value of c_{1111} grows as the ninth power of ρ. Thus, mechanical properties are not independent from the mass density, and—in accordance with expression (5.63)—the value of v_P grows with ρ, which is consistent with relation (5.62).

Exercise 5.5. Shown that—to be applied to solution (5.41)—the initial conditions must be spherically symmetric. Proceed to find the solution.

Solution. Let us write solution (5.41) in terms of its initial conditions,

$$u\,(r,t)\,|_{t=0} = \frac{1}{r}f\,(r) + \frac{1}{r}g\,(r) = \gamma$$

and

$$\left.\frac{\partial u\,(r,t)}{\partial t}\right|_{t=0} = \frac{v}{r}f'\,(r) - \frac{v}{r}g'\,(r) = \eta\,,$$

where $'$ stands for the derivative with respect to the argument, $(\cdot)$. The uniqueness of solution requires that both γ and η be function of r alone, which means that they must be spherically symmetric.

Solving the above two equations for f and g and inserting the resulting expressions into solution (5.41), we obtain the particular solution given by[47]

$$u(r,t) = \frac{1}{2\,r}\left(\gamma(r+vt) + \gamma(r-vt) + \frac{r}{v}\int\limits_{r-vt}^{r+vt}\eta(\zeta)\,\mathrm{d}\zeta\right)\,,$$

which is a function of a single spatial variable, r, even though—as a radius of a sphere—r can be interpreted in the three-dimensional space.

[47] *see also*: Slawinski (2015, Section 6.5.1)

Exercise 5.6. In view of solution (5.41), consider the change of pressure, which is proportional to $1/r$. Interpret this result in the context of solution (5.27).

Solution. According to solution (5.27), both f and g remain constant as they propagate along $\mathbb{R}^1$. To consider an analogous property in $\mathbb{R}^3$, we use the fact that the amplitude is proportional to the square of pressure. Hence, the change of amplitude with the distance from the origin is proportional to $1/r^2$. We proceed to show that, as expected from solution (5.27), amplitude remains constant if integrated over an expanding spherical wavefront.

In spherical coordinates, a surface element is $\mathrm{d}S = r^2 \sin\alpha \,\mathrm{d}\alpha \,\mathrm{d}\beta$. Herein, the amplitude over a sphere whose radius is r is

$$\iint\limits_S \frac{1}{r^2}\,\mathrm{d}S = \int\limits_0^{2\pi}\int\limits_0^{\pi} \sin\alpha \,\mathrm{d}\alpha \,\mathrm{d}\beta\,,$$

which does not depend on r, and—in view of a spherical symmetry—it is a constant: 4π.

In $\mathbb{R}^1$, since there is no geometrical spreading, the constancy of amplitude is tantamount to the constancy of pressure, but not in $\mathbb{R}^3$, since, in contrast to the amplitude, which is proportional to the surface area, the pressure is proportional to its square root.

Remark 5.1. Following the formulation presented in Section 5.2.3.3 and in view of the pattern discussed in Exercise 5.6 and exhibited by solution (5.41), we might wish to write the solution in two spatial dimensions as

$$u(r,t) = \frac{1}{\sqrt{r}} f(r+vt) + \frac{1}{\sqrt{r}} g(r-vt)\,. \tag{5.64}$$

Furthermore, since an arclength element in polar coordinates is $\mathrm{d}s = r\,\mathrm{d}\alpha$, the amplitude over a circle whose circumference is C and the radius is r is

$$\int\limits_C \frac{1}{\left(\sqrt{r}\right)^2}\,\mathrm{d}s = \int\limits_0^{2\pi} \mathrm{d}\alpha = 2\pi\,,$$

which is a constant, as required. Also, in view of solution (5.64), the radially symmetric wave equation in polar coordinates would be

$$\frac{\partial^2(\sqrt{r}\,F)}{\partial r^2} - \frac{1}{v^2}\frac{\partial^2(\sqrt{r}\,F)}{\partial t^2} = 0\,,$$

which would imply the corresponding Laplacian as

$$\nabla^2 F = \frac{1}{\sqrt{r}} \frac{\partial^2 (\sqrt{r}\, F)}{\partial r^2}\,.$$

However, to justify such a Laplacian and, in general, the approach presented in Section 5.2.3.3, a single value of n needs to result in

$$\frac{1}{r^n} \frac{\partial^2}{\partial r^2} (r^n F) = \frac{1}{r} \frac{\partial F}{\partial r} + \frac{\partial^2 F}{\partial r^2}\,.$$

Herein, the right-hand side is the Laplacian in polar coordinates for a radially symmetric scalar function, $F(r)$. Comparing it with expression (5.40), we see that it differs, with respect to spherical coordinates, by the factor of 2 in the numerator.

To search for the possible value of n, we differentiate the left-hand side to obtain

$$\frac{1}{r^n} \frac{\partial^2}{\partial r^2} (r^n F) = n(n-1) r^{-2} F + \frac{2\,n}{r} \frac{\partial F}{\partial r} + \frac{\partial^2 F}{\partial r^2}\,.$$

Comparing this result to the right-hand side, we see that there is no such value. For the first term to vanish, we need $n = 0$ or $n = 1$; for the required form of the middle term, we need $n = 1/2$.

Thus, we cannot invoke the approach discussed in Section 5.2.3.3 to obtain the solution in two spatial dimensions. Properties exhibited by the wave equation in one and three spatial dimensions are not shared by a two-dimensional case. Notably, it is a property of hyperbolic partial differential equations that the forms of their solutions in odd spatial dimensions are different from the forms in even dimensions.[48]

Exercise 5.7. Perform the Fourier transform of

$$\nabla^2 G(\mathbf{x}, t) - \frac{1}{v^2} \frac{\partial^2 G(\mathbf{x}, t)}{\partial t^2} = -\delta(\mathbf{x}) \delta(t)\,, \tag{5.65}$$

with t and ω being the transformation variables, and find its Green's function, G, for the case of a single spatial dimension.

Solution. Using properties of the Fourier transform, we write

$$\nabla^2 \widehat{G}(\mathbf{x}, \omega) + \left(\frac{\omega}{v}\right)^2 \widehat{G}(\mathbf{x}, \omega) = \delta(\mathbf{x})\,.$$

In a single spatial dimension, this partial differential equation becomes an ordinary one,

$$\frac{\mathrm{d}^2 \widehat{G}(x, \omega)}{\mathrm{d}x^2} + \kappa^2 \widehat{G}(x, \omega) = \delta(x)\,, \tag{5.66}$$

[48] *see also*: Slawinski (2015, Sections 6.5 and 6.6)

where $\kappa := \omega/v$ denotes the *spatial frequency*. For $x \neq 0$, function $g(x) = \widehat{G}(x,\omega)$ must satisfy the ordinary differential equation given by

$$\frac{\mathrm{d}^2 g}{\mathrm{d}x^2} + \kappa^2 g = 0 \tag{5.67}$$

and whose solution we write as

$$\widehat{G}(x,\omega) = A\exp(\iota\kappa x) + B\exp(-\iota\kappa x)\,. \tag{5.68}$$

To find A and B, we proceed in the following manner. To use the inverse Fourier transform to obtain G, we must deal with integrands given by

$$A\exp(\iota\kappa x)\exp(-\iota\omega t) = A\exp(-\iota(\omega t - \kappa x)) = A\exp\left(-\iota\omega\left(t - \frac{x}{v}\right)\right)$$

and

$$B\exp(-\iota\kappa x)\exp(-\iota\omega t) = B\exp(-\iota(\omega t + \kappa x)) = B\exp\left(-\iota\omega\left(t + \frac{x}{v}\right)\right),$$

which correspond to the right- and left-propagating waves.[49] Since we consider waves propagating away from the source at $x = 0$, we write solution (5.68) as

$$\widehat{G}(x,\omega) = \begin{cases} B\exp(-\iota\kappa x)\,, & x < 0 \\ A\exp(\iota\kappa x)\,, & x > 0 \end{cases}. \tag{5.69}$$

To deal with the case of $x = 0$, since the right-hand side of equation (5.66) is $\delta(x)$, we require $\widehat{G}$ to be continuous at $x = 0$ and $\mathrm{d}G/\mathrm{d}x|_{x=0}$ to be discontinuous, with discontinuity being equal to unity, as implied by Heaviside's step. Continuity of $\widehat{G}$ means that $A = B$ at $x = 0$. Discontinuity of the derivative means that

$$\left.\frac{\mathrm{d}\widehat{G}}{\mathrm{d}x}\right|_{x=0^-}^{x=0^+} = \iota\kappa B \lim_{x\to 0^-}\exp(-\iota\kappa x) + \iota\kappa A \lim_{x\to 0^+}\exp(\iota\kappa x) = 1\,.$$

Taking the limits and using the fact that $A = B$, we write

$$\left.\frac{\mathrm{d}\widehat{G}}{\mathrm{d}x}\right|_{x=0^-}^{x=0^+} = 2\iota\kappa A = 1\,;$$

hence,

$$A = B = -\frac{\iota}{2\kappa}$$

[49] *see also*: Slawinski (2015, Section 6.5.1)

and solution (5.69) is

$$\widehat{G}(x,\omega)=\begin{cases}-\dfrac{\iota}{2\kappa}\exp(-\iota\kappa x)\,, & x<0\\[2ex] -\dfrac{\iota}{2\kappa}\exp(\iota\kappa x)\,, & x>0\end{cases}\,,$$

which we write concisely as $\widehat{G}=-\iota\exp(\iota\kappa|x|)/(2\kappa)$.

To find G, we perform the inverse Fourier transform. Using $\kappa:=\omega/v$, we write

$$G(x,t)=-\int\limits_{-\infty}^{\infty}\frac{\iota v}{2\omega}\exp\left(-\iota\frac{x}{v}\omega\right)\exp\left(\iota\omega t\right)\mathrm{d}\omega=\frac{v}{2\iota}\int\limits_{-\infty}^{\infty}\frac{\exp\left(-\iota\omega\left(\frac{x}{v}-t\right)\right)}{\omega}\mathrm{d}\omega$$

and

$$G(x,t)=-\int\limits_{-\infty}^{\infty}\frac{\iota v}{2\omega}\exp\left(\iota\frac{x}{v}\omega\right)\exp\left(\iota\omega t\right)\mathrm{d}\omega=\frac{v}{2\iota}\int\limits_{-\infty}^{\infty}\frac{\exp\left(\iota\omega\left(\frac{x}{v}+t\right)\right)}{\omega}\mathrm{d}\omega\,,$$

which correspond to $x<0$ and $x>0$, respectively. These integrals can be written as

$$\int\frac{\exp(\iota a\omega)}{\omega}\mathrm{d}\omega=:Ei(\iota ax)\,,$$

where Ei denotes the *exponential integral*. However, even without such a definition, the inverse transform is[50]

$$G(x,t)=\begin{cases}0\,, & t<t_0+\dfrac{x-x_0}{v}\\[2ex] -\dfrac{1}{2v}\,, & t\geqslant t_0+\dfrac{x-x_0}{v}\end{cases}\,.$$

---

[50] *see also*: Bóna and Slawinski (2015, Appendix F.2.3)

Chapter 6

Surface, guided and interface waves

The upper part of the earth's crust is known to be made up of many different kinds of materials having a very wide range of elastic properties, so that great many internal boundaries are apt to exist in nature. The effect of these boundaries will be to complicate the wave picture beyond any reasonable hope of mathematical analysis, and any further advancement of the theory will be achieved only through the use of simple models. Ideal assumptions must be imposed upon the elastic properties of the earth materials, upon the nature and shape of the internal boundaries, and upon the physical conditions at the seismic source.

Fraser S. Grant and Gordon F. West (1965)

Preliminary remarks

In this chapter, we examine the wave equations and aspects of their solutions for the surface and interface waves. The side conditions, which describe the bounding surfaces and interfaces, play a crucial role in formulating the corresponding solutions. These conditions limit the solubility of the wave equation by imposing the constraints on its general solution.

We focus our attention on several ideal scenarios of boundaries. Also, we restrict our examination to isotropic Hookean solids.

In contrast to subjects of most chapters, where the reader is referred for details to Slawinski (2015) or Bóna and Slawinski (2015), the subject of surface, guided and interface waves is not discussed in either of these volumes. Hence, derivations herein contain more details than in other chapters.

We begin this chapter with discussions of surface and guided waves. We proceed to discuss interface waves. In each case, we examine the speed of the wave—in terms of elasticity parameters and mass densities—and the ranges of material properties within which propagation of a given wave is possible.

6.1 Introduction

It is common to say that there are two types of waves that propagate in the Earth: P and S waves. Strictly speaking, this statement is false. In view of our discussions in Section 5.1, above, the correct statement is as follows. There are two types of waves that propagate in an isotropic Hookean solid of infinite extent.

However, even though remaining within the mathematical realm, these waves serve—as analogies for seismic disturbances within the Earth—to infer physical properties of the subsurface. An example of such an inference is the lack of rigidity of the subsurface that one might infer from the absence of shear waves, since these waves do not propagate in a continuum with no resistance to a change in shape.

Applying the side conditions to the wave equations, which restrict the extent of the medium, we can infer further physical properties from the behaviour of these waves in their mathematical realm. For instance, observations of surface disturbances analogous to the SH waves, which cannot exist within an infinite halfspace unless we introduce an interface, lead us to infer that the Earth's crust is layered.

In this chapter, we focus our attention on interactions of body waves that are a consequence of boundaries within Hookean solids. The question that arises is whether or not displacements that correspond to the two waves in an unbounded medium are admissible under constraints of surfaces and interfaces.

To formulate the equations of motion at the beginning of Section 5.1 and to derive the corresponding wave equation, we consider homogeneous and isotropic Hookean solids of infinite extent. We refer to such waves as body waves, as opposed to the surface and interface waves, which are the subject of this chapter. As we demonstrate herein, for an isotropic case, the surface, guided and interface waves result from interactions of the two types of waves admissible within the corresponding unbounded Hookean solid.

As shown in Section 5.1.2, above, there are two types of waves that propagate in an unbounded homogeneous isotropic Hookean solid: P waves and S waves. Let us denote their propagation speed by α and β, respectively. P waves and S waves exhibit, respectively, displacements parallel and perpendicular to the direction of propagation; the specific orientation of the latter cannot be determined due to only two distinct eigenspaces in three dimensions.[1] This is a consequence of isotropy, where—unlike for the anisotropic Hookean solid—two *quasishear* waves are reduced to a single S wave. Furthermore, it follows from the stability conditions of Hookean solids that $\alpha > 2\beta/\sqrt{3}$.[2] Also, since the expressions for α and β are frequency-independent, these waves are nondispersive.

Remark 6.1. The three waves that appear in anisotropic Hookean solids are commonly denoted by qP, qS_1 and qS_2, where q stands for 'quasi'. They represent the quasi-P and quasi-S waves, since the displacement vectors of these waves are neither parallel nor perpendicular to the direction of propagation, but—in many cases—these displacements are almost so.[3] Subscripts $_1$ and $_2$ indicate that, in general, either quasishear wave propagates at a different speed, which is commonly referred to as shear-wave splitting.

In isotropic Hookean solids, there are two waves, P and S, and their displacement vectors are, respectively, parallel and perpendicular to their propagation directions. In isotropy, the qS_1 and qS_2 waves are reduced to a single S wave, which is a result of a double root for a cubic polynomial.

Given a Cartesian coordinate system—whose axes are viewed as vertical and horizontal—we can denote by SH the S wave whose displacement vector is horizontal, and by SV the one whose displacement vector is in the vertical plane. However, this is a convention that is not rooted in intrinsic properties of these waves, since—due to the double root—their displacement is any direction perpendicular to the direction of propagation.[4]

One can combine both terminologies. For instance, in transversely isotropic media, whose rotational symmetry axis is vertical, the shear-wave splitting results in an SH wave, whose displacement is within a horizontal plane, which is isotropic, and a qSV wave, whose displacement is within a vertical plane, which is anisotropic.

The presence of surfaces and interfaces restricts displacements of waves to particular orientations with respect to these boundaries. Furthermore, by interference, the resulting waves might exhibit hybrids of standard dis-

[1] *see also*: Slawinski (2015, Remark 9.4.1)
[2] *see also*: Slawinski (2015, Exercise 5.18)
[3] *see also*: Slawinski (2015, Figure 9.4.1)
[4] *see also*: Slawinski (2015, Remark 9.4.1)

placements, which are parallel and perpendicular to the direction of propagation. For instance, the Rayleigh waves, discussed in Section 6.2, below, exhibit an elliptical displacement within a plane perpendicular to the surface along which they propagate. Also, propagation speeds of the surface and interface waves are distinct from α and β, which are the speeds of their body-wave counterparts. Furthermore, unlike the P and S waves themselves, in certain cases, the resulting waves are dispersive.

Mathematical study of surface waves, guided waves and interface waves began in the nineteenth century and constitutes a substantial subject within mathematical physics, in general, with specific motivations from, and results applied to, seismology. In the remainder of this chapter, we examine aspects of this subject. Since the surface, guided and interface waves constitute a rich example of mathematical analogies for physical phenomena, and to foster an insightful examination, we proceed to formulate their expressions in detail, without delegating derivations to the exercises at the end of the chapter. The assumptions invoked in these derivations are the essence of mathematical concepts whose purpose is to serve as quantitative analogies for observations. If an analogy is satisfactory, in the sense of the mathematical predictions exhibiting a satisfactory agreement with physical measurements, then, in the aforecited words of Eugene Wigner (1960), we enjoy a privilege of "the unreasonable effectiveness of mathematics in the natural sciences".

6.2 Surface waves: Homogeneous elastic halfspace

Let us consider a homogeneous Hookean solid that constitutes a halfspace. For convenience, let us set the coordinate system in such an orientation that the surface limiting this halfspace coincides with the x_1x_2-plane, which we assume to be horizontal; beyond this surface, there is vacuum. Also, let us consider only waves propagating in the direction of the x_1-axis and whose displacements are contained in the vertical plane containing that axis, which is the x_1x_3-plane; in other words, we consider a propagation composed of the P and SV waves. The resulting wave is known as the *Rayleigh wave* in honour of John William Strutt, 3rd Baron Rayleigh.[5]

We need not consider the SH waves, since they are decoupled from the other two. Neither the P waves nor the SV waves propagating in the x_1x_3-

[5]Readers interested in historical documents of theoretical developments might refer to Rayleigh (1945).

plane have a displacement component in the x_2-direction, which is the only displacement direction of the SH waves propagating in that plane.[6]

In general, but remaining within isotropy, we can write the displacement function in terms of the scalar potential, $\mathcal{P}$, and vector potential, $\boldsymbol{\mathcal{S}}$, as[7]

$$\mathbf{u}(\mathbf{x},t) = \nabla\mathcal{P}(\mathbf{x},t) + \nabla \times \boldsymbol{\mathcal{S}}\,(\mathbf{x},t)\,. \tag{6.1}$$

In this expression, the use of symbols $\mathcal{P}$ and $\boldsymbol{\mathcal{S}}$ is motivated by the fact that these potentials entail the P and S waves, respectively.

As shown by Long (1967), in the context of expression (6.1), $\mathcal{P}$ and $\boldsymbol{\mathcal{S}}$ represent the entire displacement field, $\mathbf{u}$, if $\mathcal{P}$ and $\boldsymbol{\mathcal{S}}$ satisfy the corresponding wave equations.

Explicitly, we write expression (6.1) as

$$u_1 = \frac{\partial\mathcal{P}}{\partial x_1} + \frac{\partial\mathcal{S}_3}{\partial x_2} - \frac{\partial\mathcal{S}_2}{\partial x_3}\,,$$

$$u_2 = \frac{\partial\mathcal{P}}{\partial x_2} + \frac{\partial\mathcal{S}_1}{\partial x_3} - \frac{\partial\mathcal{S}_3}{\partial x_1}\,,$$

$$u_3 = \frac{\partial\mathcal{P}}{\partial x_3} + \frac{\partial\mathcal{S}_2}{\partial x_1} - \frac{\partial\mathcal{S}_1}{\partial x_2}\,.$$

Since $\boldsymbol{\mathcal{S}}$ is defined only up to the gradient of an arbitrary function, f, it can be replaced by $\tilde{\boldsymbol{\mathcal{S}}} = \boldsymbol{\mathcal{S}} + \nabla f$, where, in components, $\tilde{\mathcal{S}}_i = \mathcal{S}_i + \partial f/\partial x_i$, with $i = 1, 2, 3$. Given $\mathcal{S}$, we can find f such that $\tilde{\mathcal{S}}_1 = \tilde{\mathcal{S}}_3 = 0$.[8]

Considering displacements in the x_1x_3-plane and omitting the tilde for brevity of notation, we write the components of $\mathbf{u}$ as

$$u_1 = \frac{\partial\mathcal{P}}{\partial x_1} - \frac{\partial\mathcal{S}_2}{\partial x_3} \tag{6.2}$$

and

$$u_3 = \frac{\partial\mathcal{P}}{\partial x_3} + \frac{\partial\mathcal{S}_2}{\partial x_1}\,, \tag{6.3}$$

which correspond to both P and SV waves, where $u_2 = 0$. To express $\mathcal{P}$ and $\boldsymbol{\mathcal{S}}$ in terms of the Laplacian, we write

$$\frac{\partial u_1}{\partial x_1} = \frac{\partial^2\mathcal{P}}{\partial x_1^2} - \frac{\partial^2\mathcal{S}_2}{\partial x_1\partial x_3}\,,$$

[6]Readers interested in further study might refer to Achenbach (1973, Section 5.11), Brekhovskikh and Goncharov (1994, Section 4.3.1), Ewing *et al.* (1957, Chapter 2), Grant and West (1965, Section 3-2) and Hanyga (1984, Section 1.2.4), as well as other books on this subject, including a didactic exposition by Krebes (2004, pp. 4-1 to 4-3).

[7]*see also*: Slawinski (2015, Section 6.3)

[8]*see also*: Slawinski (2015, Section 6.3.2)

$$\frac{\partial u_1}{\partial x_3} = \frac{\partial^2 \mathcal{P}}{\partial x_1 \partial x_3} - \frac{\partial^2 \mathcal{S}_2}{\partial x_3^2}\,, \tag{6.4}$$

$$\frac{\partial u_3}{\partial x_1} = \frac{\partial^2 \mathcal{P}}{\partial x_1 \partial x_3} + \frac{\partial^2 \mathcal{S}_2}{\partial x_1^2} \tag{6.5}$$

and

$$\frac{\partial u_3}{\partial x_3} = \frac{\partial^2 \mathcal{P}}{\partial x_3^2} + \frac{\partial^2 \mathcal{S}_2}{\partial x_1 \partial x_3} \tag{6.6}$$

to write the divergence of $\mathbf{u}$ as

$$\nabla \cdot \mathbf{u} = \frac{\partial u_1}{\partial x_1} + \frac{\partial u_3}{\partial x_3} = \frac{\partial^2 \mathcal{P}}{\partial x_1^2} + \frac{\partial^2 \mathcal{P}}{\partial x_3^2} =: \nabla^2 \mathcal{P} \tag{6.7}$$

and the second component of the curl of $\mathbf{u}$ as

$$(\nabla \times \mathbf{u})_2 = -\left(\frac{\partial u_1}{\partial x_3} - \frac{\partial u_3}{\partial x_1}\right) = \frac{\partial^2 \mathcal{S}_2}{\partial x_1^2} + \frac{\partial^2 \mathcal{S}_2}{\partial x_3^2} =: \nabla^2 \mathcal{S}_2\,, \tag{6.8}$$

where ∇^2 is the Laplace differential operator. For conciseness, we denote $\mathcal{S}_2$ by $\mathcal{S}$ to write the wave equations as

$$\nabla^2 \mathcal{P} - \frac{1}{\alpha^2}\frac{\partial^2 \mathcal{P}}{\partial t^2} = 0 \tag{6.9}$$

and

$$\nabla^2 \mathcal{S} - \frac{1}{\beta^2}\frac{\partial^2 \mathcal{S}}{\partial t^2} = 0\,, \tag{6.10}$$

where α and β are the speeds of the P wave and S wave, respectively.[9] Herein, these equations can be obtained by inserting expression (6.1) into equation (5.5), using vector identities and letting $c_{1111}/\rho = \alpha^2$ and $c_{2323}/\rho = \beta^2$.[10]

To solve equations (6.9) and (6.10), let us set the ansatz, which is the term used for the trial solution, to be

$$\mathcal{P} = A(x_3)\exp(\iota(\kappa x_1 - \omega t)) \tag{6.11}$$

and

$$\mathcal{S} = B(x_3)\exp(\iota(\kappa x_1 - \omega t))\,, \tag{6.12}$$

respectively, where $\iota := \sqrt{-1}$. These expressions correspond to sinusoidal waves with frequency ω that propagate in the x_1-direction with speed ω/κ, and whose amplitude varies with the x_3-direction.

[9] *see also*: Slawinski (2015, Sections 6.1.2 and 6.1.3)
[10] *see also*: Slawinski (2015, Sections 6.3.3 and 6.3.4)

Herein, we are dealing with linear equations with real coefficients. Hence, if a complex-valued function is a solution, so are—separately—its real and imaginary parts. Thus, we can use complex-valued functions in our work, as long as we remember that we seek real solutions. Such solutions might be explicitly imposed by the real side conditions of the differential equations.

Inserting expressions (6.11) and (6.12) into equations (6.9) and (6.10), as shown in Exercise 6.1, we obtain

$$\frac{\mathrm{d}^2 A}{\mathrm{d}x_3^2} + \left(\frac{\omega^2}{\alpha^2} - \kappa^2\right) A = 0 \tag{6.13}$$

and

$$\frac{\mathrm{d}^2 B}{\mathrm{d}x_3^2} + \left(\frac{\omega^2}{\beta^2} - \kappa^2\right) B = 0 \,,$$

respectively. These are ordinary differential equations. If the term in parentheses is positive, these equations have a form of the classic expression for a simple harmonic motion, whose solution is a sinusoid, and A and B are the amplitudes of the oscillation with values to be determined by the side conditions. If that term is negative, the solution is exponential. As it becomes apparent below, we are interested in the latter case.

The general solutions are

$$A = C_1 \exp\left(-\iota\sqrt{\frac{\omega^2}{\alpha^2} - \kappa^2}\, x_3\right) + C_2 \exp\left(\iota\sqrt{\frac{\omega^2}{\alpha^2} - \kappa^2}\, x_3\right) \tag{6.14}$$

and

$$B = D_1 \exp\left(-\iota\sqrt{\frac{\omega^2}{\beta^2} - \kappa^2}\, x_3\right) + D_2 \exp\left(\iota\sqrt{\frac{\omega^2}{\beta^2} - \kappa^2}\, x_3\right) , \tag{6.15}$$

as one can verify by inserting $y = C_1 \exp(-\iota\sqrt{a}x) + C_2 \exp(\iota\sqrt{a}x)$ into $y'' + ay = 0$. Let us choose the case for which the radicands are negative; hence, the square roots are imaginary and, consequently, the exponents are real.

We justify this choice in terms of physical considerations, namely, within the bottomless halfspace, we require the diminishing of the amplitudes with increasing depth. Herein, in agreement with the seismological convention, the depth increases along the x_3-axis, which points downwards.

Thus, we write $\sqrt{\omega^2/\alpha^2 - \kappa^2}$ as $\iota\sqrt{\kappa^2 - \omega^2/\alpha^2}$ and $\sqrt{\omega^2/\beta^2 - \kappa^2}$ as $\iota\sqrt{\kappa^2 - \omega^2/\beta^2}$, in expressions (6.14) and (6.15), to get

$$A = C_1 \exp\left(\sqrt{\kappa^2 - \frac{\omega^2}{\alpha^2}}\, x_3\right) + C_2 \exp\left(-\sqrt{\kappa^2 - \frac{\omega^2}{\alpha^2}}\, x_3\right) \tag{6.16}$$

and

$$B = D_1 \exp\left(\sqrt{\kappa^2 - \frac{\omega^2}{\beta^2}}\, x_3\right) + D_2 \exp\left(-\sqrt{\kappa^2 - \frac{\omega^2}{\beta^2}}\, x_3\right),$$

where the radicands are positive numbers. The first term in either expression grows exponentially and the second one decreases exponentially.

Remark 6.2. The condition for an exponentially decaying solution, which requires positive radicands, can be obtained directly from equation (6.13). To do so, we invoke its auxiliary equation, and restrict the discriminant to positive values, which implies

$$\frac{\omega^2}{\alpha^2} - \kappa^2 < 0\,.$$

In such a case, the roots of the auxiliary equation,

$$\pm\sqrt{\kappa^2 - \frac{\omega^2}{\alpha^2}}\,,$$

are real. The same is true for β.

Note that to consider the amplitude condition, we do not examine the potentials, $\mathcal{P}$ and $\mathcal{S}$, which do not have a direct physical meaning. We examine the displacement, $\mathbf{u}$, which is directly related to energy.

Also note that our allowance of the wave propagation *ad infinitum* in the x_1-direction, albeit a mathematical idealization, is consistent with our expectation of observing such a phenomenon away from its source, even though, admittedly, with the amplitude decreasing with distance from the source. Our disallowance of the unbounded growth in the x_3-direction is tantamount to not accepting the infinite energy propagating towards the interior of the Earth. Waves whose amplitude decreases exponentially in a direction different than the direction of their propagation are called evanescent waves.

To avoid unbounded growth of amplitude with depth, we eliminate the first terms in the above expressions and keep only the second terms; thus, we set $C_1 = D_1 = 0$. Letting, for conciseness, $C := C_2 \neq 0$ and $D := D_2 \neq 0$, we write

$$A(x_3) = C \exp\left(-\sqrt{\kappa^2 - \frac{\omega^2}{\alpha^2}}\, x_3\right)$$

and

$$B(x_3) = D \exp\left(-\sqrt{\kappa^2 - \frac{\omega^2}{\beta^2}}\, x_3\right).$$

Inserting these expressions for $A(x_3)$ and $B(x_3)$ in expressions (6.11) and (6.12), respectively, we obtain

$$\mathcal{P}(\mathbf{x},t) = C\exp\left(-\sqrt{\kappa^2 - \frac{\omega^2}{\alpha^2}}\,x_3\right)\exp\left(\iota(\kappa x_1 - \omega t)\right) \tag{6.17}$$

and

$$\mathcal{S}(\mathbf{x},t) = D\exp\left(-\sqrt{\kappa^2 - \frac{\omega^2}{\beta^2}}\,x_3\right)\exp\left(\iota(\kappa x_1 - \omega t)\right), \tag{6.18}$$

respectively. In either expression, the first exponent corresponds to the decrease of amplitude with depth, along the x_3-axis, and the second exponent corresponds to propagation parallel to the x_1-axis. We can write this decrease in terms of the propagation speed, v, of the interface wave as

$$\exp\left(-\omega\sqrt{\frac{1}{v^2} - \frac{1}{\alpha^2}}\,x_3\right) \tag{6.19}$$

and

$$\exp\left(-\omega\sqrt{\frac{1}{v^2} - \frac{1}{\beta^2}}\,x_3\right),$$

where we use $v := \omega/\kappa$. As required by the decrease of amplitude with depth, the positive radicands imply $v < \alpha$ and $v < \beta$.

Herein, we introduce a condition that is required by the presence of a surface that is an interface between a Hookean solid and the vacuum. The absence of a continuum above this surface implies that there is no transmission of stresses generated in the x_1x_3-plane below that surface across the x_1x_2-plane.[11] Hence—due to continuity of traction—at that surface, which corresponds to $x_3 = 0$, we set $\sigma_{33} = 0$ and $\sigma_{13} = 0$.[12] Also, we set $\sigma_{23} = 0$, but this condition is satisfied identically, since $u_2 = 0$ and u_3 has no dependence on x_2.

Invoking Hooke's law for isotropic solids, which is equation (3.41) with the elasticity-tensor components given in matrix (3.57), we write

$$\sigma_{ij} = (C_{11} - 2C_{44})\,\delta_{ij}\sum_{k=1}^{3}\frac{\partial u_k}{\partial x_k} + C_{44}\left(\frac{\partial u_i}{\partial x_j} + \frac{\partial u_j}{\partial x_i}\right), \tag{6.20}$$

where $i,j = 1,2,3$; in this chapter, $C_{11} \equiv c_{1111}$ and $C_{44} \equiv c_{2323}$. Using expressions (6.6) and (6.7), we write a component of Hooke's law (6.20) as

$$\sigma_{33} = C_{11}\frac{\partial^2\mathcal{P}}{\partial x_3^2} + (C_{11} - 2C_{44})\frac{\partial^2\mathcal{P}}{\partial x_1^2} + 2C_{44}\frac{\partial^2\mathcal{S}}{\partial x_1\partial x_3} = 0\,.$$

[11]For further explanations, readers might refer to Chapman (2004, Section 4.3.3).

[12]*see also*: Slawinski (2015, Section 10.2.1)

Inserting into this result expressions (6.17) and (6.18), and evaluating at $x_3 = 0$, we obtain

$$\left(C_{11}\left(\kappa^2-\frac{\omega^2}{\alpha^2}\right)-(C_{11}-2C_{44})\kappa^2\right)C-2\,\iota\,\kappa\,C_{44}\sqrt{\kappa^2-\frac{\omega^2}{\beta^2}}\,D=0\,. \tag{6.21}$$

Similarly, invoking Hooke's law (6.20), we write its other component as

$$\sigma_{13}=C_{44}\left(\frac{\partial u_1}{\partial x_3}+\frac{\partial u_3}{\partial x_1}\right)=0\,,$$

which, using expression (6.4), (6.5), (6.17) and (6.18), becomes

$$2\iota\,\kappa\sqrt{\kappa^2-\frac{\omega^2}{\alpha^2}}\,C+\left(2\kappa^2-\frac{\omega^2}{\beta^2}\right)D=0\,. \tag{6.22}$$

C and D are constants related by equations (6.21) and (6.22) that result from dynamic side conditions evaluated at the surface; those relations between the constants can be applied in the entire halfspace. In general, the stress within the halfspace is not zero, but its change must be continuous.[13]

Equations (6.21) and (6.22) are a system of homogeneous linear equations to be solved for C and D. The existence of a nontrivial solution requires the determinant of the coefficient matrix to be zero,

$$\left(\frac{v}{\beta}\right)^6-8\left(\frac{v}{\beta}\right)^4+8\left(3-2\frac{\beta^2}{\alpha^2}\right)\left(\frac{v}{\beta}\right)^2+16\left(\frac{\beta^2}{\alpha^2}-1\right)=0\,, \tag{6.23}$$

where we let $v := \omega/\kappa$. This is a bicubic equation for the ratio between the speed of the surface wave, v, and β. We see that v is frequency-independent; hence, the surface wave is nondispersive, which implies that it does not change its form while propagating parallel to the x_1-axis. Its decrease along the x_3-axis is proportional to frequency, as shown in expression (6.19).

Let us examine equation (6.23) to consider several cases within the realm of a Hookean solid. To do so, we choose four values of Poisson's ratio that span its allowable range from -1 to $1/2$.[14]

The first case to consider is a Hookean solid whose Poisson's ratio is $1/4$, which is a representative value for many elastic materials and which is tantamount to the Lamé parameters being equal to one another. For this case, $\beta^2/\alpha^2 = 1/3$, and the only root of equation (6.23) that is smaller than unity is $v/\beta = 0.92$. To consider Rayleigh waves, we must choose this root. The

[13] For an illustration, readers might refer to Achenbach (1973, Figure 5.14).

[14] *see also*: Slawinski (2015, Remark 5.14.7, p. 194)

choice of $v/\beta < 1$ is required to make the radicands in expressions (6.17) and (6.18) positive, which—in turn—results in real negative exponents corresponding to amplitude decrease. The remaining two roots, 1.78 and 2.00, result in imaginary numbers. The requirement of roots being smaller than unity is tantamount to v being smaller than α and β, which—due to the stability conditions of Hookean solids—implies $v < \beta < \alpha$.

Since—for physically admissible values of α/β—equation (6.23) has exactly one root within $0 < v/\beta < 1$, there is only one propagation mode of Rayleigh waves. If it were otherwise—as is the case of body waves for which the corresponding equation has, in general, three real roots, which result in three body waves—we would have several Rayleigh waves.

For the second case, let us consider $\beta^2/\alpha^2 = 1/2$, which corresponds to Poisson's ratio being zero; the only root smaller than unity is $v/\beta = 0.87$.

For the third case, letting v/β be a ratio denoted by ζ, we can, with a certain abuse of notation, consider this ratio for the limiting case of a Hookean solid for which Poisson's ratio tends to 1/2 and $\beta \to 0$. Equation (6.23) becomes

$$\zeta^6 - 8\zeta^4 + 24\zeta^2 - 16 = 0,$$

and its only real root is $v/\beta =: \zeta = 0.96$. Since $\beta \to 0$, it follows that—to obtain a finite value of v/β—also $v \to 0$, at the same rate.

For the fourth, and the final case, let us consider equation (6.23) to calculate v/β for $\beta^2/\alpha^2 = 3/4$, which is the highest ratio allowable, according to the stability condition that allows the lowest Poisson's ratio to be -1. In such a case, the only root smaller than unity is $v/\beta = 0.69$.[15]

Let us examine the displacements of this surface wave. To do so, we use expressions (6.17) and (6.18) in expressions (6.2) and (6.3) to obtain

$$u_1 = \left(\iota C\kappa \exp\left(-\sqrt{\kappa^2 - \frac{\omega^2}{\alpha^2}}\,x_3\right) + D\sqrt{\kappa^2 - \frac{\omega^2}{\beta^2}}\exp\left(-\sqrt{\kappa^2 - \frac{\omega^2}{\beta^2}}\,x_3\right)\right) \exp(\iota(\kappa x_1 - \omega t))$$

and

$$u_3 = -\left(C\sqrt{\kappa^2 - \frac{\omega^2}{\alpha^2}}\exp\left(-\sqrt{\kappa^2 - \frac{\omega^2}{\alpha^2}}\,x_3\right) - \iota D\kappa \exp\left(-\sqrt{\kappa^2 - \frac{\omega^2}{\beta^2}}\,x_3\right)\right) \exp(\iota(\kappa x_1 - \omega t)).$$

[15]The corresponding value of expression (5.100) in Achenbach (1973) tends to infinity.

Using equation (6.22) to state D in terms of C, we write the real parts of these expressions as

$$u_1 = C\frac{\omega}{v}\left(\exp\left(-\omega\sqrt{\frac{1}{v^2}-\frac{1}{\alpha^2}}\,x_3\right)\right. \tag{6.24}$$

$$\left. -2\sqrt{\left(\frac{1}{v^2}-\frac{1}{\alpha^2}\right)\left(\frac{1}{v^2}-\frac{1}{\beta^2}\right)}\,\frac{\exp\left(-\omega\sqrt{\frac{1}{v^2}-\frac{1}{\beta^2}}\,x_3\right)}{\frac{2}{v^2}-\frac{1}{\beta^2}}\right)\sin\left(\omega\left(t-\frac{x_1}{v}\right)\right)$$

and

$$u_3 = -C\,\omega\sqrt{\frac{1}{v^2}-\frac{1}{\alpha^2}}\left(\exp\left(-\omega\sqrt{\frac{1}{v^2}-\frac{1}{\alpha^2}}\,x_3\right)\right. \tag{6.25}$$

$$\left. -\frac{2}{v^2}\,\frac{\exp\left(-\omega\sqrt{\frac{1}{v^2}-\frac{1}{\beta^2}}\,x_3\right)}{\frac{2}{v^2}-\frac{1}{\beta^2}}\right)\cos\left(\omega\left(t-\frac{x_1}{v}\right)\right),$$

where we use $\kappa = \omega/v$. Since, due to the presence of trigonometric functions, u_1 and u_3 are $\pi/2$ out of phase with one another, the resulting displacement is an ellipse. Its major axis is perpendicular to the surface and its size diminishes with depth for large values of x_3. At, and near, the surface, the displacement along this ellipse is in the sense opposite to the direction of propagation; hence, this displacement is referred to as *retrograde*.

Also, in view of the elliptical displacement, it is common to refer to the physical analogies of the Rayleigh waves as a *groundroll*. However, since we view the Earth's crust as layered, we must be careful in associating the observed groundroll with the Rayleigh waves, for which we assume a homogeneous halfspace.

To illustrate expressions (6.24) and (6.25), let us consider $\alpha^2 = 3$, $\beta^2 = 1$ and $\omega = 1$, which implies—following results on page 230—that, in such a case, $v = 0.92$. The amplitude-decrease factor for u_3 is

$$\frac{2\beta^2}{2\beta^2 - v^2}\exp(-0.43\,x_3) - \exp(-0.92\,x_3).$$

The fraction, which is positive, in general, is 1.73, herein. In this case, the amplitude of the vertical displacement increases between $x_3 = 0$ and $x_3 = 0.43$, and decreases monotonically for the remaining values, as illustrated in Figure 6.1.[16]

[16]For a graph of the other displacement and graphs of other Poisson's ratios, readers might refer to Achenbach (1973, Figure 5.13).

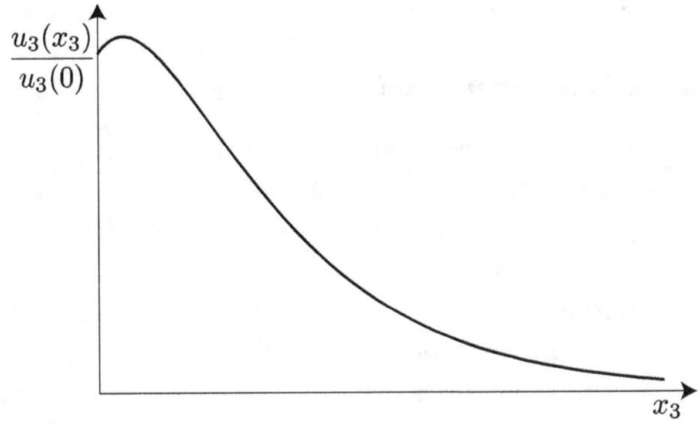

Fig. 6.1: The vertical displacement of the Rayleigh wave normalized to its amplitude at the surface. For this illustration, Poisson's ratio is $1/4$.

In Section 6.4.3, below, we examine the reason for the absence of SH waves in a homogeneous halfspace. Mathematically, the existence of a nontrivial solution that allows for the vanishing of the stress tensor at the surface implies the existence of Rayleigh waves. Since—as discussed in Section 6.4.3—this vanishing cannot be achieved with a nontrivial solution involving SH waves, we conclude that such waves cannot exist in a Hookean model examined in the present section. However, in comparing these results with observations, we must remember that we compare physical experiments with their analogues in the mathematical realm.

As examined in Section 6.3, we can modify the discussed Hookean model by introducing a horizontal interface within a halfspace, which results in a surface layer. Such a layer does not allow the existence of Rayleigh waves, which require a continuous exponential amplitude decrease with depth.[17] However, unlike the halfspace, it allows the existence of surface SH waves. Their existence is a consequence of the lack of amplitude decrease with depth and, hence, of the constructive interference of waves within the layer that are reflected from the interface and the surface; we refer to them as *guided waves*.

[17]A modification allowing for the existence of the Rayleigh-like waves in the presence of an interface is discussed in footnote 24.

6.3 Guided waves: Homogeneous layer above halfspace

6.3.1 *Elastic layer above rigid halfspace*

Let us consider a homogeneous Hookean solid that constitutes a layer above a rigid halfspace. In comparison to Section 6.2, this is tantamount to adding an interface below which the medium is rigid. We assume both the surface and the interface to be horizontal.

The waves considered in this and the following section can be viewed as surface waves, since their effects might be observed as a propagation along the surface. Such a viewpoint is common in seismological considerations. Strictly speaking, however, these are *guided waves*, whose existence requires a *waveguide*, which herein is the layer bounded by the surface and the interface.

Inserting

$$u_2(\mathbf{x},t) = U(x_3)\exp(\iota(\kappa x_1 - \omega t))\,, \tag{6.26}$$

which is analogous to expressions (6.11) and (6.12) for the Rayleigh waves, where $v = \omega/\kappa$ is the speed of a wave propagating in the x_1-direction, into the wave equation for SH waves,

$$\frac{\partial^2 u_2}{\partial x_1^2} + \frac{\partial^2 u_2}{\partial x_3^2} = \frac{1}{\beta^2}\frac{\partial^2 u_2}{\partial t^2}\,, \tag{6.27}$$

we get

$$\frac{\mathrm{d}^2 U}{\mathrm{d}\,x_3^2} + \left(\frac{\omega^2}{\beta^2} - \kappa^2\right) U = 0\,, \tag{6.28}$$

as shown in Exercise 6.1. This result is analogous to expression (6.13), for Rayleigh waves. The general solution is

$$U(x_3) = C_1 \exp\left(-\iota\sqrt{\frac{\omega^2}{\beta^2} - \kappa^2}\,x_3\right) + C_2 \exp\left(\iota\sqrt{\frac{\omega^2}{\beta^2} - \kappa^2}\,x_3\right). \tag{6.29}$$

We introduce a condition that distinguishes this expression from the analogous ones for the Rayleigh wave, in Section 6.2, and for the interface waves in Section 6.5, below. Herein, in view of the presence of an interface at a finite depth, we wish to consider the effect of interaction between the waves propagating towards the interface and the reflected waves propagating towards the surface. To do so, we assume that the amplitude does not decay below the surface but oscillates along the x_3-axis between the surface and the interface, which is a property of a guided wave.

To avoid the exponential decrease of amplitude with depth, we require the root to be real, which results in a sinusoidal solution. Also, it implies $\omega^2/\beta^2 > \kappa^2$, and hence $v > \beta$, since

$$\frac{\omega^2}{\beta^2} > \kappa^2 \Rightarrow \frac{1}{\beta^2} > \frac{\kappa^2}{\omega^2} \Rightarrow \frac{1}{\beta^2} > \frac{1}{v^2} \Rightarrow v > \beta\,;$$

the speed of the resulting wave, v, is greater than β.

Inserting U from solution (6.29) into expression (6.26), we obtain

$$u_2(\mathbf{x},t) = C_1 \exp\left(\iota\left(\kappa x_1 - \omega t - \sqrt{\frac{\omega^2}{\beta^2} - \kappa^2}\, x_3\right)\right) + C_2 \exp\left(\iota\left(\kappa x_1 - \omega t + \sqrt{\frac{\omega^2}{\beta^2} - \kappa^2}\, x_3\right)\right). \tag{6.30}$$

The first exponential term refers to the wave traveling towards the surface, and the second towards the interface; the resulting wave is an interference between the two, as explained in Section 6.3.3. The lack of amplitude decrease with depth leads to normal modes of standing waves. This expression is analogous to expressions (6.17) and (6.18), in the context of the Rayleigh wave.

Let us proceed to the side conditions. We introduce the conditions that are required by the presence of both surface and interface. Applying the condition at the surface, where $x_3 = 0$, and at the interface, where $x_3 = Z$, namely, $\sigma_{23} = C_{44}\,\partial u_2/\partial x_3 = 0$ and $u(Z,t) = 0$, since the stress tensor at the surface is zero and there is no displacement at the interface due to the *welded contact*[18] with a rigid halfspace, we obtain

$$C_1 - C_2 = 0$$

and

$$C_1 \exp\left(-\iota\sqrt{\frac{\omega^2}{\beta^2} - \kappa^2}\, Z\right) + C_2 \exp\left(\iota\sqrt{\frac{\omega^2}{\beta^2} - \kappa^2}\, Z\right) = 0\,,$$

which is a system of linear homogeneous equations for C_1 and C_2. From the first equation, we see that $C_1 = C_2$, which we denote by C. Hence, we write the second equation as

$$C\left(\exp\left(-\iota\sqrt{\frac{\omega^2}{\beta^2} - \kappa^2}\, Z\right) + \exp\left(\iota\sqrt{\frac{\omega^2}{\beta^2} - \kappa^2}\, Z\right)\right) = 0\,,$$

[18]For further explanations of a welded contact, readers might refer to Chapman (2004, Section 4.3.1).

which is

$$2\,C\cos\left(\sqrt{\frac{\omega^2}{\beta^2}-\kappa^2}\,Z\right)=0\,.$$

To obtain a nontrivial solution, we assume $C\neq 0$, which implies that

$$\sqrt{\frac{\omega^2}{\beta^2}-\kappa^2}\,Z=\pi\left(n+\frac{1}{2}\right). \tag{6.31}$$

Then, solving for κ, and inserting into $v=\omega/\kappa$, we obtain the expression of the possible phase velocities for given ω,

$$v=\frac{\omega}{\kappa}=\frac{\beta}{\sqrt{1-\left(\frac{\beta\left(n+\frac{1}{2}\right)\pi}{\omega Z}\right)^2}}\,, \tag{6.32}$$

where $n=0,1,2,\ldots$. Inserting $C_1=C_2=:C$ into solution (6.30), we obtain

$$u_2(\mathbf{x},t)=2\,C\cos\left(\sqrt{\frac{\omega^2}{\beta^2}-\kappa^2}\,x_3\right)\exp\left(\iota\left(\kappa x_1-\omega t\right)\right)\,.$$

Using equation (6.31), we obtain

$$u_2(\mathbf{x},t)=2\,C\cos\left(\pi\left(n+\frac{1}{2}\right)\frac{x_3}{Z}\right)\exp\left(\iota\left(\kappa x_1-\omega t\right)\right)\,. \tag{6.33}$$

This expression corresponds to a wave that propagates along the x_1-direction with speed (6.32), and whose amplitude oscillates in the x_3-direction. This result is consistent with ansatz (6.26), as expected.

Mathematically, the existence of a nontrivial solution is tantamount to the existence of surface SH waves. As mentioned on page 233, the reliability of this deductive argument in the context of observations depends on the pertinence of Hookean solids in modeling physical phenomena.

Since the propagation speed is a function of frequency, this wave is dispersive; herein, v is its phase speed, not its group speed, which is

$$\frac{\mathrm{d}\omega}{\mathrm{d}\kappa}=\left(\frac{1}{v}-\frac{\omega}{v^2}\frac{\mathrm{d}v}{\mathrm{d}\omega}\right)^{-1}\,.$$

Herein, following expression (6.31), we obtain

$$\frac{\mathrm{d}\omega}{\mathrm{d}\kappa}=\beta\sqrt{1-\frac{\pi^2\left(n+\frac{1}{2}\right)^2\beta^2}{\omega^2 Z^2}}\,;$$

we conclude that the dispersion disappears as Z tends to infinity.

In general, dispersion is the consequence of a variable denoting frequency being present in solubility conditions, which result in expressions for speed. This variable is introduced in a trial solution of an elastodynamic equation. Commonly, terms involving this variable disappear in the derivation process, as illustrated by expressions (6.11) and (6.12) leading to expression (6.23).[19] If the frequency dependence, as introduced in expression (6.26), remains, as in expression (6.32), the wave is dispersive. Herein, dispersion is not an intrinsic material property of a Hookean solid—as in the case of anelastic media, where the elasticity parameters are complex numbers[20]—but of the arrangement of interfaces that separate elastic solids.

For a fixed value of x_1, the wave under consideration is a standing wave whose modes are denoted by n in expression (6.32); specifically, the number of nodes between the surface and the interface is n. Examining this expression, we see that, as $Z \to \infty$, the wave ceases to be dispersive and becomes a single mode SH wave. At first sight, one might be surprised at such a conclusion, since—as mentioned in Section 6.2—there are no SH waves in homogeneous halfspaces. However, in formulating Rayleigh waves, we consider exponential solutions, which result in expressions (6.17) and (6.18). In this section, we consider sinusoidal solutions, which result in expression (6.30). In other words, the amplitude of Rayleigh waves decreases with depth and the amplitude of waves discussed herein oscillates with depth.

6.3.2 *Elastic layer above elastic halfspace*

Let us modify the case discussed in Section 6.3.1, by assuming that the halfspace below the elastic layer is elastic not rigid.[21] The SH waves within this waveguide are known as the *Love waves*, in honour of Augustus Edward Hough Love.[22]

[19] *see also*: Slawinski (2015, expression (7.2.10), which contains frequency, and expression (7.3.4), which does not)

[20] Interested readers might refer to Borcherdt (2009), for background information, and to Krebes and Slawinski (1991, expression (3), the paragraph that follows, and expression (7)).

[21] Readers interested in historical documents of theoretical developments might refer to Love (1944, Chapter XIII).

[22] Readers interested in further study might refer to Achenbach (1973, Section 6.6), Auld (1990, Vol. II, Chapter 10, Section D.1), Brekhovskikh and Goncharov (1994, Section 4.3.2 and 4.3.3), Ewing *et al.* (1957, Chapter 4), Grant and West (1965, Section 3-9), and Hanyga (1984, Section 1.2.3), as well as other books on this subject, including a didactic exposition by Krebes (2004, pp. 4-11 to 4-15).

As expected, the distinction between the Love wave and the waves formulated in Section 6.3.1 lies in the conditions associated with the interface. Thus, expression (6.33) itself, obtained in Section 6.3.1, is not a special case, *sensu stricto*, of the one to be formulated herein, since either case requires distinct side conditions, which means that—apart from the wave equation given by expression (6.27), which is common for both cases—there are other distinct equations to be considered for either case. In other words, even though both expression (6.33), above, and expression (6.38), below, refer to the surface SH wave, they satisfy different systems of equations due to distinct side conditions at the interface between the layer and the halfspace.

Herein, the stress tensor and displacement vector are nonzero; the conditions at $x_3 = Z$ are $\sigma^\ell = \sigma^h$ and $u_2^\ell = u_2^h$, where ℓ and h refer to the layer and halfspace, respectively. Since we consider the SH waves, $u_1^\ell = u_3^\ell = 0$. Invoking Hooke's law (6.20), we write

$$C_{44}^\ell \frac{\partial u_2^\ell}{\partial x_3} = C_{44}^h \frac{\partial u_2^h}{\partial x_3} \,. \tag{6.34}$$

The expressions for displacement are the same as expression (6.30), with corresponding values of κ and β that describe the properties of the layer and the halfspace, namely,

$$\begin{aligned} u_2^\ell(\mathbf{x},t) &= C_1 \exp\left(\iota\left(\kappa x_1 - \omega t - \sqrt{\frac{\omega^2}{\beta_\ell^2} - \kappa^2}\, x_3\right)\right) \\ &\quad + C_2 \exp\left(\iota\left(\kappa x_1 - \omega t + \sqrt{\frac{\omega^2}{\beta_\ell^2} - \kappa^2}\, x_3\right)\right) \end{aligned} \tag{6.35}$$

and

$$\begin{aligned} u_2^h(\mathbf{x},t) &= D_1 \exp\left(\iota\left(\kappa x_1 - \omega t - \sqrt{\frac{\omega^2}{\beta_h^2} - \kappa^2}\, x_3\right)\right) \\ &\quad + D_2 \exp\left(\iota\left(\kappa x_1 - \omega t + \sqrt{\frac{\omega^2}{\beta_h^2} - \kappa^2}\, x_3\right)\right) . \end{aligned} \tag{6.36}$$

In view of physical requirements, we need the amplitude in the halfspace to decay exponentially along the x_3-axis, even though it need not decay between the surface and the interface. The decay implies that the radicands in expression (6.36) are negative, which means that $\omega^2/\beta_h^2 < \kappa^2$.

To avoid the exponential growth, we must set $D_1 = 0$. Thus, we rewrite expression (6.36) as

$$u_2^h(\mathbf{x}, t) = D_2 \exp\left(-\sqrt{\kappa^2 - \frac{\omega^2}{\beta_h^2}}\, x_3\right) \exp\left(\iota\left(\kappa x_1 - \omega t\right)\right) . \qquad (6.37)$$

Let us consider the side conditions at $x_3 = 0$ and at $x_3 = Z$. The former is the vanishing of the stress at the surface,

$$C_{44}^{\ell} \frac{\partial u_2^{\ell}}{\partial x_3} = 0 .$$

The latter is given by expression (6.34) and, as stated above, by $u_2^{\ell} = u_2^h$: the continuity of displacement at either side of the interface. From the former, it follows that $C_1 = C_2$ in expression (6.35), which we rewrite as

$$u_2^{\ell}(\mathbf{x}, t) = 2\, C_1 \cos\left(\sqrt{\frac{\omega^2}{\beta_\ell^2} - \kappa^2}\, x_3\right) \exp\left(\iota\left(\kappa x_1 - \omega t\right)\right) . \qquad (6.38)$$

From $u_2^{\ell} = u_2^h$ it follows that the wave numbers, κ, and the frequency, ω, must be the same for both the layer and the halfspace, which we already implied by our notation in expressions (6.35) and (6.36).

Invoking the conditions at $x_3 = Z$, we write the system of two equations. Proceeding in a manner similar to the one followed in Section 6.3.1,[23] we obtain

$$\tan\left(\sqrt{\frac{\omega^2}{\beta_\ell^2} - \kappa^2}\, Z\right) = \frac{C_{44}^h \sqrt{\kappa^2 - \dfrac{\omega^2}{\beta_h^2}}}{C_{44}^{\ell} \sqrt{\dfrac{\omega^2}{\beta_\ell^2} - \kappa^2}} .$$

Recalling that $v = \omega / \kappa$, we write

$$\tan\left(\omega \sqrt{\frac{1}{\beta_\ell^2} - \frac{1}{v^2}}\, Z\right) = \frac{C_{44}^h \sqrt{\dfrac{1}{v^2} - \dfrac{1}{\beta_h^2}}}{C_{44}^{\ell} \sqrt{\dfrac{1}{\beta_\ell^2} - \dfrac{1}{v^2}}} , \qquad (6.39)$$

which is the dispersion relation for Love waves. Since both radicands must be positive, it follows—from the numerator—that $v < \beta_h$ and—from the denominator—that $v > \beta_\ell$. Hence, we conclude that $v \in (\beta_\ell, \beta_h)$. Logically, this result entails a restriction on models in which the Love waves can propagate: the S-wave speed within a layer must be smaller than within a halfspace. Otherwise, if $\beta_\ell > \beta_h$, the interval of values for v is an empty set.

[23]For details see Krebes (2004, pp. 4–11 to 4–13).

6.3.3 *Love wave as superposition of SH waves*

As suggested for expression (6.30), guided waves result from interference of body waves. Let us examine this concept in the context of Love waves and SH waves.

Consider expression (6.35), which, letting $\kappa = \omega/v$, we write as

$$u_2^\ell(x_1, x_3, t) = C_1 \exp(-\iota\kappa s_\ell x_3) \exp[\iota(\kappa x_1 - \omega t)] + C_2 \exp(\iota\kappa s_\ell x_3) \exp[\iota(\kappa x_1 - \omega t)],$$

where $s_\ell := \sqrt{(v/\beta_\ell)^2 - 1}$. Let us proceed to interpret this expression as a superposition of two SH waves within the elastic layer.

Both waves travel obliquely with respect to the surface and interface; different signs in front of x_3 means that one wave travels upwards, the other downwards. Their wave vectors are $\mathbf{k}_\pm := (\kappa, 0, \pm\kappa s_\ell)$. Considering its magnitude,

$$\|\mathbf{k}_\pm\| = \sqrt{\kappa^2 + (\kappa\, s_\ell)^2},$$

we have

$$\|\mathbf{k}_\pm\| = \kappa\sqrt{1 + (s_\ell)^2} = \kappa\, \frac{v}{\beta_\ell},$$

from which it follows that

$$\frac{\beta_\ell}{v} = \frac{\kappa}{\|\mathbf{k}_\pm\|} = \sin\theta, \tag{6.40}$$

where θ is the angle between $\mathbf{k}_\pm$ and the x_3-axis. Thus, θ is the angle between the x_3-axis and a wavefront normal, which means that it is the propagation direction of a wavefront.

For the case in which v is only slightly larger than β_ℓ, it follows that both the upgoing and downgoing SH waves propagate nearly horizontally. In other words, their propagation directions are nearly parallel to the interface. For the case in which v is much larger than β_ℓ, both waves propagate nearly vertically; their propagation directions are nearly perpendicular to the interface.

The concept of superposition allows us to understand requirements on material properties that are necessary for the existence of Love waves, which are discussed in Section 6.4.2, below. We can write these requirements as $\beta_\ell < v < \beta_h$. In view of expression (6.40), the lower limit, $v \to \beta_\ell$, is tantamount to $\sin\theta = \beta_\ell/\beta_\ell = 1$, which corresponds to waves that propagate along the x_1-axis. Hence, these waves do not exhibit any interference between one another that would be associated with the presence of the horizontal surface or interface. Formally, this limit is required by the real value of $\sin\theta$.

As shown in Exercise 6.2, the upper limit, $v \to \beta_h$, is introduced to ensure an exponential amplitude decay in the halfspace. In view of expression (6.40), this limit is tantamount to $\sin\theta = \beta_\ell/\beta_h$, which implies that, in general, the SH waves propagate obliquely. In the extreme case, if $\beta_h \to \infty$, the waves propagate vertically. Such an extreme case could be interpreted as corresponding to a rigid halfspace, $C_{44}^h \to \infty$, whose aspects are discussed in Section 6.3.1.

6.3.4 *On Rayleigh and Love waves in Earth's crust*

Let us comment on the fact that the Rayleigh waves and the Love waves cannot exist in the same model. The former requires a halfspace that is homogeneous and the latter requires an interface within that halfspace. The use of both these waves in examination of the Earth's crust, which—at first sight might appear as a contradiction—is an example of subtleties in relations between mathematical structures and seismological inferences.[24]

The solution of this apparent paradox is the fact that neither the Rayleigh waves nor the Love waves propagate within the Earth's crust. They exist in the realm of mathematics only. As such, and with certain limitations, both can be used as analogies for disturbances propagating within the crust. Explicitly, however, mathematical conclusions concerning these, and other, waves refer only to mathematical models. Inferences of physical properties based on these models do not possess the certainty of deductive arguments; these inferences are interpretations, and as such, they rely on inductive arguments, and are subject to their uncertainties and nonuniqueness.

In earthquake seismology, disturbances propagating along the surface and exhibiting purely vertical displacements are not observed. Disturbances exhibiting displacements that are horizontal are common. The Love waves exhibit displacements that are horizontal and perpendicular to the direction of propagation; the Rayleigh waves exhibit—on the surface—the dis-

[24]Readers interested in addressing this apparent contradiction by considering the existence of the Rayleigh-like waves in an elastic layer above an elastic halfspace might refer to Stacey and Davis (2008, Section 16.5) and Udías (1999, Section 10.4). Such waves result from the constructive interference of the P and SV waves within the layer. They are dispersive and—according to the dispersion equation that results from a 6×6 determinant (Udías, 1999, p. 200)—their speed is $v_r < v < \beta_h$, where v_r is the speed of a standard Rayleigh wave. Dalton *et al.* (2017) refer to these waves as *quasi-Rayleigh*.

placements that are horizontal and parallel to that direction. As stated above, the mathematical structure of these waves—with awareness of its subtleties—offers us a possibility of quantitative studies of the Earth's crust.

6.4 Existence of surface and guided waves

6.4.1 *Introduction*

At first sight, it might appear that—in our considerations in Sections 6.2 and 6.3—we address only a few special cases of waves whose propagation is observable on the surface. However, the cases considered provide us with a rather comprehensive set of mathematical analogies to examine seismological phenomena. To justify this statement, in Section 6.4.3, below, we show that the propagation of the Love waves on the surface of a homogeneous halfspace is impossible due to an inconsistency between the general solution and the side conditions. In Section 6.4.4, below, we show that the relation between the elasticity parameters imposed by the side conditions for the existence of the surface P wave is exceedingly restrictive.

However, prior to discussing these two cases, let us comment on the restrictions for the existence of waves due to material properties of the continua within the model. As we show in Section 6.4.2, below, the Love waves can exist only within a limited range of the elasticity parameters and mass densities.

6.4.2 *Elasticity parameters and mass densities*

Rayleigh waves can exist for halfspaces composed of any Hookean solid. Thus, the sole restriction on their existence is the underlying requirement for the positive definiteness of the elasticity tensor, which—in this case—is tantamount to the positive definiteness of matrix (3.57); it requires that both C_{11} and C_{44} be positive and $3C_{11} > 4C_{44}$. Contrary to the Rayleigh wave, which is based on a model consisting of a single homogeneous continuum, the other surface, guided and interface waves discussed in this book are associated with models that consist of two homogeneous continua. These waves can exist only under certain restrictive relations between the elasticity parameters and mass densities of their models.

The condition for the existence of the Love wave is a function of the speeds within the layer and the halfspace, as discussed in the context of expression (6.39) from which we conclude that $v \in (\beta_\ell, \beta_h)$. If the speed of the Love wave is outside these values, this wave cannot exist. This statement entails restrictions on the properties of the model. To analyze it, let us express these values in terms of the elasticity parameters and mass densities,

$$v \in \left(\sqrt{\frac{C_{44}^{\ell}}{\rho_{\ell}}}, \sqrt{\frac{C_{44}^{h}}{\rho_{h}}} \right) .$$

For this interval to be a nonempty set, and hence for the real value of v to exist, which is tantamount to the existence of the Love waves, it follows that $C_{44}^{\ell}/\rho_{\ell} < C_{44}^{h}/\rho_{h}$.

6.4.3 *On Love waves in homogeneous halfspace*

To show the impossibility of existence of the Love wave in a homogeneous halfspace, let us examine expression (6.37) together with the boundary condition at $x_3 = 0$, namely, $\partial u_2^h/\partial x_3 = 0$, which is

$$D_2 \left(-\sqrt{\kappa^2 - \frac{\omega^2}{\beta_h^2}} \right) \exp\left(\iota\left(\kappa x_1 - \omega t\right)\right) = 0 .$$

For a nontrivial solution, $D_2 \neq 0$, this condition requires that

$$\kappa^2 - \frac{\omega^2}{\beta_h^2} = 0 .$$

Inserting this expression into solution (6.37), we obtain

$$u_2^h(\mathbf{x}, t) = D_2 \exp\left(\iota\left(\kappa x_1 - \omega t\right)\right) .$$

However, this result has no exponential decay, which we require from physical considerations. Thus, we conclude that there are no Love waves in a homogeneous halfspace.

6.4.4 *On P waves in homogeneous halfspace*

Let us consider the P wave propagating in the x_1-direction in a homogeneous halfspace; the corresponding wave equation is

$$\nabla^2 u_1(\mathbf{x}, t) - \frac{1}{\alpha^2} \frac{\partial^2 u_1(\mathbf{x}, t)}{\partial t^2} = 0 ,$$

where α is the speed of propagation. The trial solution,

$$u_1(\mathbf{x}, t) = A(x_3) \exp\left(\iota(\kappa x_1 - \omega t)\right) , \qquad (6.41)$$

results in an ordinary differential equation,

$$\frac{\mathrm{d}^2 A}{\mathrm{d}x_3^2} - \left(\kappa^2 - \frac{\omega^2}{\alpha^2}\right) A = 0\,,$$

whose general solution is

$$A(x_3) = A_1 \exp\left(\sqrt{\kappa^2 - \frac{\omega^2}{\alpha^2}}\, x_3\right) + A_2 \exp\left(-\sqrt{\kappa^2 - \frac{\omega^2}{\alpha^2}}\, x_3\right),$$

which is equivalent to expression (6.16), and where—to avoid an unbounded growth—we set $A_1 = 0$ to obtain

$$A(x_3) = A_2 \exp\left(-\sqrt{\kappa^2 - \frac{\omega^2}{\alpha^2}}\, x_3\right). \tag{6.42}$$

Let us examine this solution in the context of the side conditions at $x_3 = 0$, which—as discussed on page 229—are $\sigma_{13} = \sigma_{33} = 0$. According to Hooke's law (6.20), we write

$$\sigma_{13} = C_{44} \frac{\partial u_1}{\partial x_3} = 0\,, \tag{6.43}$$

since $\delta_{13} = 0$, and

$$\sigma_{33} = (C_{11} - 2C_{44}) \frac{\partial u_1}{\partial x_1} = 0\,, \tag{6.44}$$

since $u_2 = u_3 = 0$; we consider only P waves and their propagation—and hence their displacement—only in the x_1-direction. This is distinct from the case of the Rayleigh wave, discussed in Section 6.2, where—due to the presence of the SV wave—displacements are both in the x_1-direction and the x_3-direction.

Inserting expression (6.41) into expressions (6.43) and (6.44), and dividing the resulting equations by the exponential term, we obtain

$$\sigma_{13} = C_{44} \left.\frac{\partial A(x_3)}{\partial x_3}\right|_{x_3=0} = 0$$

and

$$\sigma_{33} = \iota\kappa\,(C_{11} - 2C_{44})\, A(0) = 0\,.$$

To examine the first condition, we differentiate solution (6.42) to obtain

$$\frac{\partial A(x_3)}{\partial x_3} = A_2 \left(-\sqrt{\kappa^2 - \frac{\omega^2}{\alpha^2}}\right) \exp\left(-\sqrt{\kappa^2 - \frac{\omega^2}{\alpha^2}}\, x_3\right).$$

Evaluating at $x_3 = 0$, we write the first condition as

$$\sigma_{13} = C_{44}\left(-\sqrt{\kappa^2 - \frac{\omega^2}{\alpha^2}}\right) A_2 = 0\,.$$

In view of solution (6.42), the second condition is

$$\sigma_{33} = \iota\kappa\,(C_{11} - 2C_{44})\,A_2 = 0\,.$$

These two conditions are analogous to expressions (6.22) and (6.21), respectively, except for the absence of the SV wave. Herein, they constitute two homogeneous linear equations for A_2.

However, in contrast to the Rayleigh waves, which can exist in any Hookean solid, herein, the side conditions impose exceedingly restrictive requirements on the solution of this problem. Even though the first condition is satisfied trivially, since—for P waves—$\kappa = \omega/\alpha$, the second condition requires $C_{11} = 2C_{44}$, which implies that the P-wave speed must be a multiple of $\sqrt{2}$ of the S-wave speed.

The Rayleigh waves, which result from the interaction of both P and SV waves coupled by the presence of the surface, represent a better mathematical model than P waves alone for surface waves propagating in the x_1-direction in a homogeneous halfspace.

6.5 Interface waves: Homogeneous halfspaces

6.5.1 *Introduction*

[25]Considering a model similar to the one discussed in Section 6.3.2, let us examine wave propagation along the interface itself. To discuss the interface, at $x_3 = 0$, between two isotropic and elastic halfspaces, we introduce the condition at $x_3 = 0$ that is analogous to the condition for Love waves, at $x_3 = Z$, namely, the continuity of nonzero stress and displacement across the interface. To accommodate the absence of any other interface, we impose the condition of an exponential decrease of amplitude along the x_3-axis away from the interface in the same manner as for the Rayleigh wave. Commonly, all interface waves are called *Stoneley waves* in honour of Robert Stoneley.[26] Herein—to emphasize differences among waves prop-

[25]Section 6.5 benefitted from research collaborations with David Dalton, Piotr Stachura and Theodore Stanoev.

[26]Readers interested in further study might refer to Achenbach (1973, Section 5.12), Auld (1990, Vol. II, Chapter 10, Section D.3), Ewing *et al.* (1957, Chapter 3), Grant and West (1965, Section 3-3), and Hanyga (1984, Section 1.2.8), as well as other books on this subject, including a didactic exposition by Krebes (2004, pp. 4-3 to 4-4).

agating along an interface separating two solids, an interface separating a solid and a liquid and an interface separating two liquids—we restrict the term *Stoneley waves* to a solid-solid interface only.

The condition for a nontrivial solution, which is analogous to equation (6.23), is given in terms of the properties of both halfspaces stated in terms of their elasticity parameters, C_{11} and C_{44}, and mass densities, ρ.[27] If one of the halfspaces is a vacuum, this equation reduces to equation (6.23), as expected. However, in general, unlike for the Rayleigh wave, whose existence does not impose any conditions on the elasticity parameters or mass density, but similarly to the Love wave, whose existence requires the value of β in the layer to be less than in the halfspace, the existence of the Stoneley wave between two Hookean solids requires stringent conditions on ratios of the elasticity parameters and mass densities. In particular, the shear-wave speeds on either side of the interface must be similar to one another.[28] These conditions result in the Stoneley-wave speed being between the Rayleigh-wave and S-wave speeds.

Similarly to the Rayleigh-wave case, the interface model allows us to consider waves whose displacement directions are contained in a vertical plane, and which are a composition of P waves and SV waves. Waves whose displacement is horizontal cannot exist within such a model, which means that there are no interface SH waves. Also, similarly to Rayleigh waves, there is no frequency dependence of the solubility condition, which means that Stoneley waves are nondispersive. Similarities between the Rayleigh and Stoneley waves are to be expected, since the former is a particular case of the latter, where a Hookean solid is replaced by a vacuum. These similarities are also shared with the interface wave for which a Hookean solid is replaced by a liquid, as discussed in Section 6.5.2, below.

6.5.2 *Elastic and liquid halfspaces*

Let us derive expressions for the wave that propagates along the interface that separates an elastic halfspace from a liquid halfspace. Following the work of Scholte (1947), such a wave is referred to as the *Scholte wave*.

Consider an interface that coincides with the x_1x_2-plane. This interface separates the liquid halfspace, $x_3 < 0$, from an elastic one, $x_3 \geqslant 0$. To consider the wave propagation along this interface, we quantify the material

[27] Readers interested in details might refer to Grant and West (1965, equation (3-10)).

[28] Readers interested in an illustration of the Stoneley-wave-existence conditions might refer to Grant and West (1965, Figure (3-11)).

properties of, and displacements in, the liquid halfspace by

$$C_{44}^U = 0, \quad C_{11}^U \neq 0, \quad \rho^U \neq 0, \quad \mathbf{u}^U = \nabla \mathcal{P}^U : \quad u_1^U = \frac{\partial \mathcal{P}^U}{\partial x_1}, \quad u_3^U = \frac{\partial \mathcal{P}^U}{\partial x_3};$$

the superscript, U, denotes the upper halfspace; also—consistently with a seismological convention—the x_3-axis points downwards. The first two expressions are the elasticity parameters, the third expression is the mass density, the fourth is the statement of the Helmholtz theorem for the case where the scalar potential is not zero but the vector potential is zero as a consequence of no resistance of liquid to the change in shape; hence, there are no shear waves within a liquid. The last two expressions are the displacement components in the x_1x_3-plane. The wave equation in the liquid is for the pressure wave,

$$\nabla^2 \mathcal{P}^U - \frac{1}{(\alpha^U)^2} \frac{\partial^2 \mathcal{P}^U}{\partial t^2} = 0\,,$$

where $\alpha^U = \sqrt{C_{11}^U / \rho^U}$ is the propagation speed.

In the elastic halfspace, C_{44}^D, C_{11}^D and ρ^D are all nonzero. Using the same gauge-condition argument as in the Rayleigh-wave case, we write

$$\mathbf{u}^D = \nabla \mathcal{P}^D + \nabla \times \mathcal{S}^D, \quad u_1^D = \frac{\partial \mathcal{P}^D}{\partial x_1} - \frac{\partial \mathcal{S}^D}{\partial x_3}, \quad u_3^D = \frac{\partial \mathcal{P}^D}{\partial x_3} + \frac{\partial \mathcal{S}^D}{\partial x_1};$$

the superscript, D, denotes the lower halfspace. The first expression is the statement of the Helmholtz theorem with nonzero scalar and vector potentials. The last two expressions are the displacement components in the x_1x_3-plane. The wave equations of the pressure and shear waves in the elastic medium are

$$\nabla^2 \mathcal{P}^D - \frac{1}{(\alpha^D)^2} \frac{\partial^2 \mathcal{P}^D}{\partial t^2} = 0\,, \qquad \nabla^2 \mathcal{S}^D - \frac{1}{\beta^2} \frac{\partial^2 \mathcal{S}^D}{\partial t^2} = 0\,,$$

respectively, where $\beta = \sqrt{C_{44}^D / \rho^D}$ is the shear-wave propagation speed.

Since we consider the P and SV waves only, $u_2^U = u_2^D = 0$. The absence of the SH wave is justified in Section 6.6, below.

The dynamic side conditions at $x_3 = 0$ are $\sigma_{33}^U = \sigma_{33}^D$ and $\sigma_{13}^U = \sigma_{13}^D = 0$; the kinematic condition is $u_3^U = u_3^D$.[29] To express these conditions in terms of the material properties and displacements due to the wave propagation, we invoke Hooke's law (6.20). Thus, in the liquid,

$$\sigma_{13}^U = 0\,, \qquad \sigma_{33}^U = C_{11}^U \left(\frac{\partial u_1^U}{\partial x_1} + \frac{\partial u_3^U}{\partial x_3} \right);$$

[29]For further explanations, readers might refer to Chapman (2004, Section 4.3.2).

in the elastic solid,

$$\sigma_{33}^D = (C_{11}^D - 2C_{44}^D)\left(\frac{\partial u_1^D}{\partial x_1} + \frac{\partial u_3^D}{\partial x_3}\right) + 2C_{44}^D\frac{\partial u_3^D}{\partial x_3}.$$

Hence, the continuity of σ_{33} at $x_3 = 0$ implies that

$$C_{11}^U\left(\frac{\partial u_1^U}{\partial x_1} + \frac{\partial u_3^U}{\partial x_3}\right)\bigg|_{x_3=0} = \left((C_{11}^D - 2C_{44}^D)\left(\frac{\partial u_1^D}{\partial x_1} + \frac{\partial u_3^D}{\partial x_3}\right) + 2C_{44}^D\frac{\partial u_3^D}{\partial x_3}\right)\bigg|_{x_3=0}. \tag{6.45}$$

Let us consider the trial solutions for the three wave equations. We write the P wave, in the upper and lower halfspaces, as

$$\mathcal{P}^U = A^U(x_3)\exp(\iota(\kappa x_1 - \omega t)),$$

$$\mathcal{P}^D = A^D(x_3)\exp(\iota(\kappa x_1 - \omega t)),$$

respectively, and the SV wave, in the lower halfspace, as

$$\mathcal{S}^D = B(x_3)\exp(\iota(\kappa x_1 - \omega t)),$$

where ω and κ are the temporal and spatial frequencies. We consider waves that propagate in the x_1-direction and whose amplitudes change along the x_3-direction. Inserting these expressions into the corresponding wave equations, we obtain

$$\frac{\mathrm{d}^2 A^U}{\mathrm{d}\,x_3^2} + \left(\frac{\omega^2}{(\alpha^U)^2} - \kappa^2\right)A^U = 0,$$

$$\frac{\mathrm{d}^2 A^D}{\mathrm{d}\,x_3^2} + \left(\frac{\omega^2}{(\alpha^D)^2} - \kappa^2\right)A^D = 0,$$

$$\frac{\mathrm{d}^2 B}{\mathrm{d}\,x_3^2} + \left(\frac{\omega^2}{\beta^2} - \kappa^2\right)B = 0,$$

which are ordinary differential equations whose general solutions are

$$A^U = C_1\exp\left(-\iota\sqrt{\frac{\omega^2}{(\alpha^U)^2} - \kappa^2}\,x_3\right) + C_2\exp\left(\iota\sqrt{\frac{\omega^2}{(\alpha^U)^2} - \kappa^2}\,x_3\right),$$

$$A^D = C_3\exp\left(-\iota\sqrt{\frac{\omega^2}{(\alpha^D)^2} - \kappa^2}\,x_3\right) + C_4\exp\left(\iota\sqrt{\frac{\omega^2}{(\alpha^D)^2} - \kappa^2}\,x_3\right),$$

$$B = D_1\exp\left(-\iota\sqrt{\frac{\omega^2}{\beta^2} - \kappa^2}\,x_3\right) + D_2\exp\left(\iota\sqrt{\frac{\omega^2}{\beta^2} - \kappa^2}\,x_3\right).$$

In accordance with physical requirements, u_3^U and u_1^U must decay exponentially as $x_3 \to -\infty$, which forces $\kappa^2 - \omega^2/(\alpha^U)^2 > 0$ and $C_2 = 0$. Similarly, u_3^D and u_1^D must decay exponentially as $x_3 \to \infty$, and therefore, $\kappa^2 - \omega^2/(\alpha^D)^2 > 0$, $\kappa^2 - \omega^2/\beta^2 > 0$ and $C_3 = D_1 = 0$. Letting $D_2 =: D$, for brevity, we write the general solutions as

$$A^U = C_1 \exp\left(\sqrt{\kappa^2 - \frac{\omega^2}{(\alpha^U)^2}}\, x_3\right),$$

$$A^D = C_4 \exp\left(-\sqrt{\kappa^2 - \frac{\omega^2}{(\alpha^D)^2}}\, x_3\right),$$

$$B = D \exp\left(-\sqrt{\kappa^2 - \frac{\omega^2}{\beta^2}}\, x_3\right).$$

Recalling the trial solutions, we write

$$\mathcal{P}^U = C_1 \exp\left(\sqrt{\kappa^2 - \frac{\omega^2}{(\alpha^U)^2}}\, x_3\right) \exp\left(\iota(\kappa x_1 - \omega t)\right),$$

$$\mathcal{P}^D = C_4 \exp\left(-\sqrt{\kappa^2 - \frac{\omega^2}{(\alpha^D)^2}}\, x_3\right) \exp\left(\iota(\kappa x_1 - \omega t)\right),$$

$$\mathcal{S} := \mathcal{S}^D = D \exp\left(-\sqrt{\kappa^2 - \frac{\omega^2}{\beta^2}}\, x_3\right) \exp\left(\iota(\kappa x_1 - \omega t)\right), \tag{6.46}$$

where, for brevity, $\mathcal{S}^D =: \mathcal{S}$. Expressing these results in terms of the displacement-vector components, we obtain

$$u_1^U = \frac{\partial \mathcal{P}^U}{\partial x_1} = \iota\kappa C_1 \exp\left(\iota(\kappa x_1 - \omega t) + \sqrt{\kappa^2 - \frac{\omega^2}{(\alpha^U)^2}}\, x_3\right),$$

$$u_3^U = \frac{\partial \mathcal{P}^U}{\partial x_3} = \sqrt{\kappa^2 - \frac{\omega^2}{(\alpha^U)^2}}\, C_1 \exp\left(\iota(\kappa x_1 - \omega t) + \sqrt{\kappa^2 - \frac{\omega^2}{(\alpha^U)^2}}\, x_3\right),$$

$$u_1^D = \frac{\partial \mathcal{P}^D}{\partial x_1} - \frac{\partial \mathcal{S}}{\partial x_3} = \iota\kappa C_4 \exp\left(\iota(\kappa x_1 - \omega t) - \sqrt{\kappa^2 - \frac{\omega^2}{(\alpha^D)^2}}\, x_3\right)$$
$$+ \sqrt{\kappa^2 - \frac{\omega^2}{\beta^2}}\, D \exp\left(\iota(\kappa x_1 - \omega t) - \sqrt{\kappa^2 - \frac{\omega^2}{\beta^2}}\, x_3\right),$$

$$u_3^D =, \frac{\partial \mathcal{P}^D}{\partial x_3} + \frac{\partial \mathcal{S}}{\partial x_1} = -\sqrt{\kappa^2 - \frac{\omega^2}{(\alpha^D)^2}} C_4 \exp\left(\iota(\kappa x_1 - \omega t) - \sqrt{\kappa^2 - \frac{\omega^2}{(\alpha^D)^2}}\, x_3\right)$$
$$+ \iota\kappa D \exp\left(\iota(\kappa x_1 - \omega t) - \sqrt{\kappa^2 - \frac{\omega^2}{\beta^2}}\, x_3\right) .$$

Thus, from the continuity of u_3 at $x_3 = 0$, namely, $u_3^U|_{x_3=0} = u_3^D|_{x_3=0}$, we get

$$\sqrt{\kappa^2 - \frac{\omega^2}{(\alpha^U)^2}}\, C_1 = -\sqrt{\kappa^2 - \frac{\omega^2}{(\alpha^D)^2}}\, C_4 + \iota\kappa D . \tag{6.47}$$

Also, in view of the continuity of σ_{33} at $x_3 = 0$, we write

$$\sigma_{33}^U = C_{11}^U \left(\frac{\partial u_1^U}{\partial x_1} + \frac{\partial u_3^U}{\partial x_3}\right)$$

and

$$\sigma_{33}^D = (C_{11}^D - 2C_{44}^D)\left(\frac{\partial u_1^D}{\partial x_1} + \frac{\partial u_3^D}{\partial x_3}\right) + 2C_{44}^D \frac{\partial u_3^D}{\partial x_3} .$$

Letting, for brevity,

$$e_u = \exp\left(\iota(\kappa x_1 - \omega t) + \sqrt{\kappa^2 - \frac{\omega^2}{(\alpha^U)^2}}\, x_3\right) ,$$

$$e_d = \exp\left(\iota(\kappa x_1 - \omega t) - \sqrt{\kappa^2 - \frac{\omega^2}{(\alpha^D)^2}}\, x_3\right) ,$$

$$e_\beta = \exp\left(\iota(\kappa x_1 - \omega t) - \sqrt{\kappa^2 - \frac{\omega^2}{\beta^2}}\, x_3\right) ,$$

we write

$$\sigma_{33}^U|_{x_3=0} = C_{11}^U \left(-\kappa^2 C_1\, e_u + \left(\kappa^2 - \frac{\omega^2}{(\alpha^U)^2}\right) e_u\, C_1\right)\Big|_{x_3=0}$$
$$= \frac{-\omega^2}{(\alpha^U)^2} \exp\left(\iota(\kappa x_1 - \omega t)\right) C_{11}^U\, C_1$$

and

$$\sigma_{33}^D|_{x_3=0} = (C_{11}^D - 2C_{44}^D)\left(-\kappa^2 C_4 e_d + \iota\kappa\sqrt{\kappa^2 - \frac{\omega^2}{\beta^2}} D\, e_\beta \right.$$

$$\left. + \left(\kappa^2 - \frac{\omega^2}{(\alpha^D)^2}\right) C_4 e_d - \iota\kappa\sqrt{\kappa^2 - \frac{\omega^2}{\beta^2}} D e_\beta \right)\Bigg|_{x_3=0}$$

$$+ 2C_{44}^D \left(\left(\kappa^2 - \frac{\omega^2}{(\alpha^D)^2}\right) C_4 e_d - \iota\kappa\sqrt{\kappa^2 - \frac{\omega^2}{\beta^2}} D\, e_\beta\right)\Bigg|_{x_3=0}$$

$$= \left((C_{11}^D - 2C_{44}^D)\left(\frac{-\omega^2}{(\alpha^D)^2}\right) C_4 + 2C_{44}^D\left(\kappa^2 - \frac{\omega^2}{(\alpha^D)^2}\right) C_4 \right.$$

$$\left. -2C_{44}^D \iota\kappa\sqrt{\kappa^2 - \frac{\omega^2}{\beta^2}} D\right) \exp(\iota(\kappa x_1 - \omega t))\,.$$

Equating the two expressions, $\sigma_{33}^U|_{x_3=0} = \sigma_{33}^D|_{x_3=0}$, we obtain

$$\frac{-\omega^2}{(\alpha^U)^2} C_{11}^U\, C_1 = \frac{-\omega^2}{(\alpha^D)^2} C_{11}^D\, C_4 + 2\, C_{44}^D\, \kappa^2\, C_4 - 2\iota\kappa\sqrt{\kappa^2 - \frac{\omega^2}{\beta^2}}\, C_{44}^D D\,.$$

Grouping the terms on the same side and dividing by κ^2, we get

$$\frac{-v^2}{(\alpha^U)^2} C_{11}^U\, C_1 + \left(\frac{v^2}{(\alpha^D)^2} C_{11}^D - 2\, C_{44}^D\right) C_4 + 2\iota\sqrt{1 - \frac{v^2}{\beta^2}} C_{44}^D D = 0\,. \quad (6.48)$$

To complete the side-condition requirements, let us recall the continuity of σ_{13} at $x_3 = 0$, namely, $\sigma_{13}^U|_{x_3=0} = \sigma_{13}^D|_{x_3=0} = 0$.

For the upper halfspace,

$$\sigma_{13}^U = C_{44}^U\left(\frac{\partial u_1^U}{\partial x_3} + \frac{\partial u_3^U}{\partial x_1}\right) = 0\,,$$

since $C_{44}^U = 0$ in an ideal fluid. For the lower halfspace,

$$\sigma_{13}^D = C_{44}^D\left(\frac{\partial u_1^D}{\partial x_3} + \frac{\partial u_3^D}{\partial x_1}\right)$$

$$= C_{44}^D\left(-\iota\kappa\sqrt{\kappa^2 - \frac{\omega^2}{(\alpha^D)^2}} C_4 e_d - \left(\kappa^2 - \frac{\omega^2}{\beta^2}\right) D e_\beta \right.$$

$$\left. -\iota\kappa\sqrt{\kappa^2 - \frac{\omega^2}{(\alpha^D)^2}} C_4 e_d - \kappa^2 D e_\beta\right)$$

$$= 0\,.$$

Since, in the lower halfspace, $C_{44}^D \neq 0$, and, in general, $\exp(\iota(\kappa x_1 - \omega t)) \neq 0$, we evaluate σ_{13}^U and σ_{13}^D at $x_3 = 0$, and equate them to write

$$2\iota\kappa\sqrt{\kappa^2 - \frac{\omega^2}{(\alpha^D)^2}}\,C_4 + \left(2\kappa^2 - \frac{\omega^2}{\beta^2}\right)D = 0\,.$$

Dividing by κ^2 and multiplying by $-\iota$, we get

$$2\sqrt{1 - \frac{v^2}{(\alpha^D)^2}}\,C_4 - \iota\left(2 - \frac{v^2}{\beta^2}\right)D = 0\,. \tag{6.49}$$

Let us proceed to solve for C_1, C_4 and D. Dividing equation (6.47) by κ, we get

$$\sqrt{1 - \frac{v^2}{(\alpha^U)^2}}C_1 + \sqrt{1 - \frac{v^2}{(\alpha^D)^2}}C_4 - \iota D = 0\,. \tag{6.50}$$

Invoking the expressions for body-wave propagation speeds, $\alpha = \sqrt{C_{11}/\rho}$ and $\beta = \sqrt{C_{44}/\rho}$, we write

$$\frac{C_{44}^D}{C_{11}^D} = \frac{\beta^2}{(\alpha^D)^2} \qquad \text{and} \qquad \frac{C_{11}^U}{C_{11}^D} = \frac{(\alpha^U)^2\rho^U}{(\alpha^D)^2\rho^D}\,,$$

in equation (6.48), and multiplying by $(\alpha^D)^2$, we get

$$-v^2\frac{\rho^U}{\rho^D}C_1 + (v^2 - 2\beta^2)C_4 + 2\iota\beta^2\sqrt{1 - \frac{v^2}{\beta^2}}D = 0\,. \tag{6.51}$$

Thus, equations (6.50), (6.51) and (6.49) form a system to be solved for C_1, C_4 and D. A nontrivial solution requires the determinant of the coefficient matrix to be zero,

$$\det\begin{bmatrix} \sqrt{1 - \frac{v^2}{(\alpha^U)^2}} & \sqrt{1 - \frac{v^2}{(\alpha^D)^2}} & -\iota \\ -v^2\frac{\rho^U}{\rho^D} & (v^2 - 2\beta^2) & 2\iota\beta^2\sqrt{1 - \frac{v^2}{\beta^2}} \\ 0 & 2\sqrt{1 - \frac{v^2}{(\alpha^D)^2}} & -\iota\left(2 - \frac{v^2}{\beta^2}\right) \end{bmatrix} = 0\,. \tag{6.52}$$

Factoring out $-\iota$ from the last column and setting

$$S_U := \sqrt{1 - \frac{v^2}{(\alpha^U)^2}}\,, \quad S_D := \sqrt{1 - \frac{v^2}{(\alpha^D)^2}}\,, \quad S_\beta := \sqrt{1 - \frac{v^2}{\beta^2}}\,, \tag{6.53}$$

we write

$$\det\begin{bmatrix} S_U & S_D & 1 \\ -v^2\dfrac{\rho^U}{\rho^D} & v^2-2\beta^2 & -2\beta^2 S_\beta \\ 0 & 2S_D & 2-\dfrac{v^2}{\beta^2} \end{bmatrix}=0\,,$$

which, grouping similar terms and dividing by $-\beta^2$, we write as

$$\left(\frac{v}{\beta}\right)^4 S_D\frac{\rho^U}{\rho^D}-4S_\beta S_D S_U+S_U\left(2-\frac{v^2}{\beta^2}\right)^2=0\,. \tag{6.54}$$

This equation can be solved for v, which is the speed of the wave propagating along the interface. As expected, this expression is in agreement with Vinh (2013, expression (13)). Since this expression is frequency-independent, we conclude that this interface wave is nondispersive. Furthermore, this solubility condition does not impose any constraints on the relations between the values of the mass density or the elasticity parameters on either side of the interface, as stated in Section 6.5.1, above. In other words—unlike for the Stoneley waves whose existence requires conditions on relations between elasticity parameters and mass densities—the Scholte waves can exist along the interface between any Hookean solid and liquid.

Examining equation (6.54), we could conclude that the speed of the Scholte wave is less than that of the shear wave in the elastic halfspace.[30] Also, as stated by Vinh (2013, expression (14)), the necessary and sufficient conditions for the existence of the Scholte wave is equation (6.54) and

$$0<\frac{v^2}{\beta^2}<\min\left(1,\frac{(\alpha^U)^2}{\beta^2}\right),$$

which means that

$$0<v<\min(\beta,\alpha^U)\,, \tag{6.55}$$

which is expected since $\beta<\alpha^D$ and $\alpha^U<\alpha^D$; also, S_U, S_β, and S_D must be real.

Example 6.1. Let us exemplify a solution of equation (6.54). Given the specific values of all parameters, except v, we can obtain the speed of the interface wave using a numerical algorithm. In Figure 6.2, we show the plot of the left-hand side of equation (6.54), as a function of v. The used values, in the *SI* units, are $\rho^U=1025$, $\rho^D=2300$, $\alpha^U=2400$, $\alpha^D=2600$ and $\beta=2000$. These values could represent salty water over a muddy bottom.

[30]For further details, readers might refer to Vinh (2013, Table 1).

In exploration seismology, the physical disturbance analogous to that wave is called the *mudroll*, and occurs at the sea bottom. Other interfaces that lend themselves to such an analogy are the boundaries of the outer core (Dahlen and Tromp, 1998, Section 8.8.10).

As a biquadratic expression, the function is symmetric. Also, in accordance with Dong and Hovem (2011, p. 161), equation (6.54) has one positive real root for v; herein, this root is $v = 1427$.

Also, as illustrated by Figure 6.2, there is a double root at $v = 0$. To gain an insight into that root, let us examine equation (6.54), with expressions (6.53). These expressions become equal to unity at $v = 0$. Hence, the left-hand side of equation (6.54) becomes zero.

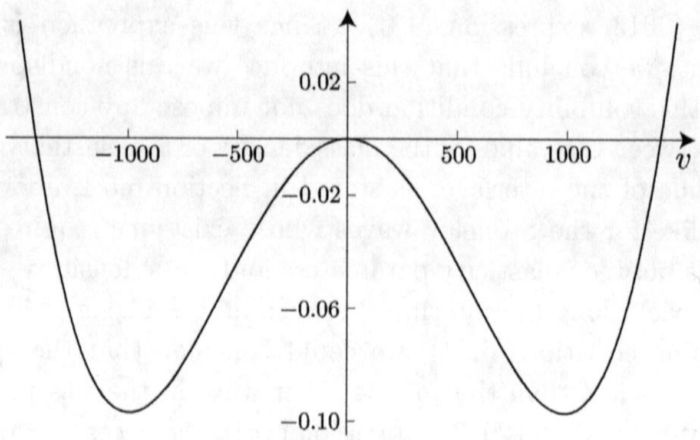

Fig. 6.2: Left-hand side of equation (6.54), as a function of v.

Example 6.2. In view of Example 6.1, let us illustrate expression (6.55). To do so, we keep all the model parameters the same as in Example 6.1, except we set $\alpha^U = 1200$ to be the slowest body-wave speed within the model. Thus, as expected from expression (6.55), the Scholte-wave speed, $v = 1148$, is slower than α^U. It is slower than the slowest body wave. Similarly, in Example 6.1, the slowest body-wave speed is β, and—in accordance with expression (6.55)—the corresponding Scholte-wave speed is slower than β.

Let us express D and C_1 in terms of C_4. Following equation (6.49), we write

$$2S_D C_4 = \iota\left(2 - \frac{v^2}{\beta^2}\right)D,$$

which means that

$$D = \frac{2S_D}{\iota\left(2 - \dfrac{v^2}{\beta^2}\right)} C_4 \,.$$

Hence, following equation (6.50), we write

$$S_U C_1 + S_D C_4 - \iota D = 0\,,$$

which leads to

$$C_1 = \frac{S_D}{S_U}\left(\frac{\dfrac{v^2}{\beta^2}}{2 - \dfrac{v^2}{\beta^2}}\right) C_4 \,.$$

Inserting D and C_1 in terms of C_4 into equation (6.51), and multiplying by $-S_U(2 - v^2/\beta^2)/\beta^2$, we obtain

$$\left(\frac{v^4}{\beta^4} S_D \frac{\rho^U}{\rho^D} + \left(2 - \frac{v^2}{\beta^2}\right)^2 S_U - 4S_\beta S_D S_U\right) C_4 = 0\,.$$

The term in parentheses is zero, since it is equation (6.54); hence, in general, $C_4 \neq 0$.

Alternatively, we can express C_4 and D in terms of C_1 to get

$$C_4 = \frac{S_U}{S_D}\left(\frac{2 - \dfrac{v^2}{\beta^2}}{\dfrac{v^2}{\beta^2}}\right) C_1 = \frac{S_U(2\beta^2 - v^2)}{S_D v^2} C_1$$

and

$$D = \frac{2S_D S_U}{S_D} \frac{\left(2 - \dfrac{v^2}{\beta^2}\right) C_1}{\iota\left(2 - \dfrac{v^2}{\beta^2}\right)\left(\dfrac{v^2}{\beta^2}\right)} = \frac{2S_U \beta^2 C_1}{\iota v^2}\,. \tag{6.56}$$

Inserting C_4 and D in terms of C_1 into the expressions for u_1^U, u_3^U, u_1^D and u_3^D and considering only the real parts of those expressions, in a manner similar to the one in the Rayleigh-wave example, we obtain

$$\begin{aligned}
\operatorname{Re}(u_1^U) &= \operatorname{Re}(\iota\kappa\, C_1 e_u) \\
&= -\kappa \exp\left(\sqrt{\kappa^2 - \frac{\omega^2}{(\alpha^U)^2}}\, x_3\right) \sin(\kappa x_1 - \omega t)\, C_1 \\
&= \frac{\omega}{v} C_1 \exp\left(\omega\sqrt{\frac{1}{v^2} - \frac{1}{(\alpha^U)^2}}\, x_3\right) \sin\left(\omega\left(t - \frac{x_1}{v}\right)\right),
\end{aligned}$$

$$
\begin{aligned}
\operatorname{Re}(u_3^U) &= \operatorname{Re}\left(\sqrt{\kappa^2 - \frac{\omega^2}{(\alpha^U)^2}}\, C_1 e_u\right) \\
&= \sqrt{\kappa^2 - \frac{\omega^2}{(\alpha^U)^2}} \exp\left(\sqrt{\kappa^2 - \frac{\omega^2}{(\alpha^U)^2}}\, x_3\right) \cos(\kappa x_1 - \omega t)\, C_1 \\
&= \frac{\omega}{v} C_1 S_U \exp\left(\omega\sqrt{\frac{1}{v^2} - \frac{1}{(\alpha^U)^2}}\, x_3\right) \cos\left(\omega\left(t - \frac{x_1}{v}\right)\right), \\
\operatorname{Re}(u_1^D) &= \operatorname{Re}\left(\iota\kappa\, C_4 e_d + \sqrt{\kappa^2 - \frac{\omega^2}{\beta^2}} D e_\beta\right) \\
&= \left(\frac{-\kappa S_U(2\beta^2 - v^2)}{S_D v^2} C_1 \exp\left(-\sqrt{\kappa^2 - \frac{\omega^2}{(\alpha^D)^2}}\, x_3\right)\right. \\
&\quad \left. +2\sqrt{\kappa^2 - \frac{\omega^2}{\beta^2}} \frac{S_U \beta^2}{v^2} C_1 \exp\left(-\sqrt{\kappa^2 - \frac{\omega^2}{\beta^2}}\, x_3\right)\right) \sin(\kappa x_1 - \omega t) \\
&= \frac{\omega}{v} C_1 \left(\frac{S_U(2\beta^2 - v^2)}{S_D v^2} \exp\left(-\omega\sqrt{\frac{1}{v^2} - \frac{1}{(\alpha^D)^2}}\, x_3\right)\right. \\
&\quad \left. -\frac{2 S_U S_\beta \beta^2}{v^2} \exp\left(-\omega\sqrt{\frac{1}{v^2} - \frac{1}{\beta^2}}\, x_3\right)\right) \sin\left(\omega\left(t - \frac{x_1}{v}\right)\right), \\
\operatorname{Re}(u_3^D) &= \operatorname{Re}\left(-\sqrt{\kappa^2 - \frac{\omega^2}{(\alpha^D)^2}}\, C_4 e_d + \iota\kappa D e_\beta\right) \\
&= \left(-\sqrt{\kappa^2 - \frac{\omega^2}{(\alpha^D)^2}} \frac{S_U(2\beta^2 - v^2)}{S_D v^2} C_1 \exp\left(-\sqrt{\kappa^2 - \frac{\omega^2}{(\alpha^D)^2}}\, x_3\right)\right. \\
&\quad \left. +\frac{2\kappa S_U \beta^2}{v^2} C_1 \exp\left(-\sqrt{\kappa^2 - \frac{\omega^2}{\beta^2}}\, x_3\right)\right) \cos(\kappa x_1 - \omega t) \\
&= \frac{\omega}{v} C_1 \left(\frac{-S_U(2\beta^2 - v^2)}{v^2} \exp\left(-\omega\sqrt{\frac{1}{v^2} - \frac{1}{(\alpha^D)^2}}\, x_3\right)\right. \\
&\quad \left. +\frac{2 S_U \beta^2}{v^2} \exp\left(-\omega\sqrt{\frac{1}{v^2} - \frac{1}{\beta^2}}\, x_3\right)\right) \cos\left(\omega\left(t - \frac{x_1}{v}\right)\right).
\end{aligned}
$$

Given specific values of α^U, α^D and β, we could solve determinant (6.54) for v, and use the above expressions to examine displacements. As discussed by Dong and Hovem (2011, pp. 163–165), based on numerical analysis and plots of particle displacements as functions of depth relative to the wavelength, these displacements—similarly to the case of Rayleigh waves examined in Section 6.2—are retrograde ellipses.

6.5.3 *Liquid halfspaces*

Let us consider a model analogous to the one discussed in Section 6.5.2, but for two liquid halfspaces. The interface, which coincides with the x_1x_2-plane, separates these halfspaces, whose coordinates along the x_3-axis are $x_3 < 0$ and $x_3 \geqslant 0$.

Herein, in contrast to equation (6.52) in Section 6.5.2, we find that there are no nontrivial solutions. Let us proceed and show the impossibility of obtaining a dispersion relation for waves propagating at the interface between two liquids.

To consider the wave propagation along this interface, we quantify the material properties of, and displacements in, the upper liquid halfspace by

$$C_{44}^U = 0, \quad C_{11}^U \neq 0, \quad \rho^U \neq 0, \quad \mathbf{u}^U = \nabla \mathcal{P}^U : \quad u_1^U = \frac{\partial \mathcal{P}^U}{\partial x_1}, \quad u_3^U = \frac{\partial \mathcal{P}^U}{\partial x_3};$$

the superscript, U, denotes the upper halfspace; also—consistently with a seismological convention—the x_3-axis points downwards. The first two expressions are the elasticity parameters, the third expression is the mass density, the fourth is the statement of the Helmholtz theorem for the case where the scalar potential is not zero but the vector potential is zero as a consequence of no resistance of liquid to the change in shape; hence, there are no shear waves within a liquid. The last two expressions are the displacement components in the x_1x_3-plane. The wave equation in this halfspace is

$$\nabla^2 \mathcal{P}^U - \frac{1}{(\alpha^U)^2} \frac{\partial^2 \mathcal{P}^U}{\partial t^2} = 0,$$

where $\alpha^U = \sqrt{C_{11}^U / \rho^U}$ is the propagation speed.

Similarly, in the lower halfspace, we have

$$C_{44}^D = 0, \quad C_{11}^D \neq 0, \quad \rho^D \neq 0, \quad \mathbf{u}^D = \nabla \mathcal{P}^D : \quad u_1^D = \frac{\partial \mathcal{P}^D}{\partial x_1}, \quad u_3^D = \frac{\partial \mathcal{P}^D}{\partial x_3};$$

the superscript, D, denotes the lower halfspace. The wave equation in this halfspace is

$$\nabla^2 \mathcal{P}^D - \frac{1}{(\alpha^D)^2} \frac{\partial^2 \mathcal{P}^D}{\partial t^2} = 0,$$

where $\alpha^D = \sqrt{C_{11}^D/\rho^D}$ is the propagation speed. Since C_{44} is zero for both halfspaces, we consider the P waves only.

The dynamic side condition at $x_3 = 0$ is $\sigma_{33}^U = \sigma_{33}^D$; the kinematic condition is $u_3^U = u_3^D$. To express these conditions in terms of the material properties and displacements due to the wave propagation, we invoke Hooke's law (6.20). Thus, in the upper halfspace,

$$\sigma_{13}^U = 0\,, \qquad \sigma_{33}^U = C_{11}^U\left(\frac{\partial u_1^U}{\partial x_1} + \frac{\partial u_3^U}{\partial x_3}\right);$$

and in the lower halfspace

$$\sigma_{13}^D = 0\,, \qquad \sigma_{33}^D = C_{11}^D\left(\frac{\partial u_1^D}{\partial x_1} + \frac{\partial u_3^D}{\partial x_3}\right).$$

Hence, at $x_3 = 0$, the continuity of σ_{13} is identically satisfied, and the continuity of σ_{33} implies that

$$C_{11}^U\left(\frac{\partial u_1^U}{\partial x_1} + \frac{\partial u_3^U}{\partial x_3}\right)\bigg|_{x_3=0} = C_{11}^D\left(\frac{\partial u_1^D}{\partial x_1} + \frac{\partial u_3^D}{\partial x_3}\right)\bigg|_{x_3=0}. \tag{6.57}$$

Let us consider the trial solutions for the two wave equations. We write the P wave in the upper and lower hemispheres as

$$\mathcal{P}^U = A^U(x_3)\exp(\iota(\kappa x_1 - \omega t))$$

and

$$\mathcal{P}^D = A^D(x_3)\exp(\iota(\kappa x_1 - \omega t))\,,$$

respectively, where ω and κ are the temporal and spatial frequencies. Each wave propagates in the x_1-direction and its amplitude changes along the x_3-direction. Inserting these expressions into the corresponding wave equations, we obtain

$$\frac{\mathrm{d}^2 A^U}{\mathrm{d}\,x_3^2} + \left(\frac{\omega^2}{(\alpha^U)^2} - \kappa^2\right)A^U = 0$$

and

$$\frac{\mathrm{d}^2 A^D}{\mathrm{d}\,x_3^2} + \left(\frac{\omega^2}{(\alpha^D)^2} - \kappa^2\right)A^D = 0\,,$$

which are ordinary differential equations whose general solutions are

$$A^U = C_1\exp\left(-\iota\sqrt{\frac{\omega^2}{(\alpha^U)^2} - \kappa^2}\,x_3\right) + C_2\exp\left(\iota\sqrt{\frac{\omega^2}{(\alpha^U)^2} - \kappa^2}\,x_3\right)$$

and

$$A^D = C_3 \exp\left(-\iota\sqrt{\frac{\omega^2}{(\alpha^D)^2} - \kappa^2}\, x_3\right) + C_4 \exp\left(\iota\sqrt{\frac{\omega^2}{(\alpha^D)^2} - \kappa^2}\, x_3\right).$$

In accordance with physical requirements, u_3^U and u_1^U must decay exponentially as $x_3 \to -\infty$, which forces $\kappa^2 - \omega^2/(\alpha^U)^2 > 0$ and $C_2 = 0$. Similarly, u_3^D and u_1^D must decay exponentially as $x_3 \to \infty$, and therefore, $\kappa^2 - \omega^2/(\alpha^D)^2 > 0$ and $C_3 = 0$. Thus, we write the general solutions as

$$A^U = C_1 \exp\left(\sqrt{\kappa^2 - \frac{\omega^2}{(\alpha^U)^2}}\, x_3\right)$$

and

$$A^D = C_4 \exp\left(-\sqrt{\kappa^2 - \frac{\omega^2}{(\alpha^D)^2}}\, x_3\right).$$

Recalling the trial solutions, we write

$$\mathcal{P}^U = C_1 \exp\left(\sqrt{\kappa^2 - \frac{\omega^2}{(\alpha^U)^2}}\, x_3\right) \exp\left(\iota(\kappa x_1 - \omega t)\right)$$

and

$$\mathcal{P}^D = C_4 \exp\left(-\sqrt{\kappa^2 - \frac{\omega^2}{(\alpha^D)^2}}\, x_3\right) \exp\left(\iota(\kappa x_1 - \omega t)\right).$$

Expressing these results in terms of the displacement-vector components, we obtain

$$u_1^U = \frac{\partial \mathcal{P}^U}{\partial x_1} = \iota\kappa\, C_1\, e_u\,,$$

$$u_3^U = \frac{\partial \mathcal{P}^U}{\partial x_3} = \sqrt{\kappa^2 - \frac{\omega^2}{(\alpha^U)^2}}\, C_1\, e_u\,,$$

$$u_1^D = \frac{\partial \mathcal{P}^D}{\partial x_1} = \iota\kappa\, C_4\, e_d\,,$$

$$u_3^D = \frac{\partial \mathcal{P}^D}{\partial x_3} = -\sqrt{\kappa^2 - \frac{\omega^2}{(\alpha^D)^2}}\, C_4\, e_d\,.$$

Thus, from the continuity of u_3 at $x_3 = 0$, namely, $u_3^U|_{x_3=0} = u_3^D|_{x_3=0}$, we get

$$\sqrt{\kappa^2 - \frac{\omega^2}{(\alpha^U)^2}}\, C_1 = -\sqrt{\kappa^2 - \frac{\omega^2}{(\alpha^D)^2}}\, C_4\,. \tag{6.58}$$

Also, returning to the continuity of σ_{33} at $x_3 = 0$, we write

$$\sigma_{33}^U = C_{11}^U \left(\frac{\partial u_1^U}{\partial x_1} + \frac{\partial u_3^U}{\partial x_3} \right)$$

and

$$\sigma_{33}^D = C_{11}^D \left(\frac{\partial u_1^D}{\partial x_1} + \frac{\partial u_3^D}{\partial x_3} \right),$$

to get

$$\begin{aligned}\sigma_{33}^U|_{x_3=0} &= C_{11}^U \left(-\kappa^2 C_1 e_u + \left(\kappa^2 - \frac{\omega^2}{(\alpha^U)^2} \right) e_u C_1 \right)\Bigg|_{x_3=0} \\ &= \frac{-\omega^2}{(\alpha^U)^2} \exp\left(\iota(\kappa x_1 - \omega t)\right) C_{11}^U C_1\end{aligned}$$

and

$$\begin{aligned}\sigma_{33}^D|_{x_3=0} &= C_{11}^D \left(-\kappa^2 C_4 e_d + \left(\kappa^2 - \frac{\omega^2}{(\alpha^D)^2} \right) e_d C_4 \right)\Bigg|_{x_3=0} \\ &= \frac{-\omega^2}{(\alpha^D)^2} \exp\left(\iota(\kappa x_1 - \omega t)\right) C_{11}^D C_4.\end{aligned}$$

Equating these two expressions, we obtain

$$\frac{-\omega^2}{(\alpha^U)^2} C_{11}^U C_1 = \frac{-\omega^2}{(\alpha^D)^2} C_{11}^D C_4.$$

Grouping the terms on the same side, dividing by ω^2, as well as replacing $C_{11}^U/(\alpha^U)^2$ by ρ^U and $C_{11}^D/(\alpha^D)^2$ by ρ^D, we get

$$\frac{-\rho^U}{\rho^D} C_1 + C_4 = 0. \tag{6.59}$$

Let us proceed to solve for C_1 and C_4. Dividing equation (6.58) by κ, we get

$$\sqrt{1 - \frac{v^2}{(\alpha^U)^2}}\, C_1 + \sqrt{1 - \frac{v^2}{(\alpha^D)^2}}\, C_4 = 0. \tag{6.60}$$

Thus, equations (6.59) and (6.60) form a system to be solved for C_1 and C_4. A nontrivial solution requires the determinant of the coefficient matrix to be zero,

$$\det \begin{bmatrix} \sqrt{1 - \frac{v^2}{(\alpha^U)^2}} & \sqrt{1 - \frac{v^2}{(\alpha^D)^2}} \\ -\frac{\rho^U}{\rho^D} & 1 \end{bmatrix} = 0, \tag{6.61}$$

which means that

$$\sqrt{1-\frac{v^2}{(\alpha^U)^2}} = -\frac{\rho^U}{\rho^D}\sqrt{1-\frac{v^2}{(\alpha^D)^2}}\,.$$

This equation cannot be solved with positive radicands, which are required by the amplitude decay away from the interface. Thus, we cannot obtain nontrivial solutions to express waves propagating along an interface between two liquids. The only solution is for a degenerate case, where the radicands are zero, and hence, $v = \alpha^U = \alpha^D$, which corresponds to a medium that is homogeneous with respect to the P-wave speed. A physical explanation for the absence of P waves at an interface between two liquids—in spite of their existence in either medium as body waves—might be an interesting question to pursue.

It is important to recognize that a hasty squaring of both sides results in

$$1-\frac{v^2}{(\alpha^U)^2} = \left(\frac{\rho^U}{\rho^D}\right)^2\left(1-\frac{v^2}{(\alpha^D)^2}\right),$$

and hence

$$v = \sqrt{\frac{1-\left(\frac{\rho^U}{\rho^D}\right)^2}{\frac{1}{(\alpha^U)^2}-\left(\frac{\rho^U}{\rho^D}\right)^2\frac{1}{(\alpha^D)^2}}}\,,$$

which—in spite of its correct appearance—is false. The final form of the expression does not betray its falsity.

Properties of the surface and interface waves are summarized in Table 6.1.

6.6 Existence of interface waves

6.6.1 *Introduction*

In a manner similar to the one used in Section 6.4, let us comment on the restrictions on the elasticity parameters and mass densities required for the existence of interface waves discussed herein. Also, let us discuss the complete impossibility of existence of certain interface waves.

Name	Location	Type	Speed	Properties
Rayleigh	Sec. 6.2	surface	$v < \beta$	evanescent
Rayleigh (modified)	footnote 24	waveguide	$v_r < v < \beta_h$	dispersive
Love (modified)	Sec. 6.3.1	waveguide	$\beta < v$	dispersive
Love	Sec. 6.3.2	waveguide	$\beta_\ell < v < \beta_h$	dispersive
Stoneley	Sec. 6.5.1	interface	$v < \min\left(\beta^U, \beta^D\right)$	evanescent
Scholte	Sec. 6.5.2	interface	$v < \min\left(\beta, \alpha^U\right)$	evanescent

Table 6.1: Properties of the surface, interface and guided waves in Hookean solids: Speed refers to the wavefront propagation. α and β are the speeds of the P and S body waves, respectively; for a given medium, $\alpha > 2\,\beta/\sqrt{3}$: a consequence of the stability condition. For Love wave, β_ℓ and β_h refer to the values from the layer and the halfspace respectively. For Stoneley wave, $\min\left(\beta^U, \beta^D\right)$ refers to the lower value from the two elastic media. For Scholte wave, $\min\left(\beta, \alpha^U\right)$ refers to the lower of the values of β or α^U .

6.6.2 *Elasticity parameters and mass densities*

In view of their definition, the interface waves are associated with models that consist of two continua, and hence can exist only under certain restrictive relations between the elasticity parameters and mass densities of their models. The range of the allowable elasticity parameters and mass densities depends on the relation among them and between the two halfspaces. This relation can be expressed as a function of the body-wave speeds within these halfspaces.

For the Stoneley waves, it is the solubility condition discussed by Achenbach (1973, Section 5.12, equation (5.110)), where it is expressed in terms of a 4×4 matrix. For the Scholte wave, this condition, given in equation (6.52), is in terms of a 3×3 matrix, and, for the interface wave between two liquids, given in equation (6.61), in terms of a 2×2 matrix.

The size of the aforementioned matrices depends on the number of body-waves allowable by a given model. In view of the P-wave and S-wave properties, there are four body waves for the interface between two isotropic solids, three waves for the interface between the solid and a liquid and two waves for the interface between two liquids.

6.6.3 *On SH waves as interface waves*

Let us turn our attention to the complete impossibility of existence of certain interface waves. Following upon the formulation presented in Section 6.5.2, let us consider an interface between an elastic solid and a liquid, and examine the possibility of the SH wave propagating therein.

Let us postulate the SH wave propagating along the interface between the elastic and liquid halfspaces. In such a case, $u_3 = u_1 = 0$ and $u_2 \neq 0$. At the interface, $x_3 = 0$, the side condition is $\sigma_{23}^U = \sigma_{23}^D = 0$, where $\sigma_{23}^U = 0$, since—for an ideal fluid—$C_{44}^U = 0$. Thus, $\sigma_{23}^D = 0$ implies

$$\left.\frac{\partial u_2}{\partial x_3}\right|_{x_3=0} = 0. \tag{6.62}$$

Consider the wave equation for SH waves

$$\frac{\partial^2 u_2}{\partial x_1^2} + \frac{\partial^2 u_2}{\partial x_3^2} = \frac{1}{\beta^2}\frac{\partial^2 u_2}{\partial t^2},$$

which is equation (6.27). Inserting the trial solution,

$$u_2(x_1, x_3, t) = U(x_3)\exp(\iota(\kappa x_1 - \omega t)),$$

we obtain

$$\frac{\mathrm{d}^2 U}{\mathrm{d}\,x_3^2} + \left(\frac{\omega^2}{\beta^2} - \kappa^2\right)U = 0,$$

whose general solution is

$$U(x_3) = C_1 \exp\left(-\iota\sqrt{\frac{\omega^2}{\beta^2} - \kappa^2}\,x_3\right) + C_2 \exp\left(\iota\sqrt{\frac{\omega^2}{\beta^2} - \kappa^2}\,x_3\right).$$

The exponential decay in the elastic halfspace requires

$$U(x_3) = C\exp\left(-\sqrt{\kappa^2 - \frac{\omega^2}{\beta^2}}\,x_3\right),$$

which means that

$$u_2(x_1, x_3, t) = C\exp\left(-\sqrt{\kappa^2 - \frac{\omega^2}{\beta^2}}\,x_3\right)\exp(\iota(\kappa x_1 - \omega t)).$$

However, condition (6.62) implies that

$$C\left(-\sqrt{\kappa^2 - \frac{\omega^2}{\beta^2}}\right)\exp(\iota(\kappa x_1 - \omega t)) = 0.$$

Thus—for a nontrivial solution, where $C \neq 0$—we require

$$\kappa^2 - \frac{\omega^2}{\beta^2} = 0,$$

which implies that

$$u_2(x_1, x_3, t) = C \exp(\iota(\kappa x_1 - \omega t)) .$$

However, similarly to the case of the homogeneous halfspace with the vacuum above it, this equation has no exponential decay, which contradicts the physical requirements. Thus, in the spirit of a proof by contradiction, the initial postulate is rejected: no SH waves are allowed in a homogeneous elastic halfspace below a homogeneous ideal liquid halfspace. Furthermore—in view of the SH waves being decoupled from the P and SV waves—no information is lost in formulating, in Section 6.5.2, the Scholte waves by considering the P and SV waves only.

Closing remarks

In this chapter, we examine surface, guided and interface waves within isotropic Hookean solids. As stated by Grant and West (1965, p. 62),

> Additional refinements to the theory are needed if the solid is æolotropic or heterogeneous.

Following this statement, Grant and West (1965) provide suggestions to formulate Rayleigh waves in transversely isotropic solids. Yet, they warn the readers that

> [d]etails of the analyses, however, require rather specific assumptions about the precise nature of the æolotropy, which in practice are difficult to make.

This warning expresses a concern about the pragmatic need for, and empirical support of, a more refined mathematical theory. Nevertheless, technological advances in more than half-a-century since the publication of Grant and West (1965) might motivate theorists to engage is such analyses to formulate adequate analogies for present-day measurements. Already, Maurycy Pius Rudzki (1912) considered such issues.

6.7 Exercises

Exercise 6.1. Show that ansatz (6.11) applied to equation (6.9) results in expression (6.13).

Solution. Differentiating

$$\mathcal{P}(\mathbf{x},t) = A(x_3)\exp(\iota(\kappa x_1 - \omega t))$$

twice with respect to each of x_1, x_3 and t, we obtain

$$\frac{\partial^2 \mathcal{P}}{\partial x_1^2} = -\kappa^2 A \exp(\iota(\kappa x_1 - \omega t)),$$

$$\frac{\partial^2 \mathcal{P}}{\partial x_3^2} = \frac{\mathrm{d}^2 A}{\mathrm{d}x_3^2}\exp(\iota(\kappa x_1 - \omega t)),$$

$$\frac{\partial^2 \mathcal{P}}{\partial t^2} = -\omega^2 A \exp(\iota(\kappa x_1 - \omega t)),$$

respectively. Inserting these expressions into equation (6.9) and factoring the exponential term, which cannot be zero, we write

$$\frac{\mathrm{d}^2 A}{\mathrm{d}x_3^2} + \left(\frac{\omega^2}{\alpha^2} - \kappa^2\right) A = 0 .$$

Similarly, ansatz (6.26) applied to equation (6.27) results in expression (6.28).

Exercise 6.2. Using equation (6.36), show that $v < \beta_h$ is required for an exponential amplitude decay within the halfspace.

Solution. Equation (6.36) models the amplitude within the halfspace. Rearranging and using $\kappa = \omega / v$, this equation can be written as

$$\begin{aligned} u_2^h(\mathbf{x},t) &= D_1 \exp\left(\iota\left(\kappa x_1 - \omega t - \iota\sqrt{\frac{\omega^2}{v^2} - \frac{\omega^2}{\beta_h^2}}\, x_3\right)\right) \\ &+ D_2 \exp\left(\iota\left(\kappa x_1 - \omega t + \iota\sqrt{\frac{\omega^2}{v^2} - \frac{\omega^2}{\beta_h^2}}\, x_3\right)\right) \\ &= D_1 \exp\left(\iota\left(\kappa x_1 - \omega t\right)\right) \exp\left(\kappa\sqrt{1 - \frac{v^2}{\beta_h^2}}\, x_3\right) \\ &+ D_2 \exp\left(\iota\left(\kappa x_1 - \omega t\right)\right) \exp\left(-\kappa\sqrt{1 - \frac{v^2}{\beta_h^2}}\, x_3\right) . \end{aligned}$$

For exponential amplitude decay, we require $D_1 = 0$, since the first term grows exponentially with x_3. For the second term, we require $v < \beta_h$, to ensure that the radicand therein is a positive number. Under such conditions, u_2^h decays exponentially with x_3.

Chapter 7

Variational principles in seismology

Peut-être que la lumière ne cherche pas à minimiser le temps de parcours, pourtant elle se comporte comme si elle le faisait. Mais de quel point de vue se placer pour faire la différence entre le comme si et la chose même?[1]

Ivar Ekeland (2000)

Preliminary remarks

Variational principles are based on an extremization of quantities, such as definite integrals, whose values depend on functions. These principles are stated in terms of the calculus of variations, which provides methods to find extremizing functions. In classical physics, variational principles provide an alternative approach to the one that relies on causality. They allow us to study the behaviour of a system, not by examining causes and effects within it but by seeking an extremum of its particular quantity. Seismology invokes at least two variational principles of classical physics: Fermat's principle of stationary traveltime and Hamilton's principle of stationary action.

We begin this chapter by introducing a few historical comments followed by an examination of Fermat's principle and an illustration of its consistency with Huygens's principle; the latter relies on causality, the former on global optimization. We generalize this consistency by discussing a proof of Fermat's principle that stems from Hamilton's ray equations.[2] We complete this chapter by discussing conserved quantities. In these dis-

---

[1] *Perhaps the light signal does not try to minimize its traveltime, even though it behaves as if it did. But from what standpoint can one distinguish between as if and the thing itself?*

[2] *see also*: Slawinski (2015, Section 13.1)

cussions, we also emphasize the distinction between the Hamiltonians and Lagrangians in mechanics and in ray theory.

To avoid any nomenclature confusion within this chapter, let us state the following. Both Fermat's and Hamilton's principles share the structure of the calculus of variations. Fermat's principle is intrinsically a part of ray theory. Hamilton's principle is intrinsically a part of Lagrangian mechanics. Hamilton's and Lagrange's equations—together with Legendre's transformation, which relates them—are mathematical formulations that allow an examination of evolution in either realm.

Legendre's transformation, named after a French mathematician, Adrien-Marie Legendre, is an abstract mathematical operation, which is used to solve partial differential equations.[3] Herein, apart from relating the Hamiltonian and Lagrangian mechanics, as well as two formalisms of ray theory, it is used to derive the thermodynamic potentials, as shown in Appendices C.2, C.4 and C.6.

7.1 Historical comments

[4]Newton revolutionized physics by expressing its laws in the language of calculus. Following the intrinsic concept of calculus with its infinitesimal displacements, he expressed the second law of motion, $F = m\,a$, as a local property connecting the instantaneous acceleration of a mass to the net force acting upon it at that instant. This was distinct from the earlier work of Ptolemy and Kepler, who described trajectories of objects by global considerations invoking geometric curves, such as epicycles, circles and ellipses. Newton's laws do not originate in the trajectories of objects, even though these trajectories can be obtained by solving the resulting equations of motion.[5]

A legendary beginning of the calculus of variations is the *Dido problem* from almost three millennia ago. Dido, the founder and first queen of Carthage, was asked to find the largest surface area of land enclosed by a constant-length boundary and a seashore. She correctly chose a semicircle. A historical beginning of the calculus of variations is the *brachistochrone problem* proposed by Johannes Bernoulli in 1696. It amounts to finding,

[3] *see also*: Slawinski (2015, Appendix B)
[4] Readers interested in history of the calculus of variations might refer to Hildebrandt and Tromba (1996).
[5] Readers interested in a historical description of Newton's work might refer to Weinberg (2015, Chapter 14).

in a constant gravitational field, the curve of fastest descent, which is a *cycloid*.

In the eighteenth century, several scientists, beginning with de Maupertuis, Euler and Lagrange, expressed mechanics in terms of global quantities. Subsequently, through work of such scholars as Jacobi and Hamilton, this global-optimization formalism reached its present form. To achieve this formalism—and motivated by the fact that an extremum of a certain quantity expressed as an integral results in a physical solution—it was necessary to develop mathematical methods to calculate the extrema of integrals; these methods constitute the calculus of variations.[6]

The lion's share of the credit for advancing variational formulations in physics belongs to Leibniz and his concepts of *vis viva*.[7] Also, in agreement with the title of Ekeland (2000), the choice of the path of stationary action is analogous to the best of all possible worlds, as discussed by Leibniz in his *Essais de théodicée sur la bonté de Dieu, la liberté de l'homme et l'origine du mal*.[8,9]

7.2 Fermat's principle

7.2.1 *Isotropic layered medium*

To introduce Fermat's principle in a model that is simple enough not to require tools of the calculus of variations, let us consider two isotropic and homogeneous media separated by an interface, as shown in Figure 7.1. Prior to discussing Fermat's principle, however, let us obtain the same result by invoking Huygens's principle for that model.

The speeds of a signal in the upper medium and lower medium are v_1 and v_2, respectively. A planar wavefront touches the interface at point A. Point C on this wavefront is to travel the distance of $v_1\,\Delta t$ to touch the interface, where Δt is the traveltime between C and B. According to Huygens's principle, A can be considered a point source. During interval Δt, the wavefront in the lower medium, generated at A, becomes a semicircle of

[6] *see also*: Slawinski (2015, Chapter 12)

[7] *see also*: Slawinski (2015, Section 13.2)

[8] Essays of Theodicy on the Goodness of God, the Freedom of Man, and the Origin of Evil

[9] Readers interested in the philosophical implications of Leibniz's theodicy and its influence on variational formulations of integral expressions in physics might refer to Ferry (2013), an audiobook in French.

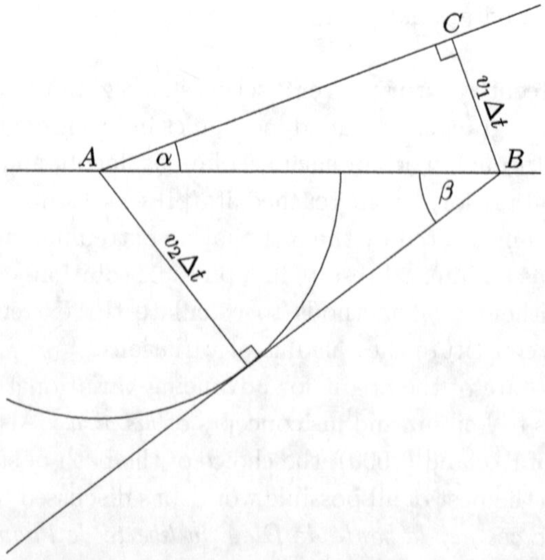

Fig. 7.1: A wavefront is in contact with an interface at point A. CB is the distance that wavefront at point C is to travel to reach the interface. Since the wavefront-propagation speed in the upper medium is v_1, this distance is $v_1\Delta t$, where Δt is the traveltime. During Δt, the wavefront generated at point A covers the distance of $v_2\Delta t$ in the lower medium. Also during Δt, each point on the interface between A and B is reached by the incident wavefront and, consequently, acts as a source resulting in the wavefront in the lower medium. The relationship between the orientations of the two wavefronts is stated in terms of v_1, v_2, α and β as Snell's law, which is a consequence of Huygens's principle.

radius $v_2\,\Delta t$. Examining Figure 7.1, we write

$$AB = \frac{v_1\,\Delta t}{\sin\alpha} \qquad \text{and} \qquad AB = \frac{v_2\,\Delta t}{\sin\beta}\,.$$

Equating the right-hand sides of these equations, we write the standard form of Snell's law, $\sin\alpha/v_1 = \sin\beta/v_2$. Since the variables on each side of the equation are evaluated in different media, it follows that each side is equal to the same constant, a concept discussed in Section 7.4.

Using the same model, let us introduce Fermat's principle. To do so, we consider source S and receiver R, as shown in Figure 7.2. The traveltime of the signal in the first medium is $\sqrt{y_S^2 + r^2}/v_1$, where r is the distance along the interface to the refraction point, and in the second, $\sqrt{y_R^2 + (D-r)^2}/v_2$, where D is the horizontal source-receiver distance. Invoking Fermat's principle—according to which traveltime of a signal between R and S is stationary, which is a global property—we differentiate the sum of these two

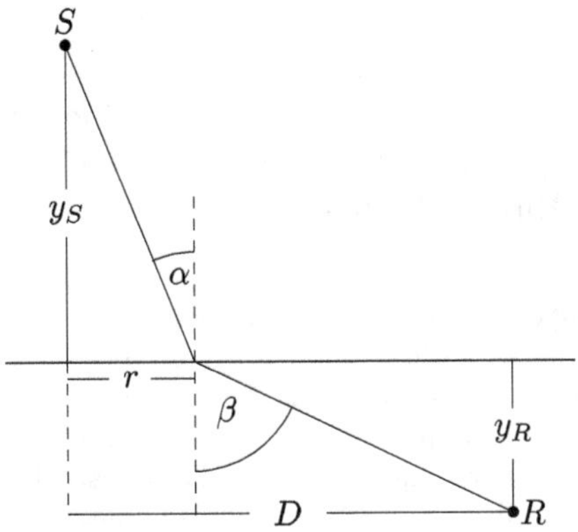

Fig. 7.2: Source S and receiver R are placed in the media for which the signal-propagation speeds are v_1 and v_2, respectively. We wish to choose the refraction point in such a manner that the traveltime of the signal between the source and the receiver is minimized. Unlike for the approach shown in Figure 7.1, where we use a geometrical construction, here we use calculus to find this point, with the figure serving as an illustration for description of traveltimes, distances and angles. As in Figure 7.1, the relationship between the orientations of the two rays is stated in terms of v_1, v_2, α and β as Snell's law, but herein it is derived by invoking Fermat's principle.

expressions with respect to r and set the resulting expression to zero. Following standard calculus operations and trigonometric relations, we obtain Snell's law, named in honour of a Dutch mathematician and astronomer, Willebrord Snel van Royen. Since—in isotropic media—rays are perpendicular to wavefronts, α is the same angle whether measured between the wavefront and the horizontal or the ray and the vertical; the same is true for β.

The results remain true if we exchange the medium of incidence with the medium of transmission or the source with the receiver; the angles, the wavefronts and the rays remain the same. This is a consequence of the signal propagation being described in terms of an even-rank tensor, which results in all properties, including the speed of propagation, being the same in the opposite directions. This statement applies to both elasticity and electromagnetism, where the corresponding tensors are of the fourth and second rank, respectively.

Both Huygens's principle, which is a local property and relies on the concept of causality, and Fermat's principle, which relies on the concept of global optimization, result in the same expression.[10]

[10]Readers interested in the philosophical importance of Fermat's principle and its influence on Descartes's rationalism might refer to Ferry (2013), an audiobook in French.

It might be interesting to note that Huygens introduced his principle in *Treatise on Light*, published in 1678. In doing so, he invoked the concept of ether, whose existence has been rejected since that time. In spite of the revision of the physical interpretation, the underlying principle remains valid. Describing Huygens's theory, Weinberg (2015, p. 220) writes,

> Just as in an ocean wave in deep water it is not the water that moves along the surface of the ocean but the disturbance of the water, so likewise in Huygens' theory it is the wave of disturbance in the particles of the ether that moves in a ray of light, not the particles themselves. Each disturbed particle acts as a new source of disturbance, which contributes to the total amplitude of the wave.

Herein—while replacing ether by an elastic continuum and a ray of light by a seismic ray—we can use the essential structure of Huygens's principle without any modifications.

7.2.2 *Isotropic continuously inhomogeneous medium*

To consider rays in a continuously inhomogeneous—as opposed to a layered—medium, where the speed of a signal is a function of position, $v(x, y)$, in the xy-plane, we must invoke the tools of the calculus of variations. We need to find a trajectory along which the traveltime is stationary. Such a statement leads to the variational expression given by $\delta\int_A^B \mathrm{d}t = 0$, where A and B are the source and receiver, respectively.

In contrast to the case discussed in Section 7.2.1, it is impossible to find point r for which the traveltime is extremal. We need to search for the entire trajectory along which the value of the traveltime integral is extremal. This trajectory is the solution of the Euler equation, whose derivation is described by Slawinski (2015, Chapter 12). Herein, we present a concise formulation of this equation, whose solution is $y(x)$ that renders

$$I = \int_A^B L(y, y'; x)\,\mathrm{d}x \tag{7.1}$$

stationary; L is a known function and the integral limits are fixed.

photo: Elena Patarini

Fig. 7.3: Monument of Luigi Lagrange in Turin, by Giovanni Albertoni

Let us assume that there exists a solution, Y, such that any infinitesimal variation, δy, of Y results in—at most—a second-order variation in I. Consequently, the first variation is zero,

$$\delta I := \delta \int\limits_A^B L(y, y'; x)\mathrm{d}x = 0\,,$$

which is tantamount to stationarity. We can write

$$\delta I = \int\limits_A^B \delta L(y, y'; x)\mathrm{d}x = \int\limits_A^B \left(\frac{\partial L}{\partial y}\delta y + \frac{\partial L}{\partial y'}\delta y'\right)\mathrm{d}x = 0\,, \tag{7.2}$$

where, for the first equality, we use the commutativity of integration and variation under the assumption that $Y \to Y + \delta y$ and $Y' \to Y' + \delta y'$, as well as no variation of integration limits. For the second equality, we use the chain rule. Using the fact that differentiations and variations commute, we write the second term as

$$\frac{\partial L}{\partial y'}\delta y' \equiv \frac{\partial L}{\partial y'}\delta\frac{\mathrm{d}y}{\mathrm{d}x} = \frac{\partial L}{\partial y'}\frac{\mathrm{d}}{\mathrm{d}x}\delta y\,,$$

which we can integrate by parts to write

$$\int\limits_A^B \frac{\partial L}{\partial y'}\frac{\mathrm{d}}{\mathrm{d}x}\delta y = \frac{\partial L}{\partial y'}\delta y\bigg|_A^B - \int\limits_A^B \delta y\frac{\mathrm{d}}{\mathrm{d}x}\left(\frac{\partial L}{\partial y'}\right)\mathrm{d}x = -\int\limits_A^B \delta y\frac{\mathrm{d}}{\mathrm{d}x}\left(\frac{\partial L}{\partial y'}\right)\mathrm{d}x\,,$$

since, at the integration limits, $\delta y = 0$. Using this result in expression (7.2), we write that expression as

$$\delta I = \int\limits_A^B \left(\frac{\partial L}{\partial y}\delta y + \frac{\partial L}{\partial y'}\delta y'\right)\mathrm{d}x = \int\limits_A^B \left(\frac{\partial L}{\partial y} - \frac{\mathrm{d}}{\mathrm{d}x}\left(\frac{\partial L}{\partial y'}\right)\right)\delta y\,\mathrm{d}x = 0\,.$$

For the integral to be zero for all δy and for any value of A and B, we require the integrand to be zero,

$$\frac{\partial L}{\partial y} - \frac{\mathrm{d}}{\mathrm{d}x}\left(\frac{\partial L}{\partial y'}\right) = 0\,; \qquad (7.3)$$

in other words, the integrand being zero is a necessary condition to render I stationary. Furthermore, this ordinary differential equation, which is the Euler equation, is also a sufficient condition because if the integrand is zero, so is the integral. Thus, the solution, $y(x)$, of this differential equation renders $\int_A^B L\,\mathrm{d}x$ stationary.

photo: Elena Patarini

Fig. 7.4: Streets named after Lagrange in Turin and Paris.

As discussed in Section 7.4.2, equation (7.3) leads to Snell's law for inhomogeneous media. Assuming that rays can be expressed as $y(x)$, L is the integrand of the right-hand side of

$$\int_A^B \mathrm{d}t = \int_A^B \frac{\sqrt{1+(y')^2}}{v(x,y)}\,\mathrm{d}x\,, \tag{7.4}$$

where $y' := \mathrm{d}y/\mathrm{d}x$, and where we could write the integrand of the intermediate step as $\mathrm{d}s/v$, where $\mathrm{d}s$ is the arclength element, $\mathrm{d}s^2 = \mathrm{d}x^2 + \mathrm{d}y^2$.

7.2.3 Global optimization and causality

As a methodology, variational principles are based on a global optimization. Their justification, as physical principles, however, relies on a causal effect.

In the context of Fermat's principle, the global optimization requires *a priori* information about conditions to be encountered by the signal. Let us address this teleological difficulty. Following Epstein and Slawinski (1999), we show that the signal trajectory between A and B in expression (7.4), even though obtained by a global optimization of equation (7.3), is actually locally determined.

Using the integrand of expression (7.4) as L for equation (7.3), we obtain

$$\frac{\partial L}{\partial y} = \frac{\partial}{\partial y}\frac{\sqrt{1+(y')^2}}{v(x,y)} = -v_y\frac{\sqrt{1+(y')^2}}{v^2}\,,$$

$$\frac{\partial L}{\partial y'} = \frac{\partial}{\partial y'}\frac{\sqrt{1+(y')^2}}{v(x,y)} = \frac{y'}{v\sqrt{1+(y')^2}}$$

and, following a laborious quotient-rule operation,

$$\frac{\mathrm{d}}{\mathrm{d}x}\frac{\partial L}{\partial y'} = \frac{\mathrm{d}}{\mathrm{d}x}\frac{y'}{v(x,y)\sqrt{1+(y')^2}} = \frac{y''v - y'(v_x + y'v_y)(1+(y')^2)}{v^2\sqrt{(1+(y')^2)^3}}\,,$$

where $y'' := \mathrm{d}^2y/\mathrm{d}x^2$, $v_x := \partial v/\partial x$ and $v_y := \partial v/\partial y$. Using these expressions, we write equation (7.3) as

$$v\,y'' + (v_y - v_x y')\left(1+(y')^2\right) = 0\,,$$

which is

$$v\,y'' + \nabla v\cdot(-y',1)\left(1+(y')^2\right) = 0\,, \tag{7.5}$$

where $\nabla v = [v_x, v_y]$ is the speed-field gradient.

Let us consider $\nabla v \cdot n$,[11] which is the directional derivative of the speed field in the direction perpendicular to the ray, where the unit ray-normal vector is

$$n = \frac{1}{\sqrt{1+(y')^2}}[-y', 1]^T, \tag{7.6}$$

with T denoting transpose.

Notation 7.1. In this chapter, vectors are not expressed in bold font. Vectorial properties can be inferred from the context.

Dividing both sides of equation (7.5) by $\sqrt{1+(y')^2}$, we obtain

$$v \frac{y''}{\sqrt{1+(y')^2}} + \nabla v \cdot n \left(1+(y')^2\right) = 0\,.$$

Dividing both sides by v and by $1+(y')^2$, we obtain

$$\frac{y''}{\sqrt{\left(1+(y')^2\right)^3}} + \frac{\nabla v \cdot n}{v} = 0\,.$$

Since the curvature of a plane curve is

$$\kappa := \frac{y''}{\sqrt{\left(1+(y')^2\right)^3}}\,,$$

we have

$$\kappa + \frac{\nabla v \cdot n}{v} = 0\,. \tag{7.7}$$

This expressions allows us to obtain the ray curvature, κ, as a function of the speed field, $v(x, y)$. As described, from an anthropocentric perspective, by Epstein and Slawinski (1999),

> the rays can be seen as constantly monitoring the gradient of the speed field across the present position, and then deciding accordingly how much to curve themselves to achieve optimality. A careful analysis of the sign conventions involved in the derivation reveals that concavity of the ray always embraces the zone of smaller speed.

Thus, the signal direction—as the ray curvature—is determined locally at each point of its trajectory, based on the local property of the speed-field gradient. The global optimization is a consequence of the infinitesimal and causal advance of the signal. A further discussion of a causal justification

[11] *see also*: Bóna and Slawinski (2015, Section 1.2)

of the calculus of variations as opposed to its teleological interpretation is presented in Section 7.2.7, below.

Examining expression (7.7), we see that, in homogeneous media, $\nabla v = 0$, which means that $\kappa = 0$, and hence, rays are straight for signals propagating in all directions. In inhomogeneous media, rays are straight for the signal propagating along the gradient direction, which is the direction of ∇v, where $\kappa = 0$, since $\nabla v \perp n$. Also, as shown in Exercise 7.1, in laterally homogeneous media with a constant speed gradient, rays are circular arcs.

7.2.4 *Stationarity versus minimization*

Colloquially, it is common to refer to Fermat's variational principle as the *least-time principle* and to Hamilton's variational principle, discussed in Section 7.3, below, as the *least-action principle.* Such a terminology is—at least partially—justified by the fact that many physical analogies of the Euler equation are in a quantitative agreement with predictions or retrodictions based on the minima of definite integrals. However, the Euler equation is a stationarity—not a minimum—condition, and there are physical analogies of the Euler equation for maxima.

To address the issue of stationarity of traveltime, let us discuss Fermat's principle in the language of the calculus of variations and comment on properties of this approach. For a ray, an infinitesimal variation of its shape results in no first-order variation in traveltime for any signal traveling along the path resulting from such a variation. It follows that the ray is a path in whose neighbourhood many paths exhibit nearly the same traveltime, hence, the property of stationarity.

For the case illustrated in Figure 7.2, above, the traveltime function is acted upon by a differential operator, $\mathrm{d}/\mathrm{d}r$, which implies infinitesimal displacements along the interface as specified by r. The variational operator, δ, on the other hand, implies a variation by modifying the entire curve that connects the source with the receiver. The former is a search for a point that extremizes the value of a function, the latter for a function that extremizes the value of an integral.

For any function, we can find an integral whose value is stationary for that function. Hence, variational principles require that the same integral be considered for all curves. Such an integral stems from an *a priori* choice of the physical principle in question. Thus, a given principle is to be chosen prior to solving the Euler equation. Analogously, in differential calculus, we consider all points for the same function.

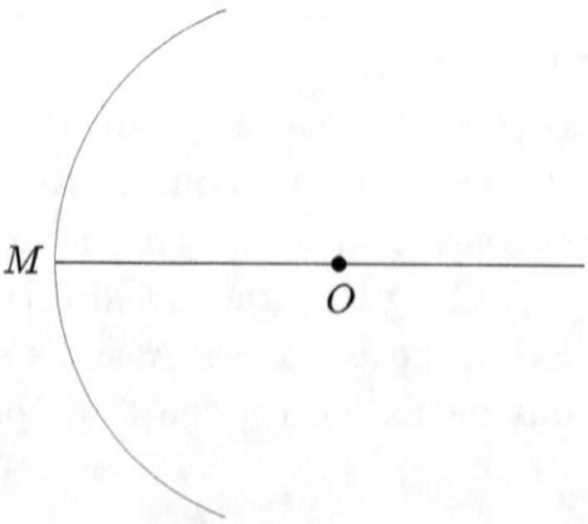

Fig. 7.5: Examining the semispherical mirror with centre at O, and considering point sources along the horizontal line, we see that, for points between O and M, the distance to and from the mirror is the least along the horizontal line; for point O, it is the same as to other points on the mirror; for points to the right of O, the distance to M and back is longer than for any other point on the mirror.

No minimum or maximum restriction is introduced in this section or in the formulation of the Euler equation in Section 7.2.2. Equation (7.3) is the condition for the stationarity of integral (7.1), not for its minimization, in a manner analogous to the first-derivative test in differential calculus. This property is consistent with the geometrical illustration given in Figure 7.5, where—due to homogeneity of the medium—distance is tantamount to traveltime. Also, this property is consistent with the macroscopic and microscopic interpretations discussed in Section 7.2.6, below.

7.2.5 *Mathematical justification*

7.2.5.1 *Fermat's principle*

Mathematically, Fermat's principle is true by virtue of a theorem.[12] To demonstrate this, we state the definition of a ray, which originates in the elastodynamic equation, and is directly contained in its bicharacteristic equations, which are Hamilton's equations. Rays are solutions of Hamilton's equations. To prove Fermat's principle is to verify that such a ray is also a solution of the Euler-Lagrange equation.[13]

[12] Readers interested in proofs of Fermat's principle might refer to Babich (1994), Epstein and Śniatycki (1992), Červený, V. (2002) and Bóna and Slawinski (2003).
[13] *see also*: Slawinski (2015, Section 13.1)

Herein, and in agreement with Slawinski (2015), Euler-Lagrange is the name given to the Euler equation upon endowing its mathematical structure with a physical interpretation by letting, for instance, L have units of slowness, as in expression (7.9), below.

Consider Hamilton's equations,

$$\dot{x}^i = \frac{\partial H}{\partial p_i}\,, \qquad \dot{p}_i = -\frac{\partial H}{\partial x^i}\,, \qquad i = 1,2,3\,, \tag{7.8}$$

where the Hamiltonian is $H = p^2v^2/2$, with $v = v(x,p)$ being a function that describes properties of the continuum, which depend on the wavefront position, x, and orientation, p.

Notation 7.2. In this chapter—in accordance with Appendix E—the indices of vectors are placed as superscripts and of one-forms as subscripts.

Rays are projections of the solutions of Hamilton's equations, which reside in the xp-space, onto the x-space; the former is called the *phase space*. They are the projected bicharacteristics of the elastodynamic equation, whose characteristics are given by the eikonal equation, whose characteristic equations, in turn, are Hamilton's equations.[14]

To prove Fermat's principle, we examine the relationship between Hamilton's and the Euler-Lagrange equations.

The first expression in system (7.8) is Legendre's transformation. The transformation of H is the Lagrangian, $L = 1/V(x,\dot{x})$, where V is the signal speed along the ray, which depends on position, x, and direction, $\dot{x}$. Equation (7.8) is transformed into

$$\frac{\partial L}{\partial x^i} + \frac{\mathrm{d}}{\mathrm{d}t}\left(\frac{\partial L}{\partial \dot{x}^i}\right) = 0\,, \qquad i = 1,2,3\,, \tag{7.9}$$

which is the solution of a variational problem given by $\delta \int L\,\mathrm{d}t = 0$.

Note that there is no need to state the variational problem *a priori*; we might view Fermat's principle as an *a posteriori* formulation following Legendre's transformation.

Legendre's transformation requires the Hamiltonian to be convex, which is tantamount to the wavefront-slowness surfaces being convex. In general, they are not; hence, there are singularities of transformation: the relationship between $\dot{x}$ and p is not one-to-one.[15] There are up to three rays for a single wavefront orientation. This is not to say that each ray might not be

[14]Readers interested in Hamilton's equations as the bicharacteristics of the elastodynamic equation might refer to Bos and Slawinski (2010).
[15]*see also*: Slawinski (2015, Section 8.4)

a stationary-traveltime path, just that we cannot prove it in the context of Legendre's transformation.

However, according to a theorem stated by Musgrave (1970, pp. 91–92), the wavefront-slowness surface is convex for the fastest among the three waves, since the wavefront-slowness surface is of degree six. Hence, any straight line can intersect it at, at most, six points. The line intersecting the innermost sheet intersects the two outer sheets twice on either side of the innermost sheet, and the innermost sheet can be intersected at, at most, two points. This implies that the innermost sheet is convex. Thus, the leading wavefronts propagate in accordance with Fermat's principle, according to the proof in the context of Legendre's transformation.

7.2.5.2 *Head waves*

Let us examine applications of Fermat's principle that reach beyond the rigours of its mathematical proof. To prove Fermat's principle, we consider rays as a consequence of Cauchy's equation of motion. According to Fermat's principle, the traveltime is stationary along any ray, which is a sufficient condition to construct a ray if the velocity function is smooth. Consider Figure 7.6. Path SR is a direct ray; path SBR a reflected one. According to Figure 7.6, a signal traveling along the interface, path $SACR$, has no intrinsic tendency to direct itself towards R at point C; along the interface, there is no change of properties at C. Thus—at the proposed level of explanation, which might be viewed as a traveltime-computation method that agrees with observations—we must impose the condition of reaching R.

In geometrical optics, where rays are postulated entities, not a consequence of Cauchy's equation of motion, we might accommodate their definition to consider $SACR$ as a ray, as long as the stationarity condition—or equivalently, Snell's law at the interface—is satisfied. Also, according to the physical mechanism described below in the context of Figure 7.7, there is no need to direct the signal towards R; in agreement with descriptions of Section 7.2.6, paths SR, SBR and $SACR$ illustrate the essence of a macroscopic description of complicated physical phenomena.

A description that allows us to predict the observed traveltime, illustrated in Figure 9.1, as opposed to an *a posteriori* computation method provided by path $SACR$, is provided by the concept of a *head wave*, illustrated in Figure 7.7. Head waves were postulated and detected by Ludger Mintrop in the first quarter of the twentieth century by examining

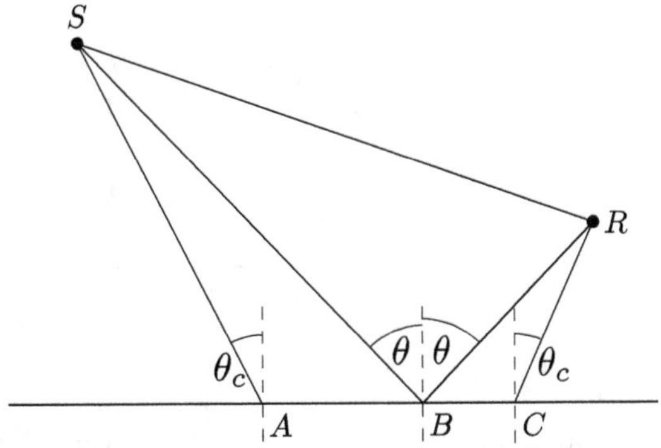

Fig. 7.6: Direct, SR, reflected, SBR, and refracted, $SACR$, signals travel between source S and receiver R along paths of least traveltime. θ_c is the critical angle, $\arcsin v_1/v_2$, where v_1 and v_2 are the signal propagation speeds in the upper and lower medium, respectively. For the refracted signal, as illustrated in this figure, the direction of propagation specified at the source is insufficient to reach R; there is no change of properties along the interface—a physical justification is provided by Figure 7.7 and described in the text. However, the concept of constructive interference illustrated in Figure 7.9 still applies.

traveltime-distance plots akin to the one illustrated in Figure 9.1. Mintrop received a patent for his formulation in the context of seismic exploration. A history of the development of a physical explanation of head waves exemplifies attempts to account for observations within the realm of a theory.[16] The presently accepted explanation as a doubly refracted wave is illustrated in Figures 7.6 and 7.7.

Consider an incident signal on a planar interface between two isotropic homogeneous media, as illustrated in Figure 7.2. Until the critical angle there are three rays that coincide at the interface and point away from it; they correspond to the incident, reflected and refracted signals; only the first and third are illustrated in Figure 7.2. The incident and reflected signals propagate with speed v_1 and the refracted signal propagates with v_2. Beyond the critical angle, the refracted signal propagates along the interface in the lower medium with speed v_2; also, from that point on, the corresponding wavefront is perpendicular to the interface. Since the two media are in a welded contact,[17] the continuity of displacement results in

[16]Readers interested in historical remarks on head waves might refer to Červený, V. and Ravindra (1971, Section 1.1).

[17]For further explanations of a welded contact, readers might refer to Chapman (2004, Section 4.3.1).

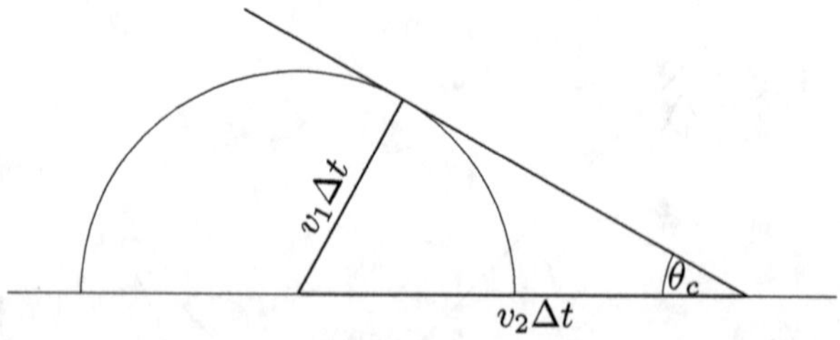

Fig. 7.7: A point source propagates along the interface with speed v_2 generating disturbances in the upper medium, where the propagation speed is v_1. The head wave is the straight-line envelope of these disturbances, as illustrated by two source locations separated by time Δt. The orientation of the wavefront is $\arcsin v_1/v_2$, which is the critical angle, θ_c. This wavefront can be viewed as being perpendicular to CR in Figure 7.6 and to paths reaching receivers R_i in Figure 9.1.

the propagation of a disturbance across the interface. Each point across the interface can be viewed as a Huygens source, which generates disturbances that, in the upper medium, propagate with speed v_1.

The envelope of the source moving with speed v_2 and generating disturbances propagating with speed v_1 is the head wave. Since v_1 and v_2 are constant, the wavefront is a straight line whose orientation is the critical angle, $\arcsin v_1/v_2$, as illustrated in Figure 7.7.[18] If the interface is not straight, the envelope of elementary wavefronts is not a straight line, as illustrated in Figure 7.8. Locally, the rays, which—under assumption of isotropy—are orthogonal to the head-wave wavefront, are at the critical angle with the interface.

We note that in the context of ray theory—in contrast to the direct, reflected and transmitted waves, which are associated with the zero-order term of the asymptotic series—the head waves are associated with higher-order terms.[19]

To complete this section, let us comment on the fact that different levels of explanation are associated with, among other considerations, the perceived aim of science. For an empiricist, whose only purpose is to account for observations, a computational method using $SACR$ in Figure 7.6 might be all that is justifiable in view of the distance-traveltime plot illustrated

[18]For further illustrations of head waves, readers might refer to Červený, V. and Ravindra (1971, Section 1.2).

[19]For a discussion of head waves and higher-order terms, readers might refer to Červený, V. (2001, Section 5.6.7).

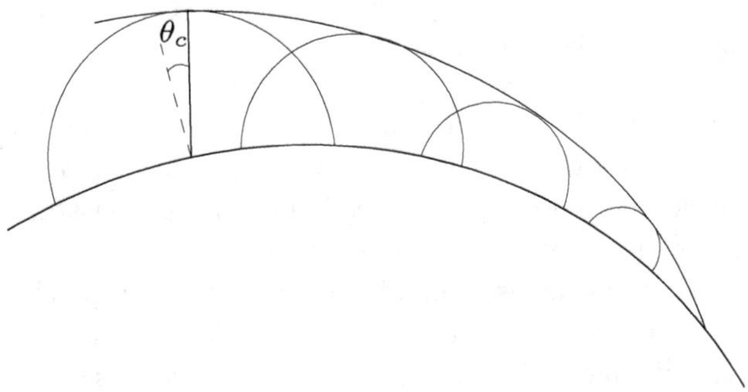

Fig. 7.8: A point source propagating along the circular interface with speed v_2 and generating disturbances in the medium outside the circle, where propagation speed is v_1. These disturbances are modeled by circular elementary wavefronts, whose envelope is the head wave. Locally, the orientation of the head wave—which globally is an arc of a spiral—is given by the orientation of the ray at the critical angle, θ_c.

in Figure 9.1. For a realist, whose purpose is to propose a physical mechanism, even if its intricacies cannot be observed, the concept of a head wave illustrated in Figures 7.7 and 7.8 might be necessary.

In Chapter 9, we examine several approaches to seismology as a science. In Section 7.2.6, to provide a realist view, we discuss both a macroscopic and a microscopic physical justification of Fermat's principle.

7.2.6 *Physical interpretation*

7.2.6.1 *Macroscopic interpretation*

Let us establish a relation between Fermat's principle, as a mathematical analogy, and macroscopic physical phenomena. To do so, we interpret this principle in terms of signals of finite wavelengths. To provide an underlying microscopic reason for the agreement between the principle and empirical information, we use an argument based on the superposition of probability amplitudes, as invoked in quantum mechanics.

Joseph-Louis Lagrange (1788), in his classic treatise, states the following.

> Nous savons aujourd'hui que la lumière se meut par *tous* les chemins, mais que ce n'est que sur celui de moindre durée de parcours que les ondes lumineuses sont renforcées en sorte que

> l'on puisse constater un résultat sensible. La lumière *semble* donc se propager suivant la ligne de moindre durée.[20]

Examining Figure 7.2, we can imagine that source S emits signals that reach every point on the interface. Following the above quote of Lagrange, we interpret Fermat's principle physically in terms of constructive interference. Consider the curve—without the arrows—in Figure 7.9. This curve illustrates the traveltime as a function of the refraction point. Waveforms arriving at the receiver at a similar traveltime along similar paths interfere constructively.[21] Since this interpretation relies on signals arriving at a similar time, it is valid for the stationary—not only the minimum—traveltime.

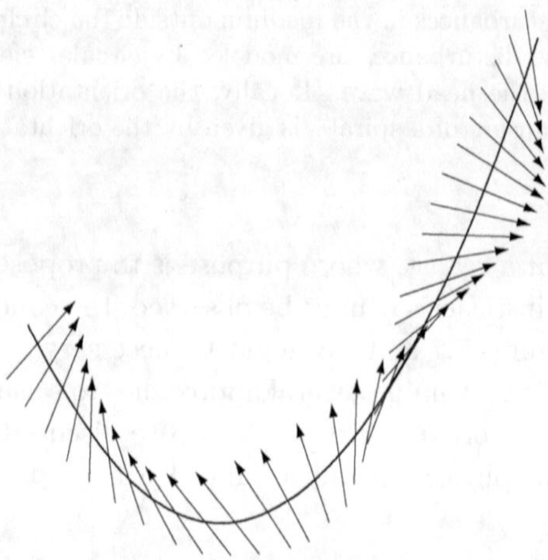

Fig. 7.9: Traveltimes of signals propagating through different paths between S and R, shown in Figure 7.2. The minimum corresponds to the refraction point. The traveltime value changes less in the vicinity of that point than in other locations, since the ray has a property that a small change to its shape results in no first-order change in traveltime for any signal traveling along the path resulting from such a change. This property justifies the constructive interference of both wavelets and probability-amplitude arrows, shown herein. The summation of arrows on a horizontal line, which denotes the interface shown in Figure 7.2, leads to a significant resultant vector near the minimum; at other points, the arrows interfere destructively.

---

[20]Today we know that the light gets through *all* paths, but it is only along the least-traveltime path that the light waves are reinforced in a manner as to render them visible. The light *appears* to propagate along the least-traveltime path.

[21]Readers interested in an insightful description according to which—as stated by Feynman—the signal, using its wavelength, *smells nearby paths, and checks them against each other*, might refer to Feynman *et al.* (1964, Vol. I, Section 26-5).

For this macroscopic interpretation, we introduce the concept of wavelength, which is not used explicitly in mathematical formulations of Fermat's principle, Huygens's principle or Snell's law; nor is it needed for mathematical consistency among them.

However, invoking the concept of wavelength and constructive interference due to similar traveltimes, is—at least macroscopically—justified, since finite wavelengths, together with Huygens's principle, provide an explanation of the bright spot of light in the centre of the shadow cast by a circular obstruction, as observed by François Jean Dominique Arago following the wave theory of light proposed by Augustin Fresnel and its mathematical formulation by Siméon Denis Poisson. This result of 1818 is a counterexample to the corpuscular theory of light and exemplifies a limitation of geometrical optics, according to which there should be no bright spot.

7.2.6.2 *Microscopic interpretation*

To provide a microscopic physical justification of Fermat's principle, we consider the probability that a signal is detected at the receiver, which is illustrated by probability-amplitude arrows in Figure 7.9.[22] The probability of observing an event is a consequence of interference of these amplitudes.[23]

Along each path, the probability of a signal is described by vector $r\exp\iota\theta$, whose phase is proportional to the traveltime and the probability to the length squared. Using the superposition principle, we add vectors corresponding to the adjacent paths to obtain the probability amplitude in their neighbourhood. For paths along which the traveltime is similar, the vectors have similar orientations, which results in this constructive interference. As stated by Basdevant (2005), *Tout porte sur la phase*: all relies on the *phase*, which results in an analogy between the physical justification and the method of stationary phase in mathematics.[24]

The number of turns per second of a probability-amplitude arrow corresponds to the frequency of the signal, which explains the reason for higher-frequency events requiring a smaller neighbourhood of the ray for the constructive interference than would be necessary for lower frequencies.

---

[22]Readers interested in an insightful description with several figures might refer to Feynman (2006, pp. 37–52).

[23]Propagation of probability amplitudes is a Green's function, which is discussed in another context in Section 5.2.5. Readers interested in this and other recent applications of Green's functions might refer to the lecture of Freeman Dyson (2001).

[24]*see also*: Bóna and Slawinski (2015, Section 5.3.3)

7.2.6.3 *Phase consideration*

Let us consider the following illustration to emphasize that Fermat's principle is a byproduct, not the cause, of a physical phenomenon. Speaking colloquially and in an anthropocentric manner, a signal does not consider the time it takes to reach a receiver. The path of least time happens to be the trajectory from which small path deviations are negligible as far as the time of arrival is concerned, and hence, signals that arrive at the receiver close to the least time arrive with a similar phase, which leads to their constructive interference.

We can rephrase this statement as follows. Consider Figure 7.2 and Snell's law written as

$$\frac{\sin\alpha}{v_1}-\frac{\sin\beta}{v_2}=0\,. \tag{7.10}$$

Herein, for the constructive interference, we assume that the signals arriving at the receiver along the ray and along its small deviations have the same phase.

Let us modify this assumption, by assuming that the phase does not change during propagation through homogeneous media but is subject to a change on the interface. Thus, we recast Snell's law by writing expression (7.10) as

$$\frac{\sin\alpha}{v_1}-\frac{\sin\beta}{v_2}=\frac{\mathrm{d}\,\phi}{\mathrm{d}\,r}\,,$$

where ϕ denotes the phase. Examining this expression, we see that—unless $\mathrm{d}\phi/\mathrm{d}r=0$—the refraction point does not coincide with the stationary-traveltime trajectory. As stated above, all relies on the phase, and the stationary traveltime is its byproduct. If there is no change in phase that depends on the location along the interface, the expression reduces to the standard form of Snell's law. To gain further insight into this issue, let us write

$$\beta=\arcsin\left(\frac{v_2}{v_1}\sin\alpha-v_2\frac{\mathrm{d}\phi}{\mathrm{d}r}\right);$$

the angle of transmission is a function of v_1, v_2 and ϕ, and hence—for a given ratio of v_1 and v_2—the transmission angle can change, which is tantamount to the refraction point moving along the interface.

This illustration indicates that Fermat's principle stated as a principle of stationary traveltime is not a fundamental law, since it can be disobeyed without leading to contradictory or physically impossible results.

7.2.7 *On teleology of Fermat's principle*

Referring to the subtitle of this book, it might be easier to formulate questions—particularly, the unasked ones—by referring to their partial answers. Many questions appear only upon a deeper examination of provisional or tentative answers. Let us exemplify such a question by examining how a preliminary interpretation might present Fermat's principle as a case for teleology, and by advancing our arguments against it.

Developments of mathematical physics, notably the work of Hamilton, superseded the need for metaphysical justifications, as entertained by Euler and, in particular, by de Maupertuis. The work of Hamilton resulted in considering variational principles as a mathematical tool only. However, developments of quantum mechanics, among them the work of Feynman discussed in Section 7.2.6.2, suggest that—while there is no need for a teleological justification of variational principles—we can interpret them by a physical argument. In doing so, we allow the concept of probability to replace teleology.

The calculus of variations does not invoke the concept of causality since it is based on definite integrals; it is a search for a global optimization. To avoid the apparent conflict between causality and global optimization, we might be tempted to use the fact that, unlike for the equation of motion, which is hyperbolic, the wellposedness of an elliptical equation requires boundary, not initial, conditions.

However, wave propagation, as formulated in mathematics, is a causal phenomenon. This means that it is a phenomenon that occurs in time and whose evolution at every instant depends on conditions that immediately precede it. There are several approaches to prove this causality, one of which is discussed in Section 7.2.3. Herein, we invoke the wellposedness of the Cauchy problem, which in the context of hyperbolic partial differential equations, requires initial conditions whose propagation exhibits finite speeds.[25]

Let us take the Fourier transform of the wave equation, with time and frequency being the transformation variables. We obtain Helmholtz's equation, $\nabla^2\hat{u} + (\iota\,\omega/v)^2\,\hat{u} = 0$.[26] The explicit time dependence vanishes and the wave equation, which is hyperbolic, becomes elliptic. However, since the bicharacteristics, which correspond to rays, are real only for hyperbolic equations, finding a ray connecting two points whose locations are given as boundary conditions is impossible in the context of an elliptic equation.

[25] *see also*: Slawinski (2015, Sections 6.5.3 and 6.5.4)

[26] *see also*: Bóna and Slawinski (2015, Appendix D)

The trajectories of the propagation of initial conditions are rays, which are intrinsic entities of the wave equation. They reside in the phase space, which is the solution space of Hamilton's equation. Hamilton's equations, in turn, are intrinsic entities of the eikonal equation, which can be viewed as an infinitesimal statement of Huygens's principle, which is tantamount to the statement of causality.

The interpretation issue arises from the fact that the propagation of initial conditions is only a partial analogue for the propagation of physical disturbances. The propagation of initial conditions asserts nothing about the mechanism of propagation of physical disturbances. It allows us to predict the traveltime between the source and receiver, which does not imply that the method by which this prediction is achieved mathematically is, in any way, analogous to the mechanical behaviour.

To present a teleological interpretation, let us examine Fermat's principle by considering Figure 7.2. According to this principle, the traveltime between source S and receiver R corresponds to the shortest time that a signal takes to travel between these points. As can be experimentally verified, this principle provides an accurate prediction or retrodiction to an aspect of the physical phenomenon in question. It does not mean, however, that it offers an explanation.

Let us offer two objections to the possibility of such an explanation. First, to choose the optimal trajectory according to Fermat's principle, the signal at S would need to be subject to its final destination prior to reaching it. In an anthropocentric vocabulary, we could say that—to make the choice of a trajectory—the signal would need an *a priori* information about its destination and properties of media through which it is to travel to reach it.

The second objection is the case against an exceedingly literal interpretation of a mathematical concept of a ray. The point source emits signals that reach each point on the horizontal interface between S and R. If we wish to illustrate it with rays, they would be radii extending from S. According to Huygens's principle, each such a point is a secondary source for a signal propagating to the receiver. The choice of a single ray, denoted—in view of Snell's law, which itself is entailed by Fermat's principle—by angles α and β, is not justified by multiple rays suggested by Huygens's principle. There is, however, no conflict, since neither principle is rooted in fundamental physics; either principle is a heuristic statement that allows us to examine macroscopic physical behaviours. Notably, the very straightness of rays is a convenient approximation that is sufficiently accurate for macroscopic considerations.[27]

Nevertheless, since the agreement between predictions of Fermat's principle and experimental measurements might tempt us to postulate a teleological interpretation, let us discuss it further by presenting two other cases against such an interpretation.

Since causality requires temporal dependence, its disappearance might be seen as supporting a teleological interpretation. At the first sight, such a disappearance results from performing the Fourier transform, where time and frequency are the transformation variables. Upon taking the transform, the wave equation, which is hyperbolic, becomes elliptic. However, this time dependence disappears for each frequency separately, and therein, the temporal dependence together with requirement of causality remain. They are contained in the specific information about frequency. The solutions of the Fourier-transformed equation satisfy the corresponding elliptic equation for a single frequency. The dependence on frequency requires a condition, which—upon performing the inverse Fourier transform and hence regaining the time variable—becomes the initial condition. As expected, the underlying requirements remain unchanged.

Another way to argue against a teleological formulation of Fermat's principle is to consider rays as solutions of Hamilton's equations, which are derived without any teleological assumption offered by the calculus of variations. In applying Legendre's transformation to these equations—which does not invoke or imply any variational concept—we obtain differential equations that are recognizable as the Euler-Lagrange equations. A scientist who does not recognize that the resulting equations might be obtained independently as a stationarity condition within the calculus of variations would not entertain any teleological questions, but would accept a formulation based on differential equations for which the evolution of the system is determined progressively at each position.[28] Be that as it may, the existence of several methods to derive a given equation together with a satisfactory physical analogy offered by that equation does not allow us to interpret the mechanism of a physical phenomenon in terms of these derivations.

Regardless of our discussion on causality within the realm of mathematics, justifying the causality in physics faces difficulties on its own. As brought to our attention by David Hume (2007), there is a fundamental

---

[27]Readers interested in a quantum-theory explanation of *why light appears to travel in straight lines* might refer to Feynman (2006, pp. 53–56).

[28]*see also*: Slawinski (2015, Chapter 11)

difficulty in verifying its existence, even though it might appear clearly as a mental idea, since we might interpret consistent empirical coincidences as being due to causality. Hume (2007) states that causality in physics cannot be asserted by deduction, only by induction, where we face *Hume's induction problem*: A deductive justification of induction itself, which would constitute its proof, is impossible due to the uncertainty of conclusions based on inductive reasoning.

In concluding the discussion on the teleological interpretation of Fermat's principle, let us emphasize the importance of careful interpretations of mathematical analogies of physical phenomena. Any relation between a mathematical analogy and its physical counterpart is only partial; it omits many aspects and it might add spurious effects. A good analogy represents the essence of the physical phenomenon in question, as it pertains to the study at hand. A fruitful use of mathematical entities in studying physical phenomena requires a competent, inquisitive, curious and humble approach prepared for both unexpected and changeable interpretations of methodologies and results. Hasty interpretations might be faulty, even though their connections with a mathematical formulation might give them an appearance of rigour. As stated by Alexander Pope in *An Essay on Criticism*,

> A little learning is a dangerous thing;
> drink deep, or taste not the Pierian spring:
> there shallow draughts intoxicate the brain,
> and drinking largely sobers us again.

Also, issues of teleology could be examined in the context of Hamilton's principle, discussed in Section 7.3, below. In general, such issues arise if motion is expressed in terms of definite integrals, not in terms of differential equations. Differential equations are statements that quantify motion at a single location or instant. Definite integrals are statements that quantify motion over a spatial or temporal interval, where the initial and final states of the system are fixed *a priori*.

7.3 Hamilton's principle

7.3.1 *Action*

In Section 7.2, we examine Fermat's principle, which belongs to ray theory. Let us examine Hamilton's principle, which belongs to mechanics.

A mechanical system is described by a Lagrangian, $L(x, \dot{x}, t)$, which depends on the position and its time derivative; it might also depend explicitly on time. For a particle of mass m in a potential field, $U(x,t)$, we have $L = m\dot{x}^2/2 - U$, the difference of the kinetic, T, and potential energies. For any trajectory, $x(t)$, the action is defined as $\int L\,\mathrm{d}t$.

Herein, $L := T - U$ is a physical quantity that underlines the concept of Lagrangian mechanics. Its units are energy. It is distinct from L used in formulating Fermat's principle whose units are the reciprocal of time.

According to Hamilton's variational principle, the trajectory describing the evolution of a physical system is stationary, $\delta\int L\,\mathrm{d}t = 0$, with the condition of stationarity given by the Euler-Lagrange equation,

$$\frac{\partial L}{\partial x^i} - \frac{\mathrm{d}}{\mathrm{d}t}\left(\frac{\partial L}{\partial \dot{x}^i}\right) = 0\,, \qquad i = 1,2,3\,. \tag{7.11}$$

The Lagrangian for a volume of a continuum is

$$L = \iiint\limits_V \mathcal{L}\,\mathrm{d}V\,,$$

where the Lagrangian density, $\mathcal{L}$, is a function of the displacement vector, $u(x,t)$, and its derivatives. The stationarity of action can be stated as

$$\delta\int\limits_{t_0}^{t} L\,\mathrm{d}t = \delta\int\limits_{t_0}^{t}\iiint\limits_V \mathcal{L}\,\mathrm{d}V\mathrm{d}t = 0\,.$$

As presented by Landau and Lifshitz (1976), classical mechanics can be formulated in terms of Lagrange's equations of motion in the context of the stationary-action principle. In classical mechanics, Hamilton's equations can be viewed as the canonical form of the equations of motion.[29] Lagrangian mechanics is contained in Hamiltonian mechanics.

The classical-mechanics Hamiltonian is convex; hence, in contrast to the ray-theory Hamiltonian, there are no singularities of transformation; the relationship between $\dot{x}$ and p is one-to-one.[30] Thus, even though, formally, the equations are identical, the relationship between the Hamilton's and Euler-Lagrange equations arising in ray theory directly from the characteristics of Cauchy's equations of motion differs from that relationship in classical mechanics. This difference arises from the definitions of H and L and from the singularities of Legendre's transformation.

[29]Readers interested in the canonical forms of the equations of motion might refer to Landau and Lifshitz (1976, Chapter VII).

[30]Readers interested in Hamilton's equations in the context of Lagrangian mechanics might refer to Arnold (1989, Chapter 3, Section 15).

7.3.2 Wave equation

Let us illustrate Hamilton's principle by using it to derive the wave equation. Also, invoking Legendre's transformation, let us show that—in the realm of mathematics—this principle is included in Hamilton's mechanics.[31]

Let us invoke the principle of stationary action to derive the wave equation in a single spatial dimension, where $x^1 \equiv x$.[32] We let

$$U = \frac{1}{2}k\int_0^\ell \left(\frac{\partial u}{\partial x}\right)^2 \mathrm{d}x\,,$$

where k is the spring constant and ℓ is the length of the string, and

$$T = \frac{1}{2}\rho\int_0^\ell \left(\frac{\partial u}{\partial t}\right)^2 \mathrm{d}x\,,$$

where ρ is the mass density. Hence, the Lagrangian density is

$$\mathcal{L} := \frac{\rho}{2}\left(\frac{\partial u}{\partial t}\right)^2 - \frac{k}{2}\left(\frac{\partial u}{\partial x}\right)^2 .$$

The corresponding Euler-Lagrange equation is

$$\frac{\partial \mathcal{L}}{\partial u} - \left(\frac{\partial}{\partial t}\left(\frac{\partial \mathcal{L}}{\partial u_t}\right) + \frac{\partial}{\partial x}\left(\frac{\partial \mathcal{L}}{\partial u_x}\right)\right) = 0\,,$$

where $u_t := \partial u/\partial t$ and $u_x := \partial u/\partial x$. Thus, we obtain

$$\frac{k}{\rho}\frac{\partial^2 u(x,t)}{\partial x^2} = \frac{\partial^2 u(x,t)}{\partial t^2}\,, \tag{7.12}$$

which is the wave equation in a single spatial dimension, analogous to equation (5.22), with $\sqrt{k/\rho} =: v$ denoting the speed of propagation.

In the case of Fermat's principle, the solution is a curve; herein the solution is a surface, $u(x,t)$, for which $\int_0^t \int_0^\ell \mathcal{L}\,\mathrm{d}x\,\mathrm{d}t$ attains a stationary value.

7.3.3 Mathematical justification

Consider Hamilton's equations of motion,

$$\dot{x}^i = \frac{\partial H}{\partial p_i} \qquad \dot{p}_i = -\frac{\partial H}{\partial x^i}\,, \qquad i = 1,2,3\,, \tag{7.13}$$

[31] Readers interested in the inclusion of Lagrange's mechanics in Hamilton's mechanics might refer to the outline of Arnold (1989).

[32] *see also*: Slawinski (2015, Section 13.2.4)

where $H(x,p) = T + U$, with T and U being the kinetic and potential energies, respectively. Legendre's transformation of these equations is equation (7.11).[33]

In spite of the same form exhibited by equations (7.8) and (7.13) and the same role of describing the evolution of a system, their Hamiltonians, H, represent different quantities. In the former case, we can write $H = p^2v^2/2$, which is a product of speed and its reciprocal, slowness; hence, in that case, H is a dimensionless quantity.[34] In the latter case, H has units of energy. To distinguish these quantities, we refer to the former as the *ray-theory Hamiltonian* and to the latter as the *classical-mechanics Hamiltonian*. To emphasize the distinction of their physical meaning, Slawinski (2015) denotes them by $\mathscr{H}$ and $\mathbb{H}$, respectively. Herein, we denote both Hamiltonians by H to emphasize that both Hamilton's equations exhibit the same mathematical structure, even though they are used as analogies for different physical scenarios; in other words, herein, we emphasize the Platonic concept within mathematical physics.

However, both sets of Hamilton's equations, given in expressions (7.8) and (7.13), are related to Cauchy's equation, which describes motions within general continua, and the elastodynamic equation, which describes this motion in Hookean solids. In spite of the same form of both sets, their relations to the elastodynamic equation differ, and so do their physical interpretations. Equations (7.8) are bicharacteristic equations of the elastodynamic equation. Equations (7.13) are analogous to the elastodynamic equation itself, but in the realm of Hamilton's mechanics.

7.3.4 *Physical interpretation*

There appears to be no immediate macroscopic justification for Hamilton's principle. Similarly to Fermat's principle discussed in Section 7.2.6.2, the microscopic justification relies on probability amplitudes. As emphasized by Bunge (1967, p. 143), if one wishes to explain continuum mechanics in terms of microphysics, one needs to perform a logical reduction to quantum mechanics. However, not achieving such a reduction does not invalidate continuum mechanics as a basic science.

The superposition of probability amplitudes allows us to calculate an amplitude in the neighbourhood of a path that describes the evolution of

[33] *see also*: Slawinski (2015, Appendix B.3)
[34] *see also*: Slawinski (2015, Section 8.2)

a physical system.[35] $L(x, \dot{x}, t)$ is an explicit function of spacetime, which allows us to view classical mechanics as the limit of quantum mechanics for macroscopic actions.

7.4 Conserved quantities

7.4.1 *Introduction*

Evolutions of physical systems follow trajectories along which certain quantities might be invariant with respect to spatial and temporal variables. If so, to examine a system, we might focus our attention on a particular conserved quantity.

Conserved quantities provide us with an insight into a given system, and they might facilitate obtaining pertinent solutions. However, there is no information provided by these quantities beyond the information contained within Lagrange's and Hamilton's equations. Conserved quantities are tantamount to integrating Lagrange's equations, and are referred to as *first integrals*.[36]

In Section 7.4.2, we examine a conserved quantity within ray theory that requires a translational symmetry of the medium. In Section 7.4.3, we examine the conservation of the Hamiltonian, H, and the Lagrangian, L, in ray theory and in classical mechanics. Given appropriate definitions, both H and L are conserved in ray theory, but only H is conserved in classical mechanics, where it stands for the total energy.

7.4.2 *Ray parameter*

7.4.2.1 *Isotropy*

Let us consider Snell's law for a series of parallel isotropic homogeneous layers. Parallel layers are tantamount to no change of properties in the direction parallel to their interfaces. Homogeneity of each layer implies that the rays are straight within each layer. Let the signal-propagation speed for layer i be denoted by v_i, and let us assume that it is independent of the propagation direction; the layers are isotropic. Since $\sin\theta_1/v_1 = \sin\theta_2/v_2$ and $\sin\theta_2/v_2 = \sin\theta_3/v_3$, it follows, from the homogeneity, that $\sin\theta_1/v_1 = \sin\theta_3/v_3$.

---

[35] Readers interested in a concise description of *how it works* might refer to Feynman *et al.* (1964, Vol. I, Section 26-6).

[36] *see also*: Slawinski (2015, Section 12.6)

Since, for each relation, the left-hand and the right-hand sides depend on variables evaluated in different media, each side must be equal to the same constant; in general, $p := \sin\theta_i/v_i$. Symbol p denotes a global conserved quantity. For a given ray, it has the same value for all layers. It is a consequence of lateral homogeneity, which is a translational invariance: the same propagation speed along directions parallel to the interfaces.

Decreasing the thickness of each layer, we remove the straightness of rays. To exemplify a conserved quantity for curved rays in a laterally homogeneous medium, let us consider equation (7.3) and assume that L is not an explicit function of y. It follows that $\partial L/\partial y = 0$; hence, since $\mathrm{d}(\cdot)/\mathrm{d}x = 0$, we see that $\partial L/\partial y'$ is constant. For the integrand in expression (7.4), this constant is

$$\frac{y'}{v\sqrt{1+(y')^2}}\,. \tag{7.14}$$

Since $y' := \mathrm{d}y/\mathrm{d}x = \cot\theta$, where θ is the angle between the ray and the vertical, we write this constant as $\cos\theta/v$, which, for a homogeneous medium, is a conserved quantity by the virtue of a constant speed; it states that the orientation of wavefronts and rays remains constant, as expected in homogeneous media. Let us contrast this result with $\sin\theta/v$ in Snell's law. The importance of $\sin\theta/v$ is its being constant, under lateral homogeneity, in inhomogeneous media.

To relate our discussion to the subject of Section 7.2.1, for which the integrand in expression (7.4) is a function of y but not of x, let us invoke Beltrami's identity,[37] discussed in Exercise 7.3, to write equation (7.3) as

$$\frac{\partial L}{\partial x} + \frac{\mathrm{d}}{\mathrm{d}x}\left(y'\frac{\partial L}{\partial y'} - L\right) = 0\,. \tag{7.15}$$

This identity is equivalent to equation (7.3). It remains true even for a system of the Euler-Lagrange equations, however—being a single equation—it is insufficient to solve such a system; hence, it is not equivalent to it.

Since, in our example, L is not an explicit function of x, the first term of expression (7.15) is zero, which implies that the term in parentheses, $y'\partial L/\partial y' - L$, is constant.[38]

[37] *see also*: Slawinski (2015, Sections 11.3 and 12.3)

[38] Also, it is common to refer to this conserved quantity, $y'\partial L/\partial y' - L = C$, as Beltrami's identity; its derivation is shown in Exercise 7.3. This nomenclature emphasizes the importance of the fact that objects follow trajectories along which certain quantities—such as the ray parameter in ray theory or linear and angular momenta and energy in mechanics—are conserved in a global sense.

photo: Elena Patarini

Fig. 7.10: Street named after Eugenio Beltrami in his native Cremona

In the case discussed herein, since

$$L(y, y') := \frac{\sqrt{1+(y')^2}}{v(y)}$$

and $y' = \cot\theta$, we can write this constant term as $-\sin\theta/v(y)$, which—up to the sign—is the conserved quantity of Snell's law.

This quantity is a consequence of translational symmetry of a continuum, which means that we can express material properties in a coordinate system with at least one component explicitly absent.[39] For instance, in the xy-plane, we might have $v(y)$: speed depends on y only. Mathematically, this conserved quantity is an example of Noether's theorem formulated by Emmy Noether (1918), and plays an important role in mathematical physics.[40,41]

If a ray is endowed with a conserved quantity due to the absence of a coordinate, we refer to this quantity as the *ray parameter*. The value of this parameter is unique for a given ray. However, if a continuum does not exhibit any translational symmetry, rays do not possess ray parameters. In such a case, rays are trajectories along which no such quantity is invariant.

7.4.2.2 *Anisotropy*

The existence of a ray parameter, which is a consequence of translational symmetry within a medium, is independent of its anisotropy. Let us examine an anisotropic case.

Since the ray parameter is a conserved quantity along the ray, intrinsically it is expressed in terms of the ray entities. These quantities are the signal position along the ray given by x, the ray orientation by $\dot{x}$ and the speed of signal propagation along the ray, $V := \sqrt{\sum_{i=1}^{3}(\dot{x}^i)^2}$.

[39] *see also*: Slawinski (2015, Chapter 14)

[40] *see also*: Slawinski (2015, Section 12.6)

[41] Readers interested in an examination of Noether's theorem might refer to Goldstein (1980, Section 12.7).

Locally, at any location, we can calculate the value of the ray parameter using the corresponding wavefront quantities, its location, x, its orientation, p, and its slowness, $\sqrt{\sum_{i=1}^{3} p_i^2}$. However, a formulation in terms of wavefronts—even though it corresponds to the same conserved quantity—is valid only locally, since there is no curve connecting wavefront normals that exhibits this quantity, except in the isotropic case, where rays are normal to wavefronts.[42]

Vector $\dot{x}$ is tangent to the ray, $x(t)$, and corresponds to the velocity of the signal along the ray at the point of tangency; we refer to it as the ray velocity, in agreement with Synge (1937, p. 12) and Born and Wolf (1999, pp. 792–795). Using the first of Hamilton's equations (7.8),

$$\dot{x}^i = \frac{\partial H}{\partial p^i}, \qquad \dot{p}_i = -\frac{\partial H}{\partial x^i}, \qquad i = 1, 2, 3,$$

where $H = p^2 v^2/2$ and $v = v(x, p)$, we write

$$\left(\dot{x}^i\right)^2 = \left(p_i\right)^2 v^4 + 2 p_i\, v \frac{\partial v}{\partial p_i} + \frac{1}{v^2}\left(\frac{\partial v}{\partial p_i}\right)^2, \qquad i = 1, 2, 3.$$

Performing the summation of the three terms, $i = 1, 2, 3$, invoking Euler's homogeneous-function theorem[43] and using the eikonal equation, $p^2 v^2 = 1$, we obtain

$$V = \sqrt{v^2 + \frac{1}{v^2}\left(\nabla_p v\right)^2},$$

where $\nabla_p v$ denotes the gradient with respect to the components of the phase-slowness vector. This is the magnitude of the speed of the signal along the ray given in terms of the magnitude of the wavefront velocity, v—which is the reciprocal of the wavefront slowness—as a function of the orientation of the wavefront given by vector p. The two magnitudes are equal to one another, $V = v$, if there is no orientation independence, which is tantamount to isotropy.

Consider the traveltime integral in an anisotropic inhomogeneous continuum for which the properties change only along the x^3-axis. Since there is no lateral change of properties, a propagating signal remains in a single plane that contains that axis. Without loss of generality, we consider the $x^1 x^3$-plane. The traveltime integral between A and B is

$$\int_A^B \frac{\sqrt{1 + \left((x^3)'\right)^2}}{V\left(x^3, (x^3)'\right)}\, \mathrm{d}x^1, \tag{7.16}$$

[42]Readers interested in a description of the importance of Fermat's principle, which entails Snell's law and *predicts new things*, might refer to the last two paragraphs of Feynman *et al.* (1964, Vol. I, Section 26-4).

[43]*see also*: Slawinski (2015, Appendix A)

where $(x^3)'$ denotes $\mathrm{d}x^3/\mathrm{d}x^1$. Using Beltrami's identity,

$$\frac{\partial L}{\partial x^1}+\frac{\mathrm{d}}{\mathrm{d}x^1}\left((x^3)'\frac{\partial L}{\partial (x^3)'}-L\right)=0\,,$$

where L is the traveltime integrand, we see that the term in parentheses is constant due to the absence of x^1 in L, which means that the first term is zero. Using the explicit expression for L in this identity, we obtain

$$\frac{\partial}{\partial (x^3)'}\left(\frac{1}{V}\right)(x^3)'\sqrt{1+((x^3)')^2}-\frac{1}{V\sqrt{1+((x^3)')^2}}=C\,,$$

where C is a constant.

Following Slawinski and Webster (1999), to compare this conserved quantity to Snell's law, we use $(x^3)'=\cot\theta$ to express this quantity in terms of the ray angle. Also, the differential operator can be stated as $\partial/\partial(x^3)'=(\partial\theta/\partial(x^3)')\,\partial/\partial\theta$. Using trigonometric identities, we obtain

$$\cos\theta\,\frac{\partial}{\partial\theta}\left(\frac{1}{V\,(x^3,\theta)}\right)+\frac{\sin\theta}{V\,(x^3,\theta)}=-\,C\,.$$

For the invariance of this quantity, the properties may vary along the x^3-axis only, but the directional dependence—hence, anisotropy—can be arbitrary. In the isotropic case, the first term is zero. Also, in such a case, there is no distinction between rays and wavefront normals, hence $V=v$; the ray parameter is identical to Snell's law.

To complete this section, let us comment on the fact that the availability of closed-form expressions for V is limited to isotropy and its immediate extension: elliptical dependence. Material properties are formulated in the context of the elastodynamic equation, where they appear originally as the mass density and the elasticity parameters of Hooke's law. Following Christoffel's equations these properties are stated in terms of the wavefront slowness or, equivalently, its reciprocal, the wavefront velocity. Their expressions in terms of the ray velocities and orientations are subject to the requirements of Legendre's transformation. To avoid the singularities of this transformation, one can calculate the value of the conserved quantity at a particular point along the ray using the Hamitonian entities, x and p, and carry that quantity along the ray.

7.4.3 *Hamiltonian and Lagrangian*

7.4.3.1 *Ray theory*

The Hamiltonian, H, and the Lagrangian, L, are conserved along the ray, but for a different reason than the ray parameter. In contrast to the ray parameter, the conservation of the Hamiltonian and the Lagrangian do not require any translational symmetry. These conserved quantities exist for all cases.

To show the invariance of H, we use the chain rule to write

$$\frac{\mathrm{d}\,H(x,p,t)}{\mathrm{d}\,t}=\sum_{i=1}^{3}\frac{\partial H}{\partial x^i}\dot{x}^i+\sum_{i=1}^{3}\frac{\partial H}{\partial p_i}\dot{p}_i+\frac{\partial H}{\partial t}\,. \tag{7.17}$$

Since we consider H that does not explicitly depend on time,

$$\frac{\mathrm{d}H}{\mathrm{d}t}=\sum_{i=1}^{3}\frac{\partial H}{\partial x^i}\dot{x}^i+\sum_{i=1}^{3}\frac{\partial H}{\partial p_i}\dot{p}_i\,,$$

and, according to equations (7.8),

$$\frac{\mathrm{d}H}{\mathrm{d}t}=-\sum_{i=1}^{3}\dot{p}_i\dot{x}^i+\sum_{i=1}^{3}\dot{x}^i\dot{p}_i=0\,,$$

which means that the value of H is constant. This value is $1/2$, which results from the definition, $H=p^2v^2/2$, where $v=v(x,p)$, and the fact that the eikonal equation, $p^2v^2=1$, must be satisfied along the ray.[44]

To show the invariance of L, we consider Legendre's transformation to write $L=\sum_{i=1}^{3}p_i\dot{x}^i-H$. Since H is homogeneous of degree 2 in the p_i and $\dot{x}_i=\partial H/\partial p_i$, it follows, by Euler's homogeneous-function theorem,[45] that $L=2H-H=H$. Thus, along each ray, the Lagrangian is also $1/2$.

To clarify the ambiguous notation for consideration of the homogeneity of $H=p^2v^2(x,p)/2$, where p appears also as an argument of v, note that its only purpose as an argument is to identify the speed, v, that corresponds to the wavefront whose normal is given by the orientation of p; this orientation dependence allows us to consider anisotropy. To avoid such an ambiguity, we could write $v(x,p/\|p\|)$.[46] In any case, as illustrated in Exercise 7.5, v itself is homogeneous of degree 0 in the p_i, and hence, H is homogeneous of degree 2.

One could use another definition of H, such as $H=p^2-1/v^2$. It satisfies the eikonal equation given by $p^2-1/v^2=0$, and remains conserved with the value of 0. However, since $H=p^2-1/v^2$ is not homogeneous of any degree, we could not infer, by Euler's homogeneous-function theorem, the conservation of L.[47]

[44] *see also*: Slawinski (2015, Section 8.2)
[45] *see also*: Slawinski (2015, Appendix A)
[46] *see also*: Slawinski (2015, see discussion between equations (A.1.2) and (A.1.3))
[47] *see also*: Slawinski (2015, Section 8.2)

7.4.3.2 *Classical mechanics*

In classical mechanics, $H := T + U$, introduced in equations (7.13), is the total energy of that system. If H does not depend on time explicitly, $\partial H/\partial t = 0$. As shown in the derivation following expression (7.17), this implies that $\mathrm{d}H/\mathrm{d}t = 0$; the Hamiltonian is conserved by its time invariance.

Let us consider conserved quantities in terms of L. In a manner analogous to expression (7.17) and assuming that L is not explicitly dependent on t, we use the chain rule to write

$$\frac{\mathrm{d}L(x,\dot{x})}{\mathrm{d}t} = \sum_{i=1}^{3} \frac{\partial L}{\partial x^i}\dot{x}^i + \sum_{i=1}^{3} \frac{\partial L}{\partial \dot{x}^i}\ddot{x}^i, \tag{7.18}$$

which, in general, is not zero. However, invoking equation (7.11), which we write as

$$\frac{\partial L}{\partial x^i} = \frac{\mathrm{d}}{\mathrm{d}t}\left(\frac{\partial L}{\partial \dot{x}^i}\right), \qquad i = 1,2,3,$$

we rewrite expression (7.18) as

$$\frac{\mathrm{d}L}{\mathrm{d}t} = \sum_{i=1}^{3}\left(\frac{\mathrm{d}}{\mathrm{d}t}\left(\frac{\partial L}{\partial \dot{x}^i}\right)\dot{x}^i + \frac{\partial L}{\partial \dot{x}^i}\ddot{x}^i\right),$$

which, in view of the product rule, we write as

$$\frac{\mathrm{d}L}{\mathrm{d}t} = \sum_{i=1}^{3}\frac{\mathrm{d}}{\mathrm{d}t}\left(\frac{\partial L}{\partial \dot{x}^i}\dot{x}^i\right),$$

and, using the linearity of the differential operator, rearrange to obtain

$$\frac{\mathrm{d}}{\mathrm{d}t}\left(\sum_{i=1}^{3}\frac{\partial L}{\partial \dot{x}^i}\dot{x}^i - L\right) = 0.$$

Since the total derivative is equal to zero, it follows that the term in parentheses is a conserved quantity. The fact that $\mathrm{d}(\,\cdot\,)/\mathrm{d}t = 0$ can also be shown directly, as illustrated in Exercise 7.6. Using Legendre's transformation, we can write this term as

$$\sum_{i=1}^{3}\frac{\partial L}{\partial \dot{x}^i}\dot{x}^i - L = H := T + U, \tag{7.19}$$

which means that it stands for the total energy of a system, whose conservation is shown in terms of H, since $\mathrm{d}H/\mathrm{d}t = 0$.

Let us consider another conserved quantity in terms of L. Examining expression (7.11),

$$\frac{\partial L}{\partial x^i} - \frac{\mathrm{d}}{\mathrm{d}t}\left(\frac{\partial L}{\partial \dot{x}^i}\right) = 0\,, \qquad i = 1, 2, 3\,,$$

we see that, if L is not explicitly dependent on x^i, then the term in parentheses is conserved for all t. This is a statement of the conservation of linear momentum in mechanics, whose components are $p_i = \partial L / \partial \dot{x}^i$. Equivalently, according to equations (7.13), the explicit absence of x^i in H results in $\dot{p} = 0$, which is the same conserved quantity. It also follows that the ray parameter is analogous to the *generalized momentum* in mechanics.

In Lagrangian mechanics, the momentum is generalized, since the same form of the equation of motion appears in different coordinates. In cylindrical coordinates, the conserved quantity is $p_i = \partial L / \partial \dot{\theta} = m\, r^2 \dot{\theta}$, where $r\,\dot{\theta}$ is the tangential speed about the x^3-axis. Hence, $mr\,(r\,\dot{\theta})$ is the angular momentum. Also, we have the conservation of the angular momentum and the ray parameter in spherical coordinates. Such a parameter is stated in expression (1.1) and used in global seismology.

To complete this section, let us comment on the fact that in ray theory, both H and L can be conserved explicitly, but not so, in mechanics. This is a consequence of their definitions, which in the former case can be chosen such that both H and L are homogeneous functions. If, in mechanics, L was a homogeneous function of degree r, according to Definition 7.1 on page 306, then—by Euler's homogeneous-function theorem—the summation term in expression (7.19) would be rL, and the left-hand side would become $(r-1)\,L$. Consequently, the conservation of H would imply the conservation of L. This is not the case, which emphasizes distinct meanings of equations (7.8) and (7.13) and their Legendre's transformations, even though their forms are identical.

For conservative systems in classical mechanics, H is conserved explicitly, but L is not. However, both $\sum_{i=1}^{3} (\partial L / \partial \dot{x}^i) \dot{x}^i - L$ and $\sum_{i=1}^{3} \partial L / \partial \dot{x}^i$ are conserved. This means that L changes in such a manner as to keep these terms constant.

Closing remarks

In this chapter, we consider two variational principles used in seismology: Fermat's principle, which resides within ray theory, and is a consequence of the characteristics and bicharacteristics of the elastodynamic equation, and Hamilton's principle, which resides within the realm of Lagrangian mechanics.

Examining these principles, we encounter the same mathematical structure for different physical concepts, which is consistent with the quote of Toretti on page 1. The richness of mathematical structures furnishes us with an inexhaustible stock of analogies for physical phenomena.

A strict reductionist could argue that, in classical physics, Lagrangian mechanics does not contribute to information contained already in Newtonian mechanics. For instance, Hamilton's principle in a classical phase space is a mathematical reformulation of Newton's laws. However, the global approach that results from examining trajectories of evolution in the phase space, rather than paths in the physical space, are significant for both the enhancement of our understanding of natural phenomena and the ease of their quantitative examination.[48] Furthermore, not surprisingly, variational approaches in quantum mechanics are not reducible to Newton's laws of motion.

7.5 Exercises

Exercise 7.1. Prove the following proposition.

Proposition 7.1. *For the laterally homogeneous media in two spatial dimensions, where* $v(x,y)=v(y)$ *, with a constant speed gradient, rays are circular arcs.*

Solution. Let us proceed with the following proof.

Proof. To prove that rays are circular arcs, it suffices to show that—for each ray—its curvature is constant for all values of the ray angle, θ. Consider expression (7.7) written as

$$\kappa=-\frac{\nabla v\cdot n}{v(y)}\,,$$

[48]Interested readers might refer to Arnold (1989).

where the ray normal is given by expression (7.6),

$$n = \frac{1}{\sqrt{1+(y')^2}}[-y',1]^T .$$

Since $\partial v/\partial x = 0$, it follows that

$$\kappa = -\frac{\frac{\partial v}{\partial y}}{v(y)\sqrt{1+(y')^2}} .$$

Since $y' \equiv \mathrm{d}y/\mathrm{d}x = \cot\theta$, where θ is measured with respect to the y-axis, we use trigonometric identities to write

$$\kappa = -\frac{\partial v}{\partial y}\frac{\sin\theta}{v(y)} .$$

where $\sin\theta/v$ is the ray parameter, which is a conserved quantity along the ray. Thus, for an arbitrary ray, if the nonzero gradient component, $\partial v/\partial y$, is constant, so is κ. Hence, rays are circular arcs, as required. □

Remark 7.1. If $\partial v/\partial y = 0$, which means that v is constant, $\kappa = 0$, and hence, rays are straight.

Remark 7.2. Proposition 7.1 and its proof have the following corollary.

Corollary 7.1. *The curvature for the solution of the brachistochrone problem,*

$$\kappa = -\frac{1}{2y\sqrt{1+(y')^2}}$$

is independent of the mass of the descending object and of the acceleration due to gravity, where both mass and acceleration are assumed to be nonzero.

Herein, y is the vertical distance between the initial and final position of the descending object. The vertical speed of this object, v, is due to gravitation, $mv^2/2 = mgy$, where m and g denote mass and the acceleration due to gravity; hence, $v = \sqrt{2gy}$.

The curve in question is a *cycloid.* Along this curve, a point mass starting from rest and moving without friction under constant acceleration, would arrive at its final position in the shortest time. The curvature of this curve is not constant. However, in accordance with Proposition 7.1, the moving point exhibits a conserved quantity, $-\sin\theta/\sqrt{2gy}$, which is akin to the ray parameter and is tantamount to the linear momentum of that point.

Exercise 7.2. Consider an isotropic Hookean layer that is 1000-metre thick and—in its middle—contains a thin and parallel one-metre layer of different mechanical properties. Using Fermat's principle, find the traveltime of a signal whose propagation speed in the bulk of this medium is 1000 m/s and in the one-metre layer is 10000 m/s. Assume that the source and the receiver are on the opposite sides of the medium and, laterally, 1000 metres apart.

Solution. According to Fermat's principle, the traveltime is the minimum of

$$2\frac{\sqrt{x^2+499.5^2}}{1000}+\frac{\sqrt{(1000-2x)^2+1}}{10000}.$$

Taking the derivative and setting it to zero, we obtain

$$\frac{x}{\sqrt{249500+x^2}}=\frac{1000-2x}{10\sqrt{1+(1000-2x)^2}}.$$

Solving numerically, we get $x\approx 50$ m, which means that the signal travels about 900 m in the thin layer. This results in a traveltime of about 1.1 s.

Remark 7.3. It is important to note that both $x\approx 50$ m and $t\approx 1.1$ s remain nearly constant, even if we keep decreasing the thickness of the thin layer, since a one-metre layer is a small portion of the entire medium in question, and, regardless of its diminishing thickness, it offers the least-time path for the travelling signal. Also, at $x\approx 50$ m, the incidence angle approaches the critical angle, since $\arctan(50/500)=\arctan(1000/10000)$. For an insight into this result, in a seismological context, see Exercise 9.1, below, and Remark 9.2, on page 357.[49]

Exercise 7.3. Given equation (7.3) and assuming that $L(y,y';x)$ does not depend explicitly on x, prove Beltrami's identity,

$$y'\frac{\partial L}{\partial y'}-L=C\,, \tag{7.20}$$

where C denotes a constant. Give a physical interpretation.

Solution. Consider the total derivative of $L(y,y';x)$ with respect to its independent variable, x,

$$\frac{\mathrm{d}L}{\mathrm{d}x}=\frac{\partial L}{\partial y}\frac{\mathrm{d}y}{\mathrm{d}x}+\frac{\partial L}{\partial y'}\frac{\mathrm{d}y'}{\mathrm{d}x}+\frac{\partial L}{\partial x}\equiv\frac{\partial L}{\partial y}y'+\frac{\partial L}{\partial y'}y''+\frac{\partial L}{\partial x}.$$

[49]Also, readers interested in such issues might refer to Dalton and Slawinski (2016).

Also, consider

$$\frac{\mathrm{d}}{\mathrm{d}x}\left(\frac{\partial L}{\partial y'}y'\right) = \frac{\partial L}{\partial y'}y'' + \frac{\mathrm{d}}{\mathrm{d}x}\frac{\partial L}{\partial y'}y' .$$

We rewrite each expression as

$$\frac{\partial L}{\partial y'}y'' = \frac{\mathrm{d}L}{\mathrm{d}x} - \frac{\partial L}{\partial y}y' - \frac{\partial L}{\partial x}$$

and

$$\frac{\partial L}{\partial y'}y'' = \frac{\mathrm{d}}{\mathrm{d}x}\left(\frac{\partial L}{\partial y'}y'\right) - \frac{\mathrm{d}}{\mathrm{d}x}\frac{\partial L}{\partial y'}y' ,$$

respectively. Equating the right-hand sides of both equations and using the fact that $\partial L/\partial x = 0$, we write

$$\frac{\mathrm{d}}{\mathrm{d}x}\left(\frac{\partial L}{\partial y'}y'\right) - \frac{\mathrm{d}L}{\mathrm{d}x} = \frac{\mathrm{d}}{\mathrm{d}x}\frac{\partial L}{\partial y'}y' - \frac{\partial L}{\partial y}y' ,$$

which, upon rearranging, we write as

$$\frac{\mathrm{d}}{\mathrm{d}x}\left(\frac{\partial L}{\partial y'}y' - L\right) = -\left(\frac{\partial L}{\partial y} - \frac{\mathrm{d}}{\mathrm{d}x}\frac{\partial L}{\partial y'}\right)y' .$$

Since the term in parenthesis on the right-hand side is zero by the virtue of equation (7.3), it follows that

$$\frac{\mathrm{d}}{\mathrm{d}x}\left(\frac{\partial L}{\partial y'}y' - L\right) = 0 ,$$

which implies that the term in parentheses is constant as required by expression (7.20).

Thus, if L does not depend explicitly on x, it follows that—along the path, $y(x)$, that extremizes the value of $\int L\,\mathrm{d}x$—there is a conserved quantity given by expression (7.20). Depending on the definition of L, C can represent such a quantity as the ray parameter or the linear momentum.

Exercise 7.4. Argue for straightness of rays in homogeneous anisotropic media.

Solution. In homogeneous media, signals propagating in any direction do not encounter any property changes. Hence, their trajectories remain unaffected. Thus, rays are straight in homogeneous media, regardless of anisotropy.

We could proceed in a manner analogous to the one presented in Section 7.2.3, and obtain κ, for anisotropic inhomogeneous Hookean solids. We conjecture that, for any anisotropic but homogeneous case, $\kappa = 0$.

Exercise 7.5. Following Slawinski (2015, Section 6.11.3), and considering two-dimensions, we write

$$\mathbf{p} = (p_1, p_2) = (p(\vartheta)\sin\vartheta, p(\vartheta)\cos\vartheta) , \tag{7.21}$$

where $p(\vartheta)$ stands for the magnitude of $\mathbf{p}$ in direction ϑ. Using equation (7.21), we express ϑ as

$$\frac{p_1}{p_2} = \frac{\sin\vartheta}{\cos\vartheta} = \tan\vartheta ,$$

which means that

$$\vartheta = \arctan\left(\frac{p_1}{p_2}\right) . \tag{7.22}$$

Use the definition of a homogeneous function[50] to show that ϑ is homogeneous of degree 0 in the p_i.

Solution. Let us invoke the following definition.

Definition 7.1. A real function, $f(x_1, \ldots, x_n)$, is homogeneous of degree r in the $x_1, \ldots, x_n$, if

$$f(ax_1, \ldots, ax_n) = a^r f(x_1, \ldots, x_n) ,$$

where $a \in \mathbb{R}$.

Hence, considering expression (7.22), we write

$$\vartheta(ap_1, ap_2) = \arctan\left(\frac{a\,p_1}{a\,p_2}\right) = \arctan\left(\frac{p_1}{p_2}\right) = \vartheta(p_1, p_2) .$$

Thus,

$$\vartheta(ap_1, ap_2) = a^0\vartheta(p_1, p_2) ,$$

which—according to the definition—means that ϑ is homogeneous of degree 0 in the p_i.

Exercise 7.6. Show that

$$\frac{\mathrm{d}}{\mathrm{d}t}\left(\sum_{i=1}^{3}\frac{\partial L}{\partial \dot{x}^i}\dot{x}^i - L\right) = 0 , \tag{7.23}$$

for all $L(x, \dot{x})$.

[50] *see also*: Slawinski (2015, Appendix A)

Solution. Using the linearity of the differential operator, we write

$$\sum_{i=1}^{3}\frac{\mathrm{d}}{\mathrm{d}t}\left(\frac{\partial L}{\partial \dot{x}^{i}}\dot{x}^{i}\right)-\frac{\mathrm{d}L}{\mathrm{d}t}.$$

Following the product and the chain rule, we obtain

$$\sum_{i=1}^{3}\left(\frac{\mathrm{d}}{\mathrm{d}t}\left(\frac{\partial L}{\partial \dot{x}^{i}}\right)\dot{x}^{i}+\frac{\partial L}{\partial \dot{x}^{i}}\ddot{x}^{i}\right)-\sum_{i=1}^{3}\left(\frac{\partial L}{\partial x^{i}}\dot{x}^{i}+\frac{\partial L}{\partial \dot{x}^{i}}\ddot{x}^{i}\right).$$

Since the second summands of the summations cancel one another, we write

$$\sum_{i=1}^{3}\left(\frac{\mathrm{d}}{\mathrm{d}t}\left(\frac{\partial L}{\partial \dot{x}^{i}}\right)\dot{x}^{i}-\frac{\partial L}{\partial x^{i}}\dot{x}^{i}\right)=\sum_{i=1}^{3}\left(\left(\frac{\mathrm{d}}{\mathrm{d}t}\left(\frac{\partial L}{\partial \dot{x}^{i}}\right)-\frac{\partial L}{\partial x^{i}}\right)\dot{x}^{i}\right),$$

where

$$\frac{\mathrm{d}}{\mathrm{d}t}\left(\frac{\partial L}{\partial \dot{x}^{i}}\right)-\frac{\partial L}{\partial x^{i}}=0,$$

by virtue of equation (7.11). Thus, equation (7.23) is satisfied for all L, as required.

Chapter 8

Gravitational and thermal effects in seismology

Consider now linearity: why does it pervade physical laws. Because the sum of two solutions of a (homogeneous) linear equation is again a solution. This property corresponds to the Principle of Superposition, exploited by Galileo: joint causes operate each as though the others were not present. [...] *Linearity is applicable to the extent, and only to the extent, that the Principle of Superposition holds, and to the extent that nature operates in a smooth, or at least piecewise smooth, manner.*

Mark Steiner (1998)

Preliminary remarks

Seismologists who consider wave propagation within the Earth—as opposed to its oscillations as a planet—ignore the effects of its gravitation and rotation. Also, they assume that the values of the elasticity-tensor components depend on positions and orientations of media, but not on the change of temperature accompanying the passage of mechanical disturbances. In other words, it is commonly assumed that the effects of gravitation and rotation on propagation of mechanical disturbances and of the temperature change due to such a propagation are negligible.

Similarly to Chapter 6 and in contrast to most chapters, where the reader is referred for further details to Slawinski (2015) or Bóna and Slawinski (2015), the subject of gravitational and thermal effects is not addressed in the first two volumes. Herein, the purpose of the presentation—apart from examining specific issues—is to gain an insight into justification for neglecting certain physical quantities in formulating theories involving these quantities. Also, as the other side of the same question, we might examine whether or not a given quantity could be measurable, and hence, support a given theory.

We begin this chapter by discussing gravitation in the context of the conservation of linear momentum. We proceed to examine the effect of gravitation within the Earth on the propagation speed of seismic waves. Also, we comment on seismic detectability of the gravitational waves propagating through interstellar space. Thermal effects are discussed in their relation to the values of elasticity parameters within Hooke's law, and hence, their effect on speeds of seismic waves.

8.1 Gravitation

8.1.1 *Body forces*

[1]To examine seismic effects of gravitation as a body force in the elastodynamic equation given in expression (5.2), let us write the conservation of linear momentum stated in expression (2.16),

$$\rho\frac{\partial^2 u_i}{\partial t^2} = \sum_{j=1}^{3}\frac{\partial\sigma_{ij}}{\partial x_j} + f_i\,, \qquad i = 1,2,3\,,$$

where ρ stands for mass density, u_i are components of the displacement vector, σ_{ij} are components of the incremental stress tensor and f_i of body forces, as

$$\rho\left(\frac{\mathrm{d}\mathbf{v}}{\mathrm{d}t} - \mathbf{F}\right) - \nabla\cdot\boldsymbol{\Sigma} = \mathbf{0}\,, \tag{8.1}$$

where $\mathbf{v}$ is the displacement-speed vector, $\mathbf{F}$ is the body force per unit volume and $\boldsymbol{\Sigma}$ is the total stress tensor, which—as discussed in Section 3.2.3.3—is associated with both the preexisting pressure and subsequent incremental elastic deformation. We consider a mass element, $\mathrm{d}m$, that is originally located at $\mathbf{r}$, and, following deformation $\mathbf{u}$, displaced to $\mathbf{p}$, where $\mathbf{p} = \mathbf{r} + \mathbf{u}$.

The Earth is a sufficiently large body to consider, in general, gravitational interactions among its parts, which is referred to as the *selfgravitation.*[2] Herein, we consider the effect of this phenomenon on the propagation of seismic disturbances.

[1]Section 8.1.1 is based on lecture notes by, and discussions with, Michael G. Rochester.
[2]Readers interested in deformations of selfgravitating terrestrial planets might refer to Müller and Weiss (2016).

Following Aki and Richards (2002, Section 8.4) and Bullen and Bolt (1987, Section 5.6.1), we write the total stress as

$$\Sigma_{ij}(\mathbf{p}) = \sigma_{ij}(\mathbf{p}) - p_0(\mathbf{r})\,\delta_{ij}\,, \tag{8.2}$$

which, following Love (1911), is referred to as Love's hypothesis. Herein, σ_{ij} is the stress tensor associated with seismic disturbances, and p_0 is the hydrostatic pressure to which the body is subjected independently of seismic phenomena; in other words, p_0 corresponds to the undeformed state of the body: its equilibrium state.

Letting $X_i = x_i - u_i$, we write the divergence operator as

$$\nabla_X \cdot = \nabla_x \cdot \; - (\nabla_x \mathbf{u}) \cdot \nabla_x\,,$$

where the coordinates of $\mathbf{r}$ and $\mathbf{p}$ are expressed in terms of X_i and x_i, respectively. These are the material and spatial descriptions discussed in Section 2.2.[3] Hence, taking the divergence of both sides of equation (8.2), we obtain

$$\nabla \cdot \boldsymbol{\Sigma} = \nabla \cdot \sigma - \nabla p_0 + (\nabla \mathbf{u})\,\nabla p_0\,; \tag{8.3}$$

herein, the right-hand side is evaluated at $\mathbf{r}$. Equation (8.3) is a vector equation. $\boldsymbol{\Sigma}$ and σ are second-rank tensors, hence, their divergences are vector fields; p_0 is a scalar field, hence, its gradient is a vector field; $\mathbf{u}$ is a vector field, hence, its gradient is a second-rank tensor, which—*contracted* with a vector field, ∇p_0—is a vector. Herein, the operation of tensor contraction is akin to a multiplication of a matrix, A_{ij}, by a vector, v_i, which results in a vector, w_j.

For a nonrotating Earth model whose undeformed configuration is one of hydrostatic equilibrium, the gravitational potential, U, and pressure, p, are related by

$$\nabla p_0 = \rho_0 \nabla U_0 = \rho_0\,\mathbf{g}_0\,, \tag{8.4}$$

where $\mathbf{g}$ is the gravitational force. The mass density of a disturbed element is $\rho(\mathbf{p}) = \rho_0(\mathbf{p}) + \rho_1(\mathbf{p})$. To evaluate the first term on the right-hand side at $\mathbf{r}$, we write the linear-term of its Taylor series as $\rho_0(\mathbf{r}) + \mathbf{u} \cdot \nabla \rho_0(\mathbf{r})$; hence,

$$\rho(\mathbf{p}) = \rho_0(\mathbf{r}) + \mathbf{u} \cdot \nabla \rho_0(\mathbf{r}) + \rho_1(\mathbf{r})\,.$$

To express the right-hand side in terms of ρ_0 only, we invoke the equation of continuity, $\rho_1 = -\rho_0 \nabla \cdot \mathbf{u} - \mathbf{u} \cdot \nabla \rho_0$, to write

$$\rho(\mathbf{p}) = \rho_0(\mathbf{r}) - \rho_0 \nabla \cdot \mathbf{u}\,.$$

[3] *see also*: Slawinski (2015, Section 1.3)

The net body force per unit mass of a disturbed element is $\nabla(U_0+U_1)$, evaluated at $\mathbf{p}$, where U_1 is the Eulerian perturbation in gravitational potential at a given spatial location due to displacement $\mathbf{u}$. To express $\nabla U_0 = \mathbf{g}_0$ at $\mathbf{p}$ using quantities evaluated at $\mathbf{r}$, we write the first-term of its Taylor series as $\mathbf{g}_0(\mathbf{r}) + \mathbf{u}\cdot\nabla\mathbf{g}_0$. Hence,

$$\nabla(U_0 + U_1) = \mathbf{g}_0(\mathbf{r}) + \mathbf{u}\cdot\nabla\mathbf{g}_0 + \nabla U_1\,,$$

which is $\mathbf{F}$, in expression (8.1).

To examine frequency effects, we let $\mathbf{u}(\mathbf{r},t) = \mathbf{u}(\mathbf{r})\cos(\omega t+\varphi)$. For a nonrotating Earth model, $\mathrm{d}\mathbf{v}/\mathrm{d}t = \partial^2\mathbf{u}/\partial t^2 = -\omega^2\mathbf{u}$. Hence, we write expression (8.1) as

$$\omega^2\,\mathbf{u} + \nabla U_1 + \nabla(\mathbf{u}\cdot\mathbf{g}_0) - \mathbf{g}_0\nabla\cdot\mathbf{u} + \frac{1}{\rho_0}\nabla\cdot\sigma = \mathbf{0}\,; \qquad (8.5)$$

in which we consider the first-order terms only, and use the fact that, according to expression (8.4), $\nabla\times\mathbf{g}_0 = \nabla\times\nabla U_0 = \mathbf{0}$, by the virtue of a vector identity.

To evaluate the effects of gravitation, we consider the conservation of the gravitational flux, as stated by *Poisson's equation*, $\nabla^2 U_1 = -4\pi G\rho_1$, where G is the *gravitational constant*. Following the equation of continuity, $\rho_1 = -\rho_0\nabla\cdot\mathbf{u} - \mathbf{u}\cdot\nabla\rho_0$, we write

$$\nabla^2 U_1 = 4\,\pi\,G\,\rho_0\,\nabla\cdot\mathbf{u} + 4\,\pi\,G\,\mathbf{u}\cdot\nabla\rho_0\,,$$

which, using the product rule, we rewrite as

$$\nabla^2 U_1 = 4\,\pi\,G\,\nabla\cdot(\rho_0\,\mathbf{u})\,;$$

all quantities are evaluated at $\mathbf{r}$. We can estimate that

$$O(\nabla U_1) \sim 4\,\pi\,G\,\bar{\rho}\,O(\mathbf{u})\,,$$

where $O(\,\cdot\,)$ is a Landau symbol[4]—named in honour of a Russian physicist Lev Davidovich Landau, and also refereed to as the big O—and $\bar{\rho}$ is the mean density below r. Since $g_0(r) = 4\,\pi\,G\,\bar{\rho}\,r/3$,

$$O(\nabla(\mathbf{u}\cdot\mathbf{g}_0)) \sim \mathbf{g}_0\,O(\nabla\cdot\mathbf{u}) \sim 4\,\pi\,G\,\bar{\rho}\,O(\mathbf{u})\,.$$

If $\omega \gg 2\sqrt{\pi\,G\,\bar{\rho}}$, which is about 0.002, the second, third and fourth terms in expression (8.5) might be neglected in comparison with the first term. Hence,

$$\rho\frac{\partial^2\mathbf{u}}{\partial t^2} = \nabla\cdot\sigma\,,$$

where $\rho_0 \equiv \rho$, and we recall that $\omega^2\mathbf{u} = -\partial^2\mathbf{u}/\partial t^2$.

Dahlen and Tromp (1998, Section 4.3.5) discuss the nongravitation limit, and conclude that

[4] *see also*: Bóna and Slawinski (2015, Section 4.3.1)

[i]n practice, gravity is routinely ignored in body-wave, regional-wave and exploration seismology, at periods shorter than 30 seconds.

8.1.2 *Wave speeds*

To examine effects of gravitation on the speed of seismic waves, let us consider equation (5.2) together with a body force that we express as the gradient of the gravitational-potential difference between the perturbed and unperturbed homogeneous continuum,

$$\rho\,\frac{\partial^2 \mathbf{u}}{\partial t^2} = (\lambda + 2\mu)\nabla\varphi - \mu\nabla\times\mathbf{\Psi} + \rho\,\nabla\left(U - U_0\right)\,; \tag{8.6}$$

the last term on the right-hand side is the body force, with $U - U_0$ denoting the difference between the perturbed—with the passage of a wave—and unperturbed gravitational potentials.

Equation (8.6) is a simplification of equation (8.5). For a general study of the gravitational effect, a formulation based on the latter one should be used. Let us examine it briefly.

Dividing both sides of equation (8.6) by ρ and including terms analogous to those of equation (8.5), we write

$$\frac{\partial^2 \mathbf{u}}{\partial t^2} = v_P^2\nabla\varphi - v_S^2\nabla\times\mathbf{\Psi} + \nabla U_1 + \nabla(\mathbf{u}\cdot\mathbf{g}_0) - \varphi\,\mathbf{g}_0\,.$$

Taking the divergence of both sides, using the equality of mixed partial derivatives together with definition (5.3), $\varphi := \nabla\cdot\mathbf{u}$,[5] the vector-calculus identity, $\nabla\cdot\nabla\times = 0$, and the definition of the Laplacian, $\nabla^2 := \nabla\cdot\nabla$, we obtain

$$\frac{\partial^2 \varphi}{\partial t^2} = v_P^2\nabla^2\varphi + \nabla^2 U_1 + \nabla^2(\mathbf{u}\cdot\mathbf{g}_0) - \nabla\cdot(\varphi\,\mathbf{g}_0)\,.$$

In accordance with Poisson's equation, the second term on the right-hand side is equal to $4\,\pi\,G\,\nabla\cdot(\rho_0\,\mathbf{u}) = 4\,\varphi\,\pi\,G\,\rho_0$. Also, applying the product rule to the last term, we write it as $\varphi\nabla\cdot\mathbf{g}_0 + \mathbf{g}_0\cdot\nabla\varphi$, where the first term is $\varphi\nabla^2 U_0$, which again can be expressed as $-4\,\varphi\,\pi\,G\,\rho_0$. Thus,

$$\frac{\partial^2 \varphi}{\partial t^2} = v_P^2\nabla^2\varphi + 8\,\varphi\,\pi\,G\,\rho_0 + \nabla^2(\mathbf{u}\cdot\mathbf{g}_0) - \mathbf{g}_0\cdot\nabla\varphi\,, \tag{8.7}$$

which is an equation of motion for P waves that includes gravitational effects. The last two terms are difficult to evaluate in terms of the body-wave speed. Since we are concerned only with an estimate of the gravitational

[5] *see also*: Slawinski (2015, Section 1.4.3, expressions (1.4.14)–(1.4.18))

effect on that speed to infer whether or not to include this effect in further seismological studies, let us return to equation (8.6).

Taking the divergence of both sides of equation (8.6), using the equality of mixed partial derivatives, definition (5.3), the vector-calculus identity and the definition of the Laplacian, we obtain

$$\rho \frac{\partial^2 \varphi}{\partial t^2} = (\lambda + 2\mu)\nabla^2 \varphi + \rho \nabla^2 (U - U_0) , \tag{8.8}$$

which is a P-wave equation, and a special case of equation (8.7).

Taking the curl of both sides of equation (8.6) eliminates the body force, since $\nabla \times \nabla = \mathbf{0}$. Thus, according to this formulation, S waves are not affected by the body force.

The influence of the difference between the perturbed and unperturbed potentials in expression (8.8) can be expressed by Poisson's equation,[6]

$$\nabla^2 (U - U_0) = -4\pi G (\rho - \rho_0) , \tag{8.9}$$

where $G(\rho - \rho_0)$ is the difference in selfgravitation that results from the change in density. In view of the conservation of mass, we express the difference in mass densities by its proportionality to the change in volume given by the dilatation, φ. We rewrite equation (8.8) as

$$\frac{\partial^2 \varphi}{\partial t^2} = v_P^2 \nabla^2 \varphi + 4\pi G \rho \varphi , \tag{8.10}$$

where v_P is the speed of the energy propagation carried by a P wave. This is a simpler form of equation (8.7).

We wish to estimate the influence of gravitation on the wavefront speed of the P wave. Consider equation (8.10) in a single spatial dimension,

$$\frac{\partial^2 \varphi}{\partial t^2} = v_P^2 \frac{\partial^2 \varphi}{\partial x^2} + 4\pi G \rho \varphi . \tag{8.11}$$

Let $\varphi = \cos(\kappa x - \omega t)$ be a solution: a harmonic function representing a monochromatic wave propagating in the positive direction; it is a particular case of solution (5.25). Since x and t are the spatial and temporal variables, respectively, it follows that κ and ω are the spatial and temporal frequencies, respectively. Their ratio, ω/κ, is the wavefront-propagation speed, v: the speed of a point that exhibits a constant phase.

Inserting the expression for φ in equation (8.11), we obtain

$$-\omega^2 \cos(\kappa x - \omega t) = -v_P^2 \kappa^2 \cos(\kappa x - \omega t) + 4\pi G \rho \cos(\kappa x - \omega t) ,$$

[6]Readers interested in effects of gravity on wave propagation might refer to Udías (1999, Section 3.5).

from which it follows that

$$v = \frac{\omega}{\kappa} = v_P \sqrt{1 - \frac{4\,\pi\,G\,\rho}{v_P^2\,\kappa^2}}\,. \tag{8.12}$$

Since $\kappa = 2\,\pi/(T\,v_P)$, where T is the period, the quotient in the radicand is $G\,\rho\,T^2/\pi$, and is illustrated in Figure 8.1. There is a limit of validity of expression (8.12), which corresponds to the radicand equal to zero. In other words, the range of periods acceptable in that expression is $T \in \left[0, \sqrt{\pi/(G\,\rho)}\right]$.

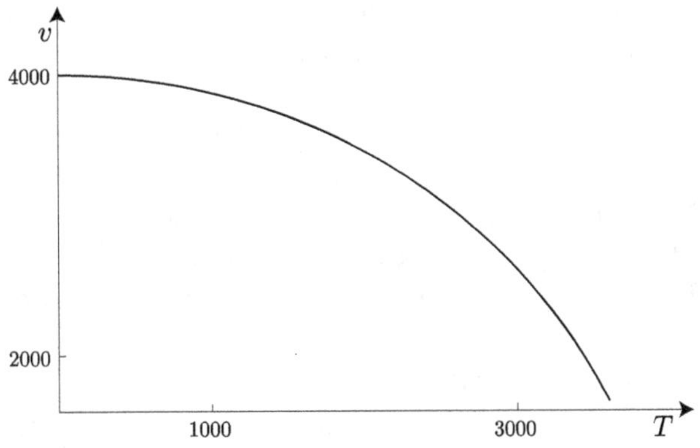

Fig. 8.1: Plot of expression (8.12); herein, $v(T) = v_P\sqrt{1 - G\,\rho\,T^2/\pi}$, where $v_P = 4000$, $\rho = 3000$, in the *SI* units. The plot extends to 3600, which is the period of one hour.

The lower limit, $T = 0$, which is equivalent to infinite frequency, is tantamount to the ray-theory formulation. The upper limit, $\sqrt{\pi/(G\,\rho)}$, is a consequence of approximations introduced in Section 8.1.1, on which the last term on the right-hand side of equation (8.6) is based, and—in particular—of replacing equation (8.7) by equation (8.10). In general, we would expect v to tend smoothly and asymptotically to zero, as $T \to \infty$; physically, there is no specific value of T for which $v = 0$. Thus, the applicable domain of expression (8.12) is limited to lower values of T. Within that domain, we can examine periods for which we can neglect the effect of gravitation in equation (8.6).

Also, referring to Stacey and Davis (2008), we estimate that—for a typical seismic wave whose period is five seconds in a Hookean solid whose mass density is $\rho = 5500\ \mathrm{kg/m^3}$, which is the average density of the Earth—the value of the root differs from unity by less than one-thousandth. The effect of gravitation decreases—by a small amount—the speed of wavefront propagation. This is consistent with the total force being accounted for by both elastic and gravitational effects; in other words, if we include gravitational effects in our consideration, the elastic effect is diminished, which results in the lowering of the speed of propagation.

Thus, the influence of gravitation on the wavefront-propagation speed is negligible for waves whose periods are of interest in seismology. Furthermore, since the characteristics of differential equations depend only on the highest derivatives of these equations, we conclude—by examining equation (8.10) and its counterpart in a single spatial dimension given in equation (8.11)—that gravitation has no effect on the shape of characteristics. Hence, the rays, which are the bicharacteristics,[7] are not altered by gravitation.

Let us evaluate the validity of our steps, and hence, the reliability of our conclusion. The gravitational-potential difference proposed in expression (8.9) is valid independent of its sources; such considerations as the rotation or the inhomogeneity of the solid are irrelevant. Furthermore, in view of differential equations (8.8) and (8.10), our formulation is in terms of local properties since the values of the mass density, elasticity parameters, and hence, the propagation speed correspond to a particular point. By letting the density and elasticity parameters be functions of position, as in equation (5.1), we would obtain similar estimates of the elastic and gravitational effects at any point, even though that equation would result in speeds that depend on both position and direction.

If we accept the difference between v_P and v as negligible, can we conclude that the influence of gravitation on wave propagation is negligible? In other words, is the difference between v_P and v all that needs to be examined? The answer—in the context of our formulation—is affirmative. Let us rewrite equation (8.11) as

$$\left(\frac{\partial}{\partial t} - v_P\frac{\partial}{\partial x}\right)\left(\frac{\partial}{\partial t} + v_P\frac{\partial}{\partial x}\right)\varphi = 4\pi G\rho\varphi\,,$$

and hence,

$$\left(\left(\frac{\partial}{\partial t},\frac{\partial}{\partial x}\right)\cdot(1,-v_P)\right)\left(\left(\frac{\partial}{\partial t},\frac{\partial}{\partial x}\right)\cdot(1,v_P)\right)\varphi = 4\pi G\rho\varphi\,, \qquad (8.13)$$

where the terms in brackets are directional derivatives in directions of vectors $(1,-v_P)$ and $(1,v_P)$, and $\mathrm{d}x/\mathrm{d}t = \pm v_P$ is the equation for the characteristics.

[7] *see also*: Bóna and Slawinski (2015, Chapter 3 and Afterword)

Using the aforementioned values, we estimate the coefficient of φ on the right-hand side of equation (8.13) to be less than 10^{-5}; the maximum amplitude of φ is equal to unity. Since this right-hand side is nearly zero, φ is almost constant along $(1, -v_P)$ and $(1, v_P)$, which are the characteristics of equation (8.13) and, hence, of equation (8.11); constancy would correspond to a Hookean solid with no gravitational influence on wave propagation, and would be expressed by the right-hand side being zero.[8] Herein, we can estimate the right-hand side to be $10^{-5}\varphi$, where φ oscillates about zero between ± 1. Thus, the amplitude of the signal exhibits minute oscillations along the characteristics, which suggests the negligibility of the gravitational effect in seismic considerations.

8.2 On weak gravitational waves

Remaining within the subject of gravitation and wave propagation—albeit in the context of general relativity, not classical mechanics—let us consider the seismic effect of gravitational waves propagating through interstellar space, which are generated by masses that change their shapes or move around a point, such as vibrating stars or binary-star systems. The 1993 Nobel prize in physics was awarded to Russell A. Hulse and Joseph H. Taylor, Jr. for the discovery of a binary pulsar that provided possibilities for the study of gravitational waves. Also, these waves can be generated by catastrophic events, such as a collision of two stars or of two black holes.[9]

In this section, we limit our attention to weak gravitational waves, whose examination is applicable to effects far from the source that produces them. This approach shares mathematical conveniences and physical limitations similar to the ones encountered in examining farfield seismic waves, which are discussed in this book.

Gravitational waves are difficult to detect. The net effect of the gravitational waves generated by a rotating pair of masses is inversely proportional to the fourth power of the distance from the source.[10] The first detection was reported by Abbott *et al.* (2016), where the source of the wave was a binary black hole merger. Thus, a century elapsed between the theoretical prediction of Einstein (1916) and experimental detection.

[8] *see also*: Slawinski (2015, Section 6.5.2, expression (6.5.15))

[9] Readers interested in an insightful description of gravitational waves might refer to Will (1986, Chapter 10).

[10] Readers interested in properties of the electromagnetic and gravitational waves might refer to Freegarde (2013, Section 1.3.3).

Gravitational waves consist of the gravitational force that propagates as waves away from the source. They are ripples in the curvature of spacetime. Their existence follows from the general theory of relativity, and is a consequence of the finite speed of propagation. In vacuum, they propagate with the speed of light.[11] As stated by Landau and Lifshitz (1975, p. 368),

> in the relativistic theory of gravitation, the finite velocity of propagation of interactions results in the possibility of the existence of free gravitational fields that are not linked to bodies—gravitational waves.

The effect of gravitational waves is undetectable in a single point of any laboratory. It requires a comparison of their effect between distinct locations. The passage of waves changes the distance between locations, with the magnitude of this change being proportional to the distance between them. This effect is represented by a time-varying spacetime strain tensor. Detecting this effect is akin to detecting the effect of a tide in the middle of the ocean, which requires that we examine relative changes in position between locations affected differently by the rising or ebbing tide.

In contrast to seismic waves, but similarly to the electromagnetic waves, gravitational waves can propagate through vacuum. Similarly to the seismic S waves and to the electromagnetic waves, the force exerted by gravitational waves is perpendicular to the direction of propagation. They are transverse waves. In contrast to the electromagnetic waves, gravitational waves interact only weakly with matter. Even dense systems are nearly transparent to gravitational waves. This property contributes to the difficulty of their detection, but allows them to penetrate systems inaccessible to the electromagnetic waves.

The seismic effect of gravitational waves—where the Earth is used as a detector—has not yet been detected. However, let us discuss its possibility.

According to Dyson (1996b, 1969, equation (2.30)), the elastodynamic equation for an isotropic inhomogeneous Hookean solid[12] that contains the

---

[11]Readers interested in propagation of gravitational waves might refer to Landau and Lifshitz (1975, Chapter 13) and to Misner *et al.* (1973, Chapter 18 and Part VIII).
[12]*see also*: Slawinski (2015, Exercise 6.21)

effect of the gravitational wave is

$$\rho\,\frac{\partial^2 u_i}{\partial t^2} = \frac{\partial}{\partial x_i}\left((c_{1111} - 2c_{2323})\sum_{j=1}^{3}\frac{\partial u_j}{\partial x_j}\right) + \sum_{j=1}^{3}\left(\frac{\partial}{\partial x_j}\left(c_{2323}\left(\frac{\partial u_i}{\partial x_j} + \frac{\partial u_j}{\partial x_i}\right)\right) - \frac{\partial\, c_{2323}}{\partial\, x_j}\, h_{ij}\right), \tag{8.14}$$

where $i = 1, 2, 3$. Herein, tensor h_{ij} represents that effect; it is referred to as the *wave metric*. Similarly to $\varepsilon_{ij} := (\partial u_i/\partial x_j + \partial u_j/\partial x_i)/2$, it is a strain tensor. As discussed by Misner *et al.* (1973, Sections 18.1, 18.2 and 35.2), this symmetric second-rank tensor corresponds to gravity within its linearized theory. An aspect of this linearization is illustrated in Exercise 8.2.

Notation 8.1. In this section, for the consistency within the chapter, we use the reciprocity of the metric tensor and its inverse—illustrated in Example F.2 on page 479—to place all indices as subscripts.

In equation (8.14), h_{ij}, with $i, j = 1, 2, 3$, is the perturbation tensor of the metric tensor, δ_{ij}, in the Galilean spacetime, discussed in Section 2.2.1 and Appendix F.1, where the line element is $\mathrm{d}s = \sqrt{\sum_{ij=1}^{3} \delta_{ij}\,\mathrm{d}x_i\,\mathrm{d}x_j}$. More generally, $h_{\alpha\beta}$, with $\alpha, \beta = 0, \ldots, 3$, is the perturbation tensor of the Minkowski metric tensor, $\eta_{\alpha\beta}$, of the special theory of relativity,

$$g_{\alpha\beta} = \eta_{\alpha\beta} + h_{\alpha\beta}\,, \qquad \alpha, \beta = 0, \ldots, 3\,, \tag{8.15}$$

where $\|\eta_{\alpha\beta}\|_F \equiv \|\mathrm{diag}(-c^2, 1, 1, 1)\|_F \gg 1$, $\|h_{\alpha\beta}\|_F \ll 1$ and $g_{\alpha\beta}$ is the metric of global frames, which—due to the presence of $h_{\alpha\beta}$—are only *nearly* inertial. Herein, c is the speed of light in vacuum and

$$\|g_{\alpha\beta}\|_F := \sqrt{\sum_{\alpha\beta=0}^{3} (\delta_{\alpha\beta}\, g_{\alpha\beta}\, g_{\alpha\beta})}$$

is the Frobenius norm, discussed in Section 4.1.3.

Expression (8.15) is a weak-field limit of general relativity. We could rewrite its spacetime metric as $g_{\alpha\beta} = \eta_{\alpha\beta} + \epsilon h_{\alpha\beta} + O(\epsilon^2)$, where ϵ is a small parameter and $O(\epsilon^2)$ is a Landau symbol, which herein stands for the neglected nonlinear terms in ϵ.[13] Spacetime remains curved in that

[13] *see also*: Bóna and Slawinski (2015, Section 4.3.1)

linearized theory of gravity, even though equations therein are stated and solved as if it was not. The Einstein Field Equation is

$$G_{\alpha\beta} + \Lambda\, g_{\alpha\beta} = \frac{8\,\pi\, G}{c^4}\, T_{\alpha\beta}\,, \qquad \alpha, \beta = 0, \ldots, 3\,,$$

where $G_{\alpha\beta}$ is the Einstein tensor, Λ is the cosmological constant, G is the gravitational constant, c is the speed of light, and $T_{\alpha\beta}$ is the stress-energy tensor. It describes the interaction of gravitation that results from the spacetime being curved by matter and energy. In the weak-field limit, this equation reduces to a wave equation for the wave metric, $\Box h_{\alpha\beta} = 0$, where $\Box$ is the d'Alembertian introduced on page 187.

Einstein (1916) shows—by solving the linearized wave equation with a source term—that the wave amplitude is proportional to the second time derivative of the quadrupole moment of the source. This solution is valid for wavelengths much longer than the size of the source.

Gravitational waves can only be generated by the time-varying quadrupole, and higher-multipole, sources. There are no monopole or dipole sources, since they result in conserved quantities of mass and angular momentum, respectively. Even catastrophic events generating great amounts of kinetic energy, but in a spherically symmetric manner, do not produce gravitational waves.

Examining the response for a model of a rotating spherical Earth, Dyson (1996b, 1969) shows the way in which a quadrupole wave differs from a monopole. In particular, Dyson (1996b, 1969) shows that a monochromatic incident wave of frequency ω is split into five modes exhibiting distinct frequencies. This split is a result of the *Doppler effect* produced by the relative motion between the detector and the source and of the rotation due to the motion of the source. Formally, the appearance of modes with distinct frequencies is akin to the case of guided waves discussed in Section 6.3, where they result from constraints of the waveguide.

Herein, the frequencies of the resulting modes are ω, $\omega \pm \Omega$ and $\omega \pm 2\Omega$, where Ω is the Earth's angular frequency. The frequencies and the amplitudes of gravitational waves are related to the motion of masses that generate them, since their sources must be undergoing dynamic changes. Otherwise, there is a steady potential field, not a wave propagation.

Dyson (1996b, 1969, equation (3.12)) derives an explicit expression of the seismic displacement induced by a gravitational wave,

$$A\,\frac{\iota\,\beta}{2\,\omega}\,\sin\theta\left(\cos\theta,\,\iota,\,-\frac{\beta}{\alpha}\sin\theta\right),$$

where A is the complex amplitude of the wave, which provides both magnitude and phase of the incident wave; θ is its incident angle and ω its frequency; α and β are the local P-wave and S-wave speeds, respectively. Dyson (1996b, 1969, equations (4.18)–(4.21)) gives complete information for the gravitational-wave displacement vector at a point with a given latitude on the rotating Earth. This vector contains the superposition of the five modes. Each mode exhibits its phase and polarization. The amplitude of the displacement vector is given relative to the incident-wave amplitude.

The *redshift* due to the Doppler effect is among the few modifications, apart from the decrease in amplitude due to geometrical spreading, to which gravitational waves are subjected through their propagation. Decrease of amplitude due to absorption by interstellar matter is extremely weak. So might be scattering and dispersion. However, as pointed out by Freeman Dyson (email comm., 2016),

> It is not true that the waves always have small scattering by interstellar material while traveling across the universe. When gravitational waves pass through massive clusters of galaxies, they are distorted by gravitational lensing just like visible light. If we were lucky enough to detect a distant gravitational wave with a cluster of galaxies in the foreground, we might see the wave split into 3 or 5 or 7 components with various magnifications and various arrival-times due to gravitational lensing, like the split optical images of a distant supernova that were recently observed (Kelly *et al.*, 2015).

According to the formulation of Dyson (1996b, 1969), within a model of a homogeneous Earth, there is no absorption; hence, there is no possible detection. As shown in equation (8.14), if c_{2323} is constant, the term that contains the gravitational-wave effect is zero. Detection is possible at interfaces between distinct rigidity moduli. The response is strongest at the solid-fluid interfaces, which are the Earth's surface and the outer-core boundaries.

As an aside, following Carroll (2004, Sections 4.1 and 4.2), let us comment on the Einstein Field Equation. Its motivation is a generalization of Poisson's equation,

$$\nabla^2\Phi = -\,4\,\pi\,G\,\rho\,,$$

to the Einsteinian spacetime. Herein, the Laplace differential operator, ∇^2, acts on on the gravitational potential, Φ, in the Galilean spacetime, with the spatial variables, x_1, x_2 and x_3. The right-hand side describes the mass distribution, where ρ is the mass density. For point masses, the classic solution is $\Phi = -GM/r$, where M is the mass and $r = \sqrt{\sum_{i=1}^{3} x_i^2}$ is the distance from the point. Heuristically, the desired generalization of this scalar equation is a tensorial form, $[\nabla^2 G]_{\alpha\beta} \propto T_{\alpha\beta}$.

In the weak-field limit, the spacetime curvature is a consequence of the perturbation tensor, $h_{\alpha\beta}$, added to the Minkowski tensor, which corresponds to the flat space, and whose illustration is presented in Exercise 8.3. The equation for a plane gravitational wave in one spatial dimension, propagating along the x_1-axis, is

$$\frac{\partial^2 h_{23}}{\partial x_1^2} - \frac{1}{c^2}\frac{\partial^2 h_{23}}{\partial t^2} = 0\,,$$

where c is the speed of light.

As discussed by Landau and Lifshitz (1975, p. 369), a plane[14] gravitational wave is a transverse wave. If it propagates along the x_1-axis, its displacements are in the x_2x_3-plane. Specifically, $h_{11} = h_{12} = h_{13} = 0$ and $h_{22} = -h_{33}$; the last equality is a consequence of h_{ij} being a traceless tensor. The solution of this one-dimensional wave equation is, as expected in view of Section 5.2, a function whose argument is $x_1 \pm ct$.

Since h_{ij} is a traceless tensor, $\sum_{ij=1}^{3} \delta_{ij} h_{ij} = 0$, it is absent from the first parentheses of expression (8.14). Also, the divergence of that tensor is zero, $\sum_{j=1}^{3} \partial h_{ij}/\partial x_j = 0$, for $i = 1, 2, 3$. In view of Section 5.1.2, this is analogous to the requirement for the existence of S waves in isotropic homogeneous solids, which means that—similarly to S waves—gravitational waves in a homogeneous medium are transverse. In inhomogeneous solids or in bounded media, discussed in Chapter 6, the P and S waves are coupled.[15]

As examined by Dyson (1996b, 1969, Section III) and discussed by Dyson (1996c, p. 35), in seismology,

> the transfer of energy between gravitational and elastic waves would occur mainly at discontinuities of the elastic properties of the earth, in particular, at the surface of the earth and at the mantle-core boundary.

Also,

> the fraction of the wave-energy transferred at the discontinuity would be proportional to the cube of the ratio of seismic velocity to light-velocity. The cube of the velocity-ratio is a number of the order of 10^{-15}, and it is this factor that makes any scheme

[14] *see also*: Slawinski (2015, Section 6.2)
[15] *see also*: Slawinski (2015, Exercise 6.21)

> for the mechanical detection of gravitational waves a heroic enterprise.

Examining equation (8.14) in view of Chapter 6, where c_{2323} is denoted by C_{44}, we infer that the effect of gravitational waves could be detectable in seismic disturbances analogous to the surface, guided and interface waves whose velocity expressions contain C_{44}, which are the Rayleigh, Love, Stoneley and Scholte waves.

Let us mention that—unlike the Rayleigh, Love, Stoneley or Scholte waves, whose existence is limited to the mathematical realm by a specific form of constitutive relations and boundary conditions—gravitational waves are physical phenomena predicted by the general theory of relativity.

Dyson (1996b, 1969, Section V) concludes that the seismic signal expected from pulsars is about five orders of magnitude below the noise level. Yet, recent increases in the data-storage capacity, which allow us to keep great volumes of recorded data, might facilitate such a detection with the help of a statistical analysis. Thus, Dyson (1996c, p. 34) states that

> [t]he gravitational signal could be integrated coherently over months or years. The use of coherent integration would increase the signal-to-noise ratio by a factor of ten thousand.

For an increase of the signal-to-noise ratio by integrating the signal over long periods, we require a periodic source of gravitational waves. Such a source could be a binary-star system, akin to the *Hulse-Taylor binary*, where two stars rotate about their common centre of mass. It could also be a rapidly rotating star with a large irregularity on its otherwise nearly spherical surface.

Also, periodic sources produce a signal of steady frequency, which lends itself to such an integration. However, such sources produce gravitational waves that are even weaker than the transient waves produced by catastrophic sources, which occur sporadically but create bursts of gravitational waves.

In view of the weakness of a signal, a possible approach for a seismic detection might be an examination—for instance, by the Bayes Information Criterion[16]—of the quantified preference between a model with and without the gravitational wave, in the context of seismic records. In other words, we could examine whether or not the accuracy of such records justifies the inclusion of h_{ij} in equation (8.14). Such an examination bears a similarity

[16]Readers interested in the model choice based on the Bayes Information Criterion might refer to Danek and Slawinski (2012).

to discussions in Section 8.1, above, and Section 8.3, below, where we justify neglecting the effects of gravitation and temperature, respectively. In this section, however, we enquire if gravitational waves could be detected.

In conclusion, let us comment on a compromise between the sophistication of a theory and its accuracy. No gravitational waves can appear in the spacetime of classical physics, which exhibits absolute time, or in the flat spacetime of the special theory of relativity. A prediction of these waves requires the curved spacetime of general relativity. Similarly, no qP, qS_1 or qS_2 waves can appear in isotropic Hookean solids; a mathematical analogy for such disturbances requires an anisotropic solid. In either case, a simpler theory results in a lesser accuracy.

8.3 Temperature

8.3.1 *Propagation and diffusion*

[17]In a manner analogous to the one presented in Section 8.1, we consider effects of temperature on parameters defining Hookean solids. In particular, we evaluate the difference between the values of the Lamé parameters obtained by an *adiabatic formulation*, which involves no heat transfer, and by an *isothermal formulation*, which involves no change in temperature. Adiabatic and isothermal processes—in the context of thermoelasticity—are discussed in Appendix C.

Neither formulation implies the values of the corresponding elasticity parameters to be independent of temperature. The values of adiabatic and isothermal elasticity parameters are different from each other, but—in both cases—they might be temperature-dependent. Examining this difference is important in considering the role of Hookean solids as mathematical entities to model seismic phenomena.

Formulating Hooke's law in Section 3.1, we do not include explicitly their temperature dependence. Expression (3.1) corresponds to a single temperature, which implicitly makes the elasticity tensor isothermal. However, the adiabatic formulation is a better analogy for rocks due to their low thermal conductivity, combined with the speed of propagation of disturbances modeled by seismic waves, which does not allow sufficient time for thermal diffusion to affect the propagation.

To illustrate the difference between the traveltime of heat flow due to

[17]Section 8.3 is based on Jeffreys (1961a, Chapter 8) and on Landau and Lifshitz (1986, Section 6).

deformation and traveltime of propagation of the disturbance itself, we follow Dahlen and Tromp (1998, Section 2.6.4). During interval Δt, the heat travels approximately the distance of $\sqrt{\kappa\,\Delta t}$; herein, κ is *thermal diffusivity*. During that interval, a wave travels approximately the distance of $v\,\Delta t$, where v is the speed of wave propagation. Let us estimate this relationship using the period of one cycle, which—in seismology—is of the order of 10^{-2} seconds.

Considering P waves in igneous rocks at 20° and following Stacey and Davis (2008), we set $\Delta t = 10^{-2}\ s$, $\kappa = 10^{-6}\ m^2/s$ and $v = 5\times 10^3\ m/s$ to get $\sqrt{\kappa\,\Delta t} = 10^{-4}\ m$ and $v\Delta t = 50\ m$. Since the distance affected by the diffusion of heat is several orders of magnitude smaller than the distance affected by deformation, we can regard such a deformation as an adiabatic process. Hence, as discussed in Exercise 8.4, solutions of inverse problems for seismic measurements provide the adiabatic values.

8.3.2 *Isothermal and adiabatic formulations*

8.3.2.1 *Lamé parameters*

Let us examine the difference between the values of the elasticity parameters for the isothermal and adiabatic cases. The stress-strain equation of an isotropic Hookean solid is[18]

$$\sigma_{ij} = \lambda\,\varphi\,\delta_{ij} + 2\mu\,\varepsilon_{ij}\,, \qquad i,j = 1,2,3\,; \tag{8.16}$$

herein, we assume that dilatation, φ, and strain, ε_{ij}, represent deformations resulting from both thermal expansion and other causes. The effect of thermal expansion is a stretch,[19] $\Delta\,\varepsilon_{ij}$, caused by the change in temperature, ΔT,

$$\Delta\varepsilon_{ij} = \tau\,\Delta T\,\delta_{ij}\,, \qquad i,j = 1,2,3\,, \tag{8.17}$$

where τ is the *coefficient of expansion*. Hence, to distinguish between thermal and nonthermal effects, we write expression (8.16) as

$$\sigma_{ij} = \lambda\left(\varphi - 3\tau\,\Delta T\right)\delta_{ij} + 2\mu\left(\varepsilon_{ij} - \tau\,\Delta T\,\delta_{ij}\right)\,, \qquad i,j = 1,2,3\,, \tag{8.18}$$

where $\Delta\,\varepsilon_{11} = \Delta\,\varepsilon_{22} = \Delta\,\varepsilon_{33}$, by the assumption of isotropic expansion adopted in expression (8.17).

To bring expressions (8.16) and (8.18) to a similar form, we rewrite the latter as

$$\sigma_{ij} = \left(\lambda\,\varphi - (3\lambda + 2\mu)\,\tau\,\Delta T\right)\delta_{ij} + 2\mu\,\varepsilon_{ij}\,, \qquad i,j = 1,2,3\,. \tag{8.19}$$

[18]*see also*: Slawinski (2015, Section 5.12.3)
[19]*see also*: Slawinski (2015, Section 1.4.3)

To obtain the coefficients of dilatation, φ, for both isothermal and adiabatic cases, we invoke the strain energy stored in a volume of a deformed continuum,[20]

$$\Delta E = \sum_{i=1}^{3}\sum_{j=1}^{3}\sigma_{ij}\,\delta\,\varepsilon_{ij}\,, \qquad i,j=1,2,3\,, \tag{8.20}$$

where $\delta\,\varepsilon_{ij}$ is the infinitesimal deformation, and where we assume that the deformation process is isothermal. To include heat in equation (8.20), we write it as

$$\Delta E = \sum_{i=1}^{3}\sum_{j=1}^{3}\sigma_{ij}\,\delta\,\varepsilon_{ij} + \Delta\,Q\,, \qquad i,j=1,2,3\,,$$

where $\Delta Q = c\,\rho\,\Delta T$, with c denoting the *specific heat* at constant strain. To relate strain to heat and temperature, we modify ΔQ to write

$$\Delta Q = c\,\rho\,\Delta\,T + \sum_{i=1}^{3}\sum_{j=1}^{3} q_{ij}\,\delta\,\varepsilon_{ij}\,, \qquad i,j=1,2,3\,, \tag{8.21}$$

where we introduce a second-rank tensor, q_{ij}. The explicit form of this tensor is deduced by Jeffreys (1961a, Chapter VIII), $q_{ij} = -T\,\partial\sigma_{ij}/\partial T$; it modifies the value of ΔQ for nonconstant strain.

In the adiabatic case, $\Delta Q = 0$, since no heat is gained or lost; hence, we write expression (8.21) as

$$\Delta T = -\frac{1}{c\rho}\sum_{i=1}^{3}\sum_{j=1}^{3} q_{ij}\,\delta\varepsilon_{ij}\,, \qquad i,j=1,2,3\,.$$

Since all terms on the right-hand side of equation (8.19) are linear in the components of the displacement vector and $\partial\Delta T/\partial\,T = 1$, we can write $\partial\,\sigma_{ij}/\partial T$ as $-(3\lambda+2\mu)\,\tau\,\delta_{ij}$. Recalling that $\Delta\varepsilon_{11} = \Delta\varepsilon_{22} = \Delta\varepsilon_{33}$, we write

$$\Delta T = -\frac{T(3\lambda+2\mu)\,\tau\,\varphi}{c\rho}\,.$$

Inserting this expression into equation (8.19), we obtain

$$\sigma_{ij} = \left(\lambda + \frac{T\,(3\lambda+2\mu)^2\,\tau^2}{c\,\rho}\right)\varphi\,\delta_{ij} + 2\mu\,\varepsilon_{ij}\,, \qquad i,j=1,2,3\,. \tag{8.22}$$

The term in parentheses is the adiabatic counterpart of λ. The value of λ, which has no explicit temperature dependence, is isothermal.

[20] *see also*: Slawinski (2015, Section 4.2, expression (4.2.3))

The value of μ is the same for both the adiabatic and isothermal cases. Thus, in view of the presented formulation, the values of P-wave speeds, which depend on both λ and μ, are different for the isothermal and adiabatic formulations. S-wave speeds, which depend on μ alone, are the same. Since these waves are equivoluminal, they are not affected by expansion.

As estimated in Exercise 8.5, the difference between the adiabatic and isothermal values of λ can be conveniently neglected for the near-surface studies. It might not be so for seismic considerations deeper within the Earth. Let us examine the latter case.

To estimate the fractional term in expression (8.22) for the Earth's mantle, we let $T = 2600\,\mathrm{K}$, $\tau = 10^{-5}$, $c = 10^3$ and $\rho = 5000$, $\lambda = 3.5\times 10^{11}$ and $\mu = 2.5\times 10^{11}$, which we assume to be isothermal. The fractional term is 1.25×10^{11}, which is of the same order of magnitude as the isothermal value of λ. Herein, this term is not negligible, and it is necessary to consider the distinction between the isothermal and adiabatic values of the Lamé parameters.

8.3.2.2 *Bulk moduli*

As discussed in Sections 8.3.1 and 8.3.2.1, as well as in Exercises 8.4 and 8.5, seismic measurements, followed by an inverse-problem solution, provide necessarily the adiabatic values—due to low thermal conductivity of rocks and insufficient time for thermal diffusion during propagation of seismic disturbances—even though either isothermal or adiabatic quantities might be used in the forward-problem formulation. Let us compare the isothermal or adiabatic quantities in terms of the *bulk modulus*.

The term $3\lambda + 2\mu$ in expression (8.22) can be written as thrice the bulk modulus, $3K$, discussed in Section 3.2.4.14, which describes the resistance of the medium to the volume change under a uniform pressure, $K := -P/\varphi$. Thus, we express the adiabatic bulk modulus, K_A, in terms of the isothermal one, K_I as

$$K_A = K_I\left(1 + \frac{9\tau^2 T\left(\alpha^2 - \frac{4}{3}\beta^2\right)}{c_0}\right)^{-1},$$

where α and β are the P-wave and S-wave speeds and c_0 is the specific heat at zero stress; c_0 is related to the specific heat at constant strain by the bulk moduli, $c_0/c = K_A/K_I$, since—for a thermal expansion at zero stress—expression (8.21) is equal to $c_0\,\rho\,\Delta T$.

Referring to Stacey and Davis (2008), we evaluate the dimensionless fraction at the radius of 4000 kilometres, where $\alpha = 13240$, $\beta = 7100$, $T = 2600$ K, $\tau = 1.27 \times 10^{-5}$ and $c_0 = 1193$; we get 0.342, which—in comparison to unity—is not a negligible quantity, as expected in view of the estimate of λ for the Earth's mantle.

Closing remarks

In this chapter, we consider effects of selfgravitation, gravitational waves and temperature on wave propagation in isotropic Hookean solids. The effects of both selfgravitation and temperature are formulated in terms of changes in volume due to, respectively, gravitational and thermal expansions or contractions. These changes affect only P waves, whose propagation is associated with changes of volume. Gravitational waves, on the other hand, affect only S waves, since the effect of gravitational waves is expressed in terms of the wave metric, which is a divergence-free tensor, in contrast to the wave function for P waves, which is $\varphi := \nabla \cdot \mathbf{u} \neq 0$.

In the elastodynamic equation, the effect of selfgravitation appears as a body force. The effects of both gravitational waves and temperature appear as contact forces. They are expressed in terms of the elasticity parameters; the former is associated with μ and the latter with λ, which are measures of rigidity and compressibility, respectively.[21]

There are other quantities that one might consider in examining descriptions of wave phenomena discussed in Chapters 5 and 6. An obvious issue stems from the Earth's rotation, as expressed by the Coriolis effect. The ratio of the magnitudes of the Coriolis force and the inertial force, which is included in the elastodynamic equation, is $2\Omega/\omega$, where Ω is the angular frequency of the Earth and ω is the angular frequency of the disturbance in question. Since for a seismic wave whose period is 1 second the ratio is of the order of 10^{-5}, we choose to ignore the influence of rotation, except we comment on it in Section 8.2. Another consequence of rotation is the ellipticity of the Earth, which one would need to include in certain aspects of global seismology, even though we ignore it in the estimates discussed in this chapter.

Furthermore, in the discussion of Section 8.3, we consider frictionless adiabatic expansion, which leaves the measure of rigidity, μ, and hence the S-wave speed, unaffected. This is equivalent to assuming that the passage

[21] *see also*: Slawinski (2015, Section 5.12.4)

of S waves is not accompanied by a temperature change. If we consider the value of μ obtained from measurements of disturbances that we interpret as S waves, we might argue for no effect of temperature change due to low thermal diffusivity in comparison to speed of wave propagation, rather than due to lack of temperature change associated with S waves themselves. Notably, the actual measurements are affected by friction.

It is important to be specific in, and to evaluate the extent of applicability of, the mathematical rigour. In examining particular cases it is important to consider other important influences. A physicist who seeks a mathematical formulation for a known aspect of a physical phenomenon might ponder the following comment of Bertrand Russell (1945) regarding St. Thomas Aquinas.

> He is not engaged in an inquiry, the result of which it is impossible to know in advance. Before he begins to philosophize he already knows the truth [...] If he can find apparently rational arguments for some parts of the faith, so much the better; if he cannot, he need only fall back on revelation. The finding of arguments for a conclusion given in advance is not philosophy, but special pleading.

Apart from specific issues addressed in this chapter, we might see its topic as an examination of hyperbolic partial differential equations in the context of the elliptic and parabolic equations. In other words, we examine the effects of gravitational potential and thermal diffusion upon wave propagation.

8.4 Exercises

Exercise 8.1. Consider equation (8.14). Examine the gravitational-wave effect for an interface between two homogeneous media that differ by the value of their rigidity parameter, $c_{2323} \equiv \mu$. Assume that the interface coincides with the x_1x_2-plane.

Solution. In equation (8.14), the effect of the gravitational-wave is expressed by

$$\sum_{j=1}^{3} \frac{\partial\, c_{2323}(x)}{\partial\, x_j}\, h_{ij}(x,t)\,, \qquad i=1,2,3\,.$$

For convenience, we let $c_{2323} \equiv \mu$ to write explicitly,

$$\frac{\partial\,\mu}{\partial\,x_1}\,h_{11} + \frac{\partial\,\mu}{\partial\,x_2}\,h_{12} + \frac{\partial\,\mu(x_3)}{\partial\,x_3}\,h_{13} = \frac{\partial\,\mu(x_3)}{\partial\,x_3}\,h_{13}(x,t)\,,$$

$$\frac{\partial\,\mu}{\partial\,x_1}\,h_{21} + \frac{\partial\,\mu}{\partial\,x_2}\,h_{22} + \frac{\partial\,\mu(x_3)}{\partial\,x_3}\,h_{23} = \frac{\partial\,\mu(x_3)}{\partial\,x_3}\,h_{23}(x,t)\,,$$

$$\frac{\partial\,\mu}{\partial\,x_1}\,h_{31} + \frac{\partial\,\mu}{\partial\,x_2}\,h_{32} + \frac{\partial\,\mu(x_3)}{\partial\,x_3}\,h_{33} = \frac{\partial\,\mu(x_3)}{\partial\,x_3}\,h_{33}(x,t)\,,$$

where we use the fact that μ is constant along the x_1-axis and the x_2-axis. The gravitational-wave effect can be written as

$$\frac{\partial\,\mu(x_3)}{\partial\,x_3}\,(h_{13}, h_{23}, h_{33})\,.$$

The interface between two homogeneous media is represented in terms of Heaviside's step, whose derivative is Dirac's delta.[22] We write the gravitational-wave effect along the x_3-axis as $\int_{-\infty}^{\infty} \delta(x_3)(h_{13}, h_{23}, h_{33})\,\mathrm{d}x_3\,.$ Hence, this effect is limited to the interface, $x_3 = 0$, where we have

$$h_{13}(x_1, x_2, 0, t), h_{23}(x_1, x_2, 0, t), h_{33}(x_1, x_2, 0, t)\,.$$

Since h_{ij} is a traceless tensor, $h_{33} = -(h_{11} + h_{22})$. If we consider plane waves propagating along the x_3-axis, $h_{33} = 0$, since gravitational waves are transverse.

Exercise 8.2. Following the definition of the strain tensor, ε_{ij}, and in view of expression (8.14), consider

$$\frac{\partial u_i}{\partial x_j} + \frac{\partial u_j}{\partial x_i} - h_{ij} := 2\varepsilon_{ij} - h_{ij}\,, \qquad i,j = 1,2,3\,.$$

Express the components of this second-rank tensor as a matrix, A, in $\mathbb{R}^2$, and examine the linear approximations of its two invariants, $\operatorname{tr} A$ and $\det A$, and of A^2 in terms of the Cayley-Hamilton theorem,

$$A^2 - (\operatorname{tr} A)A + (\det A)I = \begin{bmatrix} 0 & 0 \\ 0 & 0 \end{bmatrix}, \tag{8.23}$$

where I is the second-rank identity tensor, δ_{ij}.

[22] *see also*: Bóna and Slawinski (2015, Appendix E.2)

Solution. Since ε_{ij} is symmetric and h_{ij} is both symmetric and traceless, we write the resulting tensor as

$$A = \begin{bmatrix} 2\varepsilon_{11} - h_{11} & 2\varepsilon_{12} - h_{12} \\ 2\varepsilon_{12} - h_{12} & 2\varepsilon_{22} + h_{11} \end{bmatrix}.$$

The first invariant is $\mathrm{tr}A = 2(\varepsilon_{11} + \varepsilon_{22})$; it is not affected by the presence of h_{ij}. The second invariant is

$$\det A = -4\varepsilon_{12}^2 + 4\varepsilon_{11}\varepsilon_{22} + 2\varepsilon_{11}h_{11} - 2\varepsilon_{22}h_{11} - h_{11}^2 + 4\varepsilon_{12}h_{12} - h_{12}^2,$$

which, in view of $h_{ij} \ll 1$, we approximate by

$$\det A \approx -4\varepsilon_{12}^2 + 4\varepsilon_{11}\varepsilon_{22} + 2\varepsilon_{11}h_{11} - 2\varepsilon_{22}h_{11} + 4\varepsilon_{12}h_{12}.$$

Also,

$$A^2 = \begin{bmatrix} (h_{11} - 2\varepsilon_{11})^2 + (h_{12} - 2\varepsilon_{12})^2 & 2(\varepsilon_{11} + \varepsilon_{22})(2\varepsilon_{12} - h_{12}) \\ 2(\varepsilon_{11} + \varepsilon_{22})(2\varepsilon_{12} - h_{12}) & (2\varepsilon_{22} + h_{11})^2 + (h_{12} - 2\varepsilon_{12})^2 \end{bmatrix}$$

$$\approx \begin{bmatrix} -4h_{11}\varepsilon_{11} + 4\varepsilon_{11}^2 - 4h_{12}\varepsilon_{12} + 4\varepsilon_{12}^2 & 2(\varepsilon_{11} + \varepsilon_{22})(2\varepsilon_{12} - h_{12}) \\ 2(\varepsilon_{11} + \varepsilon_{22})(2\varepsilon_{12} - h_{12}) & 4h_{11}\varepsilon_{22} + 4\varepsilon_{22}^2 - 4h_{12}\varepsilon_{12} + 4\varepsilon_{12}^2 \end{bmatrix}.$$

Inserting these first-order approximations into expression (8.23), we obtain

$$\begin{bmatrix} 0 & 0 \\ 0 & 0 \end{bmatrix}.$$

Hence, even the linearized expressions satisfy the Cayley-Hamilton theorem. Since expression (8.23) is a function of the invariants, this function is also invariant. Herein, it remains invariant even upon linearization of the second invariant and of A^2.

Remark 8.1. An insight into the invariants of the Cayley-Hamilton theorem can be gained by considering the strain tensor, ε, associated with seismic sources, without the spacetime strain tensor, h_{ij}, associated with gravitational waves. In $\mathbb{R}^2$, we write expression (8.23) as

$$(\mathrm{tr}\,\varepsilon)\,\varepsilon - (\det\varepsilon)\,I = (\varepsilon_{11} + \varepsilon_{22})\begin{bmatrix} \varepsilon_{11} & \varepsilon_{12} \\ \varepsilon_{12} & \varepsilon_{22} \end{bmatrix} - \left(\varepsilon_{11}\varepsilon_{22} - \varepsilon_{12}^2\right)\begin{bmatrix} 1 & 0 \\ 0 & 1 \end{bmatrix},$$

which, as required, is equal to the square of the tensor—not to the square of its individual components. Herein, $\mathrm{tr}\,\varepsilon$ is the dilatation, φ; $\det\,\varepsilon$ is also an invariant, but it does not have such an immediate geometrical meaning.

Exercise 8.3. Following expression (8.15), write explicitly the expression for the corresponding quadratic form,

$$\mathrm{d}s^2 = \sum_{\alpha=0}^{3}\sum_{\beta=0}^{3}(\eta_{\alpha\beta} + h_{\alpha\beta})\,\mathrm{d}x_\alpha\,\mathrm{d}x_\beta = \sum_{\alpha=0}^{3}\sum_{\beta=0}^{3} g_{\alpha\beta}\,\mathrm{d}x_\alpha\,\mathrm{d}x_\beta\,,$$

and examine that expression in the context of $h_{\alpha\beta}$ tending to zero.

Solution. Let us write

$$\mathrm{d}s^2 = [\mathrm{d}x_0\,\mathrm{d}x_1\,\mathrm{d}x_2\,\mathrm{d}x_3]\begin{bmatrix} -c^2+h_{00} & h_{01} & h_{02} & h_{03} \\ h_{01} & 1+h_{11} & h_{12} & h_{13} \\ h_{02} & h_{12} & 1+h_{22} & h_{23} \\ h_{03} & h_{13} & h_{23} & 1+h_{33} \end{bmatrix}\begin{bmatrix} \mathrm{d}x_0 \\ \mathrm{d}x_1 \\ \mathrm{d}x_2 \\ \mathrm{d}x_3 \end{bmatrix}$$

$$\begin{aligned} =&(-c^2+h_{00})\,\mathrm{d}x_0^2+(1+h_{11})\,\mathrm{d}x_1^2+(1+h_{22})\,\mathrm{d}x_2^2+(1+h_{33})\,\mathrm{d}x_3^2 \\ &+2\Big((h_{01}\,\mathrm{d}x_1+h_{02}\,\mathrm{d}x_2+h_{03}\,\mathrm{d}x_3)\,\mathrm{d}x_0 \\ &+(h_{12}\,\mathrm{d}x_2+h_{13}\,\mathrm{d}x_3)\,\mathrm{d}x_1+h_{23}\,\mathrm{d}x_2\mathrm{d}x_3\Big)\,; \end{aligned}$$

since $h_{\alpha\beta}$ is traceless, we could write $h_{33}=-(h_{00}+h_{11}+h_{22})$.

If $h_{\alpha\beta}$, $\alpha,\beta=0,\ldots,3$, tend to zero, $\mathrm{d}s$ reduces to the case of the Minkowski metric of *special relativity*, $\mathrm{d}s^2=-c^2\,\mathrm{d}x_0^2+\mathrm{d}x_1^2+\mathrm{d}x_2^2+\mathrm{d}x_3^2$, as expected.

Exercise 8.4. Given α and β, which, respectively, are the values of the P-wave and S-wave speeds, obtained from seismic measurements, as well as assuming that other quantities of expression (8.22) are known, find the expression for the isothermal value of λ.

Solution. To proceed, we assume that physical measurements result in adiabatic values. Hence,

$$\alpha=\sqrt{\frac{\lambda_A+2\mu}{\rho}}\,;$$

consequently, $\lambda_A=\alpha^2\rho-2\mu$. Since $\mu=\beta^2\rho$ is assumed to exhibit the same value for both the adiabatic and isothermal cases, we obtain $\lambda_A=(\alpha^2-2\beta^2)\,\rho$.

Denoting the term in parentheses of expression (8.22) by λ_A, which stands for the adiabatic value of the parameter, we solve for λ, which is assumed to be isothermal, and herein denoted by λ_I. Solving a quadratic equation and considering its positive root, we obtain

$$\lambda_I=\frac{-\left(12\mu+\frac{\rho c}{T\tau^2}\right)+\sqrt{\left(12\mu+\frac{\rho c}{T\tau^2}\right)^2-36\left(4\mu^2-\frac{\rho c}{T\tau^2}\lambda_A\right)}}{18}\,,$$

which is the required expression.

Remark 8.2. In general, the value of λ might be either positive or negative, provided that—in accordance with the stability conditions—the eigenvalues of the elasticity tensor, whose components are stated in matrix (3.57), are positive. In terms of the Lamé parameters, discussed in Section 3.2.4.14, these eigenvalues are 2μ and $3\lambda + 2\mu$. Hence, the stability conditions are $\mu > 0$ and $\lambda > -2\mu/3$.

Exercise 8.5. Using measurements of the elasticity parameters for the Green-River shale (Slawinski, 2015, expression (9.4.3)), $c_{1111} = 3.13$, $c_{1133} = 0.34$, $c_{3333} = 2.25$, $c_{2323} = 0.65$ and $c_{1212} = 0.88$, estimate the fractional term in expression (8.22). Discuss the difference between the isothermal and adiabatic formulations. Assume that the values of elasticity parameters are isothermal.

In the *SI* units, the values of elasticity parameters are multiplied by 10^{10}. Also, in the *SI* units, $T = 400$ K, $\tau = 10^{-5}$, $c = 10^3$ and $\rho = 2300$.

Solution. Since the Green-River shale is represented by a transversely isotropic solid and expression (8.22) refers to an isotropic material, we use expressions (4.96) and (4.97) to find $c^{\overline{\rm iso}}_{1111} = 2.56 \times 10^{10}$ and $c^{\overline{\rm iso}}_{2323} = 0.87 \times 10^{10}$, which correspond to the closest isotropic tensor, in the sense of the Frobenius norm. Following definitions of the Lamé parameters, we obtain

$$\lambda := c^{\rm iso}_{1111} - 2c^{\rm iso}_{2323} = 0.82 \times 10^{10}$$

and

$$\mu := c^{\rm iso}_{2323} = 0.87 \times 10^{10}\,.$$

Since c_{1111}, c_{1133}, c_{3333}, c_{2323} and c_{1212} are assumed to be isothermal, so are the resulting λ and μ. Using these values in the fractional term of expression (8.22), we obtain

$$\frac{T\,(3\lambda + 2\mu)^2 \tau^2}{c\rho} \approx 3.1 \times 10^7\,,$$

which is three orders of magnitude smaller than the isothermal value of λ.

Since the difference between the adiabatic and isothermal values does not appear in $\lambda = 0.82 \times 10^{10}$, which is given up to two decimal places, we can use that value for modeling and interpretation of data in either context.

Chapter 9

Seismology as science

What is physics? One way to answer this question is to describe physics as the study of motion, energy, heat, waves, sound, light, electricity, magnetism, matter, and nuclei. This description, aside from sounding like the table of contents of a high school physics textbook, does not really specify the nature of physics. Physics is not just a study of the natural phenomena listed above but it is also a process; a process which has two distinguishable aspects.

The first one is simply the acquisition of knowledge of our physical environment. The second, and perhaps more interesting, is the creation of the worldview, which provides a framework for understanding the significance of this information. These two activities are by no means independent of each other. One requires a worldview to acquire new knowledge and vice versa one needs knowledge with which to create a worldview. But how does the process begin? Which comes first, the knowledge or the worldview?

Robert K. Logan (2010)

Preliminary remarks

In this chapter, we examine features of seismology that allow us to consider it as a science, and we comment on the type of science that it might represent. Selected issues discussed herein constitute a brief and partial overview of aspects of the philosophy of science in the context of seismology, without any claim to completeness or definiteness of presentation.

According to Sir Karl Raimund Popper (1963), the central question in the philosophy of science is the distinction between science and nonscience, and the criteria for such a distinction. Popper (1963) refers to it as the *demarcation criterion.* Another question is the purpose of science and interpretation of its results. Realists claim that science aims at truth and its theories are likely true, while instrumentalists claim that theories, albeit useful, are not necessarily true descriptions of Nature. Science relies on empirical evidence to validate theories and models; predictions should be in agreement with observations, which—in turn—are subject to interpretation. Thus, there are questions about criteria for such a validation. Also, there are questions of relations between predictions and explanations. In this chapter, we comment on these issues in the context of seismology.

We begin this chapter by discussing the concepts underlying the hypotheticodeductive formulation of seismology. Subsequently, we address aforementioned issues. We conclude by suggesting that—in spite of its intrinsic limitations, due to the notion of a continuum, and pragmatic limitations, due to such issues as accuracy of measurements—seismology is a theoretical science whose aim is a realistic description of Nature.

9.1 Hypotheticodeductive formulation

9.1.1 *Hypotheses*

Theoretical seismology follows a *hypotheticodeductive* formulation. This method is inherited from continuum mechanics (Bunge, 1967, Chapter 3, Section 2), upon which seismology is based.

In seismology, as discussed in this book, and in view of Chapters 2 and 3, the two hypotheses—from which explicitly follows the deductive argumentation to obtain quantitative predictions—are the balance of linear momentum and a constitutive equation. Implicitly, there is also the constraint imposed by the continuity equation, which originates in the balance of mass.[1]

In this book, the constitutive equation is represented by Hooke's law. Thus, seismology examines real materials using Hookean solids as their analogies. For the purpose of this section, the choice of another constitutive equation would not modify our discussion.

The balance of linear momentum is a statement that is valid for all continua. However, it does not suffice to complete the required hypotheticodeductive formulation. To relate the general theory to observations—which allows us to address the issues of theory validation—it is necessary to postulate a specific theory. Without postulating a Hookean solid, we could

[1] *see also*: Slawinski (2015, Section 2.1)

not—for instance—compute traveltimes of P and S waves, whose existence is a consequence of that postulate.

Examining the hypotheses on which the formulation of seismology is based, we can reflect on the last question in the quote of Logan on page 335: which comes first, the knowledge or the worldview? They are developing together.

In the worldview of seismology, we face the fundamental irreducibility due to the limits of continuum mechanics, whose analysis is limited to macroscopic phenomena. This limit is the reason for the existence of constitutive equations, whose purpose is to describe quantitatively such macroscopic qualities as compressibility and rigidity. This is both the strength and weakness of the subject. A seismologist's worldview is embodied within the language of continuum mechanics, which is both convenient and limiting. It provides us with predictions of phenomena whose existence is theory-dependent, such as the P and S waves, which are mathematical entities of elasticity theory. Our worldview is conditioned by the concept of continuum. However, as stated by Bunge (1967, p. 143), in spite of its mathematical nature,

> continuum mechanics is the chapter of mechanics closest to experience;

it invokes theoretical concepts, such as rigidity and compressibility, together with their sensory immediacy.

Yet, an excessive confidence in such an immediacy can be misleading, as discussed in the following quote of Ziman (1965).

> We might look at another well-known case, such as the wave theory of light which was developed at the beginning of the nineteenth century. The new hypothesis was 'light is wave-like, not particle-like'. There existed a good theory of waves in elastic solids, so it was not too difficult to write down the mathematics of such waves as if for light, and to get out some consequences, such as the slight bending of light round corners, and other observable diffraction phenomena. The point is that this was, in the first instance, only a 'toy'. Nobody knew what the waves were 'of', nor why they should be so. It took another sixty years before Clerk Maxwell showed that they were waves of electric and magnetic force, and brought them under more complete control.
>
> Nevertheless, the toy had served its purpose, for with its aid a vast range of optical phenomena had been predicted, described, or explained. Indeed, it worked so satisfactorily that it became

> too realistic. We recall Lord Kelvin suggesting that if the waves were elastic waves in a solid 'ether', then this must be enormously more rigid than steel—yet solid bodies seemed to pass through it without hindrance. Because the only sort of waves which were easily visualizable to the nineteenth-century physicists were waves in elastic solids, they could not help worrying about all the other properties that seemed to be implied by the theory—they forgot that it was only an analogy, or, as we should call it, a model.

Thus, let us not forget that Hookean solids are models. Different continua might be chosen to describe the same material, depending on the methods applied. The asthenosphere can be viewed as a Hookean solid for seismology, but not for glacial-rebound considerations, where the Maxwell model is commonly invoked.[2] The former is formulated under the assumption of the solid instantaneously regaining its shape following a deformation. The latter is formulated under the assumption of time involved in that process.

Having chosen necessary hypotheses, we deductively draw their conclusions. We do so within the structure of mathematics, where the hypotheses—albeit physically motivated—are taken as axiomatic statements within the accepted paradigm. From a formal viewpoint, a constitutive equation can be viewed as an axiom that is necessary to complete the system and obtain definite predictions based on the hypotheses of the balance of mass and the balance of linear momentum.[3]

9.1.2 *Deductive argumentation*

To emphasize the role of deductive arguments in continuum mechanics, let us refer to Truesdell (1977), who describes it as

> the part of mathematics that provides and develops logical models for the enforced changes of position and shape which we see everyday things suffer. It describes also much of what is observed or inferred in the laboratories where professional scientists produce experiments.

It is the relation between models and experiments—which, for better or worse, is a marriage between rationalism and empiricism—that distin-

[2] *see also*: Slawinski (2015, Section 3.4.4)
[3] *see also*: Slawinski (2015, Sections 3.3 and 4.4)

guishes continuum mechanics from pure mathematics. As stated by Dyson (2012),

> Over most of the territory of physics, theorists and experimenters are engaged in a common enterprise, and theories are tested rigorously by experiment. The theorists listen to the voice of nature speaking through experimental tools. This was true for the great theorists of the early twentieth century, Einstein and Heisenberg and Schrödinger, whose revolutionary theories of relativity and quantum mechanics were tested by precise experiments and found to fit the facts of nature.

We must distinguish between the reliability of deductive argumentation in mathematics and its use as a method of enquiry in natural sciences. The former—by virtue of deduction—is certain. The latter—by constraints of induction—is only a plausibility. Let us examine the following statement of Rota (1997).

> The theorems of mathematics motivate the definitions as much as the definitions motivate the theorems. A good definition is "justified" by the theorems that can be proved with it, just as the proof of the theorem is "justified" by appealing to a previously given definition.
>
> There is, thus, a hidden circularity in formal mathematical exposition. The theorems are proved starting with definitions; but the definitions themselves are motivated by the theorems that we have previously decided ought to be correct.

We can loosely paraphrase this statement in the context of continuum mechanics, in general, and quantitative seismology, in particular, by replacing definitions with hypotheses and theorems with observations.

Some philosophers call such a relation a *reflective equilibrium*. Observations suggest theories; theories predict observations. However, occasionally, a theory can overrule an observation, which is against the demarcation criterion of Popper (1963). Copernican theory tells us the Sun does not rise in the east and set in the west, even though it is manifestly obvious. We adjust and interpret a theory and observations until they are in acceptable agreement with each other. The idea of reflective equilibrium appears not only in science, but in ethics, linguistics, and other fields.

In continuum mechanics, we choose the hypotheses so as to obtain results that are in agreement with observations. Hypotheses are motivated by observations. The validity of continuum mechanics is based on its accurate description of observed phenomena, not on the physical truth of its hypotheses, which—in view of the accepted discrete, not continuous, structure of matter—are false.

The importance of deductive argumentation is its logical consistency within the theoretical formulation, not its correctness with respect to physics. A test of a physical validity of the formulation is its quantitative prediction or retrodiction of experiments. Hence, contrary to Marsden and Hughes (1983), who state that

> every beginner in elasticity theory should know that [...] Hooke's law will not be found as a basic axiom (it "really" means that you are working with the linearized theory),

we could accept this law as a basic axiom, as long as we are satisfied with the accuracy of description that it entails.

We can generalize a constitutive equation in such a manner that any form we use is a special case of a more general expression. However, a more general definition—in a mathematical sense—might not be more justifiable as a hypothesis for a theory used to infer physical properties.

Let us discuss the hidden circularity of deductive argumentation, mentioned in the above statement of Rota (1997). This circularity allows us to ensure the consistency and coherency of a formulation, but not a priority of concepts therein.

For instance, as discussed in Chapter 7, using Legendre's transformation, $L=\sum_{i=1}^{3}p_i\dot{x}^i-H$, Lagrange's equation of motion,

$$\frac{\partial L}{\partial x^i}-\frac{\mathrm{d}}{\mathrm{d}t}\frac{\partial L}{\partial \dot{x}^i}=0\,,\qquad i=1,2,3\,,$$

are obtained from Hamilton's equations of motion,[4]

$$\dot{x}^i=\frac{\partial H}{\partial p_i}\,,\qquad \dot{p}_i=-\frac{\partial H}{\partial x^i}\,,\qquad i=1,2,3\,.$$

However, postulating Hamilton's principle, $\delta\int L\,\mathrm{d}t=0$, and invoking the Euler equation of the calculus of variations, we can formulate the Lagrange equations without reference to Hamilton's equations.[5] The same is true if we postulate Fermat's principle of stationary traveltime.[6]

Subsequently—using the inverse of Legendre's transformation—we can obtain Hamilton's equations of motion, thus reversing the sequence of operations and, hence, changing the apparent priority exhibited in the former approach. Let us conclude with another statement of Rota (1997).

[4] *see also*: Slawinski (2015, Chapter 11)

[5] Such an an approach is used by Landau and Lifshitz (1976, 1986).

[6] *see also*: Slawinski (2015, Chapters 12 and 13)

> Suppose you are given two formal presentations of the same mathematical theory. The definitions of the first presentation are the theorems of the second, and *vice versa.* [...]. Which of the two presentations makes the theory "true"? Neither, evidently: what we have are two presentations of the *same* theory.

Thus, the role of deductive arguments in continuum mechanics—and, consequently, in theoretical seismology—is the rigour of their intellectual structure. The relation of theory and data is another issue.

9.2 Theory versus data

9.2.1 *Introduction*

A physical theory needs an empirical support, even though explicit empirical information might not be necessary for its original formulation. Let us consider two issues that arise within the context of an empirical examination of a theory. These issues are referred to as *theory-ladenness of data* and *underdetermination of theory by data.*

9.2.2 *Theory-ladenness of data*

To begin with the first issue, data are said to be theory-laden if they are affected by theoretical presuppositions, which means that they do not constitute information that is independent of the theory in question. Hence, the empirical support of theoretical predictions is biased by these predictions.

Seismological observations are mediated by theoretical entities. Predictions entailed by models, such as Hookean solids, raise questions about the existence of theory-independent data.

If we confine the theory in question to an isotropic Hookean solid, its predictions stem from equations of motion that contain two types of waves, P and S. In accordance with that theory and given a seismic record, such as the trace illustrated in Figure 1.2, our investigation is limited to the search for analogies of the P and S waves, among many recorded disturbances. Since an acceptable agreement between the measurement and the analogy lends the required support to a model, we might not be prompted to question our restricted theory. The same argument can be made for an anisotropic Hookean solid, whose equations of motion contain three types of waves, qP, qS_1 and qS_2.

Thomas Kuhn (1996) claims that it is impossible to isolate the hypothesis being tested from the influence of the theory to which it belongs. Kuhn (1996) argues that observations must rely on a specific paradigm, and hence, it is not possible to evaluate competing paradigms independently. However, evaluating models in seismology, we rely on an *a priori* chosen paradigm, since available models do not commonly belong to competing paradigms. An anisotropic model is an adjustment of an isotropic one; both models are within the same paradigm. The adjustment precess remains—in accordance with the structure proposed by Kuhn (1996)—within a *normal science*. Thus, even if we accept the impossibility of evaluating independently competing paradigms, we might still compare and evaluate models within a given paradigm.

Awareness of these issues might allow an alert seismologist to use seismic data, together with other information, to enhance seismic models, even though objections raised by a comparison of data and theoretical predictions might be subtle. After all, raw data illustrated in Figure 1.2 are theory-independent. A record obtained from a seismograph is independent of seismic theory.

9.2.3 *Underdetermination of theory by data*

Fundamentally, any theory about the natural world is underdetermined by data. For any data set there exist infinitely many theories that can account for it.

In seismology, the issue of underdetermination is emphasized by a pragmatic consideration of few measurements—in comparison with, say, laboratory study of materials—and of their limited accuracy. At the same time, even if an examined theory is based on a homogeneous Hookean solid, the twenty-one parameters required to describe such a solid render the choice of a model difficult. There are many possibilities without clear preferences among them. If we consider an inhomogeneous Hookean solid, where the values of $c_{ijk\ell}$ are also functions of position, there is an infinity of model parameters that can be proposed to account for the data. It is difficult to choose the appropriate model by rejecting others, if they all exhibit so many degrees of freedom.

In spite of these difficulties, we need to choose a model by confronting its theoretical predictions with actual data. Such a choice is necessary both for foundational considerations of the theory in question and for a pragmatic interpretation of given experimental information.

Ideally, we would like to apply the approach proposed by Popper (1963) and refute all conjectured models whose predictions disagree with empirical information. As a consequence of this process, we would hope that only a single model remains. However, the relation between theoretical predictions and quantitative information in seismology makes a strict application of such a criterion impossible. The number of available model parameters confronted with the limited amount of data allows us to modify a postulated model in such a manner as to prevent its refutation. As indicated by the quote on page 352, below, Popper (1963) became aware of such difficulties.[7]

To restrict available models to a handful of pertinent ones, we might use information from sciences other than seismology. In other words, we include the consistency among sciences as a model criterion. For instance, geological information about a sedimentary basin might restrict our models to horizontal layers, and—in view of the Backus (1962) average—to a homogeneous transversely isotropic medium. Thus, such disciplines as geology, mechanics, planetary sciences, as well as other branches of geophysics provide us with information that restricts the scope of theoretical models to physically pertinent ones.

Within seismology, a restricting criterion can be provided by a statistical approach, where model parameters inferred from the data are treated not as specific values but as probabilistic distributions of these values. An example of such an approach is illustrated by comparison of Figures 4.1 and 4.2, above, where—based on a better agreement within distributions and their standard deviations—we argue for the choice of an orthotropic model over a transversely isotropic one.

A statistical approach, championed by Albert Tarantola (2006) and discussed in Section 9.3, below,[8] is exemplified by *Bayesian inference*, which is an inference in which Bayes's theorem is used to update the probability of a hypothetical model as more empirical information becomes available.

9.3 Bayesian inference

Methods of, and models within, the theory of seismology might be examined by Bayesian inference, which stems from Bayes's theorem, named in honour

[7]Readers interested in adjustments of the Popper (1963) approach might refer to Bunge (1999).

[8]Readers interested in a statistical evaluation of models and their parameters might refer to Tarantola (2005).

of Thomas Bayes, who proved its special case. Laplace generalized it and used it to evaluate information in celestial mechanics.

Bayes's theorem originates in a standard statement of probability theory, given in expression (9.2), below. However, in the context of Bayesian inference, the interpretation of this theorem, and hence, its application are extended beyond its immediate meaning.[9] Within this extended interpretation—and in spite of controversies among statisticians—it is used commonly to estimate the probability of correctness of a hypothesis.

To gain an insight into the Bayesian approach, let us consider the following quote from Reid (1998), which distinguishes

> [...] between probability as a mathematical abstraction of frequencies (a view which went back to the work of Jacob Bernoulli) and probability as a measure of intensity of belief (a view identified with the name of Thomas Bayes, an English clergyman). The basic distinction between the two approaches turns on the events or statements to which each is willing to assign probability. "Frequentists" restrict their attention to those which are capable of a large number of repetitions, either in reality or in principle, while "Bayesians" are willing to assign probability to any event or statement, even though it may be unique.

The latter appears to apply more naturally to seismic considerations, where a particular set of unknown properties is considered. Assigning a probability of anisotropy to a given medium does not imply that among a large number of such media, a certain number is likely to be anisotropic.

Be that as it may, let us consider the theory of elasticity, which—for anisotropic Hookean solids—contains three types of waves. Hence, we predict that three distinct disturbances are observable. In the case of isotropy, only two waves are present. A traditional term for the appearance of the third wave, which is a consequence of anisotropy, is called *shear-wave splitting*; S wave becomes qS_1 and qS_2.

Entering the realm of anisotropy, we might wish to evaluate the usefulness of the Hookean-model prediction of shear-wave splitting. Invoking

[9] Readers interested in a classic description of Bayesian statistics might refer to Jeffreys (1961b), and for a modern approach to Jaynes (2003), the book "Dedicated to the Memory of Sir Harold Jeffreys, who saw the truth and preserved it". A concise explanation might be found in Brown (2001, pp. 73–74) and a simple seismic application in Danek and Slawinski (2012). A historical description of the development of Bayesian statistics is discussed by McGrayne (2011), where, in Chapter 2, the author describes contributions of Pierre-Simon de Laplace, and in Chapter 3, the importance of the work of Sir Harold Jeffreys.

Bayes's theorem, we write,

$$P(H|E) = P(E|H)\frac{P(H)}{P(E)}. \tag{9.1}$$

Herein, $P(H|E)$ is the confidence in the usefulness of the Hookean hypothesis, H, given the observable evidence, E, of the shear-wave splitting. $P(E|H)$ expresses the degree of confidence with which the theory predicts this phenomenon, given the background information about a material.

Since seismology is a hypotheticodeductive formulation, the anisotropy entails the splitting. However, we set $P(E|H) = 0.9$ to allow for a doubt in *a priori* information on the material in question. Both $P(H)$ and $P(E)$ are called *prior probabilities*. $P(H)$ is a degree of confidence in the usefulness of anisotropic Hookean models in view of their previous isotropic successes, which—while aware of the subjectivity of such a quantification, which is the reason for the above-mentioned controversy—we set to $P(H) = 0.7$. $P(E)$ is the probability of observing shear-wave splitting. It relies on a sufficiently accurate data set and on associating recorded disturbances with the three waves; the latter might raise epistemological concerns, since such an association requires a seismologist to formulate a theoretical entity in order to observe it.

To ensure that $P(H|E) \leqslant 1$, it is required that $P(E) \geqslant P(E|H)P(H)$, which—notably—follows from the fact that H entails E; we let this quantity be $P(E) = 0.8$. These values result in $P(H|E) = 0.7875$. If we obtain an experimental evidence of the shear-wave splitting, we update the value of $P(E)$ to, say, 0.9. We do not set $P(E) = 1$, to allow for a misinterpretation of experimental results.

If we keep other values in expression (9.1) unchanged, $P(H)$ becomes 0.7875, which is the new degree of our confidence in recognizing anisotropic properties of materials; it increased from 0.7.

The experimental evidence increases our confidence in the applicability of Hookean solids to seismological problems, and strengthens our trust in the available seismic theory. Two more examples of such an evaluation are presented in Exercises 9.2 and 9.3.

Such an approach to evaluate seismology as a science is consistent with the views of Tarantola (2006). It is also consistent with the concept of distance in the space of elasticity tensors, discussed in Section 4.1, and, in particular, with the comparison of Figures 4.1 and 4.2, according to which a Hookean solid of a particular symmetry and orientation is a good analogy for a physical material in question, while a null hypothesis of a solid exhibiting a greater symmetry is rejected within the estimate of errors.

An intuitive evaluation of $P(H)$ and $P(E)$ might raise questions regarding such an approach. Nevertheless, as stated by Jaynes (1980),

> Bayes' theorem tells us far more than intuition can. Intuition is rather good at judging what pieces of information are relevant to a question, but very unreliable in judging the relative cogency of different pieces of information. Bayes' theorem tells us quantitatively just how cogent every piece of information is.

Let us emphasize that expression (9.1), except for its interpretation discussed above, is a standard formula of the probability theory. It stems from conditional probability of both events A and B happening,

$$P(A \cap B) = P(A|B)P(B) = P(B|A)P(A)\,, \tag{9.2}$$

where $P(A)$ and $P(B)$ are the probabilities of A and B, respectively, and $P(A|B)$ and $P(B|A)$ are the probability of A, given B, and *vice versa.* Rearranging of the second equality results in expression (9.1), which was published in *Philosophical Transaction* of the Royal Society, a couple of years after Bayes's death.

Apart from the issue of subjectivity of prior probabilities, which arises in any science, the Bayesian approach to evaluate models in continuum mechanics, and hence, in seismology, suffers from a certain lack of theory-independent observables. This is not the case in evaluating specific methods therein. For instance, the appearance of Fresnel's bright spot is independent of the invoked theory or of the method applied, such as the corpuscular theory of light or geometrical optics.[10] As illustrated in Exercise 9.3, the experimental evidence contrary to predictions results in a significant weakening of geometrical optics as a comprehensive theory.

9.4 Predictions versus explanations

9.4.1 *Introduction*

Among explanations of seismological phenomena and the related issue of causality and prediction, let us consider the *covering-law model.* It has a deductive structure and requires a particular phenomenon to be covered by a general law of nature. Also, let us consider the *inference to the best explanation*, whose explanations have a nondeductive structure; they rely on a plausibility argument.

[10] Readers interested in an insightful illustration of the effect of novel predictions on Bayes's theorem might refer to Brown (2001, pp. 32-35 and 73-74).

9.4.2 Covering-law model

The covering-law model is described by Hempel (2000). In seismology, given the laws of motion and choosing a Hookean solid to represent the Earth, we deduce consequences that account for observed phenomena. According to Hempel (2000), there is a symmetry. If the phenomenon is to occur in the future, we deduce a prediction; if it occurred in the past, we deduce an explanation. In both cases, the logical structure is the same and both explanation and prediction are related to the empirical confirmation. However, there is an issue with such a structure, as described by Okasha (2002, pp. 45–47).

Imagine a flagpole that casts a shadow. Knowing the height of the pole, geometrical optics and trigonometry, we predict and explain the length of the shadow. Knowing the length of the shadow, we predict the height of the pole; however, we cannot explain this height in any manner other than, say, a carpenter's choice. The prediction-explanation symmetry breaks down.

In seismology, given the causes, we predict and explain the effects. Given the effects, we might predict or retrodict, but not explain, the causes. We need to distinguish between the causes of the measured effects due to subsurface structures and the causes for the existence of these structures within the Earth. We can explain the shadow length caused by a given height, but not the height itself.

Consider a horizontal layer whose thickness is h, and within which signal propagation speed is v_1. This layer is connected to a halfspace within which the speed is v_2, where $v_2 > v_1$. Both the model and the plot of signal traveltime, t, versus distance, x, are illustrated in Figure 9.1. Using the value of v_1, we can predict the slope of the direct-arrival line, which is v_1^{-1}. Using v_2, we can predict the slope of the refracted-arrival line, v_2^{-1}. Using h, together with v_1 and v_2, and invoking the ray theory in its geometrical-optics guise, we can predict the intercept, t_0, of the latter line. These predictions are tantamount to explanations. A given slope is a consequence of the physical property of a medium, namely, its speed; the intercept of the refracted-arrival line is

$$t_0 = \frac{2h\cos\theta_c}{v_1}, \tag{9.3}$$

where $\theta_c = \arcsin(v_1/v_2)$ is the critical angle. It is a consequence of the values of h and of ratio v_1/v_2, in the context of geometrical optics. There is a causality: preexistent geological properties of the subsurface are the cause of particular values of the slopes and the intercept.

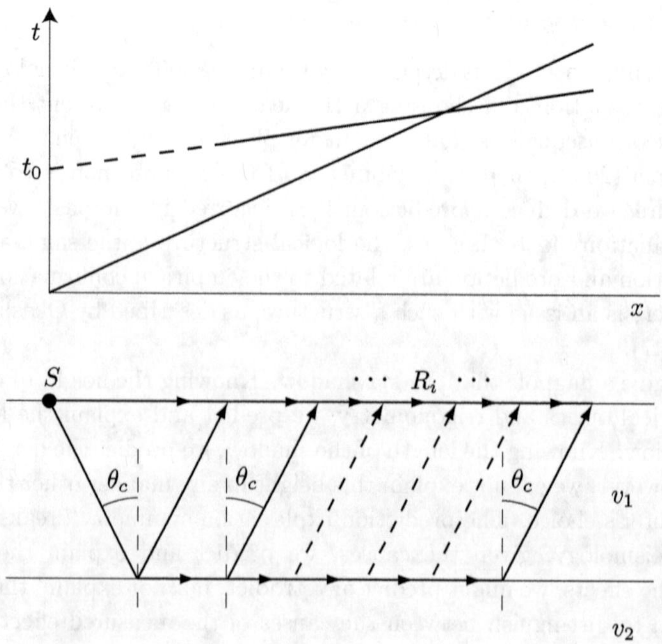

Fig. 9.1: Signals generated at the source, S, and propagating along the surface arrive directly at receivers R_i, located at the surface. Signals reaching the interface at the critical angle, θ_c, propagate along the interface with speed v_2. A physical justification for refracted signals reaching all the receivers at the critical angle is illustrated in Figure 7.7. The data set, consisting of the traveltime-versus-distance plot, can be deduced from the model in the context of geometrical optics. The slopes are v_i^{-1}, with $i = 1, 2$, and the intercept of the refracted-arrival line is given by expression (9.3). Herein, in contrast to many other cases, this data set is sufficient to infer the model parameters.

Consider only the traveltime-versus-distance plot in Figure 9.1, which is the inverse-problem information. The data consisting of the slopes and the intercept allow us—by rearranging expression (9.3)—to infer the values of v_1, v_2 and h. In certain cases, these inferences might be verified by observations; for instance, they might be verified by drilling. Notably, such an approach was used by Mintrop in the first quarter of the twentieth century. However, these inferences are not tantamount to explanations, since the properties of the subsurface are not caused by these data. They are caused by geological processes.

Obtaining the data from the model parameters is a forward problem, and obtaining the parameters from the data is an inverse problem. In the case illustrated in Figure 9.1, we have a one-to-one relationship between causes and results, which is a rare case of uniqueness for the inverse-problem solution. However—in spite of this uniqueness and in contrast to the postulate of Hempel (2000)—the predictions do not entail explanations.

In general, the lack of prediction-explanation symmetry renders the covering-law model not applicable as explanation within the realm of seismology.[11] Yet, we can consider another approach.

9.4.3 *Inference to best explanation*

Observed phenomena are explained by invoking the concepts of propagating disturbances. Quantitative properties associated with these phenomena are modelled within the realm of Hookean solids. We might accept physical interpretations of a mathematical model as the best available explanation. Such an *inference to best explanation* is in a partial agreement with Hempel (2000), in the sense that the phenomenon to be explained is deduced from general laws and particular conditions. However, there is no prediction-explanation symmetry. Explanation is an inductive inference.

For an inference to the best explanation, a hypothesis is accepted as plausible on the grounds that it explains relevant observations. For example, the hypothesis that the outer core is liquid explains a range of observations. An everyday example of such an explanation might be that, in view of an open window and torn curtains, a lamp lying broken on the floor was knocked over by the wind. It is an inductive, not deductive, inference. Still, it might be the best explanation. An issue that we do not address in this book is the choice of quantitative criteria according to which a given explanation might be considered best. A qualitative requirement for such an explanation might be its agreement with other branches of physics and geoscience. For instance, the orientation of fractures inferred from the material symmetry of a Hookean solid might be consistent with available descriptions from structural geology.

9.5 Realistic approach versus instrumental approach

A physical theory is motivated by its purpose. Let us consider two such purposes, which are distinct from one another but are not contradictory to

[11]Readers interested in the prediction-explanation symmetry and the so-called *Hempel's paradox* might refer to Ayer (1972, Chapter 3) and to Okasha (2002, Chapter 3).

one another. The corresponding approaches are referred to as *realistic* and *instrumental*. They have many commonalities, however, for the former, the postulated physical entities need to be—at least, approximately—true. For the latter approach, whose development is associated with Ernst Mach (1989), empirical adequacy is the key criterion. The postulated entities need not be physically true, provided they allow us to account for observations.

The origins of such an approach extend to Plato. In Plato's dialogues, according to the character of Timaeus of Locri, we cannot get outside to see whether the Earth goes around the Sun or the Sun goes around the Earth. Therefore, truth is not the aim of science. Its only aim is empirical adequacy. Such antirealism is common in the history of astronomy. It also appears in interpretation of quantum mechanics.

Seismological studies can be viewed as inverse problems whose goal is to describe the Earth. Thus, most seismologists are realists. In their view, the realm described by seismology is the physical world. In other words, seismic models resemble the Earth, *sensu lato*.

A common justification for realism is the success of scientific theories, such as their explanatory and predictive capabilities. A key ontological issue of realism is the status of unobservable entities that are part of scientific theories. According to realism, one can make reliable claims about certain unobservables. These unobservables have the same ontological status as observables. Other unobservables, including stress tensors and rays, are mathematical entities.

The inner core is an unobservable theory-mediated physical object. Its existence is inferred only through theory, but it exists and exhibits physical properties as does Mount Everest, which is a subject of direct experience. For seismologists, the inner core and its properties are not just parameters of a model to account for measurements on the Earth surface—as might be argued by an instrumentalist, for whom the unobservables are purely theoretical entities that allow us to account for observable phenomena—but a physical object exhibiting particular properties. For instance, the inner core is a physical material whose rigidity is greater than of the outer core.

The ontology of realism is closely related to the epistemological issue of rationalism, according to which we can appeal to reason—not only to sensory experiences—as a source of knowledge and justification. Rationalism plays an important role in hypotheticodeductive formulations, since it emphasizes a deductive, rather than sensory, criterion of truth. To avoid a confusion, rationality and rationalism are distinct notions. As explained by Kołakowski (1990, Chapter 16),

> The latter, defined in opposition to either irrationalism or empiricism, is an epistemological doctrine, a normative definition stating what does or does not have a cognitive value, whereas rationality and irrationality are characteristics of human behaviour.

Also, to avoid a confusion—in spite of important similarities, such as that scientific theories do not aim for truth about unobservable entities—we distinguish between empiricism and instrumentalism.

Returning to the ontology of realism and paraphrasing the statement of Hesse on page 103, we might postulate that a support for the physical reality of a theoretical entity lies in its possession of detectable properties in the context of distinct theories. The consistency of interpretations with respect to the outer core in both seismology and global geodynamics is an example of such a support.

To examine the instrumental approach, theories can also be considered as logical tools for pragmatic purposes only, without requiring any enquiry into the underlying mechanism or causality of observed phenomena. For instance, a traveller can use patterns in the sky for navigation without any enquiry into celestial mechanics.

An examination of earthquakes might indicate the areas of likely occurrences of future earthquakes, their extent and magnitude, and thus provide urbanists with information for city planning and civil engineers with information for building structures. Neither of these purposes requires an explanation of earthquake mechanism or of propagation of disturbances generated by such a mechanism.

In exploration seismology, the purpose of a theory is finding deposits of, say, minerals or petroleum. A symptomatic approach of detecting a particular pattern on a seismogram might suffice to predict the location of a sought deposit. In such a case, a symptomatic approach is a pertinent instrumental theory, and finding the desired deposits constitutes its validation and justification.

9.6 Coherence theory of justification

Let us comment on alternatives to epistemological difficulties faced by both rationalism and empiricism.

The central problem of epistemology consists of establishing the criteria to justify a given proposition. René Descartes (1992) argues that a person is justified in holding a proposition, if that proposition can be derived from selfevident hypotheses. Empiricists, on the other hand, argue for knowledge rooted in reliable data. The claim of either selfevident hypothesis or reliable information from data poses further problems.

Difficulties of both rationalism and empiricism, particularly in their austere forms, motivate epistemologists to seek other solutions. Among candidates, there is the *coherence theory of justification*, according to which a proposition is justified if it belongs to a system, whose coherence is based on logical consistency together with inductive relations, including explanatory ones.

A coherentist might argue that a system that is not true must—by definition—contain contradictions, and so be incoherent. On the other hand, the greater the number of phenomena explained by the system, the greater its coherence. Qualitatively, this is a variation on an epistemological view of Occam's razor. Quantitatively, the issue of coherence could be investigated within Bayesian statistics.

To examine seismological statements within the coherence theory of justification, we could use their logical consistency, which is ensured by the hypotheticodeductive structure. Also, based on models, seismologists infer a host of inductive and explanatory statements, which are in agreement with broader systems of physics and geosciences.

Closing remarks

Quantitative seismology relies on hypotheses to deduce predictable observations. As such, it appears to fit the criteria of science proposed by Popper (1963): conjectures and refutations. However, as discussed above, even remaining within Hookean solids, one might be unable to refute a model due to the plethora of available parameters, thus—in practice—failing the demarcation criterion.

As stated by Popper (1963, Chapter 1, Section III, Science: Conjectures and refutations), he became aware of such difficulties.

> Today I know, of course, that this *criterion of demarcation*—the criterion of testability, or falsifiability, or refutability—is far from obvious; for even now its significance is seldom realized. At that time, in 1920, it seemed to me almost trivial, although it solved for me an intellectual problem which had worried me deeply, and one which also had obvious practical consequences.

In spite of practical difficulties in its application, the concept of refutability remains an important criterion to consider in evaluating scientific pursuits, including seismology and its models. The Bayesian approach can be used to express refutability in a statistical context, thus accommodating certain issues, such as experimental errors. The demarcation criterion becomes a probabilistic, not an absolute, condition.

One might claim that seismology fits a covering-law model of Hempel (2000): the premises entail conclusions and contain a fundamental principle. Hence, the phenomenon to be explained is covered by a general law of nature. However, seismology fails the key requirement of Hempel (2000), since its predictions and explanations are not symmetric.

Inference to the best explanation might be a description of seismological reasoning. For instance, if the phenomenon to be explained is the detection of three distinct disturbances, one can invoke the theory of wave propagation in an anisotropic Hookean solid. An agreement between the measurements and the theoretical entailment of such properties as time differences among the three signals and their polarizations allows seismologists to validate the approach, while being aware that different or more accurate measurements might call for another type of continuum.

Even though most seismologists would consider their interpretations as corresponding to the world of physics—albeit within the limitation of continuum mechanics—few of them would be confident to state, after Weinberg (1987), that

> at the end of the day, we want to have the feeling that 'it could not have been any other way'.

Aware of the limitations of the approach, seismologists might consider that the real materials within the area of interest could be represented by a Hookean solid exhibiting certain symmetry. However, its symmetry could be different and the materials could even be represented by a continuum with different properties. Different explanations are possible, particularly in continuum mechanics, where discrete properties of matter are examined as bulk averages.

As stated by Bunge (1967), continuum mechanics is a basic theory to be studied on its own level, without explaining properties of continua in terms of underlying microscopic properties. Attempts of such explanations are beyond continuum mechanics. They belong to the realm of condensed-matter physics, and—ultimately—quantum mechanics. Even though such explanations might shed light on concepts and issues within continuum mechanics, they are not necessary to justify its scientific value.

Seismology, as a branch of continuum mechanics, examines macroscopic properties associated with disturbances propagating within the Earth. The observed phenomena are the averages of numerous causes, which are impossible to examine individually within available theory and the resolution of data. Arguably, seismology allows us to infer the best explanation for these phenomena.

9.7 Exercises

Exercise 9.1. Calculate the traveltime for the model discussed in Exercise 7.2, except without considering the one-metre layer. Compare the result with Exercise 7.2.

Solution. The traveltime is $\sqrt{1000^2 + 1000^2/1000} = \sqrt{2} \approx 1.4\ \mathrm{s}$.

The difference in traveltimes—1.1 versus 1.4 seconds—is due to the presence of the layer, where the propagation speed is tenfold greater than in the surrounding medium. Fermat's principle ensures the traveltime optimization, regardless of thickness of layers, which implicitly disregards the wavelength of a signal. In seismological studies, we would not expect a one-metre layer to affect the traveltime in a measurable manner. This issue is an example of the limitation of applicability of Fermat's principle and, in general, of geometrical optics, which relies on significant idealizations.

An attempt to accommodate that issue is offered by the Backus average, discussed in Section 4.2, above. In this average, material properties of layers are weighted by thicknesses.[12]

As illustrated in Exercise 9.3, below, a similar limitation of geometrical optics is exemplified by its failing to predict Fresnel's bright spot, whose appearance is in agreement with the wave theory but not with geometrical optics. The appearance of the bright spot supports the wave-theory approach.[13]

---

[12]Readers interested in such issues might refer to Dalton and Slawinski (2016).

[13]Readers interested in a Bayesian evaluation of wave theory versus corpuscular theory of light due to the appearance of a bright spot might refer to Brown (2001, 73-74).

Remark 9.1. Fermat's principle is a convenient mathematical analogy for quantitative examinations—within the ray theory—of observed phenomena. However, for any analogy or theory, the range of applicability is limited. A commonly observed phenomenon that is not included in predictions based on ray theory is *wavefront healing*, which results from signals penetrating into the *shadow zones* and diminishing the effects of shadow-producing obstacles.

According to ray theory, the effect of any obstacle encountered by a wavefront persists for the subsequent propagation. This is a consequence of the causality of a well-posed Cauchy problem. The effect acquired during propagation cannot disappear. Also, there can be no propagation of a signal to locations not reachable by the rays, which—as the bicharacteristics of the equation of motion—are subject to Fermat's principle.

To account for vanishing of the effect of a small obstacle, it is necessary to consider diffraction and interference, neither of which is included in Fermat's principle, in ray theory or in Cauchy problems.

Exercise 9.2. Consider the following scenario. There is a sphere of unknown composition.[14] However, there is an *a priori* conviction that it is composed of concentric layers of Hookean solids and that there are no liquid layers.

In the context of a seismic experiment—with a source generating shear waves at the surface and a receiver at its antipode, which implies normal incidence and hence, no converted waves—use Bayes's theorem to test the contrary hypothesis: there is a layer of liquid.

Assuming that the experiment results in no shear waves recorded at the receiver, estimate the *a posteriori* confidence in the existence of such a layer.

Solution. Let us endow the terms on the right-hand side of expression (9.1),

$$P(H|E) = P(E|H)\frac{P(H)}{P(E)}, \tag{9.4}$$

with their physical meanings and let us assign to them prior probabilities. $P(E|H) = 1$ is the probability of no shear waves at the antipode due to the presence of a liquid layer, since—according to the elasticity theory—shear waves do not propagate through liquids.

[14]Such a scenario resembles enquiries into properties of the Earth's interior that began a couple of centuries ago (Brush, 1996, Chapters 2.1–2.2).

$P(H)$ is the probability of the existence of a layer of liquid, which—in view of the opposite conviction—we set to 0.10; it describes the degree of confidence prior to the experiment.

$P(E)$ is the probability of not recording shear waves at the antipode of the source. In view of the above conviction, we let $P(E) = 0.11$; it is slightly larger than $P(H)$, since it is entailed by it. The existence of a liquid layer implies no recorded shear wave, $H \to E$, and we allow 0.01 for not recording a shear wave, even if it exists. However, we do not allow for misinterpreting a seismic record to include a nonexisting shear wave.

Inserting these values into expression (9.4), we obtain $P(H|E) \approx 0.91$.

Following the experiment that did not record any shear waves, we let $P_{\mathrm{new}}(E) = 0.90$. We do not set it to unity to allow for the lack of detection due to experimental limitations. In other words, we are 90% confident that no detection of shear waves implies that they are absent.

Rearranging expression (9.4), we calculate the new level of confidence in the existence of a liquid layer,

$$P_{\mathrm{new}}(H) = P_{\mathrm{new}}(E)\frac{P(H|E)}{P(E|H)} = 0.90\frac{0.91}{1} \approx 0.82\,. \tag{9.5}$$

Hence, due to the absence of shear waves on the seismic record, we increased our confidence in the existence of a liquid layer from $P(H) = 0.10$ to $P_{\mathrm{new}}(H) \approx 0.82$.

Exercise 9.3. A scientist uses geometrical optics and predicts a uniformly dark circular shape as a shadow of a disk placed between the light source and the screen. Evaluate the change in confidence for geometrical optics as a tool to predict physical phenomena if an experiment results in concentric fringes and a bright spot in the middle of the circular shape.

Solution. Since a dark circular shape is a consequence of geometrical optics, $P(E|H) = 1$. In view of prior successful predictions of geometrical optics, we set $P(H) = 0.9$. The probability of observing such a shape is estimated at $P(E) = 0.91$, where $P(E) > P(H)$, as required. Inserting these values into expression (9.4), we calculate $P(H|E) \approx 0.99$.

Following the experiment that did not result in a uniformly dark circular shadow, we set $P_{\mathrm{new}}(E) = 0.01$; we do not set it to zero to allow for a possibility of an error within the experiment. Thus, according to expression (9.5), $P_{\mathrm{new}}(H) \approx 0.01$, which emphasizes the loss of confidence in predictive capacity of geometrical optics or, more specifically, its incapacity to predict certain observable phenomena, such as diffractions. Predictions of fringes and bright spots require *wave optics*, which is a more general theory.

Remark 9.2. The Bayesian approach can be used to evaluate models and theories. In Exercise 9.2, it is used to evaluate properties of a model: existence versus nonexistence of a liquid layer. In Exercise 9.3, it is used to evaluate applicability of a method: geometrical optics versus wave optics.

The latter use of the Bayesian approach could be also exemplified in the context of Exercises 7.2 and 9.1, by evaluating the predictive capacity of ray theory and Fermat's principle. Such an evaluation would follow the pattern akin to the one presented in Exercise 9.3. Its conclusion would be a limitation of applicability of ray theory and Fermat's principle in predicting quantitative measurements. In particular, there is a necessity to account for measurements in terms of finite frequency.

Appendix A

On strains

Bodies have one distinct physical property: they occupy regions of euclidean space $\mathscr{E}$. Although a given body will occupy different regions at different times, and no one of these regions can be intrinsically associated with the body, we will find it convenient to choose such a region, $\mathscr{B}$ say, as reference, and to identify points of the body with their positions in $\mathscr{B}$.

Morton E. Gurtin (1981)

Preliminary remarks

In continuum mechanics, finite strains refer to deformations and rotations that are arbitrarily large. Consequently, since—in contrast to infinitesimal strains—the reference and the current configurations of the continuum might be significantly different from one another, it is necessary to distinguish between them to achieve an empirically adequate mathematical model.

We begin this appendix by examining the displacement-vector field and, hence, the deformation-gradient tensor. To separate deformations from rotations, which are both contained in this tensor, we formulate the Cauchy-Green strain tensors and invoke polar decomposition to obtain explicitly the rotation and stretch tensors. We also discuss transformations between the volume and surface elements in the reference and current configurations. Furthermore, we show how the infinitesimal-strain expressions are inherited from the finite-strain formulations.

Notation A.1. In this appendix, as well as in Appendices B, C and D, the quantities associated with the reference configuration are commonly denoted by the upper-case letters and the quantities associated with the current configuration are commonly denoted by the lower-case letters; not all upper-case or lower-case letters, however, refer to these configurations. Bold letters represent vectors.

A.1 Displacement-vector field

In continuum mechanics, the displacement of a body consists of two components: rigid-body motion and deformation; the former consists of translation and rotation. In the material coordinates, $\mathbf{X}$, discussed in Sections 2.2.2 and 2.2.3, we write the displacement-vector field as

$$\mathbf{U}(\mathbf{X},t)=\sum_{I=1}^{3}U_I\,\hat{\mathbf{E}}_I=\mathbf{b}(t)+\mathbf{x}(\mathbf{X},t)-\mathbf{X}\,, \tag{A.1}$$

where, as illustrated in Figure A.1, $\mathbf{b}$ is the position vector for the origin of the coordinate system for the current configuration, $\mathbf{X}$ is the position vector of a specific material point in the reference configuration and $\mathbf{x}$ is the position vector of that point in a current configuration.

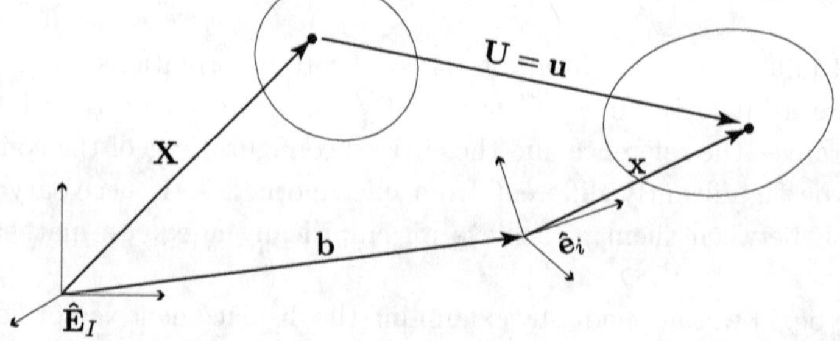

Fig. A.1: Reference and current configurations: Initially, the reference and current configurations—and, hence, the coordinates of any point in the material and spatial coordinate systems—coincide; thus, for each point, $\mathbf{X}=\mathbf{x}$, and $\mathbf{U}=\mathbf{u}=\mathbf{b}=\mathbf{0}$. Under rigid-body motion or deformation, we have, in general, $\mathbf{U}=\mathbf{u}$ and $\mathbf{X}\neq\mathbf{x}$.

In expression (A.1), with a certain abuse of notation, we let $\mathbf{x}(\mathbf{X},t)$ stand for $\phi(\mathbf{X},t)$, which is a function describing the motion of point $\mathbf{X}$ within a continuum, discussed in Section 2.2.2; also, it is referred to as function $\mathbf{f}$ in Section 3.2.3.1. The value of $\mathbf{x}(\mathbf{X},t)$ depends not only on time but also on $\mathbf{X}$; in other words, not all points within the continuum need to move together, as is the case of the rigid-body motion.

In spatial coordinates, $\mathbf{x}$, also discussed in Sections 2.2.2 and 2.2.3,

$$\mathbf{u}(\mathbf{x},t)=\sum_{i=1}^{3}u_i\,\hat{\mathbf{e}}_i=\mathbf{b}(t)+\mathbf{x}-\mathbf{X}(\mathbf{x},t)\,, \tag{A.2}$$

where $\mathbf{x}$ denotes a specific spatial location and where we let $\mathbf{X}(\mathbf{x},t)$ stand for $\Phi(\mathbf{x},t)$, discussed in Section 2.2.3, which is the inverse of ϕ. Thus, ϕ and Φ allow for the mappings between configurations.

The independent variables of expression (A.1) and, hence, of $\mathbf{U}$, are material coordinates. The independent variables of expression (A.2) and, hence, of $\mathbf{u}$, are spatial coordinates. As mentioned in Section 2.2.3, these coordinates are also said to be *referential*, or *Lagrangian*, and *Eulerian*, respectively.

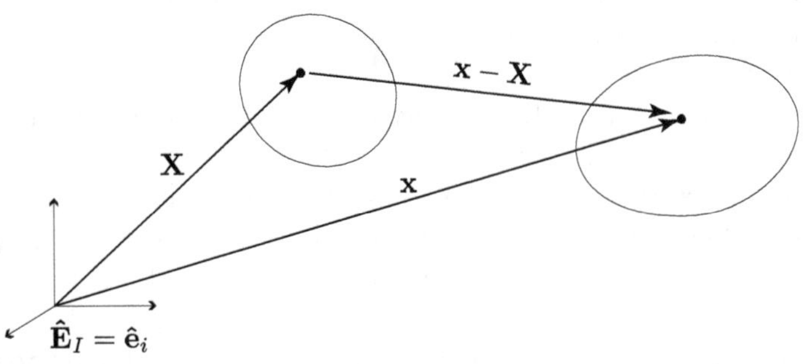

Fig. A.2: Reference and current configurations with coinciding axes, $\hat{\mathbf{E}}_I=\hat{\mathbf{e}}_i$, where $I,i=1,2,3$: If there is no deformation or rotation, $\mathbf{X}=\mathbf{x}$, otherwise, this equality need not hold.

If we choose the axes of the material and spatial coordinates to coincide for both the reference and current configurations, as illustrated in Figure A.2, we obtain[1]

$$\mathbf{U}(\mathbf{X},t)=\mathbf{x}(\mathbf{X},t)-\mathbf{X} \tag{A.3}$$

and

$$\mathbf{u}(\mathbf{x},t)=\mathbf{x}-\mathbf{X}(\mathbf{x},t)\,. \tag{A.4}$$

In such a case, $\mathbf{b}=\mathbf{0}$, even though this is not explicitly necessary, as indicated by Remark A.2 on page 364. However, the choice of the material

[1] *see also*: Slawinski (2015, Expressions 1.3.8 and 1.3.9)

and spatial coordinate systems to coincide with one another lends itself to ease of operations involving $\mathbf{X}$ and $\mathbf{x}$.

The displacement vector contains translation, rotation and deformation. The separation among these processes is discussed in Appendices A.2 and A.9.

For any given t, $\mathbf{U} = \mathbf{u}$, since $\mathbf{U}$ and $\mathbf{u}$ represent the same geometrical object: the displacement vector. Furthermore, examining expressions (A.3) and (A.4), we see that—for any given t and for the coinciding axes of the material and spatial coordinates—the values of the corresponding components of $\mathbf{U}$ and $\mathbf{u}$ are equal to one another, $U_i(\mathbf{X}, t) = u_i(\mathbf{x}, t)$, for $i = 1, 2, 3$. This is a consequence of the fact that, as illustrated in Figure A.2, the values of the components of the displacement vector, $\mathbf{U} = \mathbf{x} - \mathbf{X} = \mathbf{u}$, are the same in terms in both material and spatial coordinates.[2]

In either the material and spatial descriptions, each point of a continuum has two sets of coordinates. Its material coordinates stay with it during its motion. In other words, each point in the reference configuration is assigned a unique position vector, $\mathbf{X} = \sum_{I=1}^{3} X_I \hat{\mathbf{E}}_I$. Its spatial coordinates change as it moves; at any instant, the location occupied by this material point is assigned a unique position vector, $\mathbf{x} = \sum_{i=1}^{3} x_i \hat{\mathbf{e}}_i$.

Both the material and spatial coordinates discussed in this appendix can be expressed in terms of Cartesian coordinates.[3] Their axes are independent of deformation, as illustrated in Figures A.1 and A.2; they do not deform with the body.

An alternative method is to attach the material coordinates to the material itself, and allow them to deform together with the material, as illustrated in Figure 2.2 on page 26. Consequently, each material point has the same material-coordinate value in any configuration, and the motion is described by the change of a coordinate system. Such coordinates are referred to as the *comoving*, *convected* or *embedded* coordinates. Since the coordinates are attached to material points and deform with the body, the axes of such a system do not remain orthogonal to each other, and a curvilinear system is required.

---

[2]The fact that, for a given t, we have $U_i(\mathbf{X}, t) = u_i(\mathbf{x}, t)$, where $i = 1, 2, 3$, is generally true for finite elasticity. It is distinct from the infinitesimal-elasticity approximation, discussed by Slawinski (2015, Section 1.3.3), where $\mathbf{U}(\mathbf{X}, t) \approx \mathbf{u}(\mathbf{x}, t)$ means that—in examination of infinitesimal displacements—the distinction between the material and spatial coordinates and, hence, between the functions expressed in terms of these coordinates can be neglected.

[3]The use of spherical coordinates is illustrated in Example A.2 on page 368.

To emphasize the importance of the reference and current configurations in studying deformations, let us conclude by citing Ogden (1997, p. 83).

> Our immediate purpose is to analyse the deformation from some reference configuration $\mathcal{B}_0$ of the body B to a generic configuration $\mathcal{B}_t$ described by $\mathbf{x} = \chi(\mathbf{X}, t)$. In particular, we wish to compare the two configurations without requiring knowledge of the intermediate stages in the motion so that the time dependence in $\mathbf{x} = \chi(\mathbf{X}, t)$ is not then needed. We therefore suppress the dependence on t and replace $\mathbf{x} = \chi(\mathbf{X}, t)$ by $\mathbf{x} = \chi(\mathbf{X})$.

Marsden and Hughes (1983, p. 2) state that, in their treatise,

> [t]he words "configuration" and "deformation" are used synonymously.

In this appendix, even though we might not explicitly suppress t in $\mathbf{x}(\mathbf{X}, t)$, we focus our attention on distinct configurations, and ignore the intermediate stages, which are parametrized by t. A transformation between configurations is given by the deformation gradient, wherein time does not appear explicitly.

A.2 Deformation-gradient tensor

The deformation-gradient tensor is the fundamental measure of deformation in continuum mechanics. It originates in the gradient of the displacement-vector field.

Elastic solids, in contrast to fluids, possess a memory of their original state. Thus, to describe a solid, we specify its reference state and measure the difference between its current state and the reference. The deformation-gradient tensor bridges these two states.

Consider the gradient of expression (A.1), with respect to the material coordinates,

$$\nabla_{\mathbf{X}} \mathbf{U}(\mathbf{X}, t) = \nabla_{\mathbf{X}} \mathbf{x}(\mathbf{X}, t) - I\,,$$

where I is the identity tensor; in terms of the Cartesian components, we write

$$\frac{\partial U_I(\mathbf{X}, t)}{\partial X_J} = \frac{\partial x_i(\mathbf{X}, t)}{\partial X_J} - \delta_{iJ}\,, \qquad i, J = 1, 2, 3, \tag{A.5}$$

where δ_{iJ} is the Kronecker tensor.

Remark A.1. The Kronecker tensor and its scalar multiples are the only isotropic second-rank tensors (Jeffreys, 1961a, Chapter VII). In contrast, all scalars, which are zero-rank tensors, are isotropic, and no vectors, which are first-rank tensors, are isotropic. The Levi-Civita tensor, discussed on page 374, and its scalar multiples are the only isotropic third-rank tensors. The isotropy of a fourth-rank tensor, which is an important subject of elasticity and seismology, is discussed in Section 3.2.

In this remark, isotropy of a tensor means its invariance under all rotations in $\mathbb{R}^3$, which means that its symmetry group is $SO(3)$.

Remark A.2. Since $\mathbf{b}(t)$, which is a vector that appears in Figure A.1, is not a function of $\mathbf{X}$, expression (A.5) exhibits its form regardless of $\mathbf{b}(t) = \mathbf{0}$ or $\mathbf{b}(t) \neq \mathbf{0}$. However, it is necessary for the material and spatial coordinate systems to coincide with one another, as illustrated in Figure A.2, for functions and their arguments, $\mathbf{x}(\mathbf{X})$ and $\mathbf{X}(\mathbf{x})$, to be expressed wth respect to the same system. Such approach is commonly used.

The deformation-gradient tensor is

$$F := \nabla_{\mathbf{X}}\mathbf{x}(\mathbf{X}, t)\,, \tag{A.6}$$

which is expressed in material coordinates; its inverse is

$$F^{-1} = \nabla_{\mathbf{x}}\mathbf{X}(\mathbf{x}, t)\,,$$

which is expressed in spatial coordinates. Thus, we can write concisely the displacement-vector gradient and the deformation gradient as

$$\nabla_{\mathbf{X}}\mathbf{U} = F - I \tag{A.7}$$

and

$$\nabla_{\mathbf{x}}\mathbf{u} = I - F^{-1}\,, \tag{A.8}$$

in the material and spatial coordinates, respectively. It also follows that $\nabla_{\mathbf{x}}\mathbf{u} = \nabla_{\mathbf{X}}\mathbf{U}\,F^{-1}$.

Remark A.3. $F = I$ means that there is no deformation. However, according to the terminology used in this book,[4] $F \neq I$ does not necessarily mean that there is deformation, since a rigid-body rotation also entails $F \neq I$. The separation of deformation and rotation—both of which are contained in F—is discussed in Appendix A.9, where we study the polar decomposition of F.

[4]As indicated by quotes on page 398, below, our terminology differs from the one used in such works as Chadwick (1976) and Ogden (1997).

F and F^{-1} are *two-point tensors*: they act on vectors in one configuration to produce vectors in another configuration. In components, we write these tensors as

$$F = \sum_{i=1}^{3}\sum_{J=1}^{3} \frac{\partial x_i}{\partial X_J}\,\hat{\mathbf{e}}_i \otimes \hat{\mathbf{E}}_J\,, \tag{A.9}$$

and

$$F^{-1} = \sum_{I=1}^{3}\sum_{j=1}^{3} \frac{\partial X_I}{\partial x_j}\,\hat{\mathbf{E}}_I \otimes \hat{\mathbf{e}}_j\,,$$

where $\otimes$ denotes a tensor product, which herein results in the basis of a second-rank tensor in $\mathbb{R}^3$.

In the language of differential geometry, $F(\mathbf{X})$ is a linear map,

$$F(\mathbf{X}) \,: T_{\mathbf{X}}\mathcal{B} \to T_{\mathbf{x}}\mathcal{S}\,, \tag{A.10}$$

where $T_{\mathbf{X}}\mathcal{B}$ and $T_{\mathbf{x}}\mathcal{S}$ are the tangent spaces, discussed in Appendix E, to manifolds $\mathcal{B}$ and $\mathcal{S}$, which are, respectively, the reference and current configurations of a body. In this context, we write equation (A.9) as

$$F = \sum_{i=1}^{3}\sum_{J=1}^{3} \frac{\partial x^i}{\partial X^J}\,\hat{\mathbf{e}}_i \otimes \hat{\mathbf{E}}^J\,;$$

herein, $\hat{\mathbf{E}}^J$ is the basis of the cotangent space $T^*_{\mathbf{X}}\mathcal{B}$, which is dual to $\hat{\mathbf{E}}_J$, which means that $\hat{\mathbf{E}}^I(\hat{\mathbf{E}}_J) = \delta^I_J$.

Referring to F, Marsden and Hughes (1983, p. 71) write

> Two-point tensors are natural generalizations of vector fields and one forms over maps. Note that a two-point tensor is regarded as a function of $X \in \mathcal{B}$, and not of $x \in \mathcal{S}$. One can think of a two-point tensor as having two tensor "legs," one in $\mathcal{B}$ and one in $\mathcal{S}$. [...] The operations on tensors generalize naturally to operations on two-point tensors. These include: raising and lowering indices, tensor products, push-forward, pull-back, tilt, and contraction.

As implied by the second part of this quote, two-point tensors are not standard objects among tensors. In other words, two-point is an adjective distinguishing them from, not specifying them within, standard objects among tensors. Referring to F and F^{-1}, Maugin (1993, p. 46) writes:

> These are ***not*** tensors *per se*, but geometrical objects defined on two configurations or two bases [...].

In subsequent discussions of finite deformations, operations on two-point tensors can be avoided by introducing—in terms of the deformation-gradient tensor, which is a two-point tensor—such entities as the Cauchy-Green strain tensors, discussed in Appendix A.4, and the Second Piola-Kirchhoff and Cauchy stress tensors, discussed in Appendix B.2; these tensors, in contrast to F, are contained in a single configuration.[5] Referring to deformations, Rymarz (1993, p. 41) writes that

> [o]pis stanu deformacji za pomocą tensora rozdwojonego [...] jest ogólnie niewygodny (obliczeniowo). Dlatego praktycznie do opisu stanu deformacji stosuje się kwadratowe tensory deformacji wyrażone przez kwadraty nitek materialnych $\mathbf{dx}$, $\mathbf{dX}$ [...][6]

Also, two-point tensors disappear under the assumption of infinitesimal deformations, discussed in Appendices A.10 and B.3, for which the distinction between the reference and current configurations is negligible.

Remark A.4. F can be viewed as a $(1,1)$ tensor, whose components are F^i_J. Notably, a two-point tensor cannot be symmetric, because its upper and lower indices are not interchangeable. In orthogonal coordinate systems, however, we can move indices up or down and, hence, F can be viewed as just a second-rank tensor, whose components are F_{iJ}, which is the convention used in Appendices A–C. However, $F_{iJ} = F_{Ji}$ and $F_{iJ} \neq F_{Ji}$, which are statements that refer to their symmetry, remain meaningless. In general coordinates, discussed in Appendices E and F, a second-rank tensor is distinguished as belonging to the $(2,0)$, $(1,1)$ or $(0,2)$ type, where the numbers refer to, respectively, the contravariant and covariant ranks of a tensor. As discussed in Appendix E.3, under coordinate transformations, the $(1,1)$ type tensors are subject to mixed transformations.

[5] Readers interested in historical developments of continuum mechanics and in contributions of such scientists as Almansi, Beltrami, Jacob Bernoulli, Cauchy, Clausius, Duhem, Gibbs, Green, von Helmholtz, Kirchhoff, Lagrange, Lamé, Mohr, Nanson, Piola, Planck, de St. Venant, whose names are attached to concepts discussed in these appendices, might refer to Maugin (2013, 2014).

[6] generally, a description of the state of deformation by a two-point tensor is inconvenient (computationally). Thus, practically, to describe the state of deformation, one uses quadratic strain tensors, which are expressed by squares of material fibers, $\mathbf{dx}$, $\mathbf{dX}$ [...]

Expressing the deformation-gradient components as a 3×3 matrix, we write

$$F(\mathbf{X}, t) = \begin{bmatrix} \dfrac{\partial x_1}{\partial X_1} & \dfrac{\partial x_1}{\partial X_2} & \dfrac{\partial x_1}{\partial X_3} \\ \\ \dfrac{\partial x_2}{\partial X_1} & \dfrac{\partial x_2}{\partial X_2} & \dfrac{\partial x_2}{\partial X_3} \\ \\ \dfrac{\partial x_3}{\partial X_1} & \dfrac{\partial x_3}{\partial X_2} & \dfrac{\partial x_3}{\partial X_3} \end{bmatrix}, \tag{A.11}$$

where each entry is a coefficient associated with its basis,

$$\hat{\mathbf{e}}_1 \otimes \hat{\mathbf{E}}_1 = \begin{bmatrix} 1 \\ 0 \\ 0 \end{bmatrix} [1, 0, 0] = \begin{bmatrix} 1\,0\,0 \\ 0\,0\,0 \\ 0\,0\,0 \end{bmatrix}, \ldots, \hat{\mathbf{e}}_3 \otimes \hat{\mathbf{E}}_3 = \begin{bmatrix} 0 \\ 0 \\ 1 \end{bmatrix} [0, 0, 1] = \begin{bmatrix} 0\,0\,0 \\ 0\,0\,0 \\ 0\,0\,1 \end{bmatrix}.$$

F is a Jacobi matrix, discussed on page 28 and illustrated in Exercise 2.2.

In view of expression (A.1), F is a function of both $\mathbf{X}$ and t. Hence, to examine a particular deformation due to F, we need to specify a point, $\mathbf{X}$, in the reference configuration, and time t. If the reference configuration corresponds to $t = 0$, then—at that instant—$F = I$, since there is no deformation or rotation. Thus, the determinant of F must be positive for $t > 0$, since—as discussed in Appendix A.6, below—$\det F$, which is the change in volume of an element of continuum, must obey *local admissibility*. In other words, an element of volume cannot be reduced to zero.

Assuming a homogeneous deformation—which means that F is not a function of $\mathbf{X}$—the deformation gradient maps, for all t, a line element in the reference configuration into the line element in the current configuration, as opposed to an inhomogeneous deformation, illustrated in Figure A.3. In view of expression (A.11), a homogeneous deformation is a con-

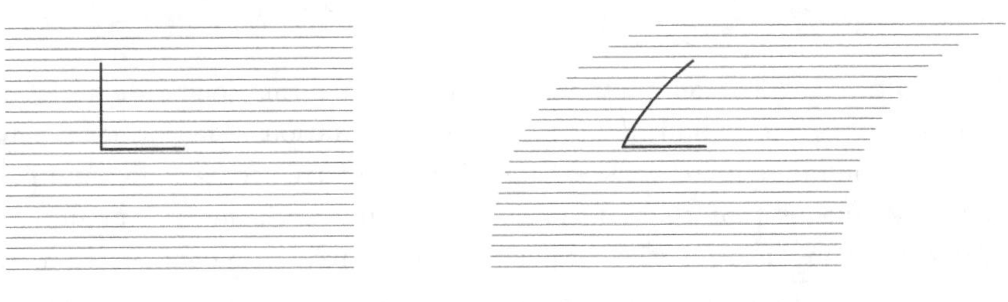

Fig. A.3: Inhomogeneous deformation; its homogeneous counterpart is illustrated in Figure A.9, on page 401.

sequence of a displacement-vector field, $\mathbf{x}(\mathbf{X}, t)$, that is not only linear—but also affine—in $\mathbf{X}$.

$\mathbf{x}(\mathbf{X})$ is a smooth map between the reference and current configurations. The derivative of $\mathbf{x}$, namely, $F(\mathbf{X})$, at a given point, is the best linear approximation of $\mathbf{x}$, near that point; such an approximation is illustrated by expression (A.35), below. This approximation can be viewed as a generalization of the total derivative of ordinary calculus to the calculus on manifolds; it pushes forward tangent vectors of one manifold to tangent vectors of another.

Herein, we have a particularly convenient context, since the maps between the reference and current configurations are diffeomorphisms, which are invertible functions whose inverses are smooth. These maps allow us to transform any differentiable tensor field between these configurations by the pushforward and pullback operations.[7] Matrix (A.11) and its inverse describe the transformation of covariant and contravariant tensors, discussed in Appendix E.

F contains information about both the deformation and the rigid-body rotation of a body. To isolate the information about deformation, we discuss, in Appendix A.9, the polar-decomposition theorem.

Example A.1. For numerical illustrations of discussed entities, let us consider the deformation gradient whose components are

$$F = \begin{bmatrix} 1 & 0.4950 & 0.5000 \\ -0.3330 & 1 & -0.2470 \\ 0.9590 & 0 & 1.5000 \end{bmatrix}; \tag{A.12}$$

it corresponds to a homogeneous deformation evaluated at a particular instant, in a manner similar to the ones illustrated in Exercises 2.1 and 2.2, on page 34. Herein, $\det F = 1.1505$, which means that, at that instant, each element of volume is expanded by about 15%, as discussed in Appendix A.6, below.

Example A.2. [8]To exemplify properties of the deformation gradient, let us consider an isochoric and inhomogeneous deformation. The former means that the volume remains constant, and is also referred to as equivoluminal; the latter means that the deformation gradient—and, hence, the amount of deformation—is a function of position in the reference configuration. In

[7]Readers interested in these operations might refer to Epstein (2010, Sections 4.9 and 4.13).

[8]Example A.2 is based on discussions with Andrey Melnikov regarding Ogden (1997, Problem 2.2.16, pp. 116–117).

this example, in which we consider a spherically symmetric deformation, the isochoric deformation requires inhomogeneity, where the deformation gradient, $F(\mathbf{X})$, must exhibit a specific form. We assume that the reference and current coordinate systems coincide; in other words, $\hat{\mathbf{E}}_I = \hat{\mathbf{e}}_i$, where $I = i = 1, 2, 3$.

To find this form, let us express a spherically symmetric deformation as

$$\mathbf{x} = f(R)\,\mathbf{X}, \tag{A.13}$$

where $R = \|\mathbf{X}\|$ is the magnitude of the radius in the reference coordinates. We write, in spherical coordinates, the deformation gradient, given expression (A.9) in Cartesian coordinates, as

$$F(\mathbf{X}) = \nabla_{\mathbf{X}}\mathbf{x} = \frac{\partial \mathbf{x}}{\partial R} \otimes \hat{\mathbf{E}}_R + \frac{1}{R}\frac{\partial \mathbf{x}}{\partial \Theta} \otimes \hat{\mathbf{E}}_\Theta + \frac{1}{R\sin\Theta}\frac{\partial \mathbf{x}}{\partial \Phi} \otimes \hat{\mathbf{E}}_\Phi, \tag{A.14}$$

where

$$\mathbf{X} = R\,\hat{\mathbf{E}}_R, \tag{A.15}$$

and, hence, expression (A.13) can be written as

$$\mathbf{x} = f(R)\,R\,\hat{\mathbf{E}}_R. \tag{A.16}$$

Taking its derivatives—using the chain rule with $\partial\hat{\mathbf{E}}_R/\partial\Theta = \hat{\mathbf{E}}_\Theta$ and $\partial\hat{\mathbf{E}}_R/\partial\Phi = \sin\Theta\,\hat{\mathbf{E}}_\Phi$—we get

$$\frac{\partial \mathbf{x}}{\partial R} = [f'(R)\,R + f(R)]\,\hat{\mathbf{E}}_R,$$

$$\frac{\partial \mathbf{x}}{\partial \Theta} = f(R)\,R\frac{\partial \hat{\mathbf{E}}_R}{\partial \Theta} = f(R)\,R\,\hat{\mathbf{E}}_\Theta$$

and

$$\frac{\partial \mathbf{x}}{\partial \Phi} = f(R)\,R\frac{\partial \hat{\mathbf{E}}_R}{\partial \Phi} = f(R)\,R\sin\Theta\,\hat{\mathbf{E}}_\Phi.$$

Using these derivatives in expression (A.14), we obtain

$$F(\mathbf{X}) = (f'(R)R + f(R))\,\hat{\mathbf{E}}_R \otimes \hat{\mathbf{E}}_R + f(R)\,\hat{\mathbf{E}}_\Theta \otimes \hat{\mathbf{E}}_\Theta + f(R)\,\hat{\mathbf{E}}_\Phi \otimes \hat{\mathbf{E}}_\Phi. \tag{A.17}$$

Using $\hat{\mathbf{E}}_R \otimes \hat{\mathbf{E}}_R + \hat{\mathbf{E}}_\Theta \otimes \hat{\mathbf{E}}_\Theta + \hat{\mathbf{E}}_\Phi \otimes \hat{\mathbf{E}}_\Phi = \mathbf{I}$ and relation (A.15), we write

$$\begin{aligned} F(\mathbf{X}) &= f'(R)R\,\hat{\mathbf{E}}_R \otimes \hat{\mathbf{E}}_R + f(R)\left(\hat{\mathbf{E}}_R \otimes \hat{\mathbf{E}}_R + \hat{\mathbf{E}}_\Theta \otimes \hat{\mathbf{E}}_\Theta + \hat{\mathbf{E}}_\Phi \otimes \hat{\mathbf{E}}_\Phi\right) \\ &= f'(R)R\left(\frac{1}{R}\mathbf{X}\right) \otimes \left(\frac{1}{R}\mathbf{X}\right) + f(R)\,\mathbf{I} = f'(R)R\,\frac{1}{R^2}\mathbf{X} \otimes \mathbf{X} + f(R)\,\mathbf{I} \\ &= \frac{1}{R}f'(R)\,\mathbf{X} \otimes \mathbf{X} + f(R)\,\mathbf{I}, \end{aligned} \tag{A.18}$$

which is the deformation gradient for a spherically symmetric isochoric deformation.

To examine properties of such a deformation, let us write

$$F = \begin{bmatrix} f'(R) + f(R) & 0 & 0 \\ 0 & f(R) & 0 \\ 0 & 0 & f(R) \end{bmatrix}. \tag{A.19}$$

Examining this matrix, we identify the principal stretches, which are $\lambda_r = f'(R)R + f(R)$ and $\lambda_\theta = \lambda_\phi = f(R)$. Since these stretches depend on R, it is an inhomogeneous deformation. Let us denote λ_θ and λ_ϕ by λ, which is the tangential stretch, $2\pi r/2\pi R$; hence, $\lambda = r/R$, where r is the radius in the current configuration. Since the deformation is isochoric, $\lambda_r\lambda^2 = 1$, and, hence, $\lambda_r = \lambda^{-2}$, which—in curvilinear orthonormal coordinates—is equivalent to the determinant being equal to unity,

$$\det F = (f'(R)R + f(R))\, f^2(R) = 1\,. \tag{A.20}$$

This condition is equivalent to

$$f'(R) + R^{-1}f(R) = R^{-1}f^{-2}(R)\,, \tag{A.21}$$

which is a Bernoulli equation, named in acknowledgement of contributions of Jacob Bernoulli, who discussed it in the last decade of the seventeenth century. In general, Bernoulli equations,

$$y'(x) + p(x)y(x) = q(x)y^n(x)\,,$$

where $n \in \mathbb{R}$, are nonlinear—for $n \neq 0, 1$—differential equations whose exact solutions are known.

To solve equation (A.21), we let

$$u(R) = f^3(R)\,, \tag{A.22}$$

which implies that $f(R) = \sqrt[3]{u}$ and $f'(R) = \frac{1}{3}u^{-\frac{2}{3}}u'(R)$. Therefore, equation (A.21) can be written as

$$u'(R) + 3\,R^{-1}\,u(R) = 3\,R^{-1}\,.$$

Multiplying both sides by R^3, we obtain

$$R^3\,u'(R) + 3\,R^2\,u(R) = 3\,R^2\,,$$

which is equivalent to

$$\frac{\mathrm{d}}{\mathrm{dR}}\left(R^3u(R)\right) = 3\,R^2\,.$$

Integrating both sides and solving for u, we obtain

$$u(R) = 1 + \frac{C}{R^3},$$

where C is the integration constant. Invoking substitution (A.22), we obtain

$$f(R) = \sqrt[3]{1 + \frac{C}{R^3}}. \tag{A.23}$$

Let us verify that, as required for isochoric deformations, expression (A.23) results—for all C—in $\det F = 1$, which is condition (A.20). Taking the derivative of expression (A.23), we obtain

$$f'(R) = \frac{1}{3}\left(1 + \frac{C}{R^3}\right)^{-\frac{2}{3}}\left(-3\frac{C}{R^4}\right) = -\frac{C}{R^4}\left(1 + \frac{C}{R^3}\right)^{-\frac{2}{3}}.$$

Inserting this result into expression (A.20), we obtain

$$\left(-\frac{C}{R^3}\left(1 + \frac{C}{R^3}\right)^{-\frac{2}{3}} + \left(1 + \frac{C}{R^3}\right)^{\frac{1}{3}}\right)\left(1 + \frac{C}{R^3}\right)^{\frac{2}{3}} = -\frac{C}{R^3} + 1 + \frac{C}{R^3} = 1,$$

as required. In other words, expression (A.23) is a solution of equation (A.21).

Let us find the expression of C. The initial volume of a spherical shell, in the reference configuration, is $\frac{4}{3}\pi\left(B^3 - A^3\right)$, where A and B are its inner and outer radii. Under a spherically symmetric deformation, in the current configuration, this volume is $\frac{4}{3}\pi\left(b^3 - a^3\right)$, where a and b are the inner and outer radii. To preserve the volume, between the initial and final configurations, we require—in accordance with an isochoric deformation—that $B^3 - A^3 = b^3 - a^3$, which means that

$$b^3 - B^3 = a^3 - A^3 \tag{A.24}$$

and, for the intermediate radii, we require that

$$r^3 - R^3 = a^3 - A^3,$$

where $a \leqslant r \leqslant b$ and $A \leqslant R \leqslant B$. Solving the latter expression for r^3, dividing by R^3 and taking the cubic root, we obtain

$$\frac{r}{R} = \sqrt[3]{1 + \frac{a^3 - A^3}{R^3}},$$

Comparing its right-hand side with solution (A.23), we conclude that

$$f(R) = \frac{r}{R} \tag{A.25}$$

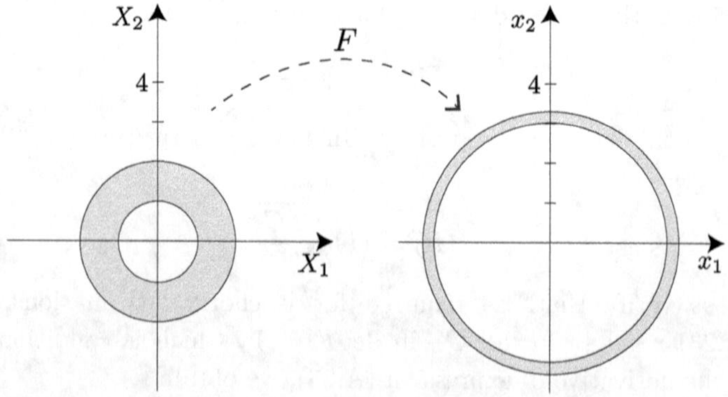

Fig. A.4: A cross-section illustrating a deformation of a spherical shell: its original radii, $A = 1$ and $B = 2$, become $a = 3$ and $b = 3.2396$; the last value is obtained using expression (A.24). Under this deformation, the volume—due to the assumption of an isochoric deformation—is preserved; it is $28\pi/3$. Even though the reference-configuration and the current-configuration coordinates are shown separately, the position of the centre of the spherical shell is not affected by the deformation; F results only in the radial displacement of each point $\mathbf{X}$ to its current location $\mathbf{x}$.

and that $C = a^3 - A^3$, which is a consequence of a boundary condition of equation (A.25), namely, $f(A) = a/A$. Also, in the context of matrix (A.19), we see that $f(R) = r/R = \lambda$ and, hence, using solution (A.23), we obtain $\lambda_r = f'(R)R + f(R) = \lambda^{-2}$, which is in agreement with condition (A.20), as required.

Let us examine further properties of a spherically symmetric isochoric deformation. Dividing both sides of equation (A.24) by A^3, we obtain

$$\left(\frac{B}{A}\right)^3 (\lambda_b^3 - 1) = \lambda_a^3 - 1, \tag{A.26}$$

where we denote a/A by λ_a and b/B by λ_b. Thus, for the case of an expanding shell, we have $\lambda_a > \lambda_b > 1$, which means that circumferential stretch at the inner boundary is greater than at the outer boundary, as illustrated in Figure A.4. For the case of a contracting shell, we have $\lambda_a < \lambda_b < 1$, which means that circumferential stretch at the inner boundary is smaller than at the outer boundary. Also, let us write expression (A.25) as

$$r = f(R)\,R,$$

which implies that

$$\mathrm{d}r = f'(R)\,R\,\mathrm{d}R + f(R)\,\mathrm{d}R = (f'(R)\,R + f(R))\,\mathrm{d}R;$$

in other words, the material line element of length $\mathrm{d}R$ aligned along the radial coordinate is stretched by $f'(R)\,R + f(R)$.

A.3 Compatibility of deformation gradient

A gradient of a vector field, which is a second-rank tensor field, is referred to as *compatible* if it allows us to obtain this vector field. For linear elasticity, the condition of compatibility was formulated at the end of the nineteenth century by Eugenio Beltrami. Let us discuss it.

Applying the gradient operator to a continuously differentiable vector field results in a unique second-rank tensor. The compatibility condition ensures that—given a tensor field that is a gradient of a vector field—we can infer this vector field. Since, in $\mathbb{R}^3$, a vector field has three components and a second-rank tensor has nine—to obtain uniquely the former from the latter—we need constraining relations among tensor components. Otherwise, we have an overdetermined system, which consists of nine partial differential equations—one for each component, $\partial x_i/\partial X_J$, where $i, J = 1, 2, 3$—for three unknowns, x_i, where $i = 1, 2, 3$.

Remark A.5. Consider the following analogy. Applying the gradient operator to a scalar field results in a unique vector field. Since a scalar field has one component and, in $\mathbb{R}^3$, the vector field has three, to obtain the former from the latter, we have to solve—for one unknown—a system that consists of three equations.

In general, given v_1, v_2, v_3, system

$$v_1 = \frac{\partial f}{\partial x_1}, \qquad v_2 = \frac{\partial g}{\partial x_2}, \qquad v_3 = \frac{\partial h}{\partial x_3}$$

is uniquely soluble—up to a constant—if and only if it is constrained by $f = g = h =: \varphi$, which means that $[v_1, v_2, v_3] = \nabla\varphi$. In other words, only a conservative field allows us to obtain the scalar field from its gradient. This is a tautology, since only gradients of scalar fields are conservative. However, in general, a vector field need not be the gradient of a scalar field, and such a vector field is not compatible, in the sense discussed in this appendix.

For finite strains in a simply connected body, the necessary and sufficient condition for a unique solution is

$$\nabla \times F = 0\,, \tag{A.27}$$

photos: Elena Patarini

Fig. A.5: A medal to commemorate Tullio Levi-Civita and Gregorio Ricci-Curbastro, professors of University of Padua, and their contributions to tensor calculus, which is instrumental in continuum mechanics and, hence, theoretical seismology.
This medal celebrates the World Year of Physics, established in recognition of the century since three key contributions were published by Einstein in 1905: on *photoelectric effect*, on *Brownian motion* and on special relativity. The last subject is a generalization of *Galilean relativity* and Galilean spacetime, discussed in Section 2.2.1; notably, Galileo was also a professor of University of Padua.

where F is the deformation gradient, discussed in Appendix A.2; herein, for conciseness, we let ∇ stand for $\nabla_{\mathbf{X}}$. Using the Levi-Civita symbol—named in honour of an Italian mathematical physicist, Tullio Levi-Civita, whose achievements are commemorated by the medal illustrated in Figure A.5—we write the left-hand side of condition (A.27) as

$$\nabla \times F = \sum_{I=1}^{3} \sum_{j=1}^{3} \epsilon_{Ijm} \frac{\partial F_{jK}}{\partial X_I} \hat{\mathbf{e}}_m \otimes \hat{\mathbf{E}}_K \,, \tag{A.28}$$

which is a second-rank two-point tensor, whose Cartesian components are $(\nabla \times F)_{mK}$. Explicitly, these components are

$$\begin{bmatrix} \frac{\partial}{\partial X_1}\left(\frac{\partial x_3}{\partial X_2}-\frac{\partial x_2}{\partial X_3}\right) & \frac{\partial}{\partial X_2}\left(\frac{\partial x_3}{\partial X_2}-\frac{\partial x_2}{\partial X_3}\right) & \frac{\partial}{\partial X_3}\left(\frac{\partial x_3}{\partial X_2}-\frac{\partial x_2}{\partial X_3}\right) \\ \frac{\partial}{\partial X_1}\left(\frac{\partial x_1}{\partial X_3}-\frac{\partial x_3}{\partial X_1}\right) & \frac{\partial}{\partial X_2}\left(\frac{\partial x_1}{\partial X_3}-\frac{\partial x_3}{\partial X_1}\right) & \frac{\partial}{\partial X_3}\left(\frac{\partial x_1}{\partial X_3}-\frac{\partial x_3}{\partial X_1}\right) \\ \frac{\partial}{\partial X_1}\left(\frac{\partial x_2}{\partial X_1}-\frac{\partial x_1}{\partial X_2}\right) & \frac{\partial}{\partial X_2}\left(\frac{\partial x_2}{\partial X_1}-\frac{\partial x_1}{\partial X_2}\right) & \frac{\partial}{\partial X_3}\left(\frac{\partial x_2}{\partial X_1}-\frac{\partial x_1}{\partial X_2}\right) \end{bmatrix}. \tag{A.29}$$

Condition (A.27) is satisfied identically for homogeneous deformations, since all terms in parentheses are constant if F does not depend on $\mathbf{X}$, which means that $\mathbf{x}(\mathbf{X})$ is a linear function of $\mathbf{X}$. Such deformations are, for instance, a simple extension, illustrated in Exercise 2.1, and a simple shear, illustrated in Exercise 2.2, where expressions (2.17) and (2.21), respectively, are linear in $\mathbf{X}$. Otherwise, in view of Proposition A.1, this condition is satisfied if $F = F^T$, which means—in view of expression (A.29)—that $\partial x_i/\partial X_J = \partial x_j/\partial X_I$, but does not mean—in view of Remark A.4—that F is a symmetric tensor. Also, $F = F^T$ is the case of a uniform dilation, illustrated in Example A.2, which is spherically symmetric.

Proposition A.1. *Given an at least twice continuously differentiable single-valued vector field,* $\mathbf{x}(\mathbf{X})$, *and its gradient,* $F = \nabla\mathbf{x}$ *—where we assume that this gradient exhibits the equality of mixed partial derivatives—it follows that*

$$\nabla \times (\nabla\mathbf{x})^T = 0\,. \tag{A.30}$$

Proof. Using the Levi-Civita symbol, we write equation (A.30) as

$$\nabla \times F^T = \sum_{I=1}^{3}\sum_{j=1}^{3} \epsilon_{Ijm}\frac{\partial F_{Kj}}{\partial X_I}\,\hat{\mathbf{e}}_m \otimes \hat{\mathbf{E}}_K = 0\,, \tag{A.31}$$

which is a second-rank two-point tensor, whose Cartesian components are $(\nabla \times F^T)_{Km}$. The components of tensor (A.31) are

$$\begin{bmatrix} \frac{\partial^2 x_1}{\partial X_2\partial X_3}-\frac{\partial^2 x_1}{\partial X_3\partial X_2} & \frac{\partial^2 x_2}{\partial X_2\partial X_3}-\frac{\partial^2 x_2}{\partial X_3\partial X_2} & \frac{\partial^2 x_3}{\partial X_2\partial X_3}-\frac{\partial^2 x_3}{\partial X_3\partial X_2} \\ \frac{\partial^2 x_1}{\partial X_1\partial X_3}-\frac{\partial^2 x_1}{\partial X_3\partial X_1} & \frac{\partial^2 x_2}{\partial X_1\partial X_3}-\frac{\partial^2 x_2}{\partial X_3\partial X_1} & \frac{\partial^2 x_3}{\partial X_1\partial X_3}-\frac{\partial^2 x_3}{\partial X_3\partial X_1} \\ \frac{\partial^2 x_1}{\partial X_1\partial X_2}-\frac{\partial^2 x_1}{\partial X_2\partial X_1} & \frac{\partial^2 x_2}{\partial X_1\partial X_2}-\frac{\partial^2 x_2}{\partial X_2\partial X_1} & \frac{\partial^2 x_3}{\partial X_1\partial X_2}-\frac{\partial^2 x_3}{\partial X_2\partial X_1} \end{bmatrix}. \tag{A.32}$$

We see that—under the equality of mixed partial derivatives—all components are identically zero, as required. □

Examining expressions (A.28) and (A.31), we see that $(F_{jK})^T = F_{Kj}$; in other words, $(F^T)^T = F$. However, $(\nabla \times F^T)^T \neq \nabla \times F$.

In orthonormal coordinate systems, F and F^T are duals of one another. They are linear maps between the reference configuration and a current configuration. In the particular case of F being an orthogonal transformation, $FF^T = I$, they are inverses of one another.

To gain an insight into expression (A.29), let us consider the following lemma.

Lemma A.1. *Given an at least twice continuously differentiable single-valued vector field, $\mathbf{x}(\mathbf{X})$, it follows that*

$$\nabla \times (\nabla \mathbf{x}) = \nabla(\nabla \times \mathbf{x}) . \tag{A.33}$$

Proof. The left-hand side of equation (A.33) is given by expression (A.29). Let us examine its right-hand side. Using the Levi-Civita symbol, we write the Cartesian components of $\nabla \times \mathbf{x}$ as

$$(\nabla \times \mathbf{x})_i = \sum_{j=1}^{3} \sum_{K=1}^{3} \epsilon_{ijK} \frac{\partial x_j}{\partial X_K} , \qquad i = 1, 2, 3 .$$

Explicitly, we write these components as

$$\left[\frac{\partial x_3}{\partial X_2} - \frac{\partial x_2}{\partial X_3}, \frac{\partial x_1}{\partial X_3} - \frac{\partial x_3}{\partial X_1}, \frac{\partial x_2}{\partial X_1} - \frac{\partial x_1}{\partial X_2} \right]^T .$$

Hence,

$$\nabla(\nabla \times \mathbf{x}) =$$

$$\begin{bmatrix} \frac{\partial}{\partial X_1}\left(\frac{\partial x_3}{\partial X_2} - \frac{\partial x_2}{\partial X_3}\right) & \frac{\partial}{\partial X_2}\left(\frac{\partial x_3}{\partial X_2} - \frac{\partial x_2}{\partial X_3}\right) & \frac{\partial}{\partial X_3}\left(\frac{\partial x_3}{\partial X_2} - \frac{\partial x_2}{\partial X_3}\right) \\ \frac{\partial}{\partial X_1}\left(\frac{\partial x_1}{\partial X_3} - \frac{\partial x_3}{\partial X_1}\right) & \frac{\partial}{\partial X_2}\left(\frac{\partial x_1}{\partial X_3} - \frac{\partial x_3}{\partial X_1}\right) & \frac{\partial}{\partial X_3}\left(\frac{\partial x_1}{\partial X_3} - \frac{\partial x_3}{\partial X_1}\right) \\ \frac{\partial}{\partial X_1}\left(\frac{\partial x_2}{\partial X_1} - \frac{\partial x_1}{\partial X_2}\right) & \frac{\partial}{\partial X_2}\left(\frac{\partial x_2}{\partial X_1} - \frac{\partial x_1}{\partial X_2}\right) & \frac{\partial}{\partial X_3}\left(\frac{\partial x_2}{\partial X_1} - \frac{\partial x_1}{\partial X_2}\right) \end{bmatrix} ,$$

which is expression (A.29), as required. □

Lemma A.1 states the commutativity of differential operators, which is a consequence of $\mathbf{x}$ being a vector field, whose derivative results in a tensor of a rank that allows for the subsequent operation. Equation (A.33) does not hold for a scalar field.

A.4 Cauchy-Green strain tensors

A.4.1 *Right Cauchy-Green strain tensor*

Consider two line elements, $\mathrm{d}\mathbf{X}^{(1)}$ and $\mathrm{d}\mathbf{X}^{(2)}$, mapped into $\mathrm{d}\mathbf{x}^{(1)}$ and $\mathrm{d}\mathbf{x}^{(2)}$, respectively. To examine the change of their lengths and of the angle between them, we consider their scalar products,

$$\mathrm{d}\mathbf{x}^{(1)} \cdot \mathrm{d}\mathbf{x}^{(2)} = \left(F\,\mathrm{d}\mathbf{X}^{(1)}\right) \cdot \left(F\,\mathrm{d}\mathbf{X}^{(2)}\right), \tag{A.34}$$

where terms in parentheses are actions of F, which is illustrated in Figure A.6. In components, we can write this action as $\mathrm{d}x_i = \sum_{J=1}^{3} F_{iJ}\,\mathrm{d}X_J$ and, invoking the definition of F, as $\mathrm{d}x_i = \sum_{J=1}^{3} \partial x_i/\partial X_J\,\mathrm{d}X_J$.

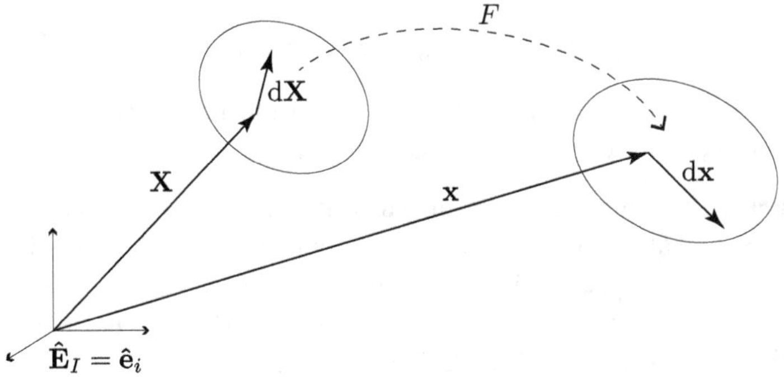

Fig. A.6: Deformation gradient acting on line element

Remark A.6. The relation between $\mathrm{d}\mathbf{x}$ and $\mathrm{d}\mathbf{X}$ is a pushforward, $\mathrm{d}\mathbf{x} = \phi_*[\mathrm{d}\mathbf{X}]$.

Example A.3. Using expression (A.12), which corresponds to a homogeneous deformation, we can calculate the action of F on vector $\mathbf{X} = [1,1,1]$. Its result is $\mathbf{x} = [1.9950, 0.4200, 2.4590]$. Both the length and orientation are affected by F.

Remark A.7. Note that the action of F, namely,

$$\mathrm{d}x_i = \sum_{J=1}^{3} \frac{\partial x_i(\mathbf{X},t)}{\partial X_J}\,\mathrm{d}X_J\,, \qquad i = 1,2,3\,,$$

is an infinitesimal quantity, and, hence, an approximation of

$$\Delta x_i = \sum_{J=1}^{3} \frac{\partial x_i(\mathbf{X},t)}{\partial X_J}\,\Delta X_J + \ldots\,, \qquad i = 1,2,3\,, \tag{A.35}$$

The terms of order $(\Delta \mathbf{X})^2$ and higher are neglected in the Taylor-series expansion represented by the right-hand side of expression (A.35). It is assumed that these terms tend to zero as the differential, $\mathrm{d}\mathbf{X}$, tends to zero. Thus, the deformation gradient—whose components are finite—quantifies these deformations in the neighbourhood of point $\mathbf{X}$; it maps an infinitesimal line element, $\mathrm{d}\mathbf{X}$, attached to $\mathbf{X}$, in the reference configuration, to the infinitesimal line element $\mathrm{d}\mathbf{x}$, attached to $\mathbf{x}$ in the current configuration.

In general, the transpose of a second-rank tensor, S, denoted by S^T, is defined by its property $(S\,\mathbf{w})\cdot\mathbf{v} = \mathbf{w}\cdot(S^T\,\mathbf{v})$, for all vectors, $\mathbf{v}$ and $\mathbf{w}$. Letting—in expression (A.34)—$\mathbf{v} := F\,\mathrm{d}\mathbf{X}^{(1)}$, $\mathbf{w} := \mathrm{d}\mathbf{X}^{(2)}$ and $S := F$, we can write its right-hand side as $(S\,\mathbf{w})\cdot\mathbf{v}$; rewriting it as $\mathbf{w}\cdot(S^T\,\mathbf{v})$, we obtain

$$\mathrm{d}\mathbf{x}^{(1)}\cdot\mathrm{d}\mathbf{x}^{(2)} = \mathrm{d}\mathbf{X}^{(2)}\cdot\left(F^T F\,\mathrm{d}\mathbf{X}^{(1)}\right); \tag{A.36}$$

herein—due to the symmetry of $F^T F$—the right-hand side remains invariant under the interchange of the superscripts, $(1)\leftrightarrow(2)$.

Remark A.8. Operations relating the right-hand sides of equations (A.34) and (A.36) treat both the reference and current configurations within the same $\mathbb{R}^3$. Otherwise, if in accordance with expression (A.10), $F(\mathbf{X}) : T_{\mathbf{X}}\mathcal{B} \to T_{\mathbf{x}}\mathcal{S}$, we need the inner product in both tangent spaces to obtain $(F(\mathbf{X}))^T : T_{\mathbf{x}}\mathcal{S} \to T_{\mathbf{X}}\mathcal{B}$, which is defined by

$$\left(F(\mathbf{X})^T\,\mathbf{w}\right)\cdot\mathbf{v} = \mathbf{w}\cdot\left(F(\mathbf{X})\,\mathbf{v}\right),$$

for any $\mathbf{v}\in T_{\mathbf{X}}\mathcal{B}$ and $\mathbf{w}\in T_{\mathbf{x}}\mathcal{S}$.

We refer to $F^T F$, which—as shown, respectively, in Lemma A.2 and Lemma A.3, below—is a symmetric and positive-definite second-rank tensor, as the *right Cauchy-Green strain tensor*, and denote it by C. Its inverse is called the *Piola strain tensor*.

Explicitly, we write

$$C := F^T F; \tag{A.37}$$

its components are

$$C_{IJ} = \sum_{k=1}^{3} F_{kI}F_{kJ}, \qquad I, J = 1, 2, 3, \tag{A.38}$$

which means that C is contained in the reference configuration, since its basis is $\hat{\mathbf{E}}_I \otimes \hat{\mathbf{E}}_J$. In other words,

$$C = \sum_{I=1}^{3}\sum_{J=1}^{3}\left(\sum_{k=1}^{3} F_{kI}F_{kJ}\right)\hat{\mathbf{E}}_I \otimes \hat{\mathbf{E}}_J. \tag{A.39}$$

Remark A.9. To show a derivation of expression (A.39), we recall expression (A.9),

$$F = \sum_{i=1}^{3} \sum_{J=1}^{3} \frac{\partial x_i}{\partial X_J} \hat{\mathbf{e}}_i \otimes \hat{\mathbf{E}}_J \,,$$

and, using the fact that, in general, $(\mathbf{a} \otimes \mathbf{b})^T = \mathbf{b} \otimes \mathbf{a}$, we write

$$F^T = \sum_{i=1}^{3} \sum_{J=1}^{3} \frac{\partial x_i}{\partial X_J} \hat{\mathbf{E}}_J \otimes \hat{\mathbf{e}}_i \,.$$

Following expression (A.37) and renaming the indices, we write

$$C = \left(\sum_{k=1}^{3} \sum_{I=1}^{3} \frac{\partial x_k}{\partial X_I} \hat{\mathbf{E}}_I \otimes \hat{\mathbf{e}}_k \right) \left(\sum_{\ell=1}^{3} \sum_{J=1}^{3} \frac{\partial x_\ell}{\partial X_J} \hat{\mathbf{e}}_\ell \otimes \hat{\mathbf{E}}_J \right) .$$

Since $(\mathbf{a} \otimes \mathbf{b})(\mathbf{c} \otimes \mathbf{d}) = (\mathbf{b} \cdot \mathbf{c})(\mathbf{a} \otimes \mathbf{d})$, and according to properties of orthonormal bases, $\hat{\mathbf{e}}_k \cdot \hat{\mathbf{e}}_\ell = \delta_{k\ell}$, we get

$$C = \sum_{k=1}^{3} \sum_{I=1}^{3} \sum_{\ell=1}^{3} \sum_{J=1}^{3} \frac{\partial x_k}{\partial X_I} \frac{\partial x_\ell}{\partial X_J} \delta_{k\ell} \hat{\mathbf{E}}_I \otimes \hat{\mathbf{E}}_J \,.$$

Using properties of the Kronecker delta, we obtain

$$C = \sum_{k=1}^{3} \sum_{I=1}^{3} \sum_{J=1}^{3} \frac{\partial x_k}{\partial X_I} \frac{\partial x_k}{\partial X_J} \hat{\mathbf{E}}_I \otimes \hat{\mathbf{E}}_J \,,$$

which is equivalent to expression (A.39).

Example A.4. For the case of tensor (A.12), we have

$$C = \begin{bmatrix} 2.0306 & 0.1620 & 2.0208 \\ 0.1620 & 1.2450 & 0.0005 \\ 2.0208 & 0.0005 & 2.5610 \end{bmatrix} , \tag{A.40}$$

which is a symmetric positive-definite tensor; its eigenvalues are 4.3376, 1.2561 and 0.2429.

Lemma A.2. *C is symmetric.*

Proof. To prove symmetry of C, it suffices to verify that $C^T = C$. Using expression (A.37), we write

$$C^T = \left(F^T F\right)^T = F^T \left(F^T\right)^T = F^T F = C \,.$$

as required. □

Remark A.10. In general, the symmetry is a property entailed by a product of a matrix and its transpose.

Lemma A.3. *C is positive-definite.*

Proof. To prove the positive definiteness of C, it suffices to verify that $\mathrm{d}\mathbf{X} \cdot (C\,\mathrm{d}\mathbf{X}) > 0$, for $\mathrm{d}\mathbf{X} \neq \mathbf{0}$. Using expression (A.37), we write

$$\mathrm{d}\mathbf{X} \cdot \left(F^T F\right) \mathrm{d}\mathbf{X} = (F\,\mathrm{d}\mathbf{X}) \cdot (F\,\mathrm{d}\mathbf{X}) = \|F\,\mathrm{d}\mathbf{X}\|^2 > 0\,, \qquad \forall\, \mathrm{d}\mathbf{X} \neq \mathbf{0}\,,$$

as required. □

Remark A.11. For the first equality in the proof of Lemma A.3, we invoke the identity used to obtain expression (A.36), namely, $\mathbf{w} \cdot S^T \mathbf{v} = S\,\mathbf{w} \cdot \mathbf{v}$. Herein,

$$\underbrace{\mathrm{d}\mathbf{X}}_{\mathbf{w}} \cdot (\underbrace{F^T}_{S^T} \underbrace{F)\,\mathrm{d}\mathbf{X}}_{\mathbf{v}} = (\underbrace{F}_{S} \underbrace{\mathrm{d}\mathbf{X}}_{\mathbf{w}}) \cdot \underbrace{(F\,\mathrm{d}\mathbf{X})}_{\mathbf{v}}\,. \tag{A.41}$$

Also, in the proof of Lemma A.3, we implicitly assume that $F\,\mathrm{d}\mathbf{X} \neq \mathbf{0}$, which means that $\mathrm{d}\mathbf{X}$ does not disappear upon deformation. In other words, a nonzero line element, $\mathrm{d}\mathbf{X} \neq \mathbf{0}$, in the reference configuration remains nonzero, $F\,\mathrm{d}\mathbf{X} = \mathrm{d}\mathbf{x} \neq \mathbf{0}$, in a current configuration, which is required by a physical consideration, namely, conservation of mass. Mathematically, $S = M^T M$, where M is an arbitrary second-rank tensor, results in a positive-semidefinite tensor, S, since, in general, $\mathbf{v} \cdot \left(M^T M\right) \mathbf{v} \geqslant 0$, for all $\mathbf{v} \neq \mathbf{0}$.

Remark A.12. In general, the positive definiteness is a property entailed by a product of a nonsingular matrix and its transpose.

A.4.2 *Left Cauchy-Green strain tensor*

In a manner similar to expression (A.34), consider

$$\mathrm{d}\mathbf{X}^{(1)} \cdot \mathrm{d}\mathbf{X}^{(2)} = \left(F^{-1}\,\mathrm{d}\mathbf{x}^{(1)}\right) \cdot \left(F^{-1}\,\mathrm{d}\mathbf{x}^{(2)}\right).$$

Since, in general, for the second-rank tensors, A and B, we have $(AB)^{-1} = B^{-1}A^{-1}$ and $A^{-T} := (A^{-1})^T = (A^T)^{-1}$, using the identity invoked in expression (A.41), we can rewrite the right-hand side as

$$\mathrm{d}\mathbf{X}^{(1)} \cdot \mathrm{d}\mathbf{X}^{(2)} = \mathrm{d}\mathbf{x}^{(1)} \cdot \left(F^{-T} F^{-1} \mathbf{dx}^{(2)}\right) =: \mathrm{d}\mathbf{x}^{(1)} \cdot \left(b^{-1} \mathbf{dx}^{(2)}\right). \tag{A.42}$$

Herein, b is a symmetric and positive-definite second-rank tensor, which we refer to as the *left Cauchy-Green strain tensor*; its inverse is called the *Cauchy strain tensor*.

The symmetry of b^{-1} could be proven in a manner analogous to the one presented in the proof of Lemma A.2. Thus, since b^{-1} is symmetric, so is b. To prove the positive definiteness of b, by the approach similar to the one presented in the proof of Lemma A.3, we follow the right-hand side of equation (A.42) to write

$$\mathrm{d}\mathbf{x} \cdot \left(b^{-1}\mathrm{d}\mathbf{x}\right) = \left(F^{-1}\,\mathrm{d}\mathbf{x}\right) \cdot \left(F^{-1}\mathrm{d}\mathbf{x}\right) = \|F^{-1}\mathrm{d}\mathbf{x}\|^2 > 0\,, \qquad \forall\ \mathrm{d}\mathbf{x} \neq \mathbf{0}\,, \tag{A.43}$$

where, as discussed in Remark A.11, to obtain a strict inequality, we assume that $F^{-1}\mathrm{d}\mathbf{x} \neq \mathbf{0}$. Thus, b^{-1} is positive-definite. If the inverse of a tensor is positive-definite, so is the tensor. To justify this statement, we use the fact that a tensor is positive-definite if and only if its eigenvalues are positive, and the fact that the eigenvalues of an inverse are reciprocals of the eigenvalues of a tensor; to complete this justification, we use the fact that a real number and its reciprocal have the same sign. The second equality of equation (A.43) can be also written as

$$\mathrm{d}\mathbf{X} \cdot \mathrm{d}\mathbf{X} = \|\mathrm{d}\mathbf{X}\|^2 > 0\,, \qquad \forall\ \mathrm{d}\mathbf{X} \neq \mathbf{0}\,,$$

which is consistent with equation (A.42) and, herein, means that the line element in the reference configuration is not zero. Otherwise, the derivation of the left Cauchy-Green strain tensor, presented in equation (A.42), would not be possible.

Explicitly, we write the left Cauchy-Green strain tensor as

$$b := FF^T\,; \tag{A.44}$$

its components are

$$b_{ij} = \sum_{K=1}^{3} F_{iK}F_{jK}\,, \qquad i,j = 1,2,3\,,$$

which means that b is contained in the current configuration. In terms of its basis, we write

$$b = \sum_{i=1}^{3}\sum_{j=1}^{3}\left(\sum_{K=1}^{3} F_{iK}F_{jK}\right)\hat{\mathbf{e}}_i \otimes \hat{\mathbf{e}}_j\,.$$

This result is analogous to expression (A.39), which is derived in Remark A.9.

Example A.5. For the case of tensor (A.12), we have

$$b = \begin{bmatrix} 1.4950 & 0.0385 & 1.7090 \\ 0.0385 & 1.1719 & -0.6898 \\ 1.7090 & -0.6898 & 3.1697 \end{bmatrix}\,, \tag{A.45}$$

which is a symmetric positive-definite tensor; its eigenvalues are the same as for C, namely, 4.3376, 1.2561 and 0.2429.

A.4.3 *Relations between right and left tensors*

As expected from their definitions, $C := F^T F$ and $b := F F^T$, which are the right and left Cauchy-Green strain tensors—whose names originate from F being to the right and to the left of F^T—are related by the deformation gradient,

$$C = F^{-1} b F \qquad \text{and} \qquad b = F C F^{-1} .$$

Since, in general, given two tensors, $\operatorname{tr}(A\,B) = \operatorname{tr}(B\,A)$, it follows that

$$\operatorname{tr} C = \operatorname{tr} b ;$$

for three, or more, tensors, the equality holds under cyclic—but not random—permutations. Thus, $\operatorname{tr}(A\,B\,C) = \operatorname{tr}(B\,C\,A) = \operatorname{tr}(C\,A\,B)$, but $\operatorname{tr}(A\,B\,C) \neq \operatorname{tr}(B\,A\,C)$. Also, in general, the eigenvalues of AB are the same as the eigenvalues of BA . For tensors in Examples A.4 and A.5, $\operatorname{tr} C = \operatorname{tr} b = 5.8366$, which is also the sum of their three eigenvalues, as discussed in Remark A.13.

Traces, which are scalar quantities, are coordinate-independent. Commonly, the strain-energy function, which is discussed in Appendix D, is formulated in terms of invariants of C and b , so that it does not depend on transformation of coordinate systems associated with the reference or current configuration. Furthermore, since these relations between C and b remain true for all powers, $\operatorname{tr} C^n = \operatorname{tr} b^n$, and the three invariants of C and b are equal to each other.

Remark A.13. In $\mathbb{R}^3$, the three invariants of a second-rank tensor, A , can be expressed in terms of its trace and the traces of A^2 and of A^3 . Also, they can be expressed in terms of its eigenvalues, λ_i , where $i = 1, 2, 3$. Explicitly, we have

$$I_A = \operatorname{tr} A = \lambda_1 + \lambda_2 + \lambda_3 ,$$

$$II_A = \tfrac{1}{2}\left((\operatorname{tr} A)^2 - \operatorname{tr} A^2 \right) = \lambda_1\lambda_2 + \lambda_2\lambda_3 + \lambda_3\lambda_1$$

and

$$III_A = \tfrac{1}{3}\left(\operatorname{tr} A^3 - \tfrac{3}{2} \operatorname{tr} A \operatorname{tr} A^2 + \tfrac{1}{2} (\operatorname{tr} A)^3 \right) = \lambda_1\lambda_2\lambda_3 .$$

The third invariant is $\det A$. These invariants result from the Cayley-Hamilton theorem.

In general, these quantities are invariant under coordinate transformations, which is tantamount to the invariance of I_A , II_A and III_A under orthogonal transformations of A, namely, $A \to A' = Q\,A\,Q^T$, where $'$

refers to the tensor expressed with respect to the transformed coordinates and $Q \in O(3)$ is an orthogonal transformation. These quantities are not invariant under such operations as interchanging rows or columns of a tensor whose components are expressed as a matrix; nor are they invariant under the inverse operation, $A \to A^{-1}$. They are, however, invariant under the transpose operation, $A \to A^T$, even though, in general, this operation cannot be expressed as an orthogonal transformation.

Example A.6. The characteristic equation of tensor (A.40) is

$$\lambda^3 - 5.8366\lambda^2 + 6.8073\lambda - 1.3234 = 0\,, \tag{A.46}$$

which is

$$\lambda^3 - I_C\lambda^2 + II_C\lambda - III_C = 0\,.$$

Equation (A.46) remains invariant under coordinate transformations of tensor (A.40). It is also invariant under the transpose operation, which—in this example—is trivial, since $C = C^T$; this property, however, is true in general. Equation (A.46) is also the characteristic equation of tensor (A.45), since $I_b = I_C$, $II_b = II_C$ and $III_b = III_C$.

In Appendix A.8—to show that C and b contain information about deformation only—we introduce the *stretch ratio*, which is a function of deformation alone. Prior to introducing the stretch ratio, however, in Appendix A.5, we revisit line elements $\mathrm{d}\mathbf{X}$ and $\mathrm{d}\mathbf{x}$ to consider their squares. In Appendix A.6, we invoke expression (A.11) to consider the transformation of the volume element between the reference and current configurations. In Appendix A.7, we consider the transformation of the oriented surface element between the two configurations.

A.5 Green-Lagrange and Almansi strain tensors

Let us consider

$$\frac{\|\mathrm{d}\mathbf{x}\|^2 - \|\mathrm{d}\mathbf{X}\|^2}{2}\,, \tag{A.47}$$

where $\|\cdot\|$ denotes the magnitude of a vector. Expression (A.47) is a measure of deformation between the neighbourhoods of $\mathbf{X}$ and $\mathbf{x}$, from which stem the corresponding line elements, $\|\mathrm{d}\mathbf{X}\|$ and $\|\mathrm{d}\mathbf{x}\|$, respectively.

Following expressions (A.36) and (A.37), we write expression (A.47) as

$$\tfrac{1}{2}\left(\mathrm{d}\mathbf{X}\cdot(C\,\mathrm{d}\mathbf{X}) - \mathrm{d}\mathbf{X}\cdot\mathrm{d}\mathbf{X}\right) = \tfrac{1}{2}\mathrm{d}\mathbf{X}\cdot((C-I)\,\mathrm{d}\mathbf{X}) =: \mathrm{d}\mathbf{X}\cdot(E\,\mathrm{d}\mathbf{X})\,,$$

where E is a positive-definite tensor expressed in the material coordinates. It is referred to as the *Green-Lagrange strain tensor*, which in terms of C and F, discussed in Appendix A.4, can be written as

$$E = \tfrac{1}{2}(C - I) \equiv \tfrac{1}{2}(F^T F - I) . \tag{A.48}$$

Similarly, in spatial coordinates, we write expression (A.47) as

$$\frac{\|\mathrm{d}\mathbf{x}\|^2 - \|\mathrm{d}\mathbf{X}\|^2}{2} = \mathrm{d}\mathbf{x} \cdot (e\, \mathrm{d}\mathbf{x}) ,$$

where

$$e = \tfrac{1}{2}(I - b^{-1}) \equiv \tfrac{1}{2}(I - F^{-T} F^{-1}), \tag{A.49}$$

which is referred to as the *Almansi strain tensor* in honour of an Italian mathematical physicist Emilio Almansi. Both E and e are convenient entities to study deformation. If there is no strain, they are equal to zero, since—in such a case—$C = b^{-1} = I$.

Remark A.14. Properties of strain tensors C, b, E and e follow from properties of the deformation gradient, F, by rules of tensor algebra. Tensors $C := F^T F$ and $b := F F^T$ are symmetric and positive-definite due to the fact that F is invertible, and, hence, $\det F \neq 0$.

Tensors $E = \frac{1}{2}(F^T F - I)$ and $e = \frac{1}{2}(I - F^{-T} F^{-1})$ are symmetric due to the fact that—for any invertible tensor, A—both $A^T A$ and $A^{-T} A^{-1}$ are symmetric, and so is I. The other property of E and e requires a more subtle approach.

In general, the eigenvalues of $A - I$ are $\lambda_i - 1$, where λ_i are eigenvalues of A. Thus, E is positive-definite if all eigenvalues of $F^T F$ are greater than unity. Also, the eigenvalues of $I - A$ are $1 - \lambda_i$. Thus, e is positive-definite if all eigenvalues of $F^{-T} F^{-1}$ are smaller than unity.

We conclude that, similarly to C and b, tensors E and e are symmetric. However, in contrast to C and b, they need not be positive-definite. In general, their eigenvalues are both positive and negative, as illustrated in Example A.7, below.

Following expressions (A.7) and (A.8), it is common to write the Green-Lagrange strain tensor and the Almansi strain tensor in terms of the displacement-gradient tensor. The former strain tensor, which is in material coordinates, is

$$\begin{aligned} E &= \frac{1}{2}\left((\nabla_{\mathbf{X}}\mathbf{U} + I)^T (\nabla_{\mathbf{X}}\mathbf{U} + I) - I\right) \\ &= \frac{1}{2}\left(\nabla_{\mathbf{X}}\mathbf{U} + (\nabla_{\mathbf{X}}\mathbf{U})^T + (\nabla_{\mathbf{X}}\mathbf{U})^T \nabla_{\mathbf{X}}\mathbf{U}\right) ; \end{aligned} \tag{A.50}$$

in components, we write it as

$$E_{IJ} = \frac{1}{2}\left(\frac{\partial U_I}{\partial X_J} + \frac{\partial U_J}{\partial X_I} + \sum_{K=1}^{3} \frac{\partial U_K}{\partial X_I}\frac{\partial U_K}{\partial X_J}\right), \qquad I, J = 1, 2, 3 .$$

The latter strain tensor, which is in spatial coordinates, is

$$\begin{aligned} e &:= \tfrac{1}{2}\left(I - F^{-T}F^{-1}\right) = \tfrac{1}{2}\left(I - (I - \nabla_{\mathbf{x}}\mathbf{u})^T(I - \nabla_{\mathbf{x}}\mathbf{u})\right) \\ &= \tfrac{1}{2}\left(\nabla_{\mathbf{x}}\mathbf{u} + (\nabla_{\mathbf{x}}\mathbf{u})^T - (\nabla_{\mathbf{x}}\mathbf{u})^T\nabla_{\mathbf{x}}\mathbf{u}\right) ; \end{aligned}$$

in components, we write it as

$$e_{ij} = \frac{1}{2}\left(\frac{\partial u_i}{\partial x_j} + \frac{\partial u_j}{\partial x_i} - \sum_{k=1}^{3} \frac{\partial u_k}{\partial x_i}\frac{\partial u_k}{\partial x_j}\right), \qquad i, j = 1, 2, 3 .$$

Remark A.15. As most entities in the reference and current configurations, E and e are related to one another. Explicitly, we can write

$$e = F^{-T} E\, F \tag{A.51}$$

and

$$E = F^T e\, F , \tag{A.52}$$

respectively.

Example A.7. For the case of expressions (A.40) and (A.45), we have

$$E = \begin{bmatrix} 0.5153 & 0.0810 & 1.0104 \\ 0.0810 & 0.1225 & 0.0003 \\ 1.0104 & 0.0003 & 0.7805 \end{bmatrix} , \tag{A.53}$$

and

$$e = \begin{bmatrix} -0.7234 & 0.4914 & 0.7666 \\ 0.4914 & -0.1868 & -0.4144 \\ 0.7666 & -0.4144 & -0.1613 \end{bmatrix} ,$$

which are related by expressions (A.51) and (A.52). Herein, both E and e are symmetric, as expected in view of Remark A.14. Their eigenvalues are both positive and negative, namely, $1.6688, -0.3785, 0.1280$ and $-1.5581, 0.3847, 0.1019$, respectively, which is consistent with Remark A.14, in the context of Example A.1, since the eigenvalues of $F^T F$ and $F^{-T}F^{-1}$ are $4.3376, 0.2429, 1.2561$ and $4.1161, 0.2305, 0.7961$, respectively; not all of them are greater or smaller than unity.

To relate explicitly the eigenvalues of $F^T F$, which we denote by λ_i, to the eigenvalues of E, we recall that $E = \frac{1}{2}(F^T F - I)$ and, hence, its eigenvalues are $\frac{1}{2}(\lambda_i - 1)$; herein, they are $1.6688, -0.3785, 0.1280$, as expected. Similarly, to relate the eigenvalues of $F^{-T} F^{-1}$ and e, we recall that $e = \frac{1}{2}(I - F^{-T} F^{-1})$ and, hence, its eigenvalues are $\frac{1}{2}(1 - \lambda_i)$, where λ_i stands for eigenvalues of $F^{-T} F^{-1}$; thus, we obtain $-1.5581, 0.3847, 0.1019$, as expected.

Example A.8. Let us obtain the *principal strains* of E.[9] To do so, we diagonalize matrix (A.53), in a manner described in Remark A.20, below, to get

$$E = \begin{bmatrix} 0.6599 & 0.7483 & -0.0679 \\ 0.0347 & -0.1206 & -0.9921 \\ 0.7505 & -0.6523 & 0.1056 \end{bmatrix}$$

$$\begin{bmatrix} 1.6688 & 0 & 0 \\ 0 & -0.3785 & 0 \\ 0 & 0 & 0.1280 \end{bmatrix} \begin{bmatrix} 0.6599 & 0.7483 & -0.0679 \\ 0.0347 & -0.1206 & -0.9921 \\ 0.7505 & -0.6523 & 0.1056 \end{bmatrix}^T .$$

The unit eigenvectors, which are the columns of the rotation matrix, are directions of principal strains, which are normal, not shearing, strains. The eigenvalues, which are the entries of the diagonalized matrix, are the corresponding magnitudes. The sign indicates the direction of strain; the positive value is extension and the negative value is compression.

A.6 Deformation of volume element

Let us consider the deformation of a volume element. Invoking the *triple product* and considering three line elements, the volume of a parallelepiped in the reference configuration can be written as

$$\mathrm{d}V = \mathrm{d}\mathbf{X}^{(1)} \cdot \left(\mathrm{d}\mathbf{X}^{(2)} \times \mathrm{d}\mathbf{X}^{(3)}\right) . \tag{A.54}$$

After the deformation, the volume of that parallelepiped is

$$\mathrm{d}v = \mathrm{d}\mathbf{x}^{(1)} \cdot \left(\mathrm{d}\mathbf{x}^{(2)} \times \mathrm{d}\mathbf{x}^{(3)}\right) .$$

Considering the action of F, we write

$$\mathrm{d}v = F\,\mathrm{d}\mathbf{X}^{(1)} \cdot \left(F\,\mathrm{d}\mathbf{X}^{(2)} \times F\,\mathrm{d}\mathbf{X}^{(3)}\right) .$$

[9] Readers interested in discussions on principal strains might refer to Timoshenko and Goodier (1987, pp. 23–24).

It follows from the properties of $\det T$—for a second-rank tensor, T, and vectors, $\mathbf{a}$, $\mathbf{b}$ and $\mathbf{c}$—that

$$(T\mathbf{a} \times T\mathbf{b}) \cdot T\,\mathbf{c} = \det T\,(\mathbf{a} \times \mathbf{b}) \cdot \mathbf{c}\,;$$

hence, we write

$$\mathrm{d}v = \det F\,\left(\mathrm{d}\mathbf{X}^{(1)} \cdot \left(\mathrm{d}\mathbf{X}^{(2)} \times \mathrm{d}\mathbf{X}^{(3)}\right)\right) = \det F\,\mathrm{d}V\,. \tag{A.55}$$

Remark A.16. Using the Levi-Civita symbol, we can write the triple product as

$$\sum_{i=1}^{3}\sum_{j=1}^{3}\sum_{k=1}^{3}\epsilon_{ijk}a_ib_jc_k\,,$$

which—in view of properties of this symbol—shows that this product is antisymmetric if we exchange any pair of vectors. For instance,

$$\mathbf{a} \cdot (\mathbf{b} \times \mathbf{c}) = -\mathbf{b} \cdot (\mathbf{a} \times \mathbf{c})\,.$$

An entity whose components—with respect to an orthonormal basis—are given by the Levi-Civita symbol is a *pseudotensor*, since, under reflections, it acquires a negative sign. Consequently, the result of a vector product is a *pseudovector*, not a vector.

Also, ϵ_{ijk} and its scalar multiples are the only isotropic third-rank tensors (Jeffreys, 1961a, Chapter VII).

If we denote the determinant of F, given in matrix (A.11) by J, we rewrite expression (A.55) as

$$\mathrm{d}v = J\,\mathrm{d}V\,, \tag{A.56}$$

where, also according to the triple product,

$$J = \frac{\partial \mathbf{x}}{\partial X_1} \cdot \left(\frac{\partial \mathbf{x}}{\partial X_2} \times \frac{\partial \mathbf{x}}{\partial X_3}\right)\,;$$

herein, each term is a column of matrix (A.11).

J is a volume ratio of an element between its reference and current configurations. For incompressible materials, $J = 1$. In general, it is a positive quantity—due to local admissibility—which means that no material element is permitted to invert itself upon deformation, which is tantamount to no penetration among elements within a continuum. It ensures that F has an inverse. Also, in view of equation (A.56), the requirement of

$$J := \det F > 0\,, \tag{A.57}$$

which is introduced in Section 3.2.3.1 and Appendix A.2, means that both $\mathrm{d}v$ and $\mathrm{d}V$ must exhibit the same sign.

Remark A.17. J, as the volume ratio, is equivalent to $\det F$, for deformation gradients, whose components are expressed as a matrix in terms of Cartesian or curvilinear orthonormal coordinates. This includes spherical coordinates, used in Example A.2, on page 368, where we set $\det F = 1$, as a requirement for an isochoric deformation. A change in volume, in terms of Cartesian coordinates, is discussed in Remark 2.3, on page 37.

In Example A.2, equation (A.56) is used in condition (A.21), which is a differential equation. For homogeneous deformations, equation (A.56) remains valid even in a form that is not differential, namely, $v = JV$. To exemplify it, let us consider Exercise 2.1, on page 34, where, at $t = 1$,

$$\det \begin{bmatrix} 1+t & 0 & 0 \\ 0 & 1 & 0 \\ 0 & 0 & 1 \end{bmatrix} = 2\,,$$

which, as illustrated in Figure 2.3, is the ratio of the current volume, at $t = 1$, which is $v = 2(1)^2$, to the reference volume, at $t = 0$, which is $V = (1)^3$. Furthermore, the determinant of matrix (2.22), in Exercise 2.2, is equal to unity, regardless of the value of α, as required for an equivoluminal deformation.

A.7 Deformation of oriented surface element

Let us consider the deformation of an oriented surface element. In the reference configuration, we can express an element of the surface area, $\mathrm{d}S$, whose unit normal is $\mathbf{N}$ and which is bounded by two line elements, as

$$\mathbf{N}\,\mathrm{d}S = \mathrm{d}\mathbf{X}^{(1)} \times \mathrm{d}\mathbf{X}^{(2)}\,.$$

In accordance with the triple product, the volume element associated with this area and an arbitrary line element, $\mathrm{d}\mathbf{X}$, is

$$\mathrm{d}V = \mathbf{N}\,\mathrm{d}S \cdot \mathrm{d}\mathbf{X}\,. \tag{A.58}$$

Similarly, in the current configuration, we can write

$$\mathrm{d}v = \mathbf{n}\,\mathrm{d}s \cdot \mathrm{d}\mathbf{x}\,.$$

In terms of the action of F, we can write

$$\mathrm{d}v = \mathbf{n}\,\mathrm{d}s \cdot F\mathrm{d}\mathbf{X}\,. \tag{A.59}$$

Invoking expressions (A.56) and (A.58), we obtain

$$\mathrm{d}v = J\mathrm{d}V = J\,\mathbf{N}\,\mathrm{d}S \cdot \mathrm{d}\mathbf{X}\,. \tag{A.60}$$

Equating the right-hand sides of equations (A.59) and (A.60) as well as using the fact that $\mathrm{d}\mathbf{X}$ is arbitrary, we obtain

$$\mathbf{n}\,\mathrm{d}s = JF^{-T}\mathbf{N}\,\mathrm{d}S\,, \tag{A.61}$$

which relates the current and the reference configurations in terms of $\mathbf{n}\,\mathrm{d}s$ and $\mathbf{N}\,\mathrm{d}S$. It is known as *Nanson's formula*. In other words, Nanson's formula relates the oriented area elements in the reference and current configurations.

A.8 Stretch ratio

Let us return to the Cauchy-Green tensors, discussed in Appendix A.4. The right and left Cauchy-Green tensors are related to the stretch ratio, which is defined in terms of a line element in the reference and current configurations,

$$\lambda := \frac{\|\mathrm{d}\mathbf{x}\|}{\|\mathrm{d}\mathbf{X}\|}\,, \tag{A.62}$$

where $\|\cdot\|$ denotes the magnitude of a vector.

Remark A.18. In a one-dimensional medium, we write the local stretch as $\mathrm{d}x/\mathrm{d}X$, which—for such a medium—is equivalent to its deformation gradient.

It follows that, in terms of the direction vectors of unit length, we can write

$$\lambda^2 = \frac{\mathrm{d}\mathbf{X}}{\|\mathrm{d}\mathbf{X}\|} \cdot \left(C\,\frac{\mathrm{d}\mathbf{X}}{\|\mathrm{d}\mathbf{X}\|} \right) \tag{A.63}$$

and

$$\lambda^{-2} = \frac{\mathrm{d}\mathbf{x}}{\|\mathrm{d}\mathbf{x}\|} \cdot \left(b^{-1}\frac{\mathrm{d}\mathbf{x}}{\|\mathrm{d}\mathbf{x}\|} \right), \tag{A.64}$$

in the material and spatial coordinates, respectively. Thus, given the orientation of a line element, tensors C and b determine its stretch. Herein, C and b provide information about deformation only.

Example A.9. Using values from Example A.3 in expression (A.62), we have

$$\lambda = \frac{\sqrt{1.9950^2 + 0.4200^2 + 2.4590^2}}{\sqrt{1^2 + 1^2 + 1^2}} = 1.8442\,.$$

Using values (A.40) in expression (A.63), with

$$\frac{\mathrm{d}\mathbf{X}}{\|\mathrm{d}\mathbf{X}\|} = \frac{[1,1,1]}{\sqrt{3}}\,,$$

we obtain 3.4010, which is λ^2, as expected.

A.9 Stretch and rotation tensors

A.9.1 *Polar decomposition*

To express explicitly the information about deformation and rotation—in contrast to both of them being included implicitly in F—we invoke the *polar-decomposition theorem.*

Theorem A.1. *Every nonsingular second-rank tensor, F, is uniquely decomposable into the product of an orthogonal tensor, R, and a symmetric positive-definite tensor, U. Also, there is a unique decomposition in the reversed order, $v\,R$, where v is also a symmetric positive-definite tensor.*

Proof. To prove Theorem A.1, we let R be orthogonal, U be positive-definite and $F = RU$. It follows that $F^T = U^T R^T = UR^{-1}$, and, hence, $F^T F = U^2$, $U = \sqrt{F^T F}$ and $R = FU^{-1}$.

Remark A.19. To take the square root of $F^T F$, we can express it in the orientation of coordinates that result in its diagonal form, take the roots of the diagonal entries, and return the tensor to its original orientation. Since $F^T F$ is a symmetric positive-definite tensor, it is diagonalizable and its main-diagonal entries, which are its eigenvalues, are real and positive.

Thus, given F that is invertible, let us define $U := \sqrt{F^T F}$ and $R := FU^{-1}$. The former is necessarily positive-definite, since F is invertible. The orthogonality of the latter is stated by the following lemma.

Lemma A.4. *R is an orthogonal tensor.*

Proof. To prove orthogonality of R, it suffices to verify that $R^T R = I$. We write

$$R^T R = \left(FU^{-1}\right)^T \left(FU^{-1}\right) = \left(U^{-1}\right)^T F^T \left(FU^{-1}\right) = \left(U^T\right)^{-1} F^T \left(FU^{-1}\right).$$

Since U is symmetric, we can write

$$R^T R = U^{-1} F^T F U^{-1} = U^{-1} U^2 U^{-1} = I\,,$$

as required. □

In the following lemma we prove the uniqueness of polar decomposition.

Lemma A.5. *Polar decomposition is unique.*

Proof. To prove uniqueness, let us suppose that

$$F = RU = \bar{R}\bar{U}\,, \tag{A.65}$$

where R and $\bar{R}$ are orthogonal and U and $\bar{U}$ are positive-definite. From equation (A.65), it follows that

$$U^2 = F^T F = \bar{U}^2\,.$$

Hence, since U and $\bar{U}$ are positive-definite, $U = \bar{U}$. Again, from equation (A.65), it follows that $R = \bar{R}$. Thus, the decomposition is unique. □

It remains to prove that $F = vR$. Since $F = RU = (RUR^T)R$, we define $v := RUR^T$, which is positive-definite. The uniqueness of this decomposition can be proven in a manner analogous to Lemma A.5.

The proof of Theorem A.1 is complete. □

A.9.2 *Right Stretch Tensor*

Following Theorem A.1, consider

$$F = RU\,, \tag{A.66}$$

where, herein, R is the rotation tensor in $\mathbb{R}^3$, and U is referred to as the *right Stretch Tensor*. R is a two-point tensor since—according to the action of F, used in expression (A.34)—we can write

$$\mathrm{d}\mathbf{x} = F\,\mathrm{d}\mathbf{X} = R\,(U\,\mathrm{d}\mathbf{X})\,. \tag{A.67}$$

Herein, the undeformed line element, $\mathrm{d}\mathbf{X}$, in the reference configuration, is first stretched by U and then rotated by R to become $\mathrm{d}\mathbf{x}$, in the current configuration.

To find U, consider definition (A.62) and the material line element, $\mathrm{d}\mathbf{X}$. In terms of λ, the action of F, and according to decomposition (A.66), it follows that

$$\lambda \frac{\mathrm{d}\mathbf{x}}{\|\mathrm{d}\mathbf{x}\|} = F \frac{\mathrm{d}\mathbf{X}}{\|\mathrm{d}\mathbf{X}\|} = RU \frac{\mathrm{d}\mathbf{X}}{\|\mathrm{d}\mathbf{X}\|}\,.$$

Taking the scalar product of $\lambda\,\mathrm{d}\mathbf{x}/\|\mathrm{d}\mathbf{x}\|$ with itself and using the fact that $R^T R = I$ and $U = U^T$, we obtain

$$\lambda^2 = \frac{\mathrm{d}\mathbf{X}}{\|\mathrm{d}\mathbf{X}\|} \cdot \left(U\,U \frac{\mathrm{d}\mathbf{X}}{\|\mathrm{d}\mathbf{X}\|} \right).$$

Hence, according to expression (A.63), it follows that

$$U = \sqrt{C}, \tag{A.68}$$

where the validity of this expression relies on the fact that any symmetric positive-definite second-rank tensor has a unique symmetric positive-definite second-rank tensor that is equal to its square.

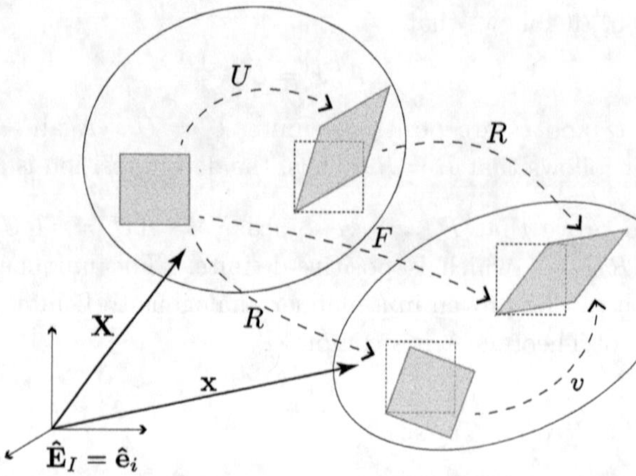

Fig. A.7: Polar decomposition of tensor F: $RU = vR$

C, unlike F, contains information about deformation only. Notably, U, and, hence, also C, contains all the information about the strain; R contains all the information about rotation. Also, since C belongs to the reference configuration, so does U. Also, as illustrated in Figure A.7, since F is a two-point tensor, it follows that R must also be the mapping between the reference configuration and a current configuration, $F_{iJ} = \sum_{K=1}^{3} R_{iK} U_{KJ}$. Hence, its geometrical interpretation as a rotation has to be considered only *sensu lato*.

Example A.10. The eigenvalues of C, which is given in expression (A.40), are

$$\Lambda_1 = 4.3376, \Lambda_2 = 1.2561, \Lambda_3 = 0.2429;$$

the corresponding unit eigenvectors are

$$\begin{aligned}\mathbf{N}_1 &= a\,[\,0.6599, 0.0347, 0.7506\,]\,,\\ \mathbf{N}_2 &= b\,[\,-0.0679, -0.9921, 0.1056\,]\,,\\ \mathbf{N}_3 &= c\,[\,-0.7483, 0.1206, 0.6523\,]\,,\end{aligned}$$

where we set $a = b = c = 1$; in general, $a, b, c \in \mathbb{R}$.

In general, $[\mathbf{N}_1, \mathbf{N}_2, \mathbf{N}_3]$ form an orthogonal system, due to the symmetry of C. Herein, each eigenvector is normalized to form an orthonormal system. All eigenvalues and eigenvectors are within the reference configuration.

Denoting the eigenvalues by Λ_I, we can write

$$C = \sum_{I=1}^{3} \Lambda_I\, \mathbf{N}_I \otimes \mathbf{N}_I\,, \tag{A.69}$$

whose values—which are independent of the choice of basis—are given in expression (A.40); for instance,

$$\begin{aligned}C_{11} &= \Lambda_1\,(\mathbf{N}_1 \otimes \mathbf{N}_1)_{11} + \Lambda_2\,(\mathbf{N}_2 \otimes \mathbf{N}_2)_{11} + \Lambda_3\,(\mathbf{N}_3 \otimes \mathbf{N}_3)_{11} \\ &= 2.0306\,,\end{aligned} \tag{A.70}$$

where the values of $(\mathbf{N}_I \otimes \mathbf{N}_I)_{11}$ are given in Remark A.21 on page 395.

Remark A.20. Expression (A.69) is equivalent to the result of the diagonalization of C. To see it, we write

$$C = \begin{bmatrix} \mathbf{N}_1\,\mathbf{N}_2\,\mathbf{N}_3 \end{bmatrix} \begin{bmatrix} \Lambda_1 & 0 & 0 \\ 0 & \Lambda_2 & 0 \\ 0 & 0 & \Lambda_3 \end{bmatrix} \begin{bmatrix} \mathbf{N}_1\,\mathbf{N}_2\,\mathbf{N}_3 \end{bmatrix}^T,$$

where

$$\begin{bmatrix} \mathbf{N}_1\,\mathbf{N}_2\,\mathbf{N}_3 \end{bmatrix} := \begin{bmatrix} 0.6599 & -0.0679 & -0.7483 \\ 0.0347 & -0.9921 & 0.1206 \\ 0.7506 & 0.1056 & 0.6523 \end{bmatrix},$$

whose determinant is -1, is the orthogonal transformation of the coordinate system that results in the diagonal form of the components of C; it consists of rotation and reflection. Explicitly, we write C as

$$\begin{bmatrix} \Lambda_1\,(\mathbf{N}_1)_1\,(\mathbf{N}_1)_1 + \Lambda_2\,(\mathbf{N}_2)_1\,(\mathbf{N}_2)_1 + \Lambda_3\,(\mathbf{N}_3)_1\,(\mathbf{N}_3)_1 \\ \Lambda_1\,(\mathbf{N}_1)_1\,(\mathbf{N}_1)_2 + \Lambda_2\,(\mathbf{N}_2)_1\,(\mathbf{N}_2)_2 + \Lambda_3\,(\mathbf{N}_3)_1\,(\mathbf{N}_3)_2 \\ \Lambda_1\,(\mathbf{N}_1)_1\,(\mathbf{N}_1)_3 + \Lambda_2\,(\mathbf{N}_2)_1\,(\mathbf{N}_2)_3 + \Lambda_3\,(\mathbf{N}_3)_1\,(\mathbf{N}_3)_3 \end{matrix}$$

$$
\begin{matrix}
\Lambda_1 (\mathbf{N}_1)_1 (\mathbf{N}_1)_2 + \Lambda_2 (\mathbf{N}_2)_1 (\mathbf{N}_2)_2 + \Lambda_3 (\mathbf{N}_3)_1 (\mathbf{N}_3)_2 \\
\Lambda_1 (\mathbf{N}_1)_2 (\mathbf{N}_1)_2 + \Lambda_2 (\mathbf{N}_2)_2 (\mathbf{N}_2)_2 + \Lambda_3 (\mathbf{N}_3)_2 (\mathbf{N}_3)_2 \\
\Lambda_1 (\mathbf{N}_1)_2 (\mathbf{N}_1)_3 + \Lambda_2 (\mathbf{N}_2)_2 (\mathbf{N}_2)_3 + \Lambda_3 (\mathbf{N}_3)_2 (\mathbf{N}_3)_3
\end{matrix}
$$

$$
\left.
\begin{matrix}
\Lambda_1 (\mathbf{N}_1)_1 (\mathbf{N}_1)_3 + \Lambda_2 (\mathbf{N}_2)_1 (\mathbf{N}_2)_3 + \Lambda_3 (\mathbf{N}_3)_1 (\mathbf{N}_3)_3 \\
\Lambda_1 (\mathbf{N}_1)_2 (\mathbf{N}_1)_3 + \Lambda_2 (\mathbf{N}_2)_2 (\mathbf{N}_2)_3 + \Lambda_3 (\mathbf{N}_3)_2 (\mathbf{N}_3)_3 \\
\Lambda_1 (\mathbf{N}_1)_3 (\mathbf{N}_1)_3 + \Lambda_2 (\mathbf{N}_2)_3 (\mathbf{N}_2)_3 + \Lambda_3 (\mathbf{N}_3)_3 (\mathbf{N}_3)_3
\end{matrix}
\right] ,
$$

where components of each eigenvector—which is denoted by $I = 1, 2, 3$—are $(\mathbf{N}_I)_J$, with $J = 1, 2, 3$. Thus, concisely,

$$
C_{JK} = \sum_{I=1}^{3} \Lambda_I (\mathbf{N}_I)_J (\mathbf{N}_I)_K = \sum_{I=1}^{3} \Lambda_I (\mathbf{N}_I \otimes \mathbf{N}_I)_{JK} , \qquad J, K = 1, 2, 3 ,
$$

which is tantamount to expression (A.69), but stated in a matrix notation. For instance,

$$
C_{11} = \Lambda_1 (\mathbf{N}_1)_1 (\mathbf{N}_1)_1 + \Lambda_2 (\mathbf{N}_2)_1 (\mathbf{N}_2)_1 + \Lambda_3 (\mathbf{N}_3)_1 (\mathbf{N}_3)_1 ,
$$

which—in terms of the tensor product—we can write as

$$
C_{11} = \Lambda_1 (\mathbf{N}_1 \otimes \mathbf{N}_1)_{11} + \Lambda_2 (\mathbf{N}_2 \otimes \mathbf{N}_2)_{11} + \Lambda_3 (\mathbf{N}_3 \otimes \mathbf{N}_3)_{11} ,
$$

which is equation (A.70), as expected.

Let us return to equation (A.68). Following expression (A.69), we have

$$
U = \sqrt{C} = \sum_{I=1}^{3} \sqrt{\Lambda_I}\, \mathbf{N}_I \otimes \mathbf{N}_I = \begin{bmatrix} 1.1880 & 0.0787 & 0.7829 \\ 0.0787 & 1.1128 & -0.0244 \\ 0.7829 & -0.0244 & 1.3956 \end{bmatrix} , \qquad \text{(A.71)}
$$

and, in accordance with Theorem A.1 and expression (A.66), we obtain

$$
R = F U^{-1} = \begin{bmatrix} 0.9143 & 0.3769 & -0.1481 \\ -0.3739 & 0.9262 & 0.0490 \\ 0.1556 & 0.0110 & 0.9878 \end{bmatrix} ,
$$

where F is given in expression (A.12). Thus, a polar decomposition of the deformation gradient illustrated in Example A.1 and stated in expression (A.12) is

$$
R U = \begin{bmatrix} 0.9143 & 0.3769 & -0.1481 \\ -0.3739 & 0.9262 & 0.0490 \\ 0.1556 & 0.0110 & 0.9878 \end{bmatrix} \begin{bmatrix} 1.1880 & 0.0787 & 0.7829 \\ 0.0787 & 1.1128 & -0.0244 \\ 0.7829 & -0.0244 & 1.3956 \end{bmatrix} .
$$

For instance, following Example A.3 and according to expression (A.67), we apply the stretch to $\mathbf{X} = [1, 1, 1]$ to obtain $U\,\mathbf{X} = [2.0500, 1.1680, 2.1550]$. Then, we rotate it to obtain $\mathbf{x} = [1.9950, 0.4200, 2.4590]$, as expected from Example A.3.

Remark A.21. Let us revisit Example A.10 to consider explicitly operations relating C and U. As implied in the context of expression (A.9), on page 365, a basis of the space of operators in $\mathbb{R}^3$, which therein appear as a second-rank tensor, $\hat{\mathbf{e}}_i \otimes \hat{\mathbf{E}}_J$, has nine elements. However, the basis of C in expression (A.69) involves only three elements, namely,

$$\mathbf{N}_1 \otimes \mathbf{N}_1 = \begin{bmatrix} 0.4354 & 0.0229 & 0.4953 \\ 0.0229 & 0.0012 & 0.0260 \\ 0.4953 & 0.0260 & 0.5634 \end{bmatrix},$$

$$\mathbf{N}_2 \otimes \mathbf{N}_2 = \begin{bmatrix} 0.0046 & 0.0674 & -0.0072 \\ 0.0674 & 0.9842 & -0.1047 \\ -0.0072 & -0.1047 & 0.0111 \end{bmatrix}$$

and

$$\mathbf{N}_3 \otimes \mathbf{N}_3 = \begin{bmatrix} 0.5600 & -0.0903 & -0.4881 \\ -0.0903 & 0.0145 & 0.0788 \\ -0.4881 & 0.0788 & 0.4255 \end{bmatrix}.$$

The fact that the linear combination of three basis elements is sufficient to express C—as could be verified by comparing the result of operations in expression (A.69) to expression (A.40)—is a consequence of the special choice of basis intrinsic to C.

Herein, each basis element possesses only one nonzero eigenvalue, which is equal to unity. Hence, the determinant of each element is zero. The orthonormal basis, $[\mathbf{N}_1, \mathbf{N}_2, \mathbf{N}_3]$, of $\mathbb{R}^3$ consists of the corresponding eigenvectors of operators $\mathbf{N}_I \otimes \mathbf{N}_I$, where $I = 1, 2, 3$. Since $\mathbf{N}_I \cdot \mathbf{N}_J = \delta_{IJ}$, products $(\mathbf{N}_I \otimes \mathbf{N}_J)(\mathbf{N}_I \otimes \mathbf{N}_J)$ are zero, if $I \neq J$, and $\mathbf{N}_I \otimes \mathbf{N}_I$, if $I = J$. The former property means that each $\mathbf{N}_I \otimes \mathbf{N}_J$, where $I \neq J$, for instance,

$$\mathbf{N}_1 \otimes \mathbf{N}_2 = \begin{bmatrix} -0.0448 & -0.6547 & 0.0697 \\ -0.0024 & -0.0344 & 0.0037 \\ -0.0500 & -0.7446 & 0.0792 \end{bmatrix},$$

is a nilpotent operator, which—if multiplied by itself—is zero. The latter property means that each $\mathbf{N}_I \otimes \mathbf{N}_I$ is an idempotent operator, which—if multiplied by itself—remains unaltered. This property is used implicitly between expressions (A.69) and (A.71), where—relating C and U—we use $(\mathbf{N}_I \otimes \mathbf{N}_I)^2 = \mathbf{N}_I \otimes \mathbf{N}_I$. The matrix interpretation of this process is described in Remark A.19.

In Example A.10, idempotent operators, $\mathbf{N}_I \otimes \mathbf{N}_I$, have traces equal to unity. In general, the trace of an idempotent operator—on a finite-dimensional space—is an integer, which is the dimension of the range of this operator.

Let us note that in expression (A.39), $\hat{\mathbf{E}}_I \otimes \hat{\mathbf{E}}_J$ are also idempotent operators, if $I = J$; otherwise, they are nilpotent. However, in contrast to Example A.10, where—due to the special choice of basis—$C_{IJ} = 0$, if $I \neq J$, in general, the nine components of C are nonzero.

A.9.3 *Left Stretch Tensor*

Following Theorem A.1, consider,

$$F = vR\,, \tag{A.72}$$

where R is the same tensor as in expression (A.66) and v is a symmetric and positive-definite second-rank tensor, and

$$\mathrm{d}\mathbf{x} = F\,\mathrm{d}\mathbf{X} = v\,(R\,\mathrm{d}\mathbf{X})\,. \tag{A.73}$$

It is referred to as the *left Stretch Tensor.* To find v, we obtain, in a manner analogous to the one discussed in Appendix A.9.2,

$$\lambda^{-2} = \frac{\mathrm{d}\mathbf{x}}{\|\mathrm{d}\mathbf{x}\|} \cdot \left(v^{-1}\, v^{-1} \frac{\mathrm{d}\mathbf{x}}{\|\mathrm{d}\mathbf{x}\|} \right).$$

Hence, according to expression (A.64), it follows that $v\,v = b$. Thus, b, unlike F, contains information about deformation only. Also, since b is contained in the current configuration, so is v. Again—as in the case of the right Stretch tensor, and as illustrated in Figure A.7—the rotation tensor, R, is the mapping between the reference configuration and a current configuration, $F_{iJ} = \sum_{k=1}^{3} v_{ik} R_{kJ}$.

Example A.11. In a manner analogous to the one presented in Example A.10, and in accordance with Theorem A.1, we can find v for expression (A.72) by considering eigenvalues and eigenvectors of tensor $b = v^2$. Also, since eigenvalues for both U and v are the same and R is the same for both decompositions, we can obtain the eigenvectors of the latter as $\mathbf{n}_i = R\mathbf{N}_I$, and, hence, construct tensor v.[10] However, since herein we have expressions for F and R, we can invoke expression (A.72) to write

$$v = FR^T = \begin{bmatrix} 1.0269 & 0.1090 & 0.6547 \\ 0.1090 & 1.0386 & -0.2852 \\ 0.6547 & -0.2852 & 1.6309 \end{bmatrix}.$$

[10] Interested readers might refer to Bonet and Wood (2008, Section 4.6).

Thus, a polar decomposition of the deformation gradient illustrated in Example A.1 and stated in expression (A.12) is

$$v\,R = \begin{bmatrix} 1.0269 & 0.1090 & 0.6547 \\ 0.1090 & 1.0386 & -0.2852 \\ 0.6547 & -0.2852 & 1.6309 \end{bmatrix} \begin{bmatrix} 0.9143 & 0.3769 & -0.1481 \\ -0.3739 & 0.9262 & 0.0490 \\ 0.1556 & 0.0110 & 0.9878 \end{bmatrix} .$$

If, following Example A.3 and expression (A.73), we apply the rotation to $\mathbf{X} = [1, 1, 1]$, we obtain $R\,\mathbf{X} = [1.1431, 0.6012, 1.1540]$. Then, if we stretch it, we obtain $\mathbf{x} = [1.9950, 0.4200, 2.4590]$, as expected from Examples A.3 and A.10.

A.9.4 *Relations between right and left Stretch Tensors*

As illustrated in Examples A.10 and A.11, the results of the actions of $R\,U$ and vR are the same, even though the order of deformation and rotation is reversed; R is the same in both cases. Also, $v = R\,U\,R^T$ and, hence, $b = R\,C\,R^T$.

As stated in Appendix A.9.3, both U and v have the same eigenvalues, and their eigenvectors are rotated by R with respect to each other. Since U is a symmetric positive-definite tensor, its three eigenvectors are orthogonal to each other. The amount of stretch is measured by the eigenvalues. If it is greater than unity, the element is expanded in the direction of the corresponding eigenvector. If it is smaller, it is contracted. The same is true for v.

Furthermore, from equation (A.55), it follows that

$$\det F = \frac{\mathrm{d}v}{\mathrm{d}V} .$$

Since U, as well as v, contain all the information about the strain, and also since a rotation of an element does not affect its volume, it follows that

$$\det F = \det U = \det v ,$$

where the latter equality is consistent with the fact that U and v have the same eigenvalues. Both equalities are consistent with the fact that $\det U$ and $\det v$ are positive by virtue of the positive definiteness of U and v, and $\det F$ is positive by virtue of local admissibility.

Also, Theorem A.1 allows us to prove two concepts discussed in Appendix A.2, namely that $F = I$ means that there is no deformation, and that $F \neq I$ does not necessarily mean that there is deformation, since a rigid-body rotation also entails $F \neq I$. For the former, the polar decompositions of $F = I$ result in $R = U = v = I$, which means that there is neither rotation nor deformation. For the latter, if $U = v = I$, which means that there is no deformation, and if $R \neq I$, which means that there is rotation, we have $F = R$.

To comment briefly on terminology, let us cite Dorfmann and Ogden (2014, p. 53), who write

> Note that if F is just a rotation R, then, since R is orthogonal, $R^T R = I$, i.e. there is no strain associated with a pure rotation.

According to Ogden (1997, p. 90)—in spite of no strain—there is, however, deformation, since one can

> regard a rigid rotation as a local deformation which corresponds to zero strain. The most general deformation which yields zero strain and is consistent with $J \equiv \det F > 0$ comprises rigid translation combined with a rigid rotation,

where $J \equiv \det F > 0$ is requirement (A.57).

It is important to be explicit, without being pedantic, about the meaning of such terms as motion, deformation and strain. Herein, we interpret the deformation of a continuum, *sensu stricto*, as a change of position of points within it, relative to each other,[11] which is distinct from the interpretation or terminology, *sensu lato*, cited above or used by Chadwick (1976, p. 52), who writes that

> [t]he term *deformation* is employed when the functions ψ and Ψ do not depend upon t and in contexts in which the time dependence is irrelevant. More precisely, a deformation is a (smooth) mapping of a reference configuration into a current (or deformed) configuration. Evidently a motion may be viewed as a one-parameter family of deformations indexed by the time.

In this appendix, functions ψ and Ψ are $\mathbf{x}(\mathbf{X}, t)$ and $\mathbf{X}(\mathbf{x}, t)$, respectively; in Chapter 2, they are denoted by ϕ and Φ.

In our terminology, we distinguish between a rigid-body motion, which consists of rotation and translation, and a deformation, which is limited to the case for which the relative distance or orientation of points within the body is changed. Motion, in our terminology, includes all of them.

In our terminology, wave propagation is a consequence of deformation, which is a change in volume or shape within a continuum. If a broader meaning of deformation is used, it is a consequence of strain.

[11] *see also*: Slawinski (2015, Section 1.4.1)

A.9.5 *Local and global rotations*

In general, polar decomposition is a mathematical statement that is applied locally. A rigid-body rotation, on the other hand, implies a global application. Yet, in general, R is not a rigid-body rotation, but a function of position within a body; in such a case, it is an intrinsic component of deformation.

Let us consider a simple shear, illustrated in Exercise 2.2, on page 36. The deformation gradient, stated in matrix (2.22), can be written as

$$F = I + \alpha(t)\,\hat{\mathbf{e}}_1 \otimes \hat{\mathbf{E}}_2\,, \tag{A.74}$$

where α is the amount of shear. At $t = 2$, which corresponds to $\alpha = 2$, the polar decomposition is

$$\underbrace{\begin{bmatrix} 1 & 2 & 0 \\ 0 & 1 & 0 \\ 0 & 0 & 1 \end{bmatrix}}_{F} = \underbrace{\begin{bmatrix} \cos\left(\frac{\pi}{4}\right) & \sin\left(\frac{\pi}{4}\right) & 0 \\ -\sin\left(\frac{\pi}{4}\right) & \cos\left(\frac{\pi}{4}\right) & 0 \\ 0 & 0 & 1 \end{bmatrix}}_{R} \underbrace{\begin{bmatrix} \frac{\sqrt{2}}{2} & \frac{\sqrt{2}}{2} & 0 \\ \frac{\sqrt{2}}{2} & \frac{3\sqrt{2}}{2} & 0 \\ 0 & 0 & 1 \end{bmatrix}}_{U}; \tag{A.75}$$

herein, R is a global rotation and U is a global stretch, as illustrated in Figure A.8, in which stretch is followed by rotation.[12] However, both result in a deformation, as illustrated in Figure A.9.

Since U is symmetric, it can be diagonalized, and the resulting polar decomposition can be expressed in terms of the principal stretches. In such a case, the rotation matrix is adjusted accordingly (Ogden, 1997, Problem 2.2.10).

In contrast to equation (A.75), for a uniaxial stretch,[13] discussed in Exercise 2.1, on page 34, we have

$$\underbrace{\begin{bmatrix} \alpha & 0 & 0 \\ 0 & 1 & 0 \\ 0 & 0 & 1 \end{bmatrix}}_{F} = \underbrace{\begin{bmatrix} 1 & 0 & 0 \\ 0 & 1 & 0 \\ 0 & 0 & 1 \end{bmatrix}}_{R} \underbrace{\begin{bmatrix} \alpha & 0 & 0 \\ 0 & 1 & 0 \\ 0 & 0 & 1 \end{bmatrix}}_{U} = \underbrace{\begin{bmatrix} \alpha & 0 & 0 \\ 0 & 1 & 0 \\ 0 & 0 & 1 \end{bmatrix}}_{v} \underbrace{\begin{bmatrix} 1 & 0 & 0 \\ 0 & 1 & 0 \\ 0 & 0 & 1 \end{bmatrix}}_{R},$$

where α is the amount of stretch. Herein, there is no rotation—globally or locally—as expected in view of Figure 2.3.[14]

[12] *see also*: Slawinski (2015, Figure 1.4.2)
[13] *see also*: Slawinski (2015, Figure 1.4.1)
[14] Readers interested in geometrical interpretations of deformations might refer to Ogden (1997, Sections 2.2.5, 2.2.6).

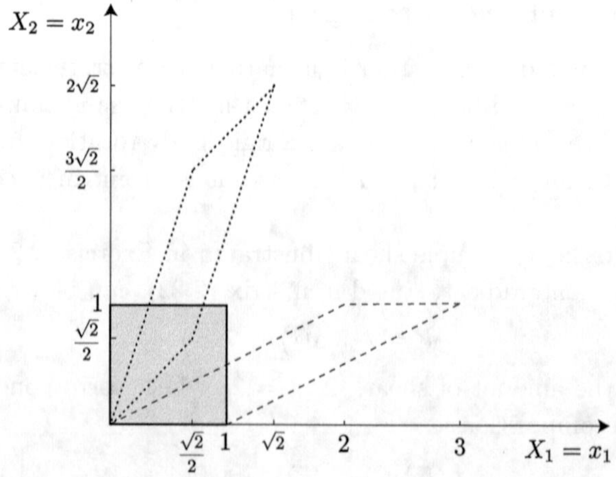

Fig. A.8: The RU polar decomposition of a simple shear, stated in expression (2.23), on page 37, for $t = 2$: a stretch, which results in dotted lines, is followed by a $\pi/4$ rotation, which results in dashed lines; in Figure 2.4, dashed lines illustrate the result of the same simple shear, for $t = 1$. For the $v\,R$ polar decomposition, the intermediate phase would be illustrated by the clockwise $\pi/4$ rotation of the square, about the origin. The final result is necessarily the same for both decompositions, since $F = RU = v\,R$.

Both the simple shear and uniaxial stretch are homogeneous deformations. The issue of local rotations—in polar decomposition—becomes particularly important for inhomogeneous deformations.

Returning to a simple shear and using equation (A.74), we write equation (A.37) as

$$C = I + \alpha(t)\left(\hat{\mathbf{E}}_1 \otimes \hat{\mathbf{E}}_2 + \hat{\mathbf{E}}_2 \otimes \hat{\mathbf{E}}_1\right) + \alpha^2(t)\left(\hat{\mathbf{E}}_2 \otimes \hat{\mathbf{E}}_2\right),$$

and equation (A.44) as

$$b = I + \alpha(t)\left(\hat{\mathbf{e}}_1 \otimes \hat{\mathbf{e}}_2 + \hat{\mathbf{e}}_2 \otimes \hat{\mathbf{e}}_1\right) + \alpha^2(t)\left(\hat{\mathbf{e}}_1 \otimes \hat{\mathbf{e}}_1\right).$$

Also, we write equation (A.48) and (A.49) as

$$E = \tfrac{1}{2}\left(\alpha(t)\left(\hat{\mathbf{E}}_1 \otimes \hat{\mathbf{E}}_2 + \hat{\mathbf{E}}_2 \otimes \hat{\mathbf{E}}_1\right) + \alpha^2(t)\left(\hat{\mathbf{E}}_2 \otimes \hat{\mathbf{E}}_2\right)\right)$$

and

$$e = \tfrac{1}{2}\left(\alpha(t)\left(\hat{\mathbf{e}}_1 \otimes \hat{\mathbf{e}}_2 + \hat{\mathbf{e}}_2 \otimes \hat{\mathbf{e}}_1\right) - \alpha^2(t)\left(\hat{\mathbf{e}}_2 \otimes \hat{\mathbf{e}}_2\right)\right),$$

respectively.

Fig. A.9: Rotation due to simple shear

Each strain-tensor expression is contained within a single configuration; also, each of them carries explicit information about local rotations. This is apparent from the definitions of these tensors, which are expressed in terms of geometric relations between line elements. Hence, C, b, E and e are affected by these rotations.

As shown in Figure A.9, we can illustrate a simple shear as a homogeneous deformation of a deck of cards. Cards slide parallel to each other in such a manner that each card slides, with respect to the card below it, in the same direction, and by the same amount, as that card slides with respect to the card below it. Let us consider two line elements prior to deformation. One line element is parallel to the cards, which we visualize as being horizontal; the other is perpendicular, and, hence, it is vertical. After deformation, the orientation of the horizontal line element remains horizontal, since since it coincides with the direction of sliding, which is the shear direction. The other line element, however, ceases to be vertical. Thus, their mutual orientation changes as a consequence of deformation, which, in this illustration, is the local rotation of the vertical line element.

A.10 Infinitesimal strain and rotation tensors

In aspects of elasticity, including many studies of seismology, we achieve a satisfactory level of empirical adequacy under assumptions of infinitesimal strains and rotations. Furthermore, resulting formulations exhibit several convenient properties.

To examine consequences of these assumptions, let us consider expression (A.50). If the displacement gradient is assumed to be infinitesimal, we can neglect the quadratic terms in this expression to write

$$E \approx \frac{1}{2}\left(\nabla_{\mathbf{X}}\mathbf{U} + (\nabla_{\mathbf{X}}\mathbf{U})^T\right) =: \varepsilon\,,$$

which is the infinitesimal-strain tensor. If we also assume that the rigid-body rotations are infinitesimal, it follows that

$$\tfrac{1}{2}\left(\nabla_{\mathbf{X}}\mathbf{U}+(\nabla_{\mathbf{X}}\mathbf{U})^{T}\right)\approx\tfrac{1}{2}\left(\nabla_{\mathbf{x}}\mathbf{u}+(\nabla_{\mathbf{x}}\mathbf{u})^{T}\right),$$

which means that the distinction between the material and spatial coordinates is negligible. For infinitesimal strains, it is common to write[15]

$$\varepsilon_{ij}=\frac{1}{2}\left(\frac{\partial u_i}{\partial x_j}+\frac{\partial u_j}{\partial x_i}\right),\qquad i,j=1,2,3\,, \tag{A.76}$$

without distinguishing between the material and spatial coordinates. Hence, we can write the components of the displacement vector as $u_i = x_i - X_i$, where $i = 1,2,3$.

For infinitesimal strains and rotations, the separation between them can be accomplished by invoking the fact that any second-rank tensor can be decomposed into its symmetric and antisymmetric parts. Hence, we write the gradient of the displacement field, $\mathbf{u}$, as

$$\nabla_{\mathbf{x}}\mathbf{u}=\tfrac{1}{2}\left(\nabla_{\mathbf{x}}\mathbf{u}+(\nabla_{\mathbf{x}}\mathbf{u})^{T}\right)+\tfrac{1}{2}\left(\nabla_{\mathbf{x}}\mathbf{u}-(\nabla_{\mathbf{x}}\mathbf{u})^{T}\right), \tag{A.77}$$

where the first summand is a symmetric part, which corresponds to an infinitesimal-strain tensor, and the second one is an antisymmetric part, which corresponds to an infinitesimal-rotation tensor. We denote them by ε and ξ, respectively. In components, these tensors are stated in expression (A.76) and by

$$\xi_{ij}=\frac{1}{2}\left(\frac{\partial u_i}{\partial x_j}-\frac{\partial u_j}{\partial x_i}\right),\qquad i,j=1,2,3\,. \tag{A.78}$$

In contrast to the deformation gradient, F, discussed in Appendix A.2, $\nabla_{\mathbf{x}}\mathbf{u}$ is a tensor, *sensu stricto*, not a two-point tensor. This is a consequence of the disappearance of distinction—under the assumption of infinitesimal deformations—between the reference and current configurations. Hence, in contrast to expression (A.6), where $\mathbf{x}(\mathbf{X})$ and $\mathbf{X}$ belong to the current and reference configurations, respectively, in expression (A.77), both the range, $\mathbf{u}(\mathbf{x})$, and the domain, $\mathbf{x}$, belong to the same configuration. Thus, within the context of infinitesimal deformations, referring to $\nabla_{\mathbf{x}}\mathbf{u}$, Thorne and Blandford (2017, p. 570) write

> [t]his tensor is a geometric object, defined independently of any coordinate system [...].

Also, Thorne and Blandford (2017, p. 570) comment on the compatibility of $\nabla_{\mathbf{x}}\mathbf{u}$,

[15] *see also*: Slawinski (2015, Expression 1.4.6)

> [i]n any small neighborhood of any point $\mathbf{x}_o$ in a deformed body, we can reconstruct the displacement vector [...] from its gradient [...] up to an additive constant.

Remark A.22. In a manner similar to the one discussed in Appendix A.3, we could establish the compatibility condition for infinitesimal strains, which ensures that—given a strain tensor—we can infer the corresponding displacement field, up to a rigid-body motion. This condition is

$$\nabla \times (\nabla \times \varepsilon) = 0 . \tag{A.79}$$

Again, as in the case of finite strains, condition (A.79) is satisfied trivially for homogeneous deformations, since all partial derivatives are zero.

Remark A.23. Expression (A.78)—multiplied by 2—can be viewed as components of the curl of a vector field, $\mathbf{u}$, which are $2\,\xi_{32}$, $2\,\xi_{13}$ and $2\,\xi_{21}$.[16] In Cartesian coordinates and using the Levi-Civita symbol, these three components are

$$(\nabla \times \mathbf{u})_i = \sum_{j=1}^{3}\sum_{k=1}^{3} \epsilon_{ijk}\frac{\partial u_k}{\partial x_j}\,, \qquad i = 1, 2, 3\,,$$

which, in view of antisymmetry of the symbol, represent the linearly independent components of the second-rank tensor that corresponds to rotation. As discussed in Section 5.1.2, the shear wave is expressed in terms of $\nabla \times \mathbf{u}$.

From expressions (A.7) and (A.8), it follows that, in general,

$$F = I + \nabla_{\mathbf{X}}\mathbf{U} = I + \nabla_{\mathbf{x}}\mathbf{u}\,F\,.$$

Hence, for infinitesimal deformations, for which we let $\mathbf{X} = \mathbf{x}$ and $\mathbf{U} = \mathbf{u}$,

$$F \approx I + \nabla_{\mathbf{x}}\mathbf{u} = I + \varepsilon + \xi\,. \tag{A.80}$$

Under the assumptions of infinitesimal strains and rotations several distinct entities become indistinguishable from each other. $C = b$, which means that there is no distinction between the right and left Cauchy-Green strain tensors; they both tend to identity. Also, using the fact that $(I + \delta)^{-1} \approx I - \delta$, it follows that $E = e$, which means that there is no distinction between the Green-Lagrange strain tensor and the Almansi strain tensor; they both tend to zero. In such a case, they are both commonly denoted by ε.

Furthermore, in general, the product of two deformation gradients is not commutative, as can be expected in view of noncommutativity of matrix

[16] *see also*: Slawinski (2015, Section 1.5 and Exercise 1.7)

multiplication and of subtleties of polar decomposition. This property can be illustrated geometrically by rotating a stretched cube versus stretching a rotated cube; in both cases, we rotate and stretch by the same amount and in the same orientation. For an infinitesimal case, however, the product becomes commutative.[17]

To relate the decomposition of a tensor into its symmetric and antisymmetric parts to the polar decomposition of a tensor, let us revisit the infinitesimal strains and rotations by considering an aspect of the *Lie algebra.* To do so, we translate the coordinate system in such a manner that we express every second-rank tensor as $T = I + \Delta T$, where I is the identity tensor at the origin. Herein, we express the infinitesimal deformation gradient as $I + \Delta F$, where I is at the origin of the reference configuration. Consequently, following expression (A.66), we write the polar decomposition as

$$I + \Delta F = (I + \Delta R)(I + \Delta U) = I + \Delta R + \Delta U + (\Delta R)(\Delta U) .$$

Neglecting the product, we obtain

$$\Delta F = \Delta R + \Delta U , \tag{A.81}$$

which means that—for infinitesimal strains and rotations—product (A.66) becomes an addition. Also, following expression (A.72), we can write $\Delta F = \Delta R + \Delta v$, which implies that, in such a case, $\Delta U = \Delta v$, and stretches commute with the rotation.

Equation (A.81) is a decomposition of a tensor into its symmetric and antisymmetric parts. To see that, consider the fact that R is an orthogonal tensor, which means that $R\,R^T = I$. If we consider infinitesimal rotations, we can write

$$(I + \Delta R)(I + \Delta R)^T = I ,$$

from which it follows—by neglecting the product, $\Delta R\,\Delta R^T$—that $\Delta R + \Delta R^T = 0$, which means that ΔR is antisymmetric, since $\Delta R = -\Delta R^T$. Furthermore—since any second-rank tensor can be decomposed into its symmetric and antisymmetric parts, and in view of expression (A.81)—it follows that

$$\Delta U = \tfrac{1}{2}(\Delta F + \Delta F^T) ,$$

which is the symmetric part of ΔF, and

$$\Delta R = \tfrac{1}{2}(\Delta F - \Delta F^T) ,$$

which is the antisymmetric part of ΔF; these expressions are tantamount to expressions (A.76) and (A.78), respectively.

[17] *see also*: Slawinski (2015, Exercise 1.3)

A.11 Objective tensors

Objectivity of a tensor means that its intrinsic properties are independent of its orientation in a current configuration, provided that we agree on the same reference configuration. As stated by Bonet and Wood (2008, Section 4.15),

> From the point of view of an observer attached to and rotating with the body many quantities describing the behavior of the solid will remain unchanged. Such quantities, [...], are said to be objective. Although the intrinsic nature of these quantities remains unchanged, their spatial description may change.

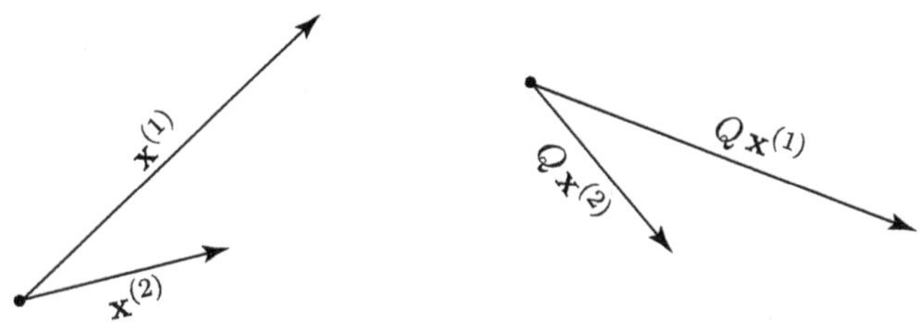

Fig. A.10: Vectors $\mathbf{x}^{(1)}$ and $\mathbf{x}^{(2)}$ in two orientations

The concept of objectivity relies on the invariance of certain quantities under orthogonal transformations, Q, as shown by the fact that orthogonal transformations preserve scalar products. To illustrate this property, let us consider a scalar product of a pair of vectors, $\mathbf{x}^{(1)}$ and $\mathbf{x}^{(2)}$, rotated by Q,

$$\mathbf{x}^{(1)} \cdot \mathbf{x}^{(2)} = \left(Q\,\mathbf{x}^{(1)}\right) \cdot \left(Q\,\mathbf{x}^{(2)}\right) . \tag{A.82}$$

To prove this identity, we write the right-hand side as $\left(Q\,\mathbf{x}^{(1)}\right)^T \left(Q\,\mathbf{x}^{(2)}\right)$, which means that we express a scalar product as a matrix multiplication. Performing the transpose operation, we write $\left(\mathbf{x}^{(1)}\right)^T Q^T Q\,\mathbf{x}^{(2)}$. Since Q is an orthogonal transformation, $Q^T Q = I$, and we obtain $\left(\mathbf{x}^{(1)}\right)^T \mathbf{x}^{(2)}$, which— in terms of a scalar product—we write as $\mathbf{x}^{(1)} \cdot \mathbf{x}^{(2)}$, which is the left-hand side of equation (A.82). The invariance of a scalar product means that both the length of a vector and the angle between vectors are not affected by the transformation, as illustrated in Figure A.10.

In general, a scalar is objective if it has the same value in the reference configuration and in the current configurations. As illustrated above, a vector, $\mathbf{v}$, is objective if its components satisfy $\mathbf{v}' = Q\,\mathbf{v}$, where $'$ denotes the rotated vector.

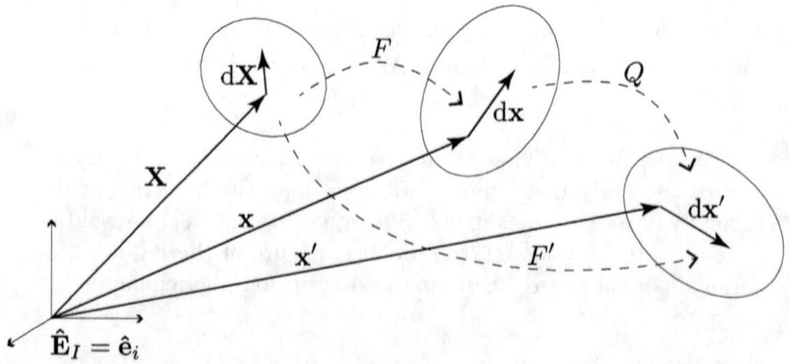

Fig. A.11: Objectivity of a two-point second-rank tensor illustrated by a deformation gradient: Composition of operations F and Q acting on $\mathrm{d}\mathbf{X}$ is $Q\,F\,\mathrm{d}\mathbf{X}$, which is tantamount to $F'\,\mathrm{d}\mathbf{X}$.

A second-rank tensor, T, expressed in spatial coordinates, is objective if $T' = Q\,T\,Q^T$. A second-rank two-point tensor, however, such as the deformation gradient, is objective if $F' = Q\,F$, which means that it is objective if and only if it maps an objective vector in the reference configuration into an objective vector in a current configuration. To explain this, following Figure A.11, let us consider the fact that $\mathrm{d}\mathbf{x}' = Q\,\mathrm{d}\mathbf{x}$, which is a line element in two current configurations. Since $\mathrm{d}\mathbf{x} = F\,\mathrm{d}\mathbf{X}$, we obtain $\mathrm{d}\mathbf{x}' = Q\,F\,\mathrm{d}\mathbf{X}$, which implies that $Q\,F = F'$, where F' is the deformation gradient relating the reference configuration to the current configuration containing $\mathrm{d}\mathbf{x}'$.

Not all physical or geometrical quantities are objective. An example of a nonobjective quantity is velocity, $\partial\mathbf{x}/\partial t$, since

$$\dot{\mathbf{x}}' = \frac{\partial(Q\,\mathbf{x})}{\partial t} = Q\,\dot{\mathbf{x}} + \dot{Q}\,\mathbf{x}, \tag{A.83}$$

where $\dot{}$ denotes the partial derivative with respect to time. Examining the right-hand side, we see that the magnitude of $\dot{\mathbf{x}}$ is not preserved under rotation due to the presence of the second summand, if $\dot{Q} \neq 0$.

Let us comment on objectivity and frame indifference, which is one of the tenets of constitutive relations. The latter is a quantity whose value does not change, such as a dilatation, stated in expression (5.3), which is a scalar given by the trace of a displacement-vector gradient. As discussed by Epstein (2010, p. 297), in general,

> [t]here is nothing sacred about being frame indifferent. In particular, the principle of material frame indifference does not claim that the quantities involved in a constitutive law must be frame indifferent. Quite to the contrary, it affirms that even though the quantities involved are in general not frame indifferent, nevertheless the constitutive functionals themselves are invariant under a change of frame.

To exemplify objectivity and frame indifference, let us examine the Cauchy-Green strain tensors, discussed in Appendix A.4. We proceed in a manner analogous to the one used to show the invariance of a scalar product from equation (A.82).

Under orthogonal transformations, the right Cauchy-Green strain tensor, given in expression (A.37), is

$$C' = F'^T F' = (QF)^T QF = F^T Q^T Q\, F = F^T\, I\, F = F^T F = C\,, \qquad \text{(A.84)}$$

which means that—in view of its being a reference-configuration entity—C is an objective tensor. Under orthogonal transformations, the left Cauchy-Green strain tensor, given in expression (A.44), is

$$b' = F' F'^T = QF\,(QF)^T = Q\,F\,F^T Q^T = Q\,b\,Q^T\,, \qquad \text{(A.85)}$$

which means that b is also objective, but is not frame-indifferent. In other words—under rigid-body rotations—its components transform in a required manner, but are not invariant under such rotations.

Analogous properties are exhibited by the Green-Lagrange strain tensor, E, given in expression (A.48), and by the Almansi strain tensor, e, given in expression (A.49), formulated in Appendix A.5. To show their properties, we can proceed—in a manner analogous to C and b—from expressions (A.48) and (A.49) stated in terms of F.

Alternatively, we can use their relations to the Cauchy-Green strain tensors, given in the same expressions. Following this approach, we write

$$E' = \tfrac{1}{2}(C' - I)\,.$$

In accordance with equation (A.84) and the fact that $I' = QIQ^T = QQ^T = I$, we conclude that $E' = E$. Also, since

$$e' = \tfrac{1}{2}\left(I - \left(b^{-1}\right)'\right)\,,$$

it follows that $e' = Q\,e\,Q^T$. This is a consequence of the linearity of tensor multiplication, namely, that—for any second-rank tensor—$Q\,(I+S)\,Q^T = I + Q\,S\,Q^T$, and—in view of expression (A.85)—of a property of orthogonal transformations, namely, that, for any invertible second-rank tensor,

$$S' = Q\,S\,Q^T \iff \left(S^{-1}\right)' = Q\,S^{-1}\,Q^T\,. \qquad \text{(A.86)}$$

In the context of objectivity, Q is viewed as a rigid-body rotation within a given coordinate system. We can do so since property (A.86) remains true if we replace $Q\,[\cdot]\,Q^T$ by $Q^T[\cdot]\,Q$.

To prove property (A.86), we use the fact that $S'\,(S^{-1})' = I$. Multiplying the corresponding sides of the equations in expression (A.86) by one another, we obtain

$$S'\,(S^{-1})' = Q\,S\,Q^T Q\,S^{-1}\,Q^T = I\,,$$

as required.

Remark A.24. $Q\,[\cdot]\,Q^T$ and $Q^T[\cdot]\,Q$ refer to a relative motion between a second-rank tensor and its spatial coordinate system. These expressions differ by the transpose of Q and, hence, they distinguish between the directions of operation. In the case of rotation, they distinguish between the clockwise and counterclockwise rotations, as can be illustrated in $\mathbb{R}^2$ by

$$Q = \begin{bmatrix} \cos(\theta) & -\sin(\theta) \\ \sin(\theta) & \cos(\theta) \end{bmatrix}$$

and

$$Q^T = \begin{bmatrix} \cos(\theta) & \sin(\theta) \\ -\sin(\theta) & \cos(\theta) \end{bmatrix} = \begin{bmatrix} \cos(-\theta) & -\sin(-\theta) \\ \sin(-\theta) & \cos(-\theta) \end{bmatrix}.$$

This property can also be interpreted as a distinction between the active and passive rotations, where the active one corresponds to a change of the orientation of a body, which is a process independent of any coordinate system, and the passive one to a rotation of a coordinate system itself.

Given a tensor in a current configuration, $S = \sum_{i=1}^3 \sum_{j=1}^3 S_{ij}\,\hat{\mathbf{e}}_i \otimes \hat{\mathbf{e}}_j$, and the coordinate system, whose basis is $\{\hat{\mathbf{e}}_1, \hat{\mathbf{e}}_2, \hat{\mathbf{e}}_3\}$, we can write the result of its active rotation as

$$S' = \sum_{i=1}^3 \sum_{j=1}^3 \sum_{k=1}^3 \sum_{\ell=1}^3 Q_{ik}\,S_{k\ell}\,Q^T_{\ell j}\,\hat{\mathbf{e}}_i \otimes \hat{\mathbf{e}}_j = \sum_{i=1}^3 \sum_{j=1}^3 S'_{ij}\,\hat{\mathbf{e}}_i \otimes \hat{\mathbf{e}}_j$$

and of its passive rotation as $S = \sum_{i=1}^3 \sum_{j=1}^3 S'_{ij}\,\hat{\mathbf{e}}'_i \otimes \hat{\mathbf{e}}'_j$. In the former, only the tensor is rotated, and its components become S'_{ij}. In the latter, only the coordinate system is rotated, $\{Q\,\hat{\mathbf{e}}_1, Q\,\hat{\mathbf{e}}_2, Q\,\hat{\mathbf{e}}_3\}$, and its basis becomes $\{\hat{\mathbf{e}}'_1, \hat{\mathbf{e}}'_2, \hat{\mathbf{e}}'_3\}$; the tensor-components are adjusted accordingly,

$$S'_{ij} = \sum_{k=1}^3 \sum_{\ell=1}^3 Q^T_{ik} S_{k\ell} Q_{\ell j}\,.$$

If to study a particular concept suffices to examine relative orientations of a tensor with respect to its coordinate system, there is no need to distinguish between active and passive rotations. Such is the case of studying material symmetry of elasticity tensors, discussed in Section 3.2 and Appendix D.3. However, objectivity is to be examined in the context of active rotations only, since it refers to the behaviour of physical properties that might be affected by rigid-body rotations, as illustrated by expression (A.83), where Q is interpreted as a rotation of a body itself, not of its spatial coordinate system; hence, $\dot{Q}$ is the speed of the rotating body. If such an interpretation is not obvious from the context, it might be convenient to include a notation that allows us to distinguish between the active and passive rotations.

Be that as it may, endowing of a mathematical structure with a physical meaning can be achieved only by analogy.[18] Herein, to examine quantitatively a change of orientation of a body, we express its properties by analogies to tensor properties with respect to its coordinate system. There is no intrinsic structure to distinguish between active and passive rotations, but such an interpretation might be facilitated by a careful notation.

Example A.12. Let us consider a vector in $\mathbb{R}^2$,

$$\mathbf{v} = \sum_{i=1}^{2} v_i\,\hat{\mathbf{e}}_i\,,$$

where $\hat{\mathbf{e}}_i$, $i = 1, 2$, constitute the basis of Cartesian coordinate system in which the components of $\mathbf{v}$ are expressed. Let us rotate $\mathbf{v}$ counterclockwise by angle θ, as illustrated in the left-hand plot of Figure A.12. The components of $\mathbf{v}$, with respect to the coordinate axes, become

$$\begin{bmatrix} v_1' \\ v_2' \end{bmatrix} = \begin{bmatrix} \cos(\theta) & -\sin(\theta) \\ \sin(\theta) & \cos(\theta) \end{bmatrix} \begin{bmatrix} v_1 \\ v_2 \end{bmatrix}; \tag{A.87}$$

the basis remains unchanged. We can interpret $\mathbf{v}' = \sum_{i=1}^{2} v_i'\,\hat{\mathbf{e}}_i$ as a vector that represents an object rotated by θ. If we rotate the basis by $-\theta$, as illustrated in the right-hand plot of Figure A.12, for the vector—expressed with respect to the new basis—to remain unaltered, we need to adjust its components by applying operation (A.87) to obtain $\mathbf{v} = \sum_{i=1}^{2} v_i'\,\hat{\mathbf{e}}_i'$, as shown explicitly in Example E.1 on page 467.

If we assume that the coordinate system in the left-hand plot of Figure A.12 is fixed, we can interpret the change of coordinates of $\mathbf{v}$, therein, as an active rotation, in contrast to the right-hand plot, which one might view as solely the rotation of coordinates. However, an *affine space* does

[18] *see also*: Brown and Slawinski (2017, Fig. A.1)

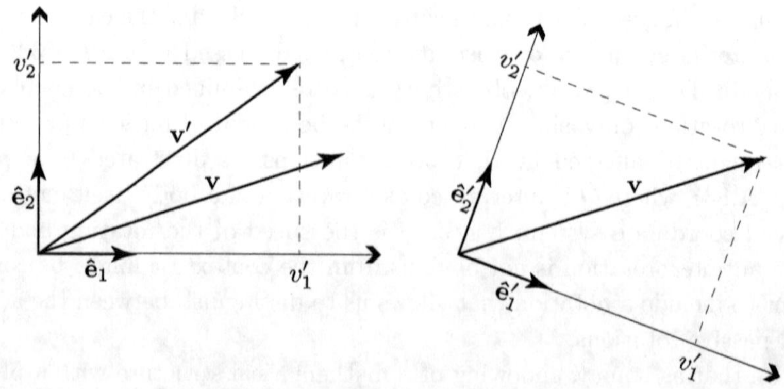

Fig. A.12: Active and passive rotations of $\mathbf{v}$ are interpretations of relations between the vector components and basis elements.

not possess a privileged immovable coordinate system within which objects move. Each *affine transformation* is only a relative motion between an object and the coordinate system.

Let us conclude with an insight of Tao (2013, p. 54).

> **Question A.1.** *Suppose one is viewing some text on a computer screen. The text is so long that one cannot display all of it at once on a screen: currently, only a middle portion of the text is visible. To see more of the text, we press the "up" arrow key (or click the "up" button). When one does so, does the text on the screen move up or down?*
>
> The answer depends on the user interface model. Most such models are *passive transformations*; the "up" command moves the *observer* up, and one's view of the text then moves *down* as a consequence. A minority of models (such as "hand" tools in various pieces of software) are instead *active transformations*; dragging a hand tool upwards causes the text to move upward (keeping the observer position fixed).

Subsequently, Tao (2013) discusses this—as he writes, somewhat informally posed—question, in a mathematical context, by emphasizing the need for a careful interpretation, whose ambiguity can be removed by careful notation, with, as he writes, "internal" and "external" labeling.

A.12 On terminology

Terminology used in finite elasticity might lead to misunderstandings, if interpreted *sensu stricto* or if endowed hastily with a geometrical meaning. The deformation-gradient tensor, F, is not the gradient in the sense of vector calculus, even though it shares with it certain properties of operations, which are expressed in terms of partial derivatives and denoted by $\nabla_{\mathbf{X}}$, with the subscript indicating their arguments.

Exhibiting operational similarities does not entail the sharing of essential properties. In the language of philosophy,[19] epistemological and ontological reductions are distinct from one another. For example, the same pattern to calculate the vector product, $\mathbf{v}\times\mathbf{w}$, and the determinant of a 3×3 matrix is only an operational similarity, since this product is not a determinant of a matrix. Furthermore, we commonly refer to the result of this product as a vector, since, similarly to a vector in $\mathbb{R}^3$, it has three linearly independent components. However, as mentioned in Appendix E, this is a consequence of its being an antisymmetric second-rank pseudotensor in $\mathbb{R}^3$, where the prefix indicates a modification in its transformation properties (e.g., Landau and Lifshitz, 1975, pp. 18–19). We might refer to $\mathbf{v}\times\mathbf{w}$ as a vector, but we must be aware of its context and subtle distinctions. For the reason of such distinctions, it is also referred to as a pseudovector, as mentioned in Remark A.16; again, the prefix indicates a modification in its transformation properties.

The issue of F lies in the fact that the independent variables, $\mathbf{X}$, are in the reference configuration and the dependent variables, $\mathbf{x}(\mathbf{X})$, which are the values of a function, are in the current configuration. In the language of differential geometry, the domain and the range of $\mathbf{x}(\mathbf{X})$ belong to different spaces. Thus, for instance, one could not interpret F in a manner analogous to vector calculus, where a gradient is the direction—within the domain of a function—along which the rate of increase of the function is the greatest.

Furthermore, for the very reason of $\mathbf{X}$ and $\mathbf{x}(\mathbf{X})$ belonging to different configurations, F is not a tensor, *sensu stricto*. It does not follow the transformation rule, described in Appendix E, which is a necessary condition to be a tensor. In view of Appendix A.11, since a two point-tensor transforms—with respect to either of its indices—as a vector, it is occasionally referred to as a *double vector*. However, the fact that two point-tensors, such as F, are not tensors does not mean that they do not share certain

[19] *see also*: Brown and Slawinski (2017, Chapter 7)

operational properties with tensors nor does it prevent their usefulness in continuum mechanics. In a manner similar to the infinitesimal-strain tensor, stated in expression (A.76), F acquires its two indices as a consequence of partial derivatives of two single-index entities; in a manner similar to the Cauchy stress tensor, stated in expression (2.15), F forms a linear relation between two single-index entities. Strain tensors discussed in this appendix, and listed in Table A.1, are formulated in terms of F.

A certain vagueness of terminology and of a common geometrical meaning is true for other two point-tensors, including the rotation tensor, R, discussed in Appendix A.9, and the First Piola-Kirchhoff stress tensor, discussed in Appendix B.2.1. Strictly speaking, neither of them is a tensor, nor can R be interpreted as a rotation, in the sense of analytical geometry. In the language of continuum mechanics, objects before and after mapping R belong to different configurations; in the language of differential geometry, they belong to different manifolds. In either case, this is illustrated by distinct bases of the coordinate system prior to, and following, the mapping. A common meaning of rotation, which requires a rotated entity to remain within the same—however abstract—space, is not preserved.

At times a misunderstanding arises in the context of, so called, active and passive transformations of tensors, illustrated in Example A.12. Therein, a mathematician might view a single vector in several guises, which are a consequence of its transformations. After all, mathematically, the essence of a tensor is the invariance of its properties under orthogonal transformations. A physicist, for whom a tensor serves as an analogy for an examination of the physical world, might view it as the same vector, $\mathbf{v}$, if its transformation is interpreted as a coordinate-transformation only, or as two distinct vectors, $\mathbf{v}$ and $\mathbf{v}'$, if it is interpreted as a change of orientation of a vector itself.

Let us conclude by a comment on deformation versus strain. In this book, we use these terms synonymously, in contrast to, for instance, Ogden (1997) and Chadwick (1976), as indicated by their quotes on page 398. Our meaning of deformation is consistent with its etymology in Latin *deformare*, which means to modify shape or size.

If necessary, the distinction between a deformed and undeformed continuum or a strained or unstrained continuum can be made obvious in the context in which these terms appear. Thus, without loss of clarity, the deformation gradient appears in Italian as *gradiente di deformazione* or in French as *tenseur gradient de la transformation*, and the strain tensor as *tensore di deformazione* or *tenseur des déformations*. Furthermore,

Strain tensor	Location	Type	Symbol
Right Cauchy-Green	Appendix A.4.1	reference configuration	C
Left Cauchy-Green	Appendix A.4.2	current configuration	b
Green-Lagrange	Appendix A.5	reference configuration	E
Almansi	Appendix A.5	current configuration	e
Right Stretch	Appendix A.9.2	reference configuration	U
Left Stretch	Appendix A.9.3	current configuration	v

Table A.1: Strain tensors

in both French and Italian, the polar decomposition results in *rotation et déformation* and *rotazione e deformazione*, respectively, where deformation is understood in the manner used in this book. In Polish, which—like English—is not a Latin language, it is common to use interchangeably *deformacja* and *odkształcenie* (e.g., Rymarz, 1993, Section 2.2), where the latter is just a Slavic translation of the former.

Closing remarks

Formulations in this appendix rely on operations involving the deformation gradient, F. F relates entities in two distinct configurations, which—in general—belong to different spaces. To simplify the operations, however, we assume that both spaces are contained in the same $\mathbb{R}^3$. Such an approach does not invoke explicitly the spaces tangent to a manifold, and, hence, restricts the generality of a formulation. Nevertheless, it allows us to examine the entities pertinent for finite strains.

The essence of this appendix is the formulation of a description that allows us to examine the deformation of a continuum, even though its motion, in general, combines a deformation with a translation and a rigid-body rotation. Deformations alone are given by the strain tensors listed in Table A.1.

Each pair of tensors in Table A.1 consists of counterparts related by the pushforward and pullback operations between the reference and current configurations. For each tensor within a pair, the deformation of a given geometric quantity is expressed with respect to only one configuration; hence, the result is not affected by coordinate transformations.

As discussed in Appendices A.4, A.5, A.9 and, in particular, in Remark A.14 on page 384, all strain tensors listed in Table A.1 are symmetric. However, not all of them are, in general, positive-definite; E and e are not. Notably, as shown in Appendix A.5, the infinitesimal-strain tensor, ε, given in expression (A.76) and omnipresent in seismology, is derived from these two tensors. As shown in Appendix A.10, it can also be derived by invoking the decomposition of the displacement gradient into its symmetric and antisymmetric parts.

There are other finite-strain measures. Notably, the *Hencky strain tensor* can be expressed in terms of the right Stretch tensor and left Stretch tensor as $H := \ln U$ and $h := \ln v$, in the material and spatial coordinates, respectively. The concept of logarithmic quantities is consistent with Zipf's law of statistics, which assumes that many types of data studied in physics can be approximated by a logarithmic distribution.

A logarithmic strain is an example of the relation between the finite and infinitesimal formulations, where the former can be viewed as the summation of increments of the latter. For a one-dimensional stretch, we write[20]

$$\varepsilon_\ell = \int\limits_0^{\varepsilon_\ell} \mathrm{d}\varepsilon = \int\limits_{x_0}^{x} \frac{\mathrm{d}\xi}{\xi} = \ln x - \ln x_0 = \ln \frac{x}{x_0} = \ln(1+\varepsilon)\,,$$

where $\mathrm{d}\xi$ is an incremental elongation, x_0 is the original length and x is the current length. Hence, in accordance with a property of the logarithm,

$$\varepsilon = \exp \varepsilon_\ell - 1\,.$$

Thus, if $\varepsilon_\ell \to 0$, so does ε, and—in the limit—the logarithmic strain becomes the infinitesimal strain.

Readers interested in strains discussed in this appendix—in the context of seismology and geodynamics—might refer to Dahlen and Tromp (1998, Chapter 2). In particular, finite strains appear in seismology in the context of wave propagation in prestressed media (e.g., Chadwick and Jarvis, 1979; Chadwick and Smith, 1977; Destrade and Saccomandi, 2007; Dey and Mukherjee, 1984; Ogden, 2001, Section 1.6).

Aspects of the finite-strain theory, discussed in this appendix, are mentioned in Section 3.2.3. The inherited infinitesimal entities appear implicitly through most of the book. In Appendix B, we discuss stress tensors that correspond to the finite strains discussed in this appendix.

[20]*see also*: Slawinski (2015, Section 1.4.3)

Appendix B

On stresses

The Cauchy stress **T** *measures the contact force per unit area in the deformed configuration. In many problems of interest—especially those involving solids—it is not convenient to work with* **T** *, since the deformed configuration is not known in advance. For this reason we introduce a stress tensor which gives the force measured per unit area in the reference configuration.*[1]

Morton E. Gurtin (1981)

Preliminary remarks

In continuum mechanics, stress and strain are mutually related, since one causes the other. Hence, finite strains, discussed in Appendix A, are associated with their corresponding stresses. Since finite strains result in significant differences between the reference and the current configurations, it is necessary to include this distinction in the context of the corresponding stresses, which map the action of forces in these configurations.

Stresses that appear in the context of finite strains do not have such direct physical interpretations as does the familiar Cauchy stress tensor, which is a measure of the force per unit area in the current configuration. However, as is the case of many abstract entities of mathematical physics, there is no need to insist on such interpretations, provided that a given entity can be related to one endowed with physical meaning.

We begin this appendix by considering the Cauchy stress tensor and, hence, introduce the two Piola-Kirchhoff stress tensors. We discuss properties of, and relations among, these three stress tensors. Notably, we show

[1]Herein, Gurtin (1981, p. 178) refers to the Second Piola-Kirchhoff stress tensor, which we discuss in Appendix B.2.2.

that both Piola-Kirchhoff stress tensors reduce to the Cauchy stress tensor for infinitesimal deformations, which explains—at least partially—its ubiquity in seismology. We conclude this appendix by showing that the three discussed stresses are objective entities.

Notation B.1. In this appendix, as well as in Appendices A, C and D, the quantities associated with the reference configuration are commonly denoted by the upper-case letters and the quantities associated with the current configuration are commonly denoted by the lower-case letters; not all upper-case or lower-case letters, however, refer to these configurations. Bold letters represent vectors.

B.1 Cauchy stress tensor

Let us consider the Cauchy stress tensor,[2] which originates in expression (2.15). In that expression, we make no distinction between the material and spatial coordinates; therein, $\mathbf{T}$ stands for $\mathbf{t}$ of this appendix. Thus, herein, $\mathbf{t} = \sigma\,\mathbf{n}$. In components, we write the stress tensor as

$$\sigma = \sum_{i=1}^{3}\sum_{j=1}^{3} \sigma_{ij}\,\hat{\mathbf{e}}_i \otimes \hat{\mathbf{e}}_j\,. \tag{B.1}$$

The basis elements of tensor (B.1) are $\hat{\mathbf{e}}_i \otimes \hat{\mathbf{e}}_j$, where $i, j = 1, 2, 3$, which means that σ is a current-configuration entity, and, hence, it is expressed in spatial coordinates.

Example B.1. For numerical illustrations of this appendix, let us consider

$$\sigma = \begin{bmatrix} 7 & 0 & -2 \\ 0 & 5 & 0 \\ -2 & 0 & 4 \end{bmatrix}, \tag{B.2}$$

whose components are real and symmetric, $\sigma_{ij} = \sigma_{ji}$.

Discussing the Cauchy stress tensor, Chou and Pagano (1992, p. 199) write that

> [a]t a given point, the stress vectors on all planes of different inclination may be conveniently represented graphically by the surface of an ellipsoid (the ellipsoid of Lamé, or the stress ellipsoid). This surface offers a graphic model of the three-dimensional state of stress at a point.

[2] *see also*: Slawinski (2015, Section 2.5)

Remark B.1. To discuss the *Lamé stress ellipsoid,*[3] let us consider the characteristic polynomial of tensor (B.2), which is $\lambda^3 - 16\lambda^2 + 79\lambda - 120$. Hence, its eigenvalues—which are the magnitudes of its *principal stresses*—are $\lambda_1 = 3$, $\lambda_2 = 5$, $\lambda_3 = 8$. Also, using these values in the three expressions of Remark A.13, we can obtain the three invariants of this tensor; they are $I_\sigma = 16$, $II_\sigma = 79$ and $III_\sigma = 120$, as expected, in view of the characteristic polynomial.

The orientation of each principal stress is given by the direction of the corresponding unit eigenvector,

$$\begin{aligned}\mathbf{n}_1 &= \left[\frac{1}{\sqrt{5}}, 0, \frac{2}{\sqrt{5}}\right],\\ \mathbf{n}_2 &= [0, 1, 0],\\ \mathbf{n}_3 &= \left[-\frac{2}{\sqrt{5}}, 0, \frac{1}{\sqrt{5}}\right].\end{aligned}$$

These eigenvectors span the *principal-stress space*; its basis components consist of three normal traction vectors,

$$\mathbf{t}_i^{(\mathbf{n})} = \lambda_i \, \mathbf{n}_i, \qquad i = 1, 2, 3. \tag{B.3}$$

This basis allows us to write—according to Hooke's law—the Cauchy traction acting on a plane with an arbitrary orientation.[4] There is no loss of generality in using $\mathbf{n}_i$, as opposed to $\hat{\mathbf{e}}_i$, where $i = 1, 2, 3$, for the basis of a coordinate system, since principal-stress directions can be determined for any stress tensor.

The equation given by

$$\hat{\mathbf{e}}_1^2 + \hat{\mathbf{e}}_2^2 + \hat{\mathbf{e}}_3^2 = 1$$

is a quadric, which is a surface expressed in terms of a second-order algebraic equation; herein, it represents a sphere whose centre is the origin of the Cartesian coordinate system. In an analogous fashion,

$$\mathbf{n}_1^2 + \mathbf{n}_2^2 + \mathbf{n}_3^2 = 1 \tag{B.4}$$

is *Cauchy's stress quadric*, which is a sphere in the principal-stress space, and whose graphical representation—in the Cartesian coordinates—is the Lamé stress ellipsoid; its centre is the point whose state of stress is represented by stress tensor σ.

[3]Readers interesting in formulations of principal stresses and the Lamé stress ellipsoid, as well as their history and derivation might refer to Timoshenko (1983, p. 116) and Timoshenko and Goodier (1987, Chapter 7).

[4]*see also*: Slawinski (2015, Section 2.5.2)

Since $[\mathbf{n}_1, \mathbf{n}_2, \mathbf{n}_3]$ coincide with the axes of Cauchy's stress quadric, principal stresses include both the minimum and the maximum normal component of traction at a point. In this context, Chou and Pagano (1992, p. 200) write that

> [t]he stress ellipsoid has the property that the radius vector from the origin to a point on its surface is equal to the stress vector on some plane [...]. Since the lengths of the semiaxes of the ellipsoid are equal to the magnitudes of the principal stresses, the two principal stresses having the largest and smallest *magnitudes* represent the stress vectors having the largest and smallest magnitudes for any plane at a given point. The stress ellipsoid by itself, however, may not be used to determine the plane on which a given stress vector acts except in certain special cases.

Denoting the three principal stresses in equation (B.3) by $\sigma_1, \sigma_2, \sigma_3$, we write equation (B.4) as

$$\mathbf{n}_1^2 + \mathbf{n}_2^2 + \mathbf{n}_3^2 = \left(\frac{\mathbf{t}_1^{(\mathbf{n})}}{\sigma_1}\right)^2 + \left(\frac{\mathbf{t}_2^{(\mathbf{n})}}{\sigma_2}\right)^2 + \left(\frac{\mathbf{t}_3^{(\mathbf{n})}}{\sigma_3}\right)^2 = 1$$

which is a condition imposed on the traction vectors by properties of $[\mathbf{n}_1, \mathbf{n}_2, \mathbf{n}_3]$ (Mase, 1970, Section 2.9-2.11).[5]

In a manner analogous to expression (A.69), we can write tensor (B.2), expressed in a tensor notation within its principal-stress space, as

$$\begin{aligned}\sigma &= \sum_{i=1}^{3} \sigma_i \, \mathbf{n}_i \otimes \mathbf{n}_i \\ &= 3\begin{bmatrix} \frac{1}{5} & 0 & \frac{2}{5} \\ 0 & 0 & 0 \\ \frac{2}{5} & 0 & \frac{4}{5} \end{bmatrix} + 5\begin{bmatrix} 0 & 0 & 0 \\ 0 & 1 & 0 \\ 0 & 0 & 0 \end{bmatrix} + 8\begin{bmatrix} \frac{4}{5} & 0 & -\frac{2}{5} \\ 0 & 0 & 0 \\ -\frac{2}{5} & 0 & \frac{1}{5} \end{bmatrix},\end{aligned}$$

where the 3×3 matrices are the basis of the coordinate system in which the components of σ are expressed. In a manner analogous to the one discussed in Remark A.20, we can write tensor (B.2), in a matrix notation within its

[5] Readers interested in a graphic representation of principal stresses within Mohr's circle might refer to Mase (1970, Section 2.12). Lamé stress ellipsoid is an alternative to Mohr's circle for the graphical representation of the stress state at a point.

principal-stress space, as

$$\sigma = \begin{bmatrix} \mathbf{n}_1\, \mathbf{n}_2\, \mathbf{n}_3 \end{bmatrix} \begin{bmatrix} \sigma_1 & 0 & 0 \\ 0 & \sigma_2 & 0 \\ 0 & 0 & \sigma_3 \end{bmatrix} \begin{bmatrix} \mathbf{n}_1\, \mathbf{n}_2\, \mathbf{n}_3 \end{bmatrix}^T$$

$$= \begin{bmatrix} \frac{1}{\sqrt{5}} & 0 & -\frac{2}{\sqrt{5}} \\ 0 & 1 & 0 \\ \frac{2}{\sqrt{5}} & 0 & \frac{1}{\sqrt{5}} \end{bmatrix} \begin{bmatrix} 3 & 0 & 0 \\ 0 & 5 & 0 \\ 0 & 0 & 8 \end{bmatrix} \begin{bmatrix} \frac{1}{\sqrt{5}} & 0 & -\frac{2}{\sqrt{5}} \\ 0 & 1 & 0 \\ \frac{2}{\sqrt{5}} & 0 & \frac{1}{\sqrt{5}} \end{bmatrix}^T ,$$

where the matrix composed of the eigenvectors is a rotation matrix, whose determinant is equal to unity; this rotation results in the orientation of the coordinate system in which the components of σ are diagonalized.

The three principal stresses, which are the stress-tensor components normal to the planes on which the traction is acting, allow us to to calculate the maximum shear-stress component, which acts tangentially on a plane. Its magnitude is half of the difference between the maximum and minimum principal stresses; herein,

$$\tau_{\max} = \frac{\sigma_3 - \sigma_1}{2} = 2.5 \,, \tag{B.5}$$

where we use symbol τ to distinguish between normal and tangential stress-tensor components. Since the minimum and maximum principal strains correspond to σ_{11} and σ_{33}, respectively, the orientation of the maximum shear is obtained by rotating the principal coordinates by $\pi/4$ in the x_1x_3-plane.

If, in formula (B.5), $\sigma_i = \sigma_j$, where $i, j = 1, 2, 3$ and $i \neq j$, then the shear-stress components are zero. This means that the stress tensor is tantamount to pressure, σI, where σ is a scalar. Similarly, in Exercise 2.1, on page 34, the shear-stress components are zero, which therein corresponds to $\sigma_{11} \neq \sigma_{22} = \sigma_{33} = 0$, as illustrated in Exercise 3.4, on page 94.

These results are analogous to the ones obtained for principal strains in Example A.8. However, therein, the principal strains are expressed in the reference configuration. Herein, the principal stresses are expressed in the current configuration. Yet, the intrinsic geometrical properties, such as the maximum and minimum values in principal directions, hold for both cases.

B.2 Piola-Kirchhoff stress tensors

B.2.1 *First Piola-Kirchhoff stress tensor*

Let us consider both the reference and the current configurations of a body and their oriented surface elements, $\mathbf{N}\,\mathrm{d}S$ and $\mathbf{n}\,\mathrm{d}s$, illustrated in Figure B.1. If a force, $\mathrm{d}\mathbf{f}$, acts on the latter, we can write

$$\mathrm{d}\mathbf{f} = \sigma\,\mathbf{n}\,\mathrm{d}s\,, \tag{B.6}$$

where σ is the Cauchy stress tensor. The First Piola-Kirchhoff stress tensor, P, is defined by

$$\mathrm{d}\mathbf{f} = P\mathbf{N}\,\mathrm{d}S\,. \tag{B.7}$$

It is a two-point tensor, since it relates the force acting in the current configuration to the surface element in the reference configuration. In components, we write

$$P = \sum_{i=1}^{3}\sum_{J=1}^{3} P_{iJ}\,\hat{\mathbf{e}}_i \otimes \hat{\mathbf{E}}_J\,. \tag{B.8}$$

Since P is a two-point tensor, its symmetry cannot be considered; components P_{iJ} and P_{Ji}—where i and J correspond to coordinate systems in different configurations—cannot be compared.

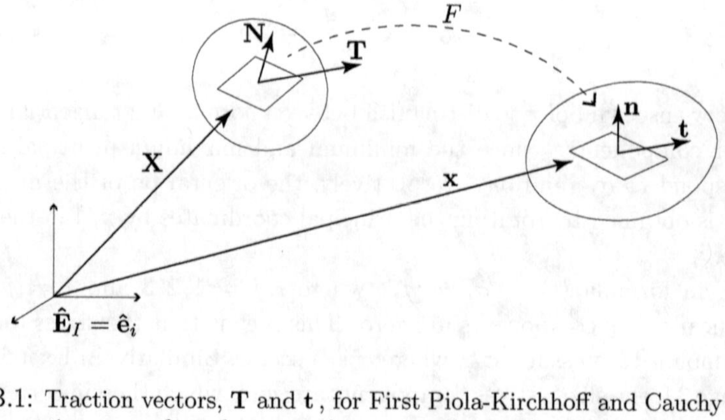

Fig. B.1: Traction vectors, $\mathbf{T}$ and $\mathbf{t}$, for First Piola-Kirchhoff and Cauchy stress tensors, acting on surface elements whose normals are $\mathbf{N}$ and $\mathbf{n}$, respectively

In a manner analogous to the Cauchy traction—where $\mathbf{t} = \mathrm{d}\mathbf{f}/\mathrm{d}s$ and $\mathbf{t} = \sigma\,\mathbf{n}$—herein, we have $\mathbf{T} = \mathrm{d}\mathbf{f}/\mathrm{d}S$ and $\mathbf{T} = P\mathbf{N}$. Since $\mathrm{d}\mathbf{f} = \mathbf{t}\,\mathrm{d}s = \mathbf{T}\,\mathrm{d}S$, it follows that $\mathbf{t}$ and $\mathbf{T}$ act in the same direction, as illustrated in Figure B.1, and that their respective magnitudes are scaled by $\mathrm{d}s$ and $\mathrm{d}S$.

In accordance with expressions (B.6) and (B.7), it follows that

$$\sigma\,\mathbf{n}\,\mathrm{d}s = P\,\mathbf{N}\,\mathrm{d}S\,. \tag{B.9}$$

Hence, using Nanson's formula (A.61), we can relate the Cauchy and the First Piola-Kirchhoff stress tensors by[6]

$$P = J\,\sigma\,F^{-T} \tag{B.10}$$

and, hence,

$$\sigma = J^{-1} P\,F^{T}\,, \tag{B.11}$$

where $J := \det F$; the determinant of a Jacobi matrix is referred to as a *Jacobian.*

Example B.2. Using expression (B.2), of Example B.1, in equation (B.10), together with tensor (A.12), we obtain

$$P = \begin{bmatrix} 11.7445 & 1.6774 & -9.0427 \\ -3.7125 & 5.1025 & 2.3735 \\ -5.4891 & -0.2033 & 6.5773 \end{bmatrix}. \tag{B.12}$$

The coordinate systems in which the components of σ and P are expressed are different from one another; hence, one cannot directly compare their components.

It is common to view the Cauchy stress tensor as a physical quantity, since—in contrast to the First Piola-Kirchhoff stress tensor—it uses the current force in a current configuration. In such a view, the First Piola-Kirchhoff stress is considered a fictitious quantity. One might argue, however, that any mathematical entity used to study a physical phenomenon is a fictitious object, and one is not more physical than another.

In view of expression (B.9), we suggest an interpretation of the First Piola-Kirchhoff stress tensor. If we write

$$\iint_s \sigma\,\mathbf{n}\,\mathrm{d}s = \iint_S P\,\mathbf{N}\,\mathrm{d}S\,,$$

then vector $P\,\mathbf{N}$ represents the force acting in a current configuration on the corresponding unit area in the reference configuration. In other words, the Cauchy stress tensor and the First Piola-Kirchhoff stress tensor are linear operators acting on vectors $\mathbf{n}$ and $\mathbf{N}$, respectively, to produce current forces acting on elements of area to which they are normal.

[6]Readers interested in details of formulation of P might refer to Bonet and Wood (2008, Sections 5.5.2).

B.2.2 *Second Piola-Kirchhoff stress tensor*

The Second Piola-Kirchhoff stress tensor is defined by[7]

$$S = J\,F^{-1}\,\sigma\,F^{-T}\,. \tag{B.13}$$

It is expressed in material coordinates,

$$S = \sum_{I=1}^{3}\sum_{J=1}^{3} S_{IJ}\,\hat{\mathbf{E}}_I \otimes \hat{\mathbf{E}}_J\,, \tag{B.14}$$

and, thus, it is contained within the reference configuration.

Remark B.2. S relates a force, $\mathrm{d}\mathbf{f} = \sigma\,\mathbf{n}\,\mathrm{d}s$, in a current configuration, to a force, $\mathrm{d}\mathbf{F} = S\,\mathbf{N}\,\mathrm{d}S$, in a reference configuration, by $\mathrm{d}\,\mathbf{F} = F^{-1}\mathrm{d}\mathbf{f}$. The middle expression is analogous to expression (B.7), except that, herein, all quantities are contained in the reference configuration.

If we assume that the Cauchy stress tensor is symmetric,[8] which is a common consideration, then so is the Second Piola-Kirchhoff stress tensor,

$$S^T = \left(J\,F^{-1}\,\sigma\,F^{-T}\right)^T = J\,\left(\sigma\,F^{-T}\right)^T\left(F^{-1}\right)^T = J\,F^{-1}\sigma^T F^{-T} = S\,,$$

where $\sigma = \sigma^T$; otherwise—according to definition (B.13)—the Second Piola-Kirchhoff stress tensor need not be symmetric. Since J is a scalar, $J = J^T$, trivially.

Remark B.3. The Second Piola-Kirchhoff stress tensor and the Cauchy stress tensor are related by the pullback and pushforward operations, discussed on page 385, and scaled by J. We can write them as

$$S = J\phi^*[\,\sigma\,]$$

and

$$\sigma = J^{-1}\phi_*[\,S\,]\,.$$

These operations are called the *Piola transformations* (Bonet and Wood, 2008, Remark 5.2). Explicitly, in view of expression (B.13), the pullback is $\phi^*[\sigma] = F^{-1}\,\sigma\,F^{-T}$, and the pushforward is $\phi_*[S] = F\,S\,F^T$.

[7]Readers interested in details of formulation of S might refer to Bonet and Wood (2008, Sections 5.5.3).

[8]*see also*: Slawinski (2015, Section 2.7)

The Second Piola-Kirchhoff stress tensor is related to the First Piola-Kirchhoff stress tensor by the deformation gradient, namely,

$$P = F\,S\,, \tag{B.15}$$

which means that

$$S = F^{-1}P\,. \tag{B.16}$$

Example B.3. Using equation (B.13), together with tensors (A.12) and (B.2)—or equation (B.16), together with tensors (A.12) and (B.12)—we obtain

$$S = \begin{bmatrix} 20.6771 & -0.9961 & -16.8789 \\ -0.9961 & 4.8946 & 0.5014 \\ -16.8789 & 0.5014 & 15.1761 \end{bmatrix}, \tag{B.17}$$

whose basis elements are $\hat{\mathbf{E}}_I \otimes \hat{\mathbf{E}}_J$, where $I, J = 1, 2, 3$. It is symmetric due to the symmetry of tensor (B.2); components of tensor (B.12) are not a symmetric matrix, but they are obtained from tensor (B.2), and, hence, they inherit certain of its properties.

Remark B.4. Let us examine tensor (B.17) in a manner presented in Remark B.1. The characteristic polynomial of this tensor is

$$\Lambda^3 - 40.7478\,\Lambda^2 + 203.14489\,\Lambda - 138.0600\,.$$

Hence, its eigenvalues, which are the magnitudes of its principal stresses, are

$$\Lambda_1 = 35.0671\,, \quad \Lambda_2 = 4.8728\,, \quad \Lambda_3 = 0.8080\,.$$

Its unit eigenvectors, which are orthogonal to each other, are

$$\begin{aligned} \mathbf{N}_1 &= [\,0.7615,\ -0.0359,\ -0.6471\,]\,, \\ \mathbf{N}_2 &= [\,0.0147,\ -0.9973,\ 0.0726\,]\,, \\ \mathbf{N}_3 &= [\,-0.6480,\ -0.0648,\ -0.7589\,]\,. \end{aligned}$$

In a manner analogous to expression (A.69), we can write tensor (B.17) as

$$S = \sum_{I=1}^{3} \Lambda_I\,\mathbf{N}_I \otimes \mathbf{N}_I = 35.0671 \begin{bmatrix} 0.5799 & -0.0273 & -0.4928 \\ -0.0273 & 0.0013 & 0.0232 \\ -0.4928 & 0.0232 & 0.4188 \end{bmatrix}$$

$$+\,4.8728 \begin{bmatrix} 0.0002 & -0.0147 & 0.0011 \\ -0.0147 & 0.9945 & -0.0724 \\ 0.0011 & -0.0724 & 0.0053 \end{bmatrix} + 0.8080 \begin{bmatrix} 0.4198 & 0.0420 & 0.4917 \\ 0.0420 & 0.0042 & 0.0492 \\ 0.4917 & 0.0492 & 0.5760 \end{bmatrix},$$

where the 3×3 matrices are the basis of its coordinate system. In a manner analogous to the one discussed in Remark A.20, we can also write it as

$$S = \begin{bmatrix} \mathbf{N}_1 \, \mathbf{N}_2 \, \mathbf{N}_3 \end{bmatrix} \begin{bmatrix} \Lambda_1 & 0 & 0 \\ 0 & \Lambda_2 & 0 \\ 0 & 0 & \Lambda_3 \end{bmatrix} \begin{bmatrix} \mathbf{N}_1 \, \mathbf{N}_2 \, \mathbf{N}_3 \end{bmatrix}^T$$

$$= \begin{bmatrix} 0.7615 & 0.0147 & -0.6480 \\ -0.0359 & -0.9973 & -0.0648 \\ -0.6471 & 0.0726 & -0.7589 \end{bmatrix} \begin{bmatrix} 35.0671 & 0 & 0 \\ 0 & 4.8728 & 0 \\ 0 & 0 & 0.8080 \end{bmatrix}$$

$$\begin{bmatrix} 0.7615 & 0.0147 & -0.6480 \\ -0.0359 & -0.9973 & -0.0648 \\ -0.6471 & 0.0726 & -0.7589 \end{bmatrix}^T ;$$

since the determinant of the matrix composed of the eigenvectors is equal to unity, its action results in rotation of a coordinate system such that S becomes diagonal.

In view of expressions (B.8), (B.1) and (B.14), the coordinate systems in which the components of σ, P and S are expressed are different from each other. Hence, one cannot compare them by directly examining their components. Also, a comparison of eigenvalues and eigenvectors requires taking into consideration the effects of F. For instance, in contrast to v and U, in Example A.11, $\mathbf{n}_i$, $i = 1, 2, 3$, of σ, are not related to $\mathbf{N}_I$, $I = 1, 2, 3$, of S, by the rotation, R, alone, since σ and S have distinct eigenvalues.

B.3 Stress tensors for infinitesimal strains

Let us consider stress tensors in the context of infinitesimal strains, discussed in Appendix A.10. To do so, we rewrite expression (A.80) as

$$F = I + \nabla_{\mathbf{x}} \mathbf{u} + o(\nabla_{\mathbf{x}} \mathbf{u}),$$

where $\mathbf{u}$ is a displacement vector and $o(\cdot)$ is the little-o symbol,[9] which herein means that the products of the displacement-vector gradient are ignored. Also, since $J := \det F$, it follows that

$$J = \det \left[I + \nabla_{\mathbf{x}} \mathbf{u} + o(\nabla_{\mathbf{x}} \mathbf{u}) \right] .$$

It follows that $J = 1 + O(\nabla_{\mathbf{x}} \mathbf{u})$, where $O(\cdot)$ is the big-O symbol. Hence, following expression (B.10), and assuming boundedness of P, we write

$$\sigma = P + O(\nabla_{\mathbf{x}} \mathbf{u})$$

[9] *see also*: Bóna and Slawinski (2015, page 138)

and, following expression (B.13),

$$\sigma = S + O(\nabla_{\mathbf{x}}\, \mathbf{u}).$$

which means that σ, P and S are equal to each other up to the first order in $\nabla_{\mathbf{x}}\, \mathbf{u}$. If $F = I$, they are explicitly equal to each other, $\sigma = P = S$.

B.4 Objective stress tensors

Following the discussion of Appendix A.11, let us examine the objectivity of the stress tensors discussed in this appendix.

Since $\mathbf{n}' = Q\,\mathbf{n}$ and $\mathbf{t}' = Q\,\mathbf{t}$ are objective vectors, it follows, for the Cauchy stress tensor, that

$$\mathbf{t} = \sigma\,\mathbf{n} \implies Q^T\mathbf{t}' = \sigma\, Q^T\mathbf{n}' \implies \mathbf{t}' = \left(Q\,\sigma\, Q^T\right)\mathbf{n}',$$

which means that σ is an objective tensor. To consider the First Piola-Kirchhoff stress tensor, we recall expression (B.10) to write

$$P' = J'\,\sigma'\,(F')^{-T} = J\,Q\,\sigma\,Q^T(QF)^{-T} = Q\left(J\,\sigma\,F^{-T}\right) = QP,$$

which is the required condition, with J, as a scalar, maintaining its value in all configurations. The First Piola-Kirchhoff stress tensor is objective in accordance with the transformation of a second-rank two-point tensor. The Second Piola-Kirchhoff stress tensor is objective, since it belongs to the reference configuration, which is not affected by any transformation.

Closing remarks

Since stress and strain are mutual related, the essence of this appendix is the formulation of a description that allows us to relate finite strains, discussed in Appendix A, to their stresses. We achieve it by discussing three stress tensors listed in Table B.1. In contrast to the strain tensors

Stress tensor	Location	Type	Symbol
Cauchy	Appendix B.2.1	current configuration	σ
First Piola-Kirchhoff	Appendix B.2.1	two-point	P
Second Piola-Kirchhoff	Appendix B.2.2	reference configuration	S

Table B.1: Stress tensors

listed in Table A.1, one of the stress tensors listed in Table B.1—namely, the First Piola-Kirchhoff stress tensor—is contained in both configurations.

There are other stress measures for finite strains, such as the *Kirchhoff stress tensor*, $\tau = J\,\sigma$, and the *corotational Cauchy stress tensor*, $\sigma' = R^T\sigma\,R$, where R is a rotation tensor. They are also objective tensors, since J is a scalar quantity and R is a tensor that accounts for the rigid-body rotation. Herein, however, we do not examine these tensors.

Since each stress tensor discussed in this appendix—including the above-mentioned Kirchhoff and corotational Cauchy stress tensors—can be related to the Cauchy stress tensor, there is no need to insist on their physical interpretations. Their physical meaning is mediated through mathematical transformations involving the Cauchy stress tensor, whose physical meaning is clear.

In general, under the assumption of infinitesimal deformations, the three distinct stress tensors become indistinguishable from each other, and are represented by the Cauchy stress tensor, except in the case of a residual stress, discussed on page 57. In general, the theory is simplified by this assumption, and its validation is sought in its empirical adequacy, which is the sufficiency of accuracy in predicting physical phenomena.

Readers interested in stresses discussed in this appendix—in the context of global seismology and geodynamics—might refer to Dahlen and Tromp (1998, Chapter 2). Also, these stresses appear in seismology in the context of wave propagation in prestressed media (e.g., Destrade and Saccomandi, 2007; Dey and Mukherjee, 1984; Ogden, 2001, Section 1.6).

Aspects of the stresses for finite strains, discussed in this appendix, are used in Sections 2.2 and 3.2.3. The inherited infinitesimal entities appear through most of the book.

Having discussed finite strains and their corresponding stresses in Appendices A and B, respectively, in Appendix C, we relate them within laws of thermodynamics. In Appendix D, we invoke the strain-energy function to relate them by formulating the stress-strain equations of hyperelastic continua.

Appendix C

On thermoelasticity

Doświadczenie uczy, że odkształcenie ciała związane jest ze zmianą zawartości w nim ciepła. Zmienne w czasie obciążenie ciała wywołuje w nim nie tylko przemieszczenia, ale i zmienny w czasie rozkład temperatury. Odwrotnie, ogrzanie ciała powoduje w nim odkształcenie i zmianę temperatury. Ruch ciała charakteryzowany jest przez wzajemne oddziaływanie na siebie pola odkształcenia i pola temperatury. Dziedzinę, zajmującą się wzajemnym oddziaływaniem tych pól, nazywamy termosprężystością.[1,2]

Witold Nowacki (1965)

Preliminary remarks

[3]In this appendix, we consider expressions that include effects of temperature in the deformation process. To do so, we consider *thermoelastic continua*, whose internal energy is solely a function of the deformation gradient and temperature.

---

[1]*Experience teaches that deformation of a body is connected with a change of its heat content. A time-dependent load on a body causes within it not only deformations, but also a time-dependent temperature distribution. Conversely, heating of a body causes within it deformation and change of temperature. A motion of a body is characterized by a reciprocal influence of the deformation and temperature fields. The domain that deals with the reciprocal influence of these fields is called thermoelasticity.*

[2]Readers interested in thermoelasticity might refer to Gurtin *et al.* (2010, Chapter 27), Holzapfel (2000, Sections 4.5, 4.6, 7.3) and Nowacki (1986); furthermore, if familiar with Polish, they might refer to Rymarz (1993, Rozdział 4). Readers interested in thermodynamics of deformable materials might refer to McLellan (1980). Readers interested in physical underpinnings might refer to Morse (1964).

[3]Appendix C is based on lectures by Andrey Melnikov, delivered at Memorial University in 2017, and discussions with Piotr Stachura during his research visit in 2018.

Apart from allowing us to address specific issues of thermoelasticity, concepts discussed in this appendix allow us to relate objects of finite and infinitesimal elasticity, discussed in Appendices A, B and D. In other words, laws of thermodynamics allow us to assemble objects that, otherwise, might appear unrelated.

We begin this appendix by discussing the first and second laws of thermodynamics, in the context of continuum mechanics. Subsequently, we invoke these laws to examine thermodynamic processes and potentials. Then, we derive stress tensors, formulated in Appendix B, in terms of thermal potentials expressed either per unit volume or per unit mass. We conclude this appendix by exemplifying thermodynamic concepts in the context of isothermal infinitesimal strains.

Notation C.1. In this appendix, as well as in Appendices A, B and D, the quantities associated with the reference configuration are commonly denoted by the upper-case letters and the quantities associated with the current configuration are commonly denoted by the lower-case letters; not all upper-case or lower-case letters, however, refer to these configurations. Bold letters represent vectors.

C.1 First law of thermodynamics

To formulate the first law of thermodynamics, we consider energy. An elastic continuum that possesses not only the mechanical—but also the thermal—energy is called thermoelastic. Thermal energy is transferred within it as heat by the virtue of a temperature gradient. Thermal power, in terms of spatial and material quantities, can be expressed as

$$Q(t) = \iint\limits_{s} q_n \, \mathrm{d}s + \iiint\limits_{v} r \, \mathrm{d}v = \iint\limits_{S} Q_N \, \mathrm{d}S + \iiint\limits_{V} R \, \mathrm{d}V, \tag{C.1}$$

where q_n and Q_N are heat fluxes in a current and reference configurations, respectively; r and R are heat sources per unit volume.

Let us consider *Stokes's heat flux-theorem*, which can be written in spatial and material coordinates as

$$q_n = -\mathbf{q} \cdot \mathbf{n}\,, \quad \text{and} \quad Q_N = -\mathbf{Q} \cdot \mathbf{N}\,, \tag{C.2}$$

respectively, where $\mathbf{n}$ and $\mathbf{N}$ are outward unit normals to the body boundary and, hence, the negative signs imply that influxes have a positive value if heat enters the body; the value is negative, if heat leaves the body.

To relate the heat flux in the reference and current configurations, we write

$$\iint_S \mathbf{Q} \cdot \mathbf{N} \, \mathrm{d}S = \iint_s \mathbf{q} \cdot \mathbf{n} \, \mathrm{d}s = \iint_S \mathbf{q} \cdot J \mathrm{F}^{-\mathrm{T}} \mathbf{N} \, \mathrm{d}S \,, \tag{C.3}$$

where, for the second equality, we use Nanson's formula (A.61). Rearranging the terms of the integrand, we write

$$\iint_S J\mathbf{N} \cdot \mathrm{F}^{-1} \mathbf{q} \, \mathrm{d}S = \iint_S J(\mathrm{F}^{-1}\mathbf{q}) \cdot \mathbf{N} \, \mathrm{d}S \,.$$

Comparing this result to the left-hand side of equation (C.3), we see that

$$\mathbf{Q} = J \mathrm{F}^{-1} \mathbf{q} \,, \tag{C.4}$$

which is the sought relation.

Let us consider the balance of thermal energy,

$$P_{\mathrm{int}}(t) + Q(t) = \frac{\mathrm{d}}{\mathrm{d}t} \mathcal{E}(t) \,, \tag{C.5}$$

where $P_{\mathrm{int}}(t)$ is the rate of work done by internal stresses, $Q(t)$ is thermal power and $\mathrm{d}\mathcal{E}(t)/\mathrm{d}t$ is the rate of change of the internal energy. We define the internal energy as the integral of energy densities, which can be written in the current and reference configurations, respectively, as

$$\mathcal{E}(t) = \iiint_v e \, \mathrm{d}v = \iiint_V e_0 \, \mathrm{d}V \,. \tag{C.6}$$

Let us also include mechanical energy in the postulated balance equation to write

$$\frac{\mathrm{d}}{\mathrm{d}t} T(t) + P_{\mathrm{int}}(t) = P_{\mathrm{ext}}(t) \,,$$

where $T(t)$ is the kinetic energy of the body and $P_{\mathrm{ext}}(t)$ is the rate of external mechanical work. It is the first law of thermodynamics, expressed in spatial coordinates, as the balance of mechanical and thermal energies per unit time. Let us examine it in terms of continuum-mechanics quantities.

Using expression (C.5), we write

$$\frac{\mathrm{d}}{\mathrm{d}t} T(t) + \frac{\mathrm{d}}{\mathrm{d}t} \mathcal{E}(t) = P_{\mathrm{ext}}(t) + Q(t) \,, \tag{C.7}$$

which can be written as

$$\frac{\mathrm{d}}{\mathrm{d}t} \iiint_v \left(\frac{1}{2} \rho \mathbf{v}^2 + e \right) \mathrm{d}v = \iint_s (\mathbf{t} \cdot \mathbf{v} + q_n) \, \mathrm{d}s + \iiint_v (\mathbf{b} \cdot \mathbf{v} + r) \, \mathrm{d}v, \tag{C.8}$$

where we use the expression for kinetic energy, external power—namely,

$$P_{\mathrm{ext}}(t) = \iint\limits_{s} \mathbf{t}\cdot\mathbf{v}\,\mathrm{d}s + \iiint\limits_{v} \mathbf{b}\cdot\mathbf{v}\,\mathrm{d}v\,, \tag{C.9}$$

where $\mathbf{t}$, $\mathbf{b}$, $\mathbf{v}$ represent traction, body force and velocity, respectively—and expressions (C.1) and (C.6), in spatial coordinates. $P_{\mathrm{ext}}(t)$ is the sum of surface and volume powers.

Let us return to expression (C.5) and rewrite its right-hand side as

$$P_{\mathrm{int}}(t) + Q(t) = \frac{\mathrm{d}}{\mathrm{d}t}\mathcal{E}(t) = \frac{\mathrm{d}}{\mathrm{d}t}\iiint\limits_{V} e_0\,\mathrm{d}V = \iiint\limits_{V} \dot{e}_0\,\mathrm{d}V\,, \tag{C.10}$$

where—to commute differentiation and integration—we use the fact that the region of integration, V, does not depend on time, since e_0 is in the reference configuration.

Remark C.1. The commutativity of differentiation and integration is a particular case of the *Leibniz integral rule*,

$$\frac{\mathrm{d}}{\mathrm{d}t}\int\limits_{a(t)}^{b(t)} f(x,t)\,\mathrm{d}x = \int\limits_{a(t)}^{b(t)} \frac{\partial f}{\partial t}\,\mathrm{d}x + f(b(t),t)\,\frac{\partial b}{\partial t} - f(a(t),t)\,\frac{\partial a}{\partial t}\,,$$

for which $a(t)$ and $b(t)$ are constants. In general, this rule can be extended to volume integrals, which—in continuum mechanics—is known as the *Reynolds transport theorem.*

In expression (C.10),

$$P_{\mathrm{int}} = \iiint\limits_{V} (P : \dot{F})\,\mathrm{d}V\,, \tag{C.11}$$

where P is the First Piola-Kirchhoff stress tensor and $\dot{F}$ is the temporal rate of change of the deformation gradient (Holzapfel, 2000, equation (4.101)); operator $:$ is the double-dot product and $\dot{}$ is the time derivative; in other words,

$$P : \dot{F} := \sum_{i=1}^{3}\sum_{J=1}^{3} P_{iJ}\,\frac{\mathrm{d}F_{iJ}}{\mathrm{d}t} = P_{11}\,\frac{\mathrm{d}F_{11}}{\mathrm{d}t} + \cdots + P_{33}\,\frac{\mathrm{d}F_{33}}{\mathrm{d}t}\,,$$

which is a scalar composed of nine summands.

Remark C.2. For two second-rank tensors, S and T, the double-dot product can be expressed as

$$S : T = \mathrm{tr}\left(S^T\,T\right)\,.$$

The double-dot product is a scalar product for second-rank tensors. In general, it lowers the rank from r to $r-2$. For instance, for Hooke's law, $\sigma = c : \varepsilon := \sum_{k=1}^{3}\sum_{\ell=1}^{3} c_{ijk\ell}\,\varepsilon_{k\ell} = c_{ij11}\,\varepsilon_{11} + \cdots + c_{ij33}\,\varepsilon_{33} = \sigma_{ij}$, where $i,j = 1,2,3$.

Following expressions (C.1) and (C.2), and invoking the divergence theorem, we write the thermal power, in terms of the material coordinates, as

$$Q(t) = -\iint\limits_{S} \mathbf{Q}\cdot\mathbf{N}\,\mathrm{d}S + \iiint\limits_{V} R\,\mathrm{d}V = \iiint\limits_{V} \left(-\nabla_{\mathbf{X}}\cdot\mathbf{Q} + R\right)\mathrm{d}V\,. \tag{C.12}$$

Using this result and expression (C.10), we write

$$P_{\mathrm{int}}(t) + \iiint\limits_{V} \left(-\nabla_{\mathbf{X}}\cdot\mathbf{Q} + R\right)\mathrm{d}V = \iiint\limits_{V} \dot{e}_0\,\mathrm{d}V\,.$$

Recalling expression (C.11), we have

$$\iiint\limits_{V} \left(P : \dot{F} - \nabla_{\mathbf{X}}\cdot\mathbf{Q} + R\right)\mathrm{d}V = \iiint\limits_{V} \dot{e}_0\,\mathrm{d}V\,.$$

Requiring the validity of this equation for an arbitrary volume,[4] we obtain

$$\dot{e}_0 = P : \dot{F} - \nabla_{\mathbf{X}}\cdot\mathbf{Q} + R\,, \tag{C.13}$$

which is the balance of energy stated as a differential equation.

In thermoelasticity, the balance of energy, stated by equation (C.12) or (C.13), must be included—together with the conservation of mass and balance of linear momentum—among the field equations, discussed in Section 2.3. These equations provide constraints required to formulate a thermodynamic theory. In a purely mechanical case, equations (C.12) and (C.13) are trivially satisfied.

C.2 Second law of thermodynamics

To formulate the second law of thermodynamics, we consider entropy. The entropy of a continuum—in the reference and current configurations—is

$$S(t) = \iiint\limits_{V} \eta_0\,\mathrm{d}V = \iiint\limits_{v} \eta\,\mathrm{d}v\,, \tag{C.14}$$

[4]*see also*: Slawinski (2015, Exercise 2.1)

respectively. Its units are the units of energy divided by the units of absolute temperature, and densities, η and η_0, are defined per unit volume. Herein, entropy is a measure of disorder within a macroscopic system, and—according to the second law of thermodynamics—it cannot decrease in an isolated system, which spontaneously evolves towards thermodynamic equilibrium: the state of maximum entropy.

In a manner analogous to expression (C.1), we define the rate of entropy entering the body as

$$\tilde{Q}(t) = -\iint\limits_{s} \mathbf{h}\cdot\mathbf{n}\,\mathrm{d}s + \iiint\limits_{V} \tilde{r}\,\mathrm{d}v = -\iint\limits_{S} \mathbf{H}\cdot\mathbf{N}\,\mathrm{d}S + \iiint\limits_{V} \tilde{R}\,\mathrm{d}V\,, \tag{C.15}$$

where we use the same convention for the unit normal as in expression (C.2); herein, we refer to $\mathbf{h}$ as the Cauchy entropy flux and to $\mathbf{H}$ as the Piola-Kirchhoff entropy flux.

According to the second law of thermodynamics, the rate of entropy production is nonnegative and is obtained as the difference between the rate of change of the entropy possessed by a given body and the rate of entropy entering that body,

$$\Gamma(t) = \frac{\mathrm{d}}{\mathrm{d}t}S(t) - \tilde{Q}(t) \geqslant 0\,. \tag{C.16}$$

Rearranging, we see that the rate of change of entropy possessed by a body is balanced by the rate of entropy input and entropy production. Using spatial expressions for $S(t)$ and $\tilde{Q}(t)$,

$$\Gamma(t) = \frac{\mathrm{d}}{\mathrm{d}t}\iiint\limits_{v} \eta\,\mathrm{d}v + \iint\limits_{s} \mathbf{h}\cdot\mathbf{n}\,\mathrm{d}s - \iiint\limits_{v} \tilde{r}\,\mathrm{d}v \geqslant 0\,, \tag{C.17}$$

which is the second law of thermodynamics expressed in spatial coordinates.

The rate of entropy input is related to the rate of thermal work. It is assumed that entropy fluxes, $\mathbf{h}$ and $\mathbf{H}$, are related to heat fluxes, $\mathbf{q}$ and $\mathbf{Q}$, by a coefficient of proportionality, $1/\Theta$, where Θ is absolute temperature, measured in kelvins. The same relations apply for entropy sources, $\tilde{r}$ and $\tilde{R}$, and for heat sources, r and R, so we can postulate

$$\mathbf{h} = \frac{\mathbf{q}}{\Theta}, \quad \tilde{r} = \frac{r}{\Theta} \tag{C.18}$$

and

$$\mathbf{H} = \frac{\mathbf{Q}}{\Theta}, \quad \tilde{R} = \frac{R}{\Theta}\,. \tag{C.19}$$

Following expressions (C.17) and (C.18), we obtain

$$\Gamma(t) = \frac{\mathrm{d}}{\mathrm{d}t}\iiint\limits_{v} \eta\,\mathrm{d}v + \iint\limits_{s} \frac{\mathbf{q}}{\Theta}\cdot\mathbf{n}\,\mathrm{d}s - \iiint\limits_{v} \frac{r}{\Theta}\,\mathrm{d}v \geqslant 0\,, \tag{C.20}$$

which is the Clausius-Duhem inequality, in spatial coordinates.

In thermoelasticity, the Clausius-Duhem inequality, which is a continuum-mechanics statement of the second law of thermodynamics, must be included among the field equations, together with the conservation of mass, balance of linear momentum and balance of energy, discussed in Sections 2.3.2, 2.3.3 and Appendix C.1, respectively. These equations provide constraints required to formulate a thermodynamic theory. In a purely mechanical case, inequality (C.20) is trivially satisfied.

Let us return to the second law of thermodynamics, stated in equation (C.16), and let us modify it by using material expressions for $S(t)$ and $\tilde{Q}(t)$. We introduce γ, which is the rate of entropy production per unit time and unit volume, in the reference configuration; thus, $\Gamma(t) = \iiint_V \gamma \,\mathrm{d}V$. Using expressions (C.14), (C.15), (C.18), (C.19), the second law of thermodynamics can be expressed as

$$\begin{aligned}\iiint\limits_V \gamma \,\mathrm{d}V &= \frac{\mathrm{d}}{\mathrm{d}t}\iiint\limits_V \eta_0 \,\mathrm{d}V + \iint\limits_S \mathbf{H}\cdot\mathbf{N}\,\mathrm{d}S - \iiint\limits_V \tilde{R}\,\mathrm{d}V \\ &= \frac{\mathrm{d}}{\mathrm{d}t}\iiint\limits_V \eta_0 \,\mathrm{d}V + \iint\limits_S \frac{\mathbf{Q}}{\Theta}\cdot\mathbf{N}\,\mathrm{d}S - \iiint\limits_V \frac{R}{\Theta}\,\mathrm{d}V \geqslant 0 .\end{aligned}$$

Since the region of integration, V, does not depend on time, the operations of differentiation and integration commute; hence, we write

$$\iiint\limits_V \gamma \,\mathrm{d}V = \iiint\limits_V \frac{\mathrm{d}\,\eta_0}{\mathrm{d}t}\,\mathrm{d}V + \iint\limits_S \frac{\mathbf{Q}}{\Theta}\cdot\mathbf{N}\,\mathrm{d}S - \iiint\limits_V \frac{R}{\Theta}\,\mathrm{d}V \geqslant 0 .$$

Invoking the divergence theorem and the linearity of integration, we write

$$\begin{aligned}\iiint\limits_V \gamma \,\mathrm{d}V &= \iiint\limits_V \frac{\mathrm{d}\,\eta_0}{\mathrm{d}t}\,\mathrm{d}V + \iiint\limits_V \nabla_{\mathbf{X}}\cdot\left(\frac{\mathbf{Q}}{\Theta}\right)\mathrm{d}V - \iiint\limits_V \frac{R}{\Theta}\,\mathrm{d}V \\ &= \iiint\limits_V \left(\frac{\mathrm{d}\,\eta_0}{\mathrm{d}t} + \nabla_{\mathbf{X}}\cdot\left(\frac{\mathbf{Q}}{\Theta}\right) - \frac{R}{\Theta}\right)\mathrm{d}V \geqslant 0 .\end{aligned}$$

Since the equation must be valid for an arbitrary volume, we require the integrand itself to be zero; hence, we write

$$\gamma = \dot{\eta}_0 + \nabla_{\mathbf{X}}\cdot\left(\frac{\mathbf{Q}}{\Theta}\right) - \frac{R}{\Theta} \geqslant 0 , \tag{C.21}$$

where $\dot{\eta}_0 := \mathrm{d}\eta_0/\mathrm{d}t$, which is the Clausius-Duhem inequality stated as a differential equation.

Let us consider the middle summand of expression (C.21). Differentiating, we obtain

$$\nabla_{\mathbf{x}} \cdot \left(\frac{\mathbf{Q}}{\Theta}\right) = \frac{1}{\Theta} \nabla_{\mathbf{x}} \cdot \mathbf{Q} - \frac{1}{\Theta^2} \mathbf{Q} \cdot \nabla_{\mathbf{x}} \Theta .$$

Since absolute temperature is nonnegative, we use this result, with $\Theta \geqslant 0$, to write expression (C.21) as

$$\gamma \Theta = \dot{\eta}_0 \Theta + \nabla_{\mathbf{x}} \cdot \mathbf{Q} - \frac{1}{\Theta} \mathbf{Q} \cdot \nabla_{\mathbf{x}} \Theta - R \geqslant 0 . \tag{C.22}$$

Using equation (C.13) to express R in terms of P and $\dot{F}$, we write equation (C.22) as

$$\gamma \Theta = P : \dot{F} - \dot{e}_0 + \dot{\eta}_0 \Theta - \frac{1}{\Theta} \mathbf{Q} \cdot \nabla_{\mathbf{x}} \Theta \geqslant 0 .$$

In view of the middle summands, we introduce a new function,

$$\psi = e_0 - \Theta \eta_0 , \tag{C.23}$$

which is the *Helmholtz free energy*, named in acknowledgement of contributions of a German physicist Hermann Ludwig Ferdinand von Helmholtz. Herein, it is Legendre's transformation of $e_0(F, \eta_0)$. Differentiating and rearranging, we obtain

$$\underbrace{\gamma \Theta}_{\text{rate of dissipation of energy}} = \underbrace{P : \dot{F}}_{\text{mechanical energy}} \underbrace{- \dot{\Theta} \eta - \frac{1}{\Theta} \mathbf{Q} \cdot \nabla_{\mathbf{x}} \Theta}_{\text{thermal energy}} - \underbrace{\dot{\psi}}_{\text{free energy}} \geqslant 0 , \tag{C.24}$$

where energies are expressed per unit time.

Remark C.3. It might be convenient to define a new variable, with a corresponding new function, to examine information contained in an old function; Legendre's transformation allows us to do so. Let us exemplify it by considering transformation (C.23), where we replace the old function, $e_0(F, \eta_0)$, by a new function, $\psi(F, \Theta)$, using η_0 and Θ as transformation variables, to write

$$\underbrace{\psi(F, \Theta)}_{\text{new function}} = \underbrace{e_0(F, \eta_0)}_{\text{old function}} - \underbrace{\frac{\partial e_0(F, \eta_0)}{\partial \eta_0}}_{\text{new variable: } \Theta} \underbrace{\eta_0}_{\text{old variable}} . \tag{C.25}$$

On the right-hand side, the quantities are expressed in terms of the old variable, η_0. If $\Theta = \partial e_0(F, \eta_0)/\partial \eta_0$ is soluble for η_0 as an explicit function of F and Θ, we can use it in equation (C.25), to express ψ in terms of F and Θ, which is the left-hand side.

To return from the new function, $\psi(F,\Theta)$, to the old function, $e_0(F,\eta_0)$, let us consider the inverse transformation. Rearranging equation (C.25), we obtain

$$\underbrace{e_0(F,\eta_0)}_{\text{old function}} = \underbrace{\psi(F,\Theta)}_{\text{new function}} - \underbrace{\frac{\partial\psi}{\partial\Theta}}_{\text{old variable: }\eta_0} \underbrace{\Theta}_{\text{new variable}} . \tag{C.26}$$

Equations (C.25) and (C.26) can be rewritten as $e_0 - \psi = \eta_0\,\Theta$ and $\psi - e_0 = \eta_0\,\Theta$, respectively. For this reason, we refer to Legendre's transformation used in Appendices C.2 and C.4 as antisymmetric. In contrast, Legendre's transformation used in Chapter 7, above, Appendix C.6, below, and discussed by Slawinski (2015, Appendix B), where $H + L = \mathbf{p}\cdot\dot{\mathbf{x}}$, is symmetric.

Remark C.4. Energy per unit time is equivalent to power. It is expressed in watts, which are joules per second, and—in base units—are $\mathrm{kg\,m^2\,s^{-3}}$.

If thermal effects are negligible, $\dot\Theta = 0$ and $\nabla_{\mathbf{X}}\Theta = 0$, the right-hand side of equation (C.24) reduces to

$$\gamma\,\Theta = P : \dot F - \dot\psi \geqslant 0\,.$$

In general, inequality (C.24) implies that the free energy is not balanced by the mechanical and thermal energies. To balance free energy, we would need to take into account the dissipation of energy per unit time, which is equal to the production of entropy multiplied by the absolute temperature, $\gamma\,\Theta$.

To endow inequality (C.24) with a physical meaning, let us consider its integral form. To do so, we integrate it—and also add and subtract

$$\frac{\mathrm{d}}{\mathrm{d}t}\iiint\limits_V \frac{1}{2}\rho_0\mathbf{V}^2\mathrm{d}V\,,$$

which is the temporal rate of change of kinetic energy—to obtain

$$\iiint\limits_V \gamma\,\Theta\,\mathrm{d}V = \iiint\limits_V P : \dot F\,\mathrm{d}V + \frac{\mathrm{d}}{\mathrm{d}t}\iiint\limits_V \frac{1}{2}\rho_0\,\mathbf{V}^2\,\mathrm{d}V \tag{C.27}$$

$$- \iiint\limits_V \left(\dot\Theta\eta_0 + \frac{1}{\Theta}\mathbf{Q}\cdot\nabla_{\mathbf{X}}\Theta\right)\mathrm{d}V - \frac{\mathrm{d}}{\mathrm{d}t}\iiint\limits_V \left(\frac{1}{2}\rho_0\mathbf{V}^2 + \psi\right)\mathrm{d}V \geqslant 0\,,$$

where we use the fact that, in a reference configuration, V is time independent and, hence,

$$\iiint_V \dot{\psi}\,\mathrm{d}V := \iiint_V \frac{\mathrm{d}\psi}{\mathrm{d}t}\,\mathrm{d}V = \frac{\mathrm{d}}{\mathrm{d}t}\iiint_V \psi\,\mathrm{d}V\,.$$

The sum of the first and second terms on the right-hand side of equation (C.27) can be viewed as external power. Thus, in view of equation (C.9)—which is stated in terms of spatial coordinates—we write

$$\iiint_V P:\dot{F}\,\mathrm{d}V + \frac{\mathrm{d}}{\mathrm{d}t}\iiint_V \frac{1}{2}\rho_0\mathbf{V}^2\,\mathrm{d}V = \iint_S \mathbf{T}\cdot\mathbf{V}\,\mathrm{d}S + \iiint_V \mathbf{B}\cdot\mathbf{V}\,\mathrm{d}V\,,$$

and, hence, inequality (C.27) can be written as

$$\underbrace{\iiint_V \gamma\,\Theta\,\mathrm{d}V}_{\text{power dissipation}} = \underbrace{\iint_S \mathbf{T}\cdot\mathbf{V}\mathrm{d}S + \iiint_V \mathbf{B}\cdot\mathbf{V}\,\mathrm{d}V}_{\text{external mechanical power}}$$
$$\underbrace{-\iiint_V \left(\dot{\Theta}\,\eta_0 + \frac{1}{\Theta}\mathbf{Q}\cdot\nabla_{\mathbf{X}}\Theta\right)\mathrm{d}V}_{\text{thermal power}}$$
$$-\underbrace{\frac{\mathrm{d}}{\mathrm{d}t}\iiint_V \left(\frac{1}{2}\rho_0\mathbf{V}^2 + \psi\right)\mathrm{d}V}_{\text{rate of change kinetic and free energy}} \geqslant 0\,.$$

Examining this expression, we see that—due to power dissipation—the temporal rate of change of kinetic and free energy of the body is not balanced by the externally applied mechanical and thermal powers, which means that processes under consideration are irreversible.

> Wspominamy tylko, że nieodwracalność jest zjawiskiem makroskopowym, które jest efektem wypadkowym oddziaływania ogromnej liczby mikrociał. Same zjawiska fizyczne na poziomie mikroskopowym są bowiem odwracalne. Wytłumaczeniem związku odwracalnych zjawisk mikroskopowych z nieodwracalnymi makroskopowymi zajmuje się fizyka statystyczna [...]. (Rymarz, 1993, p. 137)[5]

[5]We only mention that irreversibility is a macroscopic phenomenon, which is the resultant effect of the action of a large number of microbodies, and the physical phenomena themselves, at the microscopic level, are reversible. The explanation of the relation between reversible microscopic and irreversible macroscopic processes belongs to statistical physics.

C.3 Heat conduction

Let us consider the heat-conduction inequality,

$$\mathbf{Q} \cdot \nabla_{\mathbf{X}} \Theta \leqslant 0 \tag{C.28}$$

and

$$\mathbf{q} \cdot \nabla_{\mathbf{x}} \Theta \leqslant 0 \,, \tag{C.29}$$

written, respectively, in terms of the material and spatial coordinates. These inequalities state that heat flows against a temperature gradient; in other words, heat flows from regions with high temperatures to the regions with low temperatures.

To examine the heat-conduction inequality, let us consider a case to which we refer as *Duhamel's heat-conduction law*—in honour of a French mathematical physicist, Jean-Marie Constant Duhamel, whose study of differential equations entail this result—and which, in spatial coordinates, we write as

$$\mathbf{q} = -K \, \nabla_{\mathbf{x}} \Theta \,, \tag{C.30}$$

where K is a thermal-conductivity tensor, which is a symmetric. Equation (C.30) is a linear relation between heat flux and temperature gradient.

K plays a role analogous to that of the elasticity tensor in Hooke's law, and can also be anisotropic. However, K relates two first-rank tensors; hence, it is a second-rank tensor, and, thus, exhibits fewer symmetry classes. Also, in a manner analogous to the elasticity tensor, it can depend on position, which implies thermal inhomogeneity.

Substituting expression (C.30) in inequality (C.29), we obtain

$$\nabla_{\mathbf{x}} \Theta \cdot (K \, \nabla_{\mathbf{x}} \Theta) \geqslant 0 \,. \tag{C.31}$$

which implies that K is a positive-semidefinite tensor. This is in contrast to the elasticity tensor, which is positive definite.

If a medium is thermally isotropic, which means that there is no preferred direction for a heat flow, $K = k\, I$, where k is a scalar and I is the identity tensor. Since K is positive-semidefinite, its eigenvalues are $k_1 = k_2 = k_3 =: k \geqslant 0$,[6] and expression (C.30) becomes

$$\mathbf{q} = -k \, \nabla_{\mathbf{x}} \Theta \,. \tag{C.32}$$

Condition (C.31) and the negative sign ensure that the heat cannot flow in the direction of increasing temperature.

[6]Readers interested in a proof might refer to Chadwick (1976, p. 26).

Equation (C.32) is commonly referred to as *Fourier's heat-conduction law*, which was proposed, in 1804, by Jean-Baptiste Biot, and, in 1822, examined empirically by Jean-Baptiste Joseph Fourier. It can be stated as follows. The temporal rate of heat transfer is proportional to the negative gradient of the temperature and to the area—perpendicular to that gradient—through which the heat flows. Ohm's law is the electrical analogue of Fourier's law.

To consider the material counterpart of expression (C.30), we let $K_0 = JK$, which is the corresponding thermal-conductivity tensor. Since $\Theta = \Theta(\mathbf{x}(\mathbf{X}))$, it follows—according to the chain rule—that

$$\underbrace{\frac{\partial\Theta}{\partial\mathbf{X}}}_{\nabla_{\mathbf{X}}\Theta} = \underbrace{\frac{\partial\Theta}{\partial\mathbf{x}}}_{\nabla_{\mathbf{x}}\Theta}\underbrace{\frac{\partial\mathbf{x}}{\partial\mathbf{X}}}_{F},$$

where F is the deformation gradient; hence,

$$\nabla_{\mathbf{x}}\Theta = F^{-T}\nabla_{\mathbf{X}}\Theta\,. \tag{C.33}$$

According to expression (C.4),

$$\mathbf{q} = J^{-1}F\mathbf{Q}\,, \tag{C.34}$$

which, if used—sequentially—in expressions (C.33) and (C.30), results in

$$\mathbf{Q} = -F^{-1}K_0\,F^{-T}\nabla_{\mathbf{X}}\Theta\,, \tag{C.35}$$

which is Duhamel's law of heat conduction, expressed in material coordinates.

Remark C.5. Inequality (C.28) can be obtained from inequality (C.29) if we substitute expressions (C.33) and (C.34) into inequality (C.29) and use the fact that $J > 0$. In an analogous manner, we can obtain inequality (C.29) from inequality (C.28).

C.4 Thermodynamic processes and potentials

C.4.1 *Processes*

Let us discuss types of thermodynamics processes in the context of continuum mechanics.[7]

An adiabatic process means that there is no heat flux nor are there heat sources or sinks. Thus, according to relation (C.15), there is no entropy

[7] Readers interested in thermodynamic potentials in physics might refer to Morse (1964, Chapter 8).

input and the second law of thermodynamics is reduced to $\dot{S} \geqslant 0$, which means that the total entropy, S, cannot decrease with time.

An isothermal process means that the temperature remains constant, $\dot{\Theta} = 0$, in time. An *isentropic process* means that the total entropy remains constant, $\dot{S} = 0$, in time.

If the temporal rate of dissipation of energy in inequality (C.24) is zero, we have an equality. In such a case, we deal with a *reversible process*.

Let us consider an adiabatic and reversible process. Since it is adiabatic, the rate of entropy input is equal to zero, $\tilde{Q} = 0$. Since it is reversible, according to expression (C.16), we conclude that $\dot{S} = 0$. Let us also consider an isentropic reversible process. Since it is isentropic, $\dot{S} = 0$. Since it is reversible, according to expression (C.16), we conclude that $\tilde{Q} = 0$. Therefore, an adiabatic reversible process and an isentropic reversible process are identical to one another. In other words, they are the same thermodynamic processes.

Prior to discussing thermodynamic potentials, in Appendices C.4.2–C.4.5, let us refer again to Rymarz (1993, p. 141).

> Potencjały termodynamiczne wykazują jednak ograniczone zastosowanie. Mają one bowiem tylko sens dla procesów równowagowych lub/i odwracalnych. Są to jednak najczęściej procesy bardzo powolne. Niekiedy stosuje się je bezpodstawnie dla procesów nieodwracalnych (nierównowagowych), tzn. bardzo szybkich.
>
> Należy również zauważyć, że potencjały termodynamiczne są na ogół zależne od zmiennej $\mathbf{X}$ (bądź $\mathbf{x}$) [...]. Oznacza to wtedy, że mamy do czynienia z ośrodkiem termodynamicznie niejednorodnym [...]. W takim przypadku wykorzystywanie potencjałów termodynamicznych dla procesów bardzo szybkich staje się tym bardziej nieuzasadnione.[8]

To proceed, we need to postulate a generic form of the equation of state. To do so, we postulate that thermodynamic potentials are functions of the deformation gradient, F, and temperature, Θ, or entropy, η_0.

[8]Yet, thermodynamic potentials exhibit limited applicability. They can be used only for balanced and/or reversible processes. Most commonly, however, these are very slow processes. Occasionally, these potentials are applied unjustifiably to irreversible (unbalanced), i.e., very fast processes. One should also notice that thermodynamic potentials are, in general, dependent on variable $\mathbf{X}$ (or $\mathbf{x}$) [...]. This means that, in such a case, we are dealing with a thermodynamically inhomogeneous medium [...]. In such a case, using thermodynamic potentials for very fast processes becomes even more unjustified.

C.4.2 *Helmholtz free energy as potential*

Let us return to the Clausius-Duhem inequality, which is stated by inequality (C.24). Invoking inequality (C.28), we obtain a stronger form of inequality (C.24),

$$P : \dot{F} - \dot{\Theta}\,\eta_0 - \dot{\psi} \geqslant 0\,, \tag{C.36}$$

which is the Clausius-Planck statement of the second law of thermodynamics. If we assume the dissipation of energy to be zero, then—following inequality (C.36)—we obtain

$$\dot{\psi} = P : \dot{F} - \dot{\Theta}\,\eta_0\,. \tag{C.37}$$

Since Helmholtz free energy, ψ, is a function of both the deformation gradient, F, and temperature, Θ, with both arguments being functions of time—in view of the chain rule—we write

$$\dot{\psi}(F,\Theta) = \left(\frac{\partial\psi(F,\Theta)}{\partial F}\right)_{\Theta} : \dot{F} + \left(\frac{\partial\psi(F,\Theta)}{\partial\Theta}\right)_{F}\dot{\Theta}\,. \tag{C.38}$$

where subscripts mean that a given variable is kept constant. Comparing the right-hand sides of equations (C.37) and (C.38), we see that

$$P = \left(\frac{\partial\psi(F,\Theta)}{\partial F}\right)_{\Theta}\,, \qquad \eta_0 = -\left(\frac{\partial\psi(F,\Theta)}{\partial\Theta}\right)_{F}\,. \tag{C.39}$$

Thus, Helmholtz free energy is a potential whose derivatives are the First Piola-Kirchhoff stress and entropy.

C.4.3 *Internal energy as potential*

If we consider the temporal derivative of transformation (C.23), we can write inequality (C.36) as

$$P : \dot{F} - \dot{e}_0 + \Theta\,\dot{\eta}_0 \geqslant 0\,, \tag{C.40}$$

which—for reversible processes—becomes

$$\dot{e}_0 = P : \dot{F} + \Theta\,\dot{\eta}_0\,. \tag{C.41}$$

Since the internal energy, e_0 is a function of both the deformation gradient, F, and entropy, η_0, with both arguments being functions of time—in view of the chain rule—we write

$$\dot{e}_0(F,\eta_0) = \left(\frac{\partial e_0(F,\eta_0)}{\partial F}\right)_{\eta_0} : \dot{F} + \left(\frac{\partial e_0(F,\eta_0)}{\partial\eta_0}\right)_{F}\dot{\eta}_0\,. \tag{C.42}$$

Comparing the right-hand sides of equations (C.41) and (C.42), we conclude that

$$P = \left(\frac{\partial e_0(F,\eta_0)}{\partial F}\right)_{\eta_0}\,, \qquad \Theta = \left(\frac{\partial e_0(F,\eta_0)}{\partial\eta_0}\right)_{F}\,. \tag{C.43}$$

Thus, internal energy is a potential whose derivatives are the First Piola-Kirchhoff stress and temperature.

C.4.4 *Gibbs free energy as potential*

Let us introduce the Gibbs free energy, $g(P,\Theta)$, named in recognition of work by an American scientist Josiah Willard Gibbs. To do so, we perform Legendre's transformation of the Helmholtz free energy, $\psi(F,\Theta)$. Using the first of expressions (C.39), we write

$$g = \psi - \frac{\partial\psi}{\partial F} : F = \psi - P : F \tag{C.44}$$

and its time derivative, $\dot{g} = \dot{\psi} - \dot{P} : F - P : \dot{F}$. Thus, using $\dot{\psi} - P : \dot{F} = -\dot{\Theta}\,\eta_0$, and the fact that $\dot{P} : F = F : \dot{P}$, we obtain

$$\dot{g} = -F : \dot{P} - \eta_0\,\dot{\Theta}\,.$$

Since Gibbs free energy is a function of P and Θ, we write

$$\dot{g}(P,\theta) = \left(\frac{\partial g(P,\Theta)}{\partial P}\right)_\Theta : \dot{P} + \left(\frac{\partial g(P,\Theta)}{\partial \Theta}\right)_P : \dot{\Theta}\,.$$

In view of equation (C.44), we see that

$$F = -\left(\frac{\partial g(P,\Theta)}{\partial P}\right)_\Theta, \qquad \eta_0 = -\left(\frac{\partial g(P,\Theta)}{\partial \Theta}\right)_P. \tag{C.45}$$

Thus, Gibbs free energy is a potential whose derivatives are the deformation gradient and entropy.

C.4.5 *Enthalpy as potential*

In a manner analogous to the introduction of the Gibbs free energy, we can introduce enthalpy, $h_0(P,\eta_0)$, as Legendre's transformation of internal energy, $e_0(\mathrm{F},\eta_0)$. Letting $P = \partial e_0/\partial F$, we write

$$h_0 = e_0 - \frac{\partial e_0}{\partial F} : F = e_0 - P : F \tag{C.46}$$

and its time derivative $\dot{h}_0 = \dot{e}_0 - \dot{P} : F - P : \dot{F}$. According to equation (C.41), $\dot{e}_0 - P : \dot{F} = \Theta\,\dot{\eta}_0$, and, hence, we write

$$\dot{h}_0 = -F : \dot{P} + \Theta\,\dot{\eta}_0\,, \tag{C.47}$$

where we use the fact that $\dot{P} : F = F : \dot{P}$. Since enthalpy is a function of a first Piola-Kirchhoff stress tensor and a local entropy $h_0(P,\eta_0)$, we can write

$$\dot{h}_0(P,\eta_0) = \left(\frac{\partial h_0(\mathrm{P},\eta_0)}{\partial P}\right)_{\eta_0} : \dot{P} + \left(\frac{\partial h_0(P,\eta_0)}{\partial \eta_0}\right)_P \dot{\eta}_0. \tag{C.48}$$

Comparing the right-hand sides of equations (C.47) and (C.48), we conclude that

$$F = -\left(\frac{\partial h_0(P,\eta_0)}{\partial P}\right)_{\eta_0}, \qquad \Theta = \left(\frac{\partial h_0(P,\eta_0)}{\partial \eta_0}\right)_P. \tag{C.49}$$

Thus, enthalpy is a potential whose derivatives are the deformation gradient and temperature.

C.5 Stress tensors in terms of potentials

For finite strains, stress tensors—in the theory of thermoelasticity—are functions of the derivative of the Helmholtz free energy, ψ, with respect to the deformation gradient, F, where ψ is a function of F and temperature, Θ.

As discussed by Epstein (2010, pp. 300–301),

> [t]he second law of thermodynamics is a restriction that Nature imposes on the observable phenomena: Certain things simply cannot happen.

From this law, we can derive the Cauchy stress tensor. To do so, let us recall expression (C.37),

$$\dot{\psi} = P : \dot{F} - \dot{\Theta}\,\eta_0\,,$$

and expression (B.10),

$$P = J\,\sigma\,F^{-T}\,,$$

to write

$$\dot{\psi} = J\,\sigma\,F^{-T} : \dot{F} - \dot{\Theta}\,\eta_0\,.$$

Comparing this result with expression (C.38), we recognize that

$$J\sigma F^{-T} = \left(\frac{\partial\psi(F,\Theta)}{\partial F}\right)_{\Theta}\,,$$

which results in

$$\sigma = J^{-1}\,\frac{\partial\psi}{\partial F}\,F^{T}\,. \tag{C.50}$$

[9]In the infinitesimal case, the deformation gradient tends to identity and expression (C.50) reduces to expression (C.64), below.

From the thermodynamic restriction, it also follows—in the context of thermoelasticity—that

$$\frac{\partial\psi}{\partial(\nabla_{\mathbf{x}}\Theta)} = 0\,,$$

[9]Readers interested in a derivation of expression (C.50) might also refer to Epstein (2010, Appendices A.9 and A.10.2), where the author also reminds us that

> [a]n important conceptual element in Continuum Mechanics is the presence of an arrow of time, that is, the natural irrevocable direction of phenomena prescribed by the second law of thermodynamics.

which means that the free energy is not explicitly dependent on the temperature gradient, as it would be for a more general theory. In other words, for thermoelasticity, $\psi = \psi(F, \Theta)$, even though, in general, $\psi(F, \Theta, \nabla_{\mathbf{x}}\Theta)$.

Let us return to expressions for stress tensors. Using expression (C.50) and invoking expressions (B.11), namely, $\sigma = J^{-1}PF^T$, and expression (B.16), namely, $S = F^{-1}P$, we can relate the free energy to the First and Second Piola-Kirchhoff stress tensors. We write the First Piola-Kirchhoff stress tensor as

$$P = J\,\sigma\,F^{-T} = \frac{\partial\psi}{\partial F}\,. \tag{C.51}$$

Remark C.6. It is possible to examine the free-energy function as a quantity per unit mass, not per unit volume. To do so, we use $\rho_0 = J\,\rho$, which is a statement of the balance of mass; herein, ρ_0 and ρ denote mass density in the reference and current configurations, respectively.

Using the fact that, as discussed in Appendix A.6 and Remark A.17, in orthonormal coordinates, the determinant of the deformation gradient, denoted by J, is a measure of the change of volume for an element of continuum subject to deformation, $J = \mathrm{d}v/\mathrm{d}V$, we have $\rho_0\,\mathrm{d}V = \rho\,\mathrm{d}v$, as expected. Several other relations can be interpreted geometrically. For instance, since the change of volume depends only on deformation, and not on rotation, we conclude that $\det F = \det U$, as discussed in Appendix A.9.4, where U is the right Stretch Tensor.

Let us consider expression (C.51) to derive the Piola-Kirchhoff stress tensors from the Helmholtz free energy per unit mass, Ψ. For the First Piola-Kirchhoff stress, we write

$$P = \frac{\partial\,(\rho_0\Psi)}{\partial F} = \rho_0\frac{\partial\Psi}{\partial F}\,.$$

For the Second Piola-Kirchhoff stress tensor, we invoke expression (B.16), to write

$$S = \rho_0\,F^{-1}\,\frac{\partial\Psi}{\partial F}\,.$$

Also, following expression (B.11),

$$\sigma = J^{-1}\rho_0\,\frac{\partial\Psi}{\partial F}\,F^T\,.$$

Thus, we have the three stress tensors discussed in Appendix B. Their counterparts—per unit volume—are

$$\begin{aligned} P &= \frac{\partial \psi}{\partial F}, \\ S &= F^{-1}\frac{\partial \psi}{\partial F}, \\ \sigma &= J^{-1}\frac{\partial \psi}{\partial F}F^T. \end{aligned} \tag{C.52}$$

Examining the free energy in the reference and current configurations, we see that

$$\iiint_V \psi\,\mathrm{d}V = \iiint_v \rho_0 J^{-1}\Psi\,\mathrm{d}v = \iiint_v \rho\Psi\,\mathrm{d}v,$$

as expected in view of the balance of energy. The value of Ψ, formulated per unit mass, is the same regardless of configurations in which it is specified due to conservation of mass.

For the isothermal case, which means that there is no change of temperature, the free energy per unit volume in the reference configuration, ψ, becomes the strain energy, W. Consequently, following expressions (D.3), (D.1) and (D.2), the three stress tensors are those given in expressions (C.52). Herein, $\psi \equiv W$, which is not the case in Appendix C.5, in spite of the general similarity of forms in terms of partial derivatives.

Also, for the adiabatic case, which means that the heat flux is zero, $\psi \equiv W$, and the general form of the stress-tensor expressions is the same as for the isothermal case.

In both cases, the dissipation inequality becomes an equality, which means that there is no increase of entropy and, hence, the deformation process is reversible.

Remark C.7. Let us express the right-hand side of equation (C.50) in terms of the Green-Lagrange strain tensor, E, discussed in Appendix A.5. To do so, we assume that ψ depends on F solely through E; hence, according to the chain rule, we obtain

$$\frac{\partial \psi}{\partial F} = \frac{\partial \psi}{\partial E} : \frac{\partial E}{\partial F}. \tag{C.53}$$

Following definition (A.48), which, in components, we write as

$$E_{IJ} = \frac{1}{2}\left(\sum_{j=1}^{3}(F_{Ij})^T F_{jJ} - \delta_{IJ}\right) = \frac{1}{2}\left(\sum_{j=1}^{3}F_{jI}F_{jJ} - \delta_{IJ}\right)$$

we apply the product rule, to get

$$\frac{\partial E}{\partial F}=\frac{1}{2}\left(\frac{\partial F^T}{\partial F}F+F^T\frac{\partial F}{\partial F}\right), \tag{C.54}$$

which, in components, we write as

$$\frac{\partial E_{IJ}}{\partial F_{iK}}=\frac{1}{2}\sum_{j=1}^{3}\left(\frac{\partial F_{jI}}{\partial F_{iK}}F_{jJ}+F_{jI}\frac{\partial F_{jJ}}{\partial F_{iK}}\right).$$

Let us express $\partial F^T/\partial F$ and $\partial F/\partial F$ in terms of components; they are, respectively,

$$\frac{\partial F_{jI}}{\partial F_{iK}}=\delta_{ij}\delta_{IK} \qquad \text{and} \qquad \frac{\partial F_{jJ}}{\partial F_{iK}}=\delta_{ij}\delta_{JK}\,, \qquad i,j,k,I,J,K=1,2,3\,.$$

Hence, using the property of the Kronecker delta, we get

$$\frac{\partial E_{IJ}}{\partial F_{iK}}=\frac{1}{2}\sum_{j=1}^{3}\left(\delta_{ij}\delta_{IK}F_{jJ}+F_{jI}\delta_{ij}\delta_{JK}\right)=\frac{1}{2}\left(\delta_{IK}F_{iJ}+\delta_{JK}F_{iI}\right),$$

where $i,I,J,K=1,2,3$. Inserting these results into equation (C.53), we obtain

$$\frac{\partial\psi}{\partial F_{iK}}=\frac{1}{2}\sum_{I=1}^{3}\sum_{J=1}^{3}\left(\frac{\partial\psi}{\partial E_{IJ}}\left(\delta_{IK}F_{iJ}+\delta_{JK}F_{iI}\right)\right), \qquad i,K=1,2,3\,.$$

To insert this result into equation (C.50), we write that equation in terms of components,

$$\sigma_{ij}=J^{-1}\sum_{K=1}^{3}\frac{\partial\psi}{\partial F_{iK}}F_{jK} \qquad i,j=1,2,3\,.$$

Thus,

$$\sigma_{ij}=\frac{1}{2}J^{-1}\sum_{I=1}^{3}\sum_{J=1}^{3}\sum_{K=1}^{3}\left(\frac{\partial\psi}{\partial E_{IJ}}\left(\delta_{IK}F_{iJ}+\delta_{JK}F_{iI}\right)F_{jK}\right), \qquad i,j=1,2,3,$$

which, using the property of the Kronecker delta, we write as

$$\sigma_{ij}=\frac{1}{2}J^{-1}\sum_{K=1}^{3}\left(\left(\sum_{J=1}^{3}\frac{\partial\psi}{\partial E_{KJ}}F_{iJ}+\sum_{I=1}^{3}\frac{\partial\psi}{\partial E_{IK}}F_{iI}\right)F_{jK}\right), \qquad i,j=1,2,3\,.$$

Considering the summands in the inner parentheses, we write

$$\sigma_{ij}=\frac{1}{2}J^{-1}\sum_{K=1}^{3}\left(\left(\sum_{J=1}^{3}F_{iJ}\frac{\partial\psi}{\partial E_{JK}}+\sum_{I=1}^{3}F_{iI}\frac{\partial\psi}{\partial E_{KI}}\right)F_{jK}\right),$$

where $i,j = 1,2,3$. Using the fact that, herein, I and J are summation indices, we get

$$\sigma_{ij} = \frac{1}{2} J^{-1} \sum_{K=1}^{3} \left(\sum_{I=1}^{3} \left(F_{iI} \left(\frac{\partial \psi}{\partial E_{IK}} + \frac{\partial \psi}{\partial E_{KI}} \right) \right) F_{jK} \right), \qquad i,j = 1,2,3.$$

Since E is symmetric, $E_{KI} = E_{IK}$, the summands in the innermost parentheses are equal to one another, which is tantamount to $(\partial\psi/\partial E)^T = \partial\psi/\partial E$; thus, renaming the summation indices, we write

$$\sigma_{ij} = J^{-1} \sum_{I=1}^{3} \sum_{J=1}^{3} F_{iI} \frac{\partial \psi}{\partial E_{IJ}} F_{jJ} = J^{-1} \sum_{I=1}^{3} \sum_{J=1}^{3} F_{iI} \frac{\partial \psi}{\partial E_{IJ}} \left(F_{Jj} \right)^T,$$

where $i,j = 1,2,3$. Expressing this result in a coordinate-free notation, we write

$$\sigma = J^{-1} F \frac{\partial \psi}{\partial E} F^T, \tag{C.55}$$

which is equation (C.50) expressed, as required, in terms of the Green-Lagrange strain tensor.

Examining equation (C.55), in the context of equation (B.11), we see that $F\,\partial\psi/\partial E = P$, where P is the First Piola-Kirchhoff stress tensor. Also, in accordance with equation (B.15), we have $P = FS$, where S is the Second Piola-Kirchhoff stress tensor. Hence, $S = \partial\psi/\partial E$. Thus, we write expression (C.55) as

$$\sigma = J^{-1} F\, S\, F^T,$$

which is consistent with equation (B.13), solved for σ.

For infinitesimal strains, F tends to identity and, hence, its determinant, J, tends to unity; thus, $\sigma = S$, as expected in accordance with Section B.3. For isothermal processes, $\psi(F,\Theta) \equiv W(F)$. Also, as discussed on page 402, E becomes ε. Thus, for isothermal infinitesimal strains, we have $\sigma = \partial W/\partial\varepsilon$, as expected in view of expression (D.5), which is a general form for constitutive relations in hyperelasticity. Its particular case is Hooke's law, which results from W given in expression (D.6).

These discussions illustrate the fact that the second law of thermodynamics, expressed as the Clausius-Duhem inequality, imposes restrictions on the form of constitutive equations. It also illustrates relations among S, $\dot{E}$ and $\dot{C}$, since, in accordance with the chain rule,

$$\dot{\psi} = \frac{\partial \psi}{\partial E} : \dot{E} \equiv S : \dot{E} \tag{C.56}$$

and, invoking expression (A.48),

$$S : \dot{E} = \frac{1}{2} S : \dot{C}\,, \tag{C.57}$$

Also, by analogy with expression (C.56), we write

$$\dot{\psi} = \frac{\partial \psi}{\partial C} : \dot{C}\,.$$

Equating the right-hand side of this equation with the right-hand side of equation (C.57), where $S = \partial\psi/\partial E$, we see that $\partial\psi/\partial E = 2\partial\psi/\partial C$. All these relations are expressed in terms of potential ψ, and, as such, are the *work conjugates* of each other.[10]

C.6 Isothermal infinitesimal strains

C.6.1 *Introduction*

To relate the concepts that are discussed in Appendices C.1–C.5 to isothermal infinitesimal strains, which is a common assumption in seismology, we formally write the temporal rate of change of energy per unit volume as

$$\frac{\mathrm{d}W}{\mathrm{d}t} = \sigma : \frac{\mathrm{d}\varepsilon}{\mathrm{d}t}\,; \tag{C.58}$$

herein—since σ and $\mathrm{d}\varepsilon/\mathrm{d}t$ are second-rank tensors—the result of this product is a scalar.

In equation (C.58), we assume that—as a consequence of isothermal infinitesimal strains—the stress tensor is the derivative of the free energy with respect to the strain-tensor components, as shown in expression (C.64), below. Herein, W is a pure mechanical strain-energy function, which can also be viewed as the Helmholtz free energy with a fixed value of temperature.

[10]Readers interested in insightful descriptions of these relations and the formulation of Hooke's law in terms of S and E, not σ and ε, which means that the corresponding elasticity tensor is in the reference, not current, configuration, might refer to Bonet and Wood (2008, Sections 6.2 and 6.3). Formally, such a model, which is referred to as the *St. Venant-Kirchhoff material* is valid for finite strains. However, as mentioned by Bonet and Wood (2008, p. 158),

> [u]nfortunately, this St. Venant-Kirchhoff material has been found to be of little practical use beyond the small strain regime.

This practical limitation is tantamount to the necessity of constraining Hookean solids to infinitesimal strains to maintain their empirical adequacy as mathematical models of physical phenomena, beyond the coherency of their intrinsic mathematical structure.

In components, we write equation (C.58) as

$$\frac{\mathrm{d}W}{\mathrm{d}t} = \sum_{i=1}^{3}\sum_{j=1}^{3}\sigma_{ij}\frac{\mathrm{d}\varepsilon_{ij}}{\mathrm{d}t}. \tag{C.59}$$

If the stress-strain relation is linear, $\sigma_{ij} = \sum_{k=1}^{3}\sum_{\ell=1}^{3} c_{ijk\ell}\varepsilon_{k\ell}$, we obtain the expression derived in Example D.2,

$$\frac{\mathrm{d}W}{\mathrm{d}t} = \sum_{i=1}^{3}\sum_{j=1}^{3}\sum_{k=1}^{3}\sum_{\ell=1}^{3} c_{ijk\ell}\,\varepsilon_{k\ell}\,\frac{\mathrm{d}\varepsilon_{ij}}{\mathrm{d}t},$$

which is a particular case of expression (C.59).

Remark C.8. Note that Hooke's law, stated in expression (3.1) on page 43, could also be written as $\sigma = c : \varepsilon$; since c is a fourth-rank tensor, the result of this product, σ, is a second-rank tensor.

C.6.2 *Single spatial dimension*

Let us examine a special case of thermoelasticity by considering isothermal infinitesimal strains in a single spatial dimension. Using the chain rule, we write the one-dimensional case of equation (C.58) as

$$\frac{\mathrm{d}W(\varepsilon)}{\mathrm{d}t} = \sigma\,\frac{\mathrm{d}\varepsilon}{\mathrm{d}t}, \tag{C.60}$$

where—having only single components—the double-dot product does not appear. Also, using the chain rule, we write

$$\frac{\mathrm{d}W(\varepsilon)}{\mathrm{d}t} = \frac{\partial W(\varepsilon)}{\partial\varepsilon}\frac{\mathrm{d}\varepsilon}{\mathrm{d}t}. \tag{C.61}$$

Equating the right-hand sides of equations (C.60) and (C.61), we write

$$\sigma\,\frac{\mathrm{d}\varepsilon}{\mathrm{d}t} = \frac{\partial W(\varepsilon)}{\partial\varepsilon}\frac{\mathrm{d}\varepsilon}{\mathrm{d}t}.$$

Since this equation is valid for arbitrary $\mathrm{d}\varepsilon/\mathrm{d}t$, it follows that

$$\sigma = \frac{\mathrm{d}W(\varepsilon)}{\mathrm{d}\varepsilon}, \tag{C.62}$$

where, in view of a single independent variable, we make no distinction between the partial and total derivatives. Expression (C.62) is equivalent to expression (D.5).

In view of expression (C.62), let us find a dual of W, such that its derivative with respect to σ results in ε. To do so, we apply Legendre's transformation to function $W(\varepsilon)$. In general, differentiating W with respect to ε, we obtain a new variable to be the argument of the transformed function, which we denote by ξ. Herein, this derivative is stated in expression (C.62) and, hence, the new variable is σ. According to Legendre's transformation,

$$\xi = \sigma\,\varepsilon - W(\varepsilon)\,. \tag{C.63}$$

To express ξ solely as a function of σ, we would need to have function W and, hence, invert expression (C.62) to obtain ε in terms of σ to be used in expression (C.63).

Differentiating expression (C.63) with respect to σ, we obtain

$$\varepsilon = \frac{\mathrm{d}\xi(\sigma)}{\mathrm{d}\sigma}\,;$$

ξ and W are duals of one another, since expression (C.63) and

$$W(\varepsilon) = \sigma\varepsilon - \xi(\sigma)$$

are related by Legendre's transformation.[11] Thus, in view of Appendices C.4.2 and C.4.4, ξ and W are analogous to the Gibbs and Helmholtz free energies, for the case of infinitesimal strains and a fixed temperature. In Appendices C.4.2 and C.4.4, we use antisymmetric Legendre's transformation; herein, we use a symmetric one.

C.6.3 *Three spatial dimensions*

Let us rewrite expression (C.59) as

$$\frac{\mathrm{d}W}{\mathrm{d}t} = \sum_{i=1}^{3}\sum_{j=1}^{3}\sigma_{ij}\dot{\varepsilon}_{ij}\,.$$

Also, according to the chain rule,

$$\frac{\mathrm{d}W}{\mathrm{d}t} = \sum_{i=1}^{3}\sum_{j=1}^{3}\frac{\partial W}{\partial \varepsilon_{ij}}\dot{\varepsilon}_{ij}\,.$$

Equating the right-hand sides of these equations, and using the fact that the resulting equality must be valid for arbitrary $\dot{\varepsilon}_{ij}$, we obtain

$$\sigma_{ij} = \frac{\partial W}{\partial \varepsilon_{ij}}\,, \qquad i,j = 1,2,3\,, \tag{C.64}$$

as expected from expression (C.62).

[11] *see also*: Slawinski (2015, Appendix B)

In view of expression (C.64), let us find a dual of W, such that its derivative with respect to σ_{ij} results in ε_{ij}. To do so we perform Legendre's transformation of $W(\varepsilon_{ij})$,

$$\xi(\sigma_{ij}) = \sum_{i=1}^{3}\sum_{j=1}^{3}\sigma_{ij}\varepsilon_{ij} - W(\varepsilon_{ij}), \tag{C.65}$$

which means that W and ξ are duals of one another, since, by the inverse transformation,

$$W(\varepsilon_{ij}) = \sum_{i=1}^{3}\sum_{j=1}^{3}\sigma_{ij}\varepsilon_{ij} - \xi(\sigma_{ij}).$$

Differentiating expression (C.65) with respect to σ_{ij}, we obtain

$$\varepsilon_{ij} = \frac{\partial \xi}{\partial \sigma_{ij}}, \qquad i,j = 1,2,3,$$

as expected, in view of the one-dimensional case, discussed in Appendix C.6.2.

Closing remarks

The thermal aspect of thermoelasticity requires, in contrast to elasticity, an additional field equation, which is the first law of thermodynamics. In continuum mechanics, field equations are expressed in the form discussed in Section 2.3.1. They relate the rate of change of a quantity—mass, momentum, energy—within a volume to the flow of this quantity through the surface of this volume and to its amount produced by the sources, which can be on the surface and within the volume. These equations govern the motion within a solid.

The theory of thermoelasticity also requires an inequality to express the second law of thermodynamics, which is a statement of irreversibility of natural processes. In a manner analogous to the balance of mass, momentum and energy, we assume that there is a flux of entropy and its source. This inequality imposes restrictions on the form of constitutive equations by establishing relations between the deformation gradient and stress tensors. Since strain tensors can be expressed in terms of this gradient, we obtain stress-strain equations.

Thermoelastic and hyperelastic materials share many properties. Notably, their behaviour depends solely on the current state of deformation.

For the former, this dependence is a function of both the deformation-gradient tensor and temperature. For the latter, it depends on the deformation-gradient tensor alone, as discussed in Appendix D.

Let us conclude this appendix with a statement that relates it to the subject discussed in Appendix D.2. As written by Epstein (2010, p. 301),

> If we should limit the processes that a thermoelastic material undergoes to isothermally homogeneous processes (namely, with a constant temperature in time and in space), we would arrive at the conclusion that the resulting elastic material is *hyperelastic*, namely, it is completely characterized by a stress potential, which is obtained as the restriction of the free energy density to the pre-established temperature.

In such a case, expressions (D.5) and (C.62) are equivalent to one another, which means that W and ψ are equivalent to one another.

Adiabatic and isothermal processes, discussed in this appendix, are used—albeit with a different approach—in Section 8.3.

Appendix D

On hyperelasticity

The first concept to be introduced is that of material response functional. *Since the geometrical structure we are interested in exposing will emerge already in the context of elasticity, we will confine our attention to a purely elastic response.* [...] *Without loss of geometrical generality, we will also assume that the (purely mechanical) material response is entirely characterized by a single scalar-valued functional:*

$$W : \mathcal{Q} \longrightarrow \mathbb{R},$$

whose physical meaning is the elastic energy stored in the body at a given configuration. In a more technical terminology, the body is said to be hyperelastic.[1]

Marcelo Epstein (2010)

Preliminary remarks

In this appendix, as in Appendix C, we study materials that regain their original form upon the removal of the load that causes a deformation. We relate the strain and stress tensors, discussed in Appendices A and B, respectively, by the strain-energy function, which is a particular case of energies discussed in Appendix C. Herein, we consider deformations for which the temperature is assumed to be constant.

Thus, in contrast to energies discussed in Appendix C, which depend on both the deformation gradient and temperature, the strain-energy function depends only on the former. Consequently, stress tensors are related to strain tensors by derivatives of energy with respect to the deformation gradient only. Bodies endowed with the strain-energy function are called

[1]Herein, $\mathcal{Q}$ is a configuration space (Epstein, 2010, p. 123).

hyperelastic, as opposed to thermoelastic bodies, which are discussed in Appendix C. In spite of this limiting assumption, resulting models are empirically adequate in many branches of mechanical engineering and seismology.

We begin this appendix by discussing the strain-energy function and, hence, examining hyperelastic materials. In that context, we discuss the concept of material symmetry. We conclude this appendix with an explicit material-symmetry condition for infinitesimal deformations of linearly elastic solids, which constitute a particular case of hyperelastic materials.

Notation D.1. In this appendix, as well as in Appendices A, B and C, the quantities associated with the reference configuration are commonly denoted by the upper-case letters and the quantities associated with the current configuration are commonly denoted by the lower-case letters; not all upper-case or lower-case letters, however, refer to these configurations. Bold letters represent vectors.

D.1 Strain-energy function

Strain-energy function accounts for a potential energy stored in a deformed material. It is a scalar-valued function, which can be written as

$$W = W(F)\,,$$

where F is the deformation gradient discussed in Appendix A.2. It is the energy equivalent to the work performed in deforming the material.

The units of W are newtons per squared metres, which means that, strictly speaking, W is a density of a function per unit of volume. Upon integration over that volume, we obtain newton metres, which are equivalent to joules: the units of energy and work.

For example, in Hookean solids, $W = \sigma\,\varepsilon/2$, where σ is the Cauchy stress tensor and ε is the strain tensor.[2]

Remark D.1. The density of the strain-energy function can be also defined per unit of mass. Such an approach is exemplified on page 444.

[2] *see also*: Slawinski (2015, Section 4.1)

D.2 Hyperelastic materials

D.2.1 *Introduction*

Hyperelastic materials are objects whose stress tensor, at each point, can be derived solely from the strain-energy function in the current configuration. Since the strain energy—as its name suggests—is stored in the strained material, it is implicitly assumed that all work spent on deformation is contained therein. In other words, no work is dissipated during a deformation process, in such forms as heat, nor any permanent damage is made to the material during its deformation. The entire energy is available to do work.

Hyperelastic materials, which are also referred to as the *Green elastic materials*, constitute a special case of the *Cauchy elastic materials*, whose stress at each point is determined only by the current configuration with respect to an arbitrary reference configuration. Hyperelastic materials appear in the case of both finite and infinitesimal strains, as well as for both linear and nonlinear formulations.

D.2.2 *Finite strains*

For hyperelastic materials, we can express stress tensors in terms of the strain-energy function, W, only. The form of W must be such that the stress-tensor expressions of Appendix B are satisfied. The First Piola-Kirchhoff stress tensor is

$$P = \frac{\partial W}{\partial F}, \tag{D.1}$$

where the units agree, as required. $F = \nabla_{\mathbf{X}} \mathbf{x}$ is unitless, and hence P and W share the same units, which—as expected—are the units of both the stress tensor and the energy density per unit volume. In components, we write P as

$$P_{iJ} = \frac{\partial W}{\partial F_{iJ}}, \qquad i, J = 1, 2, 3.$$

We can also express P in terms of other derivatives of W; for instance, if W depends on F solely through E,

$$P = F \frac{\partial W}{\partial E},$$

where E is the Green-Lagrange strain tensor. In components,

$$P_{iJ} = \sum_{K=1}^{3} F_{iK} \frac{\partial W}{\partial E_{JK}}, \qquad i, J = 1, 2, 3.$$

Remark D.2. The dependence of W on F—solely through E—means that W depends only on $F^T F = U^2$, where $F = RU$ is the polar decomposition discussed in Appendix A.9.1. In other words, W does not depend on rotation.

In terms of the right Cauchy-Green strain tensor, if W depends on F solely through $F^T F$, we have

$$P = 2F\frac{\partial W}{\partial C}$$

and

$$P_{iJ} = 2\sum_{K=1}^{3} F_{iK}\frac{\partial W}{\partial C_{JK}}, \qquad i, J = 1, 2, 3.$$

To express the Second Piola-Kirchhoff stress tensor in terms of W, we recall equation (B.16), namely, $S = F^{-1}P$, and expression (D.1), to write

$$S = F^{-1}\frac{\partial W}{\partial F}; \tag{D.2}$$

in components, we write

$$S_{IJ} = \sum_{k=1}^{3} F_{Ik}^{-1}\frac{\partial W}{\partial F_{kJ}}, \qquad I, J = 1, 2, 3.$$

Also,[3]

$$S = \frac{\partial W}{\partial E},$$

which, in components, we write as

$$S_{IJ} = \frac{\partial W}{\partial E_{IJ}}, \qquad I, J = 1, 2, 3.$$

In terms of the right Cauchy-Green strain tensor, we have

$$S = 2\frac{\partial W}{\partial C}$$

and

$$S_{IJ} = 2\frac{\partial W}{\partial C_{IJ}}, \qquad I, J = 1, 2, 3.$$

[3]Readers interested in the motivation for this formulation of S might refer to Bonet and Wood (2008, Expression (6.7a,b)).

To express the Cauchy stress tensor in terms of W, we recall equation (B.11), namely, $\sigma = J^{-1}PF^T$, to write the following expressions, which correspond to the expressions for P.

$$\sigma = \frac{1}{J}\frac{\partial W}{\partial F}F^T\,, \tag{D.3}$$

where $J := \det F$, which, in components, we write as

$$\sigma_{ij} = \frac{1}{J}\sum_{K=1}^{3}\frac{\partial W}{\partial F_{iK}}F_{jK}\,, \qquad i,j = 1,2,3\,.$$

Also, in terms of the Green-Lagrange strain tensor,

$$\sigma = \frac{1}{J}F\frac{\partial W}{\partial E}F^T\,, \tag{D.4}$$

which, in components, we write as

$$\sigma_{ij} = \frac{1}{J}\sum_{K=1}^{3}\sum_{L=1}^{3}F_{iK}\frac{\partial W}{\partial E_{KL}}F_{jL}\,, \qquad i,j = 1,2,3\,.$$

In terms of the right Cauchy-Green strain tensor, we have

$$\sigma = \frac{2}{J}F\frac{\partial W}{\partial C}F^T$$

and

$$\sigma_{ij} = \frac{2}{J}\sum_{K=1}^{3}\sum_{L=1}^{3}F_{iK}\frac{\partial W}{\partial C_{KL}}F_{jL}\,, \qquad i,j = 1,2,3\,.$$

D.2.3 *Infinitesimal strains*

Combining the assumption of hyperelasticity with the assumption of infinitesimal strains, discussed in Appendices A.5 and B.3, let us examine expression (D.4). Since F tends to identity and, hence, J tends to unity, we write

$$\sigma(E) = \frac{\partial W}{\partial E}\,.$$

In view of expression (A.76), we rewrite it as

$$\sigma(\varepsilon) = \frac{\partial W(\varepsilon)}{\partial \varepsilon}\,. \tag{D.5}$$

If, furthermore, we consider linear elasticity, we can write, in components, expression (3.1) on page 43,[4]

$$\sigma_{ij}(\varepsilon) = \sum_{k=1}^{3}\sum_{\ell=1}^{3}c_{ijk\ell}\varepsilon_{k\ell}\,, \qquad i,j = 1,2,3\,,$$

[4] *see also*: Slawinski (2015, Section 4.1)

which—by integration with respect to the strain tensor—defines the corresponding strain-energy function,

$$W(\varepsilon) = \frac{1}{2}\sum_{i=1}^{3}\sum_{j=1}^{3}\sigma_{ij}\varepsilon_{ij} = \frac{1}{2}\sum_{i=1}^{3}\sum_{j=1}^{3}\sum_{k=1}^{3}\sum_{\ell=1}^{3}c_{ijk\ell}\varepsilon_{ij}\varepsilon_{k\ell}\,, \tag{D.6}$$

where we assume that—for unstrained continua— $W = 0$.

Example D.1. To verify result (D.6), let us differentiate this expression with respect to the strain-tensor components. Using the product rule, we write

$$\frac{\partial W(\varepsilon)}{\partial\varepsilon_{pq}} = \frac{1}{2}\left(\sum_{i=1}^{3}\sum_{j=1}^{3}\sum_{k=1}^{3}\sum_{\ell=1}^{3}c_{ijk\ell}\frac{\partial\varepsilon_{ij}}{\partial\varepsilon_{pq}}\varepsilon_{k\ell} + \sum_{i=1}^{3}\sum_{j=1}^{3}\sum_{k=1}^{3}\sum_{\ell=1}^{3}c_{ijk\ell}\varepsilon_{ij}\frac{\partial\varepsilon_{k\ell}}{\partial\varepsilon_{pq}}\right),$$

where we assume that $c_{ijk\ell}$ is not a function of ε_{pq}. Using the Kronecker delta, we write

$$\frac{\partial\varepsilon_{ij}}{\partial\varepsilon_{pq}} = \delta_{ip}\delta_{jq}\,, \qquad i,j,p,q = 1,2,3\,,$$

and

$$\frac{\partial\varepsilon_{k\ell}}{\partial\varepsilon_{pq}} = \delta_{kp}\delta_{\ell q}\,, \qquad k,\ell,p,q = 1,2,3\,.$$

Hence,

$$\frac{\partial W(\varepsilon)}{\partial\varepsilon_{pq}} = \frac{1}{2}\sum_{i=1}^{3}\sum_{j=1}^{3}\sum_{k=1}^{3}\sum_{\ell=1}^{3}c_{ijk\ell}\delta_{ip}\delta_{jq}\varepsilon_{k\ell} + \frac{1}{2}\sum_{i=1}^{3}\sum_{j=1}^{3}\sum_{k=1}^{3}\sum_{\ell=1}^{3}c_{ijk\ell}\varepsilon_{ij}\delta_{kp}\delta_{\ell q}\,.$$

Since $\sum_{i=1}^{3}\sum_{j=1}^{3}c_{ijk\ell}\delta_{ip}\delta_{jq} = c_{pqk\ell}$ and $\sum_{k=1}^{3}\sum_{\ell=1}^{3}c_{ijk\ell}\delta_{kp}\delta_{\ell q} = c_{ijpq}$, we write

$$\frac{\partial W}{\partial\varepsilon_{pq}} = \frac{1}{2}\sum_{k=1}^{3}\sum_{\ell=1}^{3}c_{pqk\ell}\varepsilon_{k\ell} + \frac{1}{2}\sum_{i=1}^{3}\sum_{j=1}^{3}c_{ijpq}\varepsilon_{ij}\,.$$

Assuming $c_{ijpq} = c_{pqij}$, we rewrite the second sum as

$$\frac{\partial W(\varepsilon)}{\partial\varepsilon_{pq}} = \frac{1}{2}\sum_{k=1}^{3}\sum_{\ell=1}^{3}c_{pqk\ell}\varepsilon_{k\ell} + \frac{1}{2}\sum_{i=1}^{3}\sum_{j=1}^{3}c_{pqij}\varepsilon_{ij}\,.$$

Renaming the summation indices in the second sum, we obtain

$$\frac{\partial W(\varepsilon)}{\partial\varepsilon_{pq}} = \sum_{k=1}^{3}\sum_{\ell=1}^{3}c_{pqk\ell}\varepsilon_{k\ell} = \sigma_{pq}\,,$$

which is the Cauchy stress tensor, as expected. This result confirms that expression (D.6) is the integral of expression (3.1), with respect to the strain-tensor components

Example D.2. In this example, let us differentiate expression (D.6) with respect to time. Such temporal derivatives appear in Appendix C.

We assume that ε, but not c, depends on time. Hence, using the product rule, we write

$$\frac{\mathrm{d}W(\varepsilon(t))}{\mathrm{d}t}=\frac{1}{2}\sum_{i=1}^{3}\sum_{j=1}^{3}\sum_{k=1}^{3}\sum_{\ell=1}^{3}c_{ijk\ell}\,\frac{\mathrm{d}\varepsilon_{ij}}{\mathrm{d}t}\,\varepsilon_{k\ell}+\frac{1}{2}\sum_{i=1}^{3}\sum_{j=1}^{3}\sum_{k=1}^{3}\sum_{\ell=1}^{3}c_{ijk\ell}\,\varepsilon_{ij}\,\frac{\mathrm{d}\varepsilon_{k\ell}}{\mathrm{d}t}\,.$$

Using the index symmetry of the elasticity tensor that allows us to interchange ij and $k\ell$, we can rewrite the second summand as

$$\frac{\mathrm{d}W(\varepsilon(t))}{\mathrm{d}t}=\frac{1}{2}\sum_{i=1}^{3}\sum_{j=1}^{3}\sum_{k=1}^{3}\sum_{\ell=1}^{3}c_{ijk\ell}\,\frac{\mathrm{d}\varepsilon_{ij}}{\mathrm{d}t}\,\varepsilon_{k\ell}+\frac{1}{2}\sum_{i=1}^{3}\sum_{j=1}^{3}\sum_{k=1}^{3}\sum_{\ell=1}^{3}c_{k\ell ij}\,\frac{\mathrm{d}\varepsilon_{k\ell}}{\mathrm{d}t}\,\varepsilon_{ij}\,.$$

Also, since ij and $k\ell$ are summation indices, we can rewrite the second summand as

$$\frac{\mathrm{d}W(\varepsilon(t))}{\mathrm{d}t}=\frac{1}{2}\sum_{i=1}^{3}\sum_{j=1}^{3}\sum_{k=1}^{3}\sum_{\ell=1}^{3}c_{ijk\ell}\,\frac{\mathrm{d}\varepsilon_{ij}}{\mathrm{d}t}\,\varepsilon_{k\ell}+\frac{1}{2}\sum_{i=1}^{3}\sum_{j=1}^{3}\sum_{k=1}^{3}\sum_{\ell=1}^{3}c_{ijk\ell}\,\frac{\mathrm{d}\varepsilon_{ij}}{\mathrm{d}t}\,\varepsilon_{k\ell}\,.$$

Since the two summands are the same, we obtain

$$\frac{\mathrm{d}W(\varepsilon(t))}{\mathrm{d}t}=\sum_{i=1}^{3}\sum_{j=1}^{3}\sum_{k=1}^{3}\sum_{\ell=1}^{3}c_{ijk\ell}\,\frac{\mathrm{d}\varepsilon_{ij}}{\mathrm{d}t}\,\varepsilon_{k\ell}\,,$$

which means that the temporal rate of change of the strain-energy function depends on both the strain itself and its rate of change.

D.3 Material symmetry

D.3.1 *Rotation invariance*

Material symmetry of a continuum can be studied in terms of the strain-energy function. Let us examine Figure D.1, which illustrates two reference configurations, $\mathbf{X}$ and $\mathbf{X}'$. The deformation gradient relating these two configurations is $Q:=\nabla_{\mathbf{X}}\mathbf{X}'$. Each reference configuration is related to the current configuration, $\mathbf{x}$, by $F=\nabla_{\mathbf{X}}\mathbf{x}$ and $F'=\nabla_{\mathbf{X}'}\mathbf{x}$, where $F=F'\,Q$.

Remark D.3. There is a similarity between Figures A.11 and D.1. Algebraically, this similarity is due to $F=F'Q$, in Figure D.1, and $F'=QF$, in Figure A.11. Both figures refer to a response of a tensor to a rigid-body rotation.

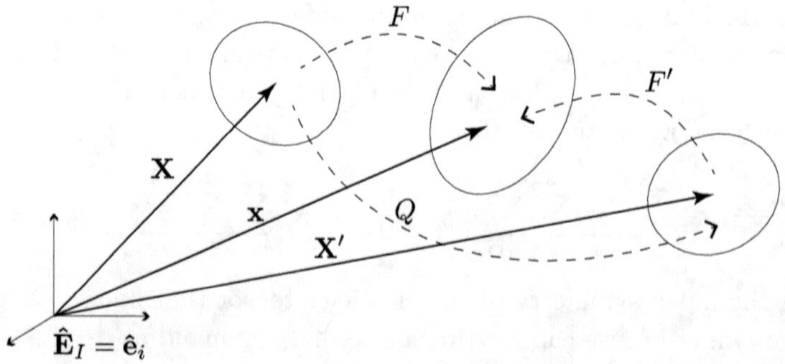

Fig. D.1: Material symmetry as transformations Q that result in $W(F) = W(F')$
Modified from Dorfmann and Ogden (2014)

In general, the value of a strain-energy function is different depending on the orientation of a continuum subjected to deformation. In other words, the amount of work spent on a deformation is direction-dependent. Hence, a rotation of the material in a reference configuration, prior to a deformation, affects the value of this function.

Consider the strain energy in the configuration whose coordinates are denoted by $\mathbf{x}$ that results from the configuration whose coordinates are $\mathbf{X}$; we denote it by W. Similarly, we denote the strain energy resulting from $\mathbf{X}'$ by W'. In general, $W \neq W'$, however, if $Q = \nabla_{\mathbf{X}}\mathbf{X}' = I$, it follows that $W = W'$; in other words, $W(F) = W(F')$. For certain materials, the same property exists even if $Q = \nabla_{\mathbf{X}}\mathbf{X}' \neq I$.

To consider such cases, let $Q \in SO(3)$, which are rotations in three-dimensional Euclidean space. The material-symmetry group is the set of rotations between configurations whose coordinates are denoted by $\mathbf{X}$ and $\mathbf{X}'$, for which the strain-energy function in the current configuration is the same. In other words,

$$W(F\,Q) = W(F)\,, \tag{D.7}$$

for all F, which means that the strain-energy function at $\mathbf{x}$ is not affected by these rotations. If W is not affected by any rotation, we say that the medium is isotropic.

Since, for any hyperelastic material, its stress tensor can be derived from the strain-energy function, we can also write the material-symmetry condition as $\sigma(F\,Q) = \sigma(F)$, for all F, where σ is the Cauchy stress tensor.

D.3.2 *Isotropy*

Isotropy can be stated by

$$W(F\,Q) = W(F)\,,$$

for any Q that belongs to $SO(3)$. In other words, $W(F)$ is invariant under all rotations.

To illustrate simplifications that result from the assumption of isotropy, let us recall polar decomposition, discussed in Appendix A.9.1, and let $Q = R^T$, to write

$$W(v\,R\,R^T) = W(v)\,;$$

this implies that W is an isotropic function of v, which is the left Stretch tensor, and of $b = v^2$, which is the left Cauchy-Green strain tensor. The function is isotropic since its value is orientation-independent. It also follows that W depends only on the invariants of v and b, which, for the former, are $\operatorname{tr} v$, $\frac{1}{2}\left((\operatorname{tr} v)^2 - \operatorname{tr} v^2\right)$ and $\det v$.

Also, considering the Cauchy stress tensor, the isotropy implies that $\sigma(W) = \sigma(v)$, and, hence, $\sigma(W) = \sigma(b)$. We can verify that objectivity is satisfied, as discussed in Appendices A.11 and B.4. Since $b' = Q\,b\,Q^T$, we write the arguments of σ as

$$\sigma(b') = \sigma\left(Q\,b\,Q^T\right)\,;$$

hence, $\sigma(b') = Q\,\sigma(b)\,Q^T$ is the required condition for the objectivity of σ. This condition is also satisfied if we replace b by v.

D.3.3 *Infinitesimal strain*

For finite elasticity, the material-symmetry condition is given in expression (3.22), which—in the form analogous to expression (D.7)—we write as

$$\sigma(FQ) = \sigma(F)\,,$$

for all F. In other words, the material-symmetry group of σ consists of transformations Q, which are a subset of $SO(3)$ that satisfies this condition.

For the infinitesimal linear elasticity, discussed in Appendix D.2.3, and in view of expression (3.23), this condition becomes, in components,

$$c_{ijk\ell} = \sum_{m=1}^{3}\sum_{n=1}^{3}\sum_{o=1}^{3}\sum_{p=1}^{3} A_{im}A_{jn}A_{ko}A_{\ell p}\,c_{mnop}\,, \qquad i,j,k,\ell = 1,2,3\,, \tag{D.8}$$

where $A \in SO(3)$. This means that—under the assumptions of the infinitesimal linear elasticity—the symmetry of σ_{ij} is tantamount to the symmetry of $c_{ijk\ell}$.[5]

An elasticity tensor can be viewed as a four-linear map, $c: \mathbb{R}^3 \times \mathbb{R}^3 \times \mathbb{R}^3 \times \mathbb{R}^3 \to \mathbb{R}$, that acts on four vectors in $\mathbb{R}^3$ to produce a real number, $\mathbb{R}$. In other words, $c(\mathbf{a}, \mathbf{b}, \mathbf{c}, \mathbf{d}) \in \mathbb{R}$. Using this four-linear map, condition (D.8) can be rewritten as

$$c(\mathbf{a}, \mathbf{b}, \mathbf{c}, \mathbf{d}) = c(A\,\mathbf{a}, A\,\mathbf{b}, A\,\mathbf{c}, A\,\mathbf{d}), \qquad \forall\, \mathbf{a}, \mathbf{b}, \mathbf{c}, \mathbf{d}, \tag{D.9}$$

where $A \in SO(3)$. If this condition is satisfied for all $A \in SO(3)$, then the tensor is isotropic.

To exemplify the action of an isotropic tensor on four vectors, let us consider

$$c(\mathbf{a}, \mathbf{b}, \mathbf{c}, \mathbf{d}) = (\mathbf{a} \cdot \mathbf{d})(\mathbf{b} \cdot \mathbf{c}) + (\mathbf{a} \cdot \mathbf{c})(\mathbf{b} \cdot \mathbf{d}) \tag{D.10}$$

and

$$c(\mathbf{a}, \mathbf{b}, \mathbf{c}, \mathbf{d}) = (\mathbf{a} \cdot \mathbf{b})(\mathbf{c} \cdot \mathbf{d}). \tag{D.11}$$

These expressions satisfy the required index symmetries,

$$c(\mathbf{a}, \mathbf{b}, \mathbf{c}, \mathbf{d}) = c(\mathbf{b}, \mathbf{a}, \mathbf{c}, \mathbf{d}) = c(\mathbf{a}, \mathbf{b}, \mathbf{d}, \mathbf{c}) = c(\mathbf{c}, \mathbf{d}, \mathbf{a}, \mathbf{b}).$$

To obtain a general form of an isotropic elasticity tensor, we use the fact that any elasticity tensor can be written in terms of a Cartesian basis as

$$c_{ijk\ell} = c(\hat{\mathbf{e}}_i, \hat{\mathbf{e}}_j, \hat{\mathbf{e}}_k, \hat{\mathbf{e}}_\ell), \qquad i, j, k, \ell = 1, 2, 3,$$

where the right-hand side represents components with respect to the basis,

$$\sum_{i=1}^{3}\sum_{j=1}^{3}\sum_{k=1}^{3}\sum_{\ell=1}^{3} c_{ijk\ell}\, \hat{\mathbf{e}}_i \otimes \hat{\mathbf{e}}_j \otimes \hat{\mathbf{e}}_k \otimes \hat{\mathbf{e}}_\ell.$$

Considering expressions (D.10) and (D.11), we write

$$(\hat{\mathbf{e}}_i \cdot \hat{\mathbf{e}}_\ell)(\hat{\mathbf{e}}_j \cdot \hat{\mathbf{e}}_k) + (\hat{\mathbf{e}}_i \cdot \hat{\mathbf{e}}_k)(\hat{\mathbf{e}}_j \cdot \hat{\mathbf{e}}_\ell), \qquad i, j, k, \ell = 1, 2, 3, \tag{D.12}$$

and

$$(\hat{\mathbf{e}}_i \cdot \hat{\mathbf{e}}_j)(\hat{\mathbf{e}}_k \cdot \hat{\mathbf{e}}_\ell), \qquad i, j, k, \ell = 1, 2, 3, \tag{D.13}$$

respectively. It can be shown that all isotropic tensors consist of the linear combinations of expressions (D.12) and (D.13), whose sum—using the

[5] *see also*: Slawinski (2015, Section 4.2.1)

orthonormality of the basis vectors—we write, using the Kronecker delta, as

$$c^{\mathrm{iso}}_{ijk\ell} = \lambda\,\delta_{ij}\,\delta_{k\ell} + \mu\,(\delta_{i\ell}\,\delta_{jk} + \delta_{ik}\,\delta_{j\ell})\,, \qquad i,j,k,\ell = 1,2,3\,, \tag{D.14}$$

where λ and μ are the Lamé parameters. Thus, an isotropic elasticity tensor has only two linearly independent parameters, which—in terms of the tensor components—are c_{1111} and c_{2323}.

Remark D.4. In general, an isotropic fourth-rank tensor in $\mathbb{R}^3$ has three linearly independent components (Jeffreys, 1961a, Chapter VII).[6] The elasticity tensor has only two due to its index symmetries.

Inserting tensor (D.14) into expression (3.1), we write

$$\sigma_{ij} = \lambda\,\delta_{ij}\sum_{k=1}^{3}\sum_{\ell=1}^{3}\,\delta_{k\ell}\,\varepsilon_{k\ell} + \mu\sum_{k=1}^{3}\sum_{\ell=1}^{3}(\delta_{i\ell}\,\delta_{jk} + \delta_{ik}\,\delta_{j\ell})\,\varepsilon_{k\ell}\,, \qquad i,j = 1,2,3\,.$$

Using properties of the Kronecker delta, we obtain

$$\sigma_{ij} = \lambda\,\delta_{ij}\sum_{k=1}^{3}\varepsilon_{kk} + 2\,\mu\,\varepsilon_{ij}\,, \qquad i,j = 1,2,3\,,$$

which is the stress-strain equation for an isotropic Hookean solid.

Closing remarks

The assumption of no dissipation of energy upon a deformation, which—together with the constancy of temperature—leads to the concept of hyperelasticity, is a convenient simplification that, in many cases, results in empirically adequate modelling. In particular, such an approach lends itself to infinitesimal strains, in which neither heat production nor permanent damage of material are likely to play an important role.

The strain-energy function relates the stress and strain tensors discussed, respectively, in Appendices A and B. It follows from the fact that a stress tensor can be expressed in terms of that function, together with a deformation gradient and a strain tensor. Also, material symmetry is formulated in terms of the strain-energy function. Furthermore, one of the index symmetries of a Hookean solid, namely, $c_{ijk\ell} = c_{k\ell ij}$, is a consequence of that function.

[6] *see also*: Slawinski (2015, Exercise 5.12)

The concept of material symmetry discussed in this appendix is used explicitly in Section 3.2. Hyperelastic entities appear implicitly through most of the book, since Hookean solids, whose study constitutes the bulk of this book, are hyperelastic materials.

Let us conclude Appendices A–D with a remark by Irgens (2008, p. vi), which allows us to see continuum mechanics within a broader scientific context.

> The study of distributions of energy, matter, and other physical quantities under circumstances where their discrete nature is unimportant, and where they may be regarded as continuous functions of position and time, is called Classical Field Theory or Continuum Physics.[7] This branch of Physics presumes nothing directly regarding the structure of the materials, although the structure of a material may be reflected in the way the physical quantities are described. When the physical quantities are purely of mechanical or thermodynamical nature, the subject is called Continuum Mechanics.

[7]Readers interested in this subject might refer to such textbooks as Landau and Lifshitz (1975) and Hertel (2012).

Appendix E

On covariant and contravariant transformations: One-forms and vectors

The facts of mathematics are verified and presented by the axiomatic method. One must guard, however, against confusing the presentation of mathematics with the content of mathematics. An axiomatic presentation of a mathematical fact differs from the fact that is being presented as medicine differs from food. It is true that this particular medicine is necessary to keep the mathematician at a safe distance from the self-delusions of the mind. Nonetheless, understanding mathematics means being able to forget the medicine and enjoy the food. Confusing mathematics with the axiomatic method for its presentation is as preposterous as confusing the music of Johann Sebastian Bach with the techniques for counterpoint in the Baroque age.

Gian-Carlo Rota (1997)

Preliminary remarks

The fact that certain quantities of interest in continuum mechanics, in general, and in seismology, in particular, cannot be conveniently expressed other than in terms of curvilinear coordinates leads to subtleties of coordinate transformations. There are two types of such transformations: *covariant* and *contravariant*.[1] First-rank tensors that transform covariantly are called one-forms; first-rank tensors that transform contravariantly are called vectors. This distinction is absent if we remain within Cartesian coordinates.

We begin this appendix with discussions of contravariant and covari-

[1]Readers interested in further examination of the covariant and contravariant entities and their transformations might refer to Misner *et al.* (1973, Chapter 2) and to Schutz (1980, Chapter 2).

ant transformations of first-rank tensors. We complete it by discussing higher-rank tensors, which—as combinations of first-rank tensors—might exhibit mixed transformations. Also, in this appendix, we comment on Cartesian coordinates, where the distinction between covariant and contravariant transformations disappears.

Notation E.1. In this appendix and in Appendix F, the rank of a tensor is the number of its indices, which is the sum of the number of its superscripts and subscripts.

E.1 Contravariant transformations

The archetype of an entity that exhibits a contravariant transformation are components of a position vector, r. In physics, contravariant objects are commonly associated with position and its temporal derivatives; notably, ray velocity is a seismological example.

Let us consider two curvilinear coordinate systems, ξ and ξ'; their bases are $(\hat{e}_1, \ldots, \hat{e}_n)$ and $(\hat{e}_{1'}, \ldots, \hat{e}_{n'})$, respectively; each $\hat{e}_i$ and $\hat{e}_{i'}$ stands for a basis vector. Explicitly, $\hat{e}_i = \partial r / \partial \xi^i$ and $\hat{e}_{i'} = \partial r / \partial \xi^{i'}$.

Notation E.2. The prime on the index refers to the symbol, not to the index; $\hat{e}_{i'}$ stands for $\hat{e}'_i$, and $\xi^{i'}$ for ξ'^i. Thus, for instance, the use of $\xi^{i'}$ and ξ^i is to distinguish ξ' from ξ, not i' from i. In other words, herein, we refer to distinct coordinate systems, not to different numbering schemes.

Following the chain rule,

$$\hat{e}_{i'} = \frac{\partial r}{\partial \xi^{i'}} = \sum_{j=1}^{n} \frac{\partial r}{\partial \xi^j} \frac{\partial \xi^j}{\partial \xi^{i'}} = \sum_{j=1}^{n} \frac{\partial \xi^j}{\partial \xi^{i'}} \hat{e}_j \,. \tag{E.1}$$

To distinguish from covariant entities, discussed in Appendix E.2, $\xi^1, \xi^2, \ldots, \xi^n$ stand for components of ξ; they are not powers of ξ. The occurrence of the same index—positioned once as superscript and once as a subscript—implies the summation over the range of that index; within this convention, a superscript in the denominator is tantamount to a subscript, $1/x^i \equiv y_i$. Herein, we specify the summations explicitly by $\sum$. In the spirit of the quote from Rota (1997), on page 465, we assure the reader that clear notational formalisms are necessary to allow examinations of intrinsic properties of geometry and physics, but they are not their essence.

Components of ξ' and ξ are related by $\xi^{i'} = \xi^{i'}(\xi^1,\ldots,\xi^n)$ and $\xi^i = \xi^i(\xi^{1'},\ldots,\xi^{n'})$. Their partial derivatives form the $n \times n$ Jacobi matrix,

$$P = \begin{bmatrix} \dfrac{\partial \xi^{1'}}{\partial \xi^1} & \cdots & \dfrac{\partial \xi^{1'}}{\partial \xi^n} \\ \vdots & \ddots & \vdots \\ \dfrac{\partial \xi^{n'}}{\partial \xi^1} & \cdots & \dfrac{\partial \xi^{n'}}{\partial \xi^n} \end{bmatrix}, \tag{E.2}$$

which serves as the *transition matrix* between the bases. Following matrix (E.2), equation (E.1) can be written as

$$(\hat{e}_{1'},\ldots,\hat{e}_{n'}) = (\hat{e}_1,\ldots,\hat{e}_n)\,P^{-1}. \tag{E.3}$$

Equation (E.3) is a system of n equations. Each of them relates the components of vector $\hat{e}_{i'}$ to the components of vector $\hat{e}_i$.

On the other hand, the components, r^i, of vector r, transform according to

$$\begin{bmatrix} r^{1'} \\ \vdots \\ r^{n'} \end{bmatrix} = P \begin{bmatrix} r^1 \\ \vdots \\ r^n \end{bmatrix}, \tag{E.4}$$

which is opposite to the transformation of the basis vectors. Hence, r is contravariant; its components vary in a manner contrary to the basis of the coordinate system. By convention, these components are marked with superscripts.

Example E.1. Let us revisit Example A.12. Therein, for a vector to remain invariant, its components vary in the opposite direction to the basis elements. The rotation matrix,

$$P = \begin{bmatrix} \cos(\theta) & -\sin(\theta) \\ \sin(\theta) & \cos(\theta) \end{bmatrix},$$

that transforms the vector components is the inverse of the matrix that transforms the basis vectors. As a result, the vector remains unchanged,

$$V^{1'}\hat{e}_{1'} + V^{2'}\hat{e}_{2'} = V^1\hat{e}_1 + V^2\hat{e}_2\,, \tag{E.5}$$

where $\hat{e}_i$ and $\hat{e}_{i'}$ stand, respectively, for vectors $\mathbf{\hat{e}}_i$ and $\mathbf{\hat{e}}'_i$, in Example A.12; also, V^i and $V^{i'}$ stand for v_i and v'_i.

Letting $\hat{e}_1 = [1,0]^T$ and $\hat{e}_2 = [0,1]^T$, and following equation (E.3), we obtain

$$\begin{bmatrix} \hat{e}_{1'} & \hat{e}_{2'} \end{bmatrix} = \begin{bmatrix} 1 & 0 \\ 0 & 1 \end{bmatrix} \begin{bmatrix} \cos(\theta) & \sin(\theta) \\ -\sin(\theta) & \cos(\theta) \end{bmatrix} = \begin{bmatrix} \cos(\theta) & \sin(\theta) \\ -\sin(\theta) & \cos(\theta) \end{bmatrix},$$

where the transition matrix is P^{-1}. Hence, $\hat{e}_{1'} = [\cos(\theta), -\sin(\theta)]^T$ and $\hat{e}_{2'} = [\sin(\theta), \cos(\theta)]^T$. If $\theta = 0$, $\hat{e}_{1'} = \hat{e}_1$ and $\hat{e}_{2'} = \hat{e}_2$, as required.

Thus, following equation (E.4), we can write the left-hand side of equation (E.5) as

$$\underbrace{\left(V^1\cos(\theta) - V^2\sin(\theta)\right)}_{V^{1'}}\underbrace{[\cos(\theta), -\sin(\theta)]^T}_{\hat{e}_{1'}} + \underbrace{\left(V^1\sin(\theta) + V^2\cos(\theta)\right)}_{V^{2'}}\underbrace{[\sin(\theta), \cos(\theta)]^T}_{\hat{e}_{2'}}.$$

Multiplying, we obtain

$$\begin{bmatrix} V^1\cos^2(\theta) - V^2\sin(\theta)\cos(\theta) \\ \\ -V^1\sin(\theta)\cos(\theta) + V^2\sin^2(\theta) \end{bmatrix} + \begin{bmatrix} V^1\sin^2(\theta) + V^2\sin(\theta)\cos(\theta) \\ \\ V^1\sin(\theta)\cos(\theta) + V^2\cos^2(\theta) \end{bmatrix},$$

which—upon summing the corresponding terms—results in

$$\begin{bmatrix} V^1 \\ V^2 \end{bmatrix} = V^1\begin{bmatrix} 1 \\ 0 \end{bmatrix} + V^2\begin{bmatrix} 0 \\ 1 \end{bmatrix} = V^1\hat{e}_1 + V^2\hat{e}_2,$$

which is the right-hand side of equation (E.5), as expected.

Remark E.1. In the language of differential geometry, a vector,

$$V^1\frac{\partial}{\partial\xi^1} + \cdots + V^n\frac{\partial}{\partial\xi^n} =: \sum_{i=1}^{n} V^i\frac{\partial}{\partial\xi^i},$$

is a linear combination of the tangents along the coordinate axes of the manifold, associated to the local coordinates, $(\xi^1, \ldots, \xi^n)$. Herein, $\partial/\partial\xi^i$, where $i = 1, \ldots, n$, are directional derivatives that constitute a basis in the space tangent to the manifold. This basis corresponds to $(\hat{e}_1, \ldots, \hat{e}_n)$, in our notation.

E.2 Covariant transformations

The archetype of an entity that exhibits a covariant transformation is the components of the gradient of a scalar field. In physics, covariant objects commonly contain the units of the reciprocal of distance. In seismology, a typical example is the wavefront slowness.

Geometrically, the gradient behaves differently from the position vector, which is the archetype of a contravariant entity, even though—in a manner analogous to a vector—it is described by the same number of components for a given spatial dimension; herein, these components are $\partial f/\partial\xi^i$, where f is a scalar field. The gradient is an example of a one-form, rather than of a vector. This fact becomes important if we use curvilinear coordinates, but can be ignored in Cartesian coordinates, which illustrates the necessity of revising commonly accepted structures upon examining a broader context.

To express the gradient of f with respect to the ξ' coordinates, which is tantamount to obtaining $\partial f/\partial\xi^{i'}$, we apply the chain rule to $f(\xi)$, where $\xi^i = \xi^i(\xi^{1'},\ldots,\xi^{n'})$, to write

$$\frac{\partial f}{\partial\xi^{i'}} = \sum_{j=1}^{n} \frac{\partial\xi^j}{\partial\xi^{i'}}\frac{\partial f}{\partial\xi^j},$$

where $\partial\xi^j/\partial\xi^{i'}$ form the transition matrix,

$$Q = \begin{bmatrix} \dfrac{\partial\xi^1}{\partial\xi^{1'}} & \cdots & \dfrac{\partial\xi^n}{\partial\xi^{1'}} \\ \vdots & \ddots & \vdots \\ \dfrac{\partial\xi^1}{\partial\xi^{n'}} & \cdots & \dfrac{\partial\xi^n}{\partial\xi^{n'}} \end{bmatrix} = \left(P^{-1}\right)^T, \tag{E.6}$$

where P is matrix (E.2).

The gradient is an example of an entity whose components, $V_i := \partial f/\partial\xi^i$ and $V_{i'} := \partial f/\partial\xi^{i'}$, transform as

$$[V_{1'},\ldots,V_{n'}] = [V_1,\ldots,V_n]\,P^{-1}, \tag{E.7}$$

which has the same form as equation (E.3). However, in equation (E.3), $\hat{e}_i$ and $\hat{e}_{i'}$ are vectors; herein, V_i and $V_{i'}$ are components. In other words, equation (E.7) is analogous to a single equation of system (E.3), which is composed of n such equations.

Thus, in view of equations (E.3) and (E.7), the components of a one-form transform in the same manner as the basis vectors of a coordinate system. For this reason, such entities are called *covariant*; by convention, the components of covariant entities are written with subscripts, rather than superscripts.

Remark E.2. In the language of differential geometry, a one-form, which is also known as *covector*,

$$V_1\,\mathrm{d}\xi^i + \cdots + V_n\,\mathrm{d}\xi^i =: \sum_{i=1}^{n} V_i\,\mathrm{d}\xi^i,$$

is a linear combination of the differentials associated to the local coordinates, $(\xi^1,\ldots,\xi^n)$, of the manifold. Herein, the differentials, $\mathrm{d}\xi^i$, where $i=1,\ldots,n$, constitute the basis in the dual space of the tangent space; it is dual to the basis therein, $(\partial/\partial\xi^1,\ldots,\partial/\partial\xi^n)$, introduced in Remark E.1.

If we remain within Cartesian coordinates, the distinction between the covariant and contravariant transformations disappears. In such a case, P is an orthogonal matrix and, hence, its transpose is its inverse; consequently, $Q = P$.

Example E.2. It is common to illustrate a vector as an arrow, which emphasizes the direction of the vector and indicates its magnitude by the length of the arrow. There is a similar illustration for a one-form (Misner *et al.* (1973, Section 2.5), Schutz (1980, Section 2.16–2.19)).

In ray theory, the ray velocity, $\dot{x} := \mathrm{d}x/\mathrm{d}t$, which is a vector, illustrated in the left-hand plot of Figure E.1, is the dual of the wavefront slowness, $\nabla\psi$, which—as a real-valued linear function—is a one-form, illustrated in the right-hand plot; herein, ψ stands for the eikonal function.[2] The direction of the increase of the one-form is indicated by progressive increasing length of dashes forming the lines. Its magnitude is indicated by the density of lines, which is constant.

Since the ray velocity is $\mathrm{d}x/\mathrm{d}t$, where x stands for distance and t for time, it is expressed in units of speed, say, km/s, which is the number of kilometres covered by a wave in one second. Since ψ is expressed in units of time, $\nabla\psi$ is expressed in units of slowness, say, s/km, which is the number of seconds needed to cover one kilometre.

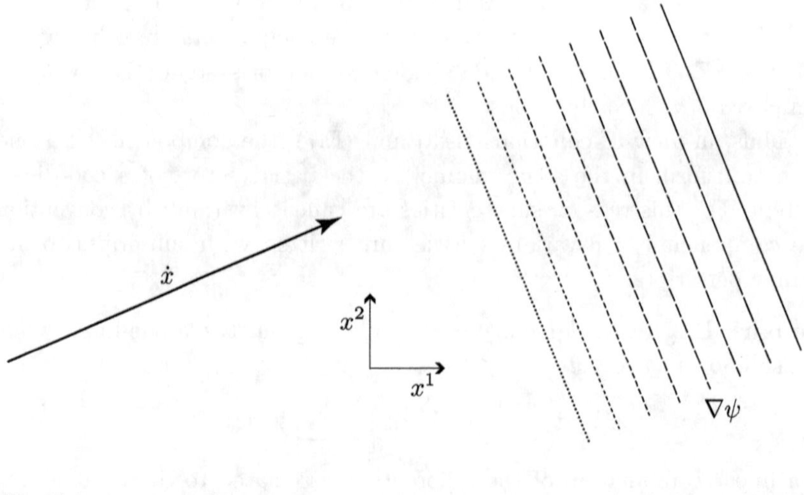

Fig. E.1: Ray velocity and wavefront slowness as vector and one-form, which—in ray theory—are related by Legendre's transformation.

[2] *see also*: Slawinski (2015, Section 7.3)

In $\mathbb{R}^3$, the arrow remains an arrow and the lines, which remain the level sets of the function, become planes. In $\mathbb{R}^n$, they are hyperplanes; their dimension is $n-1$ (e.g., Penrose, 2016, Appendix A.4).

In the language of differential geometry, $\dot{x}$ belongs to the space tangent to a manifold at point x, whereas $\nabla\psi$ belongs to the dual space of the tangent space, which is known as the cotangent space at point x. This manifold represents a medium within which waves propagate.

In general, even though the spaces containing vectors and one-forms are duals of one another, there is no relation between individual elements of these spaces. In other words, in general, there is no correspondence between vectors and one-forms. To define such a correspondence, one needs an additional structure, such as an inner product. As discussed in Section 7.2.5.1 on page 278, the procedure connecting ray velocity and wavefront slowness is Legendre's transformation.[3] Herein, it relates the Lagrangian and Hamiltonian formulations of ray theory. Thus, the relation between the ray velocity and the corresponding wavefront slowness is a property of dynamics, not a property of the manifold alone, which represents kinematics only.

The correspondence between ray velocity and wavefront slowness is one-to-one if the Hamiltonian, H, is regular; in other words, if $\partial^2 H/\partial p_i \partial p_j \neq 0$, where $p_i := \partial\psi/\partial x^i$, which is tantamount to the convexity of the wavefront (Bucataru and Slawinski, 2009b). Otherwise, up to three ray velocities correspond to a single wavefront slowness.[4]

E.3 Mixed transformations

We can consider objects that result from a linear combination of tensor products of the covariant and contravariant entities. For instance, T^i_j, where $i = 1, \ldots, n$ and $j = 1, \ldots, n$ is a second-rank tensor of the $(1,1)$ type, which has n^2 components.

The transformation rules of a higher-rank tensor are inherited from the first-rank tensors,

$$T^{i'}_{j'} = \sum_{j=1}^{n}\sum_{k=1}^{n} \frac{\partial \xi^{i'}}{\partial \xi^{j}} \frac{\partial \xi^{k}}{\partial \xi^{j'}} T^j_k \,, \tag{E.8}$$

[3] *see also*: Slawinski (2015, Appendix B.3)
[4] *see also*: Slawinski (2015, Sections 8.4, 13.1)

where the partial derivatives are stated explicitly in matrices (E.2) and (E.6). In a similar manner, we would apply twice either of these transformation matrices for a second-rank tensor of the $(0,2)$ and $(2,0)$ type, which are contravariant and covariant second-rank tensors, respectively. The same pattern applies to all higher-rank tensors, where the number of partial derivatives, analogous to two that appear in expression (E.8), is equal to the rank of a tensor. Such a tensor is referred to as the (p,q)-type tensor, depending on the number of covariant and contravariant indices.

In continuum mechanics, the deformation-gradient tensor and the First Piola-Kirchhoff stress tensor, discussed in Appendices A and B, respectively, share several similarities with mixed second-rank tensors. However, they are two-point tensors, which means that their indices refer to different spaces, which is not allowed in expression (E.8).

Closing remarks

The existence of two types of transformation, and hence the distinction between a vector, which is the contravariant entity, and a *one-form*, which is the covariant entity, is another example of essential subtleties that appear in quantitative seismology if we reach beyond conventional contexts. Notably, *vector* is a term used to denote both these entities in a conventional setting; for instance, in Cartesian coordinates, the gradient is viewed as a vector, even though—in general—it is a one-form. A seismological example of such a distinction is the wavefront velocity and slowness, which are a vector and one-form, respectively. In Example F.2 of Appendix F.1, we introduce the method of expressing a vector as a one-form, and *vice versa*.

Marsden and Hughes (1983, p. 46) write that

> [i]n classical tensor analysis, a vector and one-form are indistinguishable in the sense that we have the representations
>
> $$\mathbf{v} = v^a\,\mathbf{e}_a = v_a\,\mathbf{e}^a\,, \qquad \text{and} \qquad \alpha = \alpha_a\,\mathbf{e}^a = \alpha^a\,\mathbf{e}_a\,.$$
>
> In tensor analysis on manifolds one avoids confusing vectors and one-forms.

Since—in Cartesian coordinates—there is no distinction between the covariant and contravariant entities, it is common to use subscripts for their components. Objects resulting from combining such entities are *Cartesian tensors* (Jeffreys, 1961a), with components denoted by subscripts. Their transformation rules are the same as for general tensors, except that there is no distinction between the covariant and contravariant transformations.

As an aside, following Remark A.16, and remaining within the subject of tensors, it might be interesting to note that a result of the vector product, in $\mathbb{R}^3$, is neither a vector nor a one-form. It is an antisymmetric second-rank pseudotensor that—in Cartesian coordinates in $\mathbb{R}^3$—behaves like a pseudovector.

Extending our interests beyond the conventional setting requires notational intricacies to accommodate the existence of subtle properties. Sir Harold Jeffreys (1924, Preface, pp. viii–ix) writes that

> there is no way of obtaining quantitative results without mathematics. [...] In short, if geophysics requires mathematics for its treatment, it is the earth that is responsible, not the geophysicist.

If we accept the latter part of this quote literally, we follow the claim of *Pythagoreanism*, according to which the world is mathematics; there is nothing but mathematics.[5] By contrast—according to *Platonism*[6]—there are two realms: abstract and physical. Mathematical objects belong to the former, even though they might be used as models for the latter. Be that as it may, herein, the essence of the quote is the fact that we are incapable of formulating sufficiently sophisticated models without subtleties of mathematics. Note that, in the above quote, we choose to interpret the meaning of *quantitative* as *theoretical*; if it meant *numerical*, then the statement would be tautological, since—by definition—we cannot obtain numerical results without mathematics.

[5]Readers interested in a particular view of Pythagoreanism might refer to Tegmark (2014).

[6]Readers interested in Platonism might refer to Penrose (1997).

Appendix F

On covariant derivatives

The aviator pursuing his great circle route from Peking to Vancouver finds himself early going north, but later going south, although he is navigating the straightest route that is at all open to him (geodesic). The apparent change in direction indicates a turning, not in his route, but in the system of coordinates with respect to which his route is described.

Charles W. Misner, Kip S. Thorne, John Archibald Wheeler (1973)

Preliminary remarks

In spite of its closeness to experience, discussed on page 19, continuum mechanics, which is a study of deformable materials, must invoke many subtleties and much sophistication of mathematics to achieve insightful descriptions of nature. Tensor calculus and differential geometry are mathematical subjects whose methods appear in such a description. Notably, continuum mechanics shares the need for these subjects with the general theory of relativity, as indicated by such treatises as Misner *et al.* (1973) and Epstein (2010).

In this appendix, we discuss an extension of the partial derivative that allows us to apply it beyond Cartesian coordinates. The need for such an extension appears in quantitative seismology in the description of deformable materials and their deformations.[1]

We begin this appendix by introducing the *metric tensor*. Subsequently, we formulate the Christoffel symbol, whose purpose is to ensure the consistency of derivative with a given metric. We conclude this appendix with the discussion on the *covariant derivative*, which allows us to obtain the

[1]Readers interested in further examination of the covariant derivative might refer to Misner *et al.* (1973, Chapter 10) and to Schutz (1980, Chapter 6).

intrinsic properties of a function regardless of the coordinates in which it is expressed.

Notation F.1. In this appendix and in Appendix E, the rank of a tensor is the number of its indices, which is the sum of the number of its superscripts and subscripts. However, as illustrated by the Christoffel symbols, not every entity with indices is a tensor.

F.1 Metric tensor

The apparent contradiction presented in the quote of Misner *et al.* (1973) is a consequence of different requirements in accounting for distance on a plane and on the surface of a sphere. To motivate—beyond the apparent contradiction between the straight and turning routes in the above quote—the concept of a derivative that accommodates such issues, let us consider the following example.

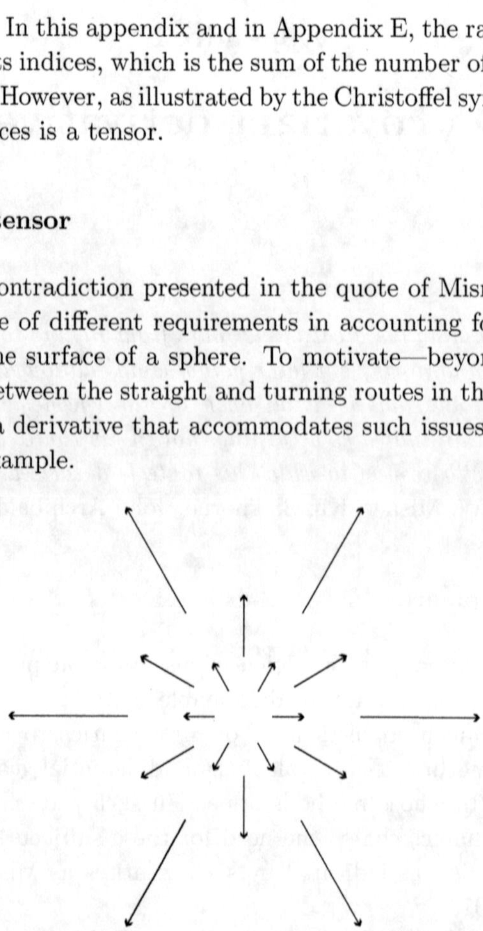

Fig. F.1: Radially symmetric vector field, whose amplitude increases linearly with the distance from the origin; it is a geometrical entity, which can be described using a variety of coordinates.

Example F.1. Let us examine quantitatively the vector field in Figure F.1. To do so, we need to express it in terms of coordinates. Let us consider Cartesian and polar coordinates to write the former as $[V^x, V^y]$, and the latter as $[V^r, V^\theta]$. Choosing the centre of this field as the origin of the coordinate systems, we write the coordinates as $[x, y]$ and $[r, 0]$, respectively.

In these coordinate systems, let us consider partial derivatives of the components of the vector field in Figure F.1. In Cartesian coordinates,

$$\begin{bmatrix} \dfrac{\partial V^x}{\partial x} & \dfrac{\partial V^x}{\partial y} \\ \dfrac{\partial V^y}{\partial x} & \dfrac{\partial V^y}{\partial y} \end{bmatrix} = \begin{bmatrix} 1 & 0 \\ 0 & 1 \end{bmatrix} . \tag{F.1}$$

In polar coordinates,

$$\begin{bmatrix} \dfrac{\partial V^r}{\partial r} & \dfrac{\partial V^r}{\partial \theta} \\ \dfrac{\partial V^\theta}{\partial r} & \dfrac{\partial V^\theta}{\partial \theta} \end{bmatrix} = \begin{bmatrix} 1 & 0 \\ 0 & 0 \end{bmatrix} . \tag{F.2}$$

Expressions (F.1) and (F.2) are not components of a second-rank tensor in two coordinate systems; they cannot be connected by relation (E.8). However, as for a scalar field, we wish the derivative to measure intrinsic changes within the field, regardless of coordinates in which this field is expressed, which requires that such a derivative be a tensor.

The disagreement between results (F.1) and (F.2) is a consequence of ignoring the difference in the measure of distance expressed in Cartesian and polar coordinates. The element of distance in Cartesian coordinates is

$$\mathrm{d}s^2 = \mathrm{d}x^2 + \mathrm{d}y^2 ,$$

which we can formally write as

$$\mathrm{d}s^2 = \sum_{i=1}^{2}\sum_{j=1}^{2} \delta_{ij}\, \mathrm{d}x^i\, \mathrm{d}x^j ;$$

in polar coordinates,

$$\mathrm{d}s^2 = \mathrm{d}r^2 + r^2(\mathrm{d}\theta)^2 ,$$

as we justify in expression (F.4), below. Formally, we can write

$$\mathrm{d}s^2 = \sum_{i=1}^{2}\sum_{j=1}^{2} g_{ij}(\xi)\, \mathrm{d}\xi^i\, \mathrm{d}\xi^j ,$$

where ξ, ξ^i and ξ^j refer to a curvilinear coordinate system. In matrix notation, in Cartesian coordinates, we have

$$\mathrm{d}s^2 = \begin{bmatrix} \mathrm{d}x \, , \mathrm{d}y \end{bmatrix} \begin{bmatrix} 1 & 0 \\ 0 & 1 \end{bmatrix} \begin{bmatrix} \mathrm{d}x \\ \mathrm{d}y \end{bmatrix}$$

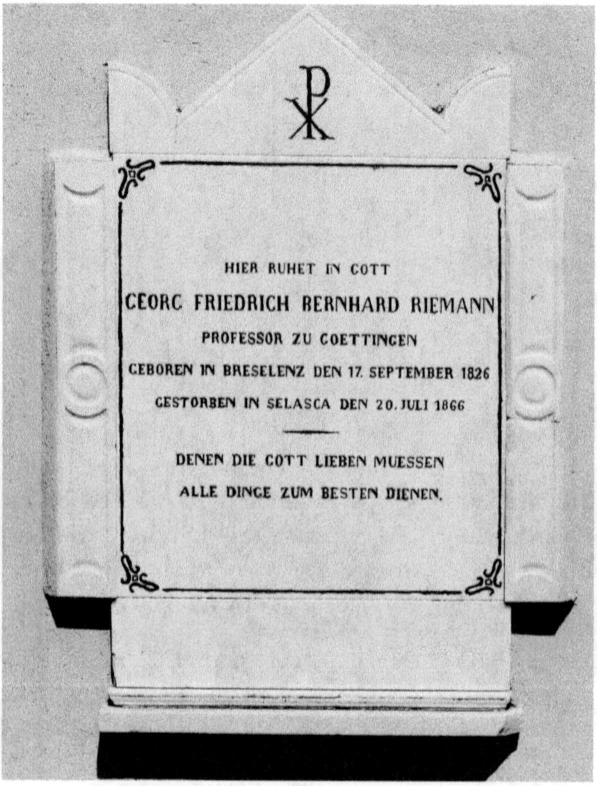

photo: Elena Patarini

Fig. F.2: Georg Friedrich Bernhard Riemann's grave in Biganzolo di Selasca, near Verbania, Italy.

and, in polar coordinates,

$$\mathrm{d}s^2 = \begin{bmatrix} \mathrm{d}r \,, \mathrm{d}\theta \end{bmatrix} \begin{bmatrix} 1 & 0 \\ 0 & r^2 \end{bmatrix} \begin{bmatrix} \mathrm{d}r \\ \mathrm{d}\theta \end{bmatrix} , \tag{F.3}$$

where the entries of the square matrices are denoted by g_{ij}; they are the components of the *metric tensor*.

For Cartesian coordinates, the metric tensor is constant; it is given by the Kronecker delta, δ_{ij}. For polar—and all non-Cartesian—coordinates, the corresponding metric tensor is not constant; it is a function of coordinates. This entity is referred to as the *Riemannian metric*, in honour of Georg Friedrich Bernhard Riemann, whose grave is shown in Figure F.2. A differentiable manifold endowed with this metric is a *Riemannian manifold*.

The applicability of Cartesian coordinates is limited. In many cases, we consider curvilinear coordinates, and hence, their metric tensors are not constant. In general, objects in flat space, such as a plane or a volume in $\mathbb{R}^3$, can be described by both Cartesian and curvilinear coordinates; however, a space with an intrinsic curvature, such as the surface of a sphere, does not lend itself to Cartesian coordinates, even if the latitudes and longitudes might be—in a manner similar to Cartesian coordinates—orthogonal to each other, as is the case for the Mercator projection of the surface of the Earth. However, unlike the metric for Cartesian coordinates, the Mercator metric is not constant, as can be seen from the large size of Greenland in comparison to Africa. Cartesian coordinates are the only coordinates whose metric tensor is constant.

For any coordinate system that is expressible in terms of Cartesian coordinates, the metric tensor can be obtained by invoking its relation to these coordinates. The relations between the polar and Cartesian coordinates are $x = r\cos\theta$ and $y = r\sin\theta$. The position vectors, such as $[x, y]$ and $[r, \theta]$, are contravariant tensors of rank one. Hence, according to the contravariant transformation discussed in Appendix E.1,

$$\begin{bmatrix} \mathrm{d}x \\ \mathrm{d}y \end{bmatrix} = \begin{bmatrix} \dfrac{\partial x}{\partial r} & \dfrac{\partial x}{\partial \theta} \\[2ex] \dfrac{\partial y}{\partial r} & \dfrac{\partial y}{\partial \theta} \end{bmatrix} \begin{bmatrix} \mathrm{d}r \\ \mathrm{d}\theta \end{bmatrix} = \begin{bmatrix} \cos\theta & -r\sin\theta \\ \sin\theta & r\cos\theta \end{bmatrix} \begin{bmatrix} \mathrm{d}r \\ \mathrm{d}\theta \end{bmatrix}.$$

The corresponding metric tensor, which is a covariant symmetric second-rank tensor, is

$$g_{ij} = \begin{bmatrix} \cos\theta & -r\sin\theta \\ \sin\theta & r\cos\theta \end{bmatrix}^T \begin{bmatrix} \cos\theta & -r\sin\theta \\ \sin\theta & r\cos\theta \end{bmatrix} = \begin{bmatrix} 1 & 0 \\ 0 & r^2 \end{bmatrix} =: \begin{bmatrix} g_{rr} & g_{r\theta} \\ g_{r\theta} & g_{\theta\theta} \end{bmatrix}, \tag{F.4}$$

where T denotes the transpose. This result justifies expression (F.3). Herein, the inverse metric is

$$g^{ij} \equiv (g_{ij})^{-1} = \begin{bmatrix} 1 & 0 \\ 0 & r^{-2} \end{bmatrix} =: \begin{bmatrix} g^{rr} & g^{r\theta} \\ g^{r\theta} & g^{\theta\theta} \end{bmatrix}, \tag{F.5}$$

which is a contravariant symmetric second-rank tensor. We can refer to g_{ij} and its inverse, g^{ij}, as the covariant and contravariant metric tensor, respectively.

Example F.2. The reciprocity of the metric tensor and its inverse, $\sum_{k=1}^{n} g_{ik}g^{k\ell} = \delta_i^\ell$, allows us to form a covariant object from a contravariant one, and *vice versa*. For example, for vectors and one-forms in n dimensions, we have

$$V^i = \sum_{k=1}^{n} g^{ik} V_k \qquad \text{and} \qquad V_i = \sum_{k=1}^{n} g_{ik} V^k ,$$

where $i = 1, \ldots, n$. For a general tensor,

$$\sum_{i_1 \ldots i_n = 1}^{N} \sum_{j_1 \ldots j_m = 1}^{M} g_{i_1 p_1} \cdots g_{i_n p_n} g^{j_1 q_1} \cdots g^{j_m q_m} T^{i_1 \ldots i_n}_{j_1 \ldots j_m} = T^{q_1 \ldots q_m}_{p_1 \ldots p_n} .$$

Thus, the raising and lowering of indices is an important role of the metric tensor in tensor algebra; the existence of the metric allows us to interchange the covariant and contravariant properties of a tensor.

F.2 Christoffel symbol

An entity whose purpose is to ensure the consistency of derivatives involving distance, for all coordinate systems, is the *Christoffel symbol*. Its components are expressed in terms of components of the metric tensor corresponding to a given system,

$$\Gamma^i_{jk} := \frac{1}{2} \sum_{\ell=1}^{n} g^{i\ell} \left(\frac{\partial g_{k\ell}}{\partial \xi^j} + \frac{\partial g_{\ell j}}{\partial \xi^k} - \frac{\partial g_{jk}}{\partial \xi^\ell} \right) , \qquad i, j, k = 1, \ldots, n , \tag{F.6}$$

where $\xi^1, \ldots, \xi^n$ are the curvilinear coordinates. In n dimensions, Γ^i_{jk} is an array of real numbers, $n \times n \times n$, where n is the dimension of the space. In spite of each of its components being denoted by three indices, it is not a tensor of rank three; it is not a tensor at all, even though it behaves like a tensor under linear coordinate transformations.[2] In a manner akin to a tensor, we can express the symmetry of the Christoffel symbol as $\Gamma^i_{jk} = \Gamma^i_{kj}$, and its number of components, which is n^3, reduces to $n^2(n+1)/2$ distinct components, due to the lower-index symmetry.

According to expression (F.6), the Christoffel symbol—which is defined in terms of derivatives of the metric tensor—is zero for Cartesian coordinates, since their metric is constant. For polar coordinates, the Christoffel symbol is not zero. Using expressions (F.4) and (F.5), in expression (F.6),

[2] This property of the Christoffel symbol is used in formulations presented by Bóna *et al.* (2007a).

we write

$$\begin{aligned}\Gamma^1_{11} =: \Gamma^r_{rr} =& \frac{1}{2} g^{rr}\left(\frac{\partial g_{rr}}{\partial r} + \frac{\partial g_{rr}}{\partial r} - \frac{\partial g_{rr}}{\partial r}\right) \\ &+ \frac{1}{2} g^{r\theta}\left(\frac{\partial g_{r\theta}}{\partial r} + \frac{\partial g_{\theta r}}{\partial r} - \frac{\partial g_{rr}}{\partial \theta}\right) \\ =& \frac{1}{2}(1)(0+0-0) + \frac{1}{2}(0)(0+0-0) = 0\,;\end{aligned} \tag{F.7}$$

continuing in a similar manner,

$$\Gamma^1_{22} =: \Gamma^r_{\theta\theta} = -r\,, \tag{F.8}$$

$$\Gamma^1_{21} = \Gamma^1_{12} =: \Gamma^r_{\theta r} = \Gamma^r_{r\theta} = 0\,, \tag{F.9}$$

$$\Gamma^2_{11} =: \Gamma^\theta_{rr} = 0\,, \tag{F.10}$$

$$\Gamma^2_{12} = \Gamma^2_{21} =: \Gamma^\theta_{r\theta} = \Gamma^\theta_{\theta r} = \frac{1}{r}\,, \tag{F.11}$$

$$\Gamma^2_{22} =: \Gamma^\theta_{\theta\theta} = 0\,. \tag{F.12}$$

Herein, there are eight components, six of which are distinct and two among them are nonzero. Components of the Christoffel symbol are zero if the metric is constant or if there are common symmetries between the metric tensor and the coordinate system.

Remark F.1. Let us consider Cartesian coordinates, x^i, and curvilinear coordinates, ξ^i, whose basis vectors are $\hat{e}_i$ and $\hat{\varepsilon}_i$, respectively. The Christoffel symbol in ξ^i can be expressed as a derivative of $\hat{\varepsilon}_i$, with respect to ξ^i,

$$\frac{\partial \hat{\varepsilon}_i}{\partial \xi^j} = \sum_{k=1}^{n} \Gamma^k_{ij}\,\hat{\varepsilon}_k\,, \qquad i,j = 1,\ldots,n\,. \tag{F.13}$$

In particular, for polar coordinates, $(\xi^1, \xi^2) = (r, \theta)$, and $\hat{\varepsilon}_1 = \hat{e}_r$ and $\hat{\varepsilon}_2 = \hat{e}_\theta$; explicitly,

$$\hat{e}_r = \begin{bmatrix}\cos\theta \\ \sin\theta\end{bmatrix}, \qquad \hat{e}_\theta = \begin{bmatrix}-r\,\sin\theta \\ r\,\cos\theta\end{bmatrix}.$$

We see that $\partial \hat{e}_r/\partial r = 0$. Thus, following expression (F.13),

$$\frac{\partial \hat{e}_r}{\partial r} = \Gamma^r_{rr}\,\hat{e}_r + \Gamma^\theta_{rr}\,\hat{e}_\theta = 0\,.$$

Thus—since $\hat{e}_r$ and $\hat{e}_\theta$ are nonzero linearly independent vectors—it follows that $\Gamma^r_{rr} = 0$ and $\Gamma^\theta_{rr} = 0$, as expected, in accordance with expressions (F.7) and (F.10). Also,

$$\frac{\partial \hat{e}_r}{\partial \theta} = \frac{\partial}{\partial \theta}\begin{bmatrix}\cos\theta\\ \sin\theta\end{bmatrix} = \begin{bmatrix}-\sin\theta\\ \cos\theta\end{bmatrix} = \frac{1}{r}\begin{bmatrix}-r\sin\theta\\ r\cos\theta\end{bmatrix} = \frac{1}{r}\,\hat{e}_\theta\,.$$

Thus, following expression (F.13), for polar coordinates,

$$\frac{\partial \hat{e}_r}{\partial \theta} = \Gamma^r_{r\theta}\,\hat{e}_r + \Gamma^\theta_{r\theta}\,\hat{e}_\theta = \frac{1}{r}\,\hat{e}_\theta\,.$$

Hence, it follows that $\Gamma^r_{r\theta} = 0$ and $\Gamma^\theta_{r\theta} = 1/r$, as expected, in accordance with expressions (F.9) and (F.11). Similar calculations would result in the remaining Christoffel-symbol expressions.

The formulation of the Christoffel symbol presented in expression (F.13) leads to a more general concept of covariant derivative, which—in contrast to expression (F.6)—does not require a metric, but instead the notion of a *parallel transport.*[3]

F.3 Covariant derivative

Let us use results (F.7)–(F.12) to remove the discrepancy between results (F.1) and (F.2). To do so, we introduce the definition of the *covariant derivative.* In general, in $\mathbb{R}^2$, the covariant derivative of a vector field, V, is defined as

$$\begin{bmatrix}\nabla_1 V^1 & \nabla_2 V^1\\ \nabla_1 V^2 & \nabla_2 V^2\end{bmatrix} := \begin{bmatrix}\dfrac{\partial V^1}{\partial \xi^1} & \dfrac{\partial V^1}{\partial \xi^2}\\[2ex] \dfrac{\partial V^2}{\partial \xi^1} & \dfrac{\partial V^2}{\partial \xi^2}\end{bmatrix} + \begin{bmatrix}\Gamma^1_{11}V^1 + \Gamma^1_{12}V^2 & \Gamma^1_{21}V^1 + \Gamma^1_{22}V^2\\ \Gamma^2_{11}V^1 + \Gamma^2_{12}V^2 & \Gamma^2_{21}V^1 + \Gamma^2_{22}V^2\end{bmatrix}. \tag{F.14}$$

This expression uses 2^3 Christoffel symbols; the same pattern is followed in $\mathbb{R}^3$, with 3^3 Christoffel symbols. The conventional partial derivative, is extended—by the Christoffel symbol—to curvilinear coordinates. Commonly, several among these symbols are zero.

The covariant derivatives of a vector field, expressed in polar coordinates, $V = [V^r(r,\theta), V^\theta(r,\theta)]$, are[4]

$$\nabla_r V^r(r,\theta) = \frac{\partial V^r}{\partial r} + \Gamma^r_{rr}V^r + \Gamma^r_{r\theta}V^\theta\,, \tag{F.15}$$

[3]Readers interested in such formulations might refer to Carroll (2004, Section 3.3).
[4]Readers interested in a detailed formulation might refer to Carroll (2004, Section 3.2).

$$\nabla_\theta V^r(r,\theta) = \frac{\partial V^r}{\partial \theta} + \Gamma^r_{\theta r} V^r + \Gamma^r_{\theta\theta} V^\theta \,, \tag{F.16}$$

$$\nabla_r V^\theta(r,\theta) = \frac{\partial V^\theta}{\partial r} + \Gamma^\theta_{rr} V^r + \Gamma^\theta_{r\theta} V^\theta \,, \tag{F.17}$$

$$\nabla_\theta V^\theta(r,\theta) = \frac{\partial V^\theta}{\partial \theta} + \Gamma^\theta_{\theta r} V^r + \Gamma^\theta_{\theta\theta} V^\theta \,. \tag{F.18}$$

Their general form is given in expression (F.26), below.

In the context of Example F.1, let us comment on the notation, which resembles the conventional notation for the gradient. In Cartesian coordinates, we have $\nabla_x V^x = \partial V^x/\partial x$, where ∇_x stands for the covariant derivative with respect to x, and V^x denotes the first component of V; in polar coordinates, however, we have $\nabla_r V^r = \partial V^r/\partial r + \Gamma^r_{rr} V^r + \Gamma^r_{r\theta} V^\theta$, with an analogous pattern to follow for other components and derivatives.

Indeed, we can write the covariant derivatives for the vector field of Example F.1, in polar coordinates, as

$$\begin{bmatrix} \nabla_r V^r & \nabla_\theta V^r \\ \nabla_r V^\theta & \nabla_\theta V^\theta \end{bmatrix} = \begin{bmatrix} \dfrac{\partial V^r}{\partial r} & \dfrac{\partial V^r}{\partial \theta} \\ \dfrac{\partial V^\theta}{\partial r} & \dfrac{\partial V^\theta}{\partial \theta} \end{bmatrix} + \begin{bmatrix} 0 & -rV^\theta \\ \dfrac{1}{r}V^\theta & \dfrac{1}{r}V^r \end{bmatrix} , \tag{F.19}$$

which are components of a second-rank tensor, where the entries in the second matrix on the right-hand side take into account the effects of the curvilinearity of coordinates. In Cartesian coordinates, the second matrix is zero.

Since the vector-field components, in polar coordinates, are $[V^r, V^\theta] = [r, 0]$, using expression (F.19), we write the covariant derivatives as

$$\begin{bmatrix} \nabla_r V^r & \nabla_\theta V^r \\ \nabla_r V^\theta & \nabla_\theta V^\theta \end{bmatrix} = \begin{bmatrix} \dfrac{\partial r}{\partial r} & \dfrac{\partial r}{\partial \theta} \\ \dfrac{\partial 0}{\partial r} & \dfrac{\partial 0}{\partial \theta} \end{bmatrix} + \begin{bmatrix} 0 & 0 \\ 0 & \dfrac{1}{r} r \end{bmatrix} = \begin{bmatrix} 1 & 0 \\ 0 & 0 \end{bmatrix} + \begin{bmatrix} 0 & 0 \\ 0 & 1 \end{bmatrix} = \begin{bmatrix} 1 & 0 \\ 0 & 1 \end{bmatrix} , \tag{F.20}$$

which—in accordance with relation (E.8)—agrees with result (F.1), as required.

Remark F.2. The gradient of a vector field, V, which is a linear map from vectors to vectors, is a second-rank tensor. In Cartesian coordinates, it is

$$\nabla V = \sum_{i=1}^n \sum_{j=1}^n \sum_{k=1}^n \delta^{jk} \frac{\partial V^i}{\partial x^j} \, \hat{e}_i \otimes \hat{e}_k .$$

In curvilinear coordinates, it is

$$\nabla V = \sum_{i=1}^{n}\sum_{j=1}^{n}\sum_{k=1}^{n} g^{jk}\left(\frac{\partial V^i}{\partial \xi^j} + \sum_{\ell=1}^{n}\Gamma^i_{j\ell}V^\ell\right)\hat{e}_i\otimes\hat{e}_k\,, \tag{F.21}$$

where $\hat{e}_i\otimes\hat{e}_k$, with $i,k=1,\ldots,n$, being the corresponding basis.

In contrast to the deformation-gradient tensor, F, discussed in Appendix A.2, ∇V is not a two-point tensor, since both the values of its components and their arguments belong to the same space.

In general, if a tensor is expressed in Cartesian coordinates, there is no need to invoke explicitly covariant derivatives, since the corresponding Christoffel symbols are zero. Also, the covariant differentiation of a scalar field, f, is just a partial differentiation, even if the field is expressed in curvilinear coordinates; in such a case,

$$\nabla f = \sum_{i=1}^{n}\sum_{j=1}^{n} g^{ij}\frac{\partial f}{\partial \xi^j}\,\hat{e}_i\,.$$

In Cartesian coordinates, $g^{ij}=\delta^{ij}$, and we obtain $\nabla f = \sum_{i=1}^{n}\left(\partial f/\partial x^i\right)\hat{e}_i$, as expected.

The necessity for the explicit use of covariant derivatives derivatives can be expressed as follows. ∇f is a first-rank tensor; hence, it transforms as such, due to the chain rule, alone. However, expression (F.21)—without the Christoffel symbol—is not a second-rank tensor; the partial derivatives, by themselves, do not transform as components of a tensor. Each of the three expressions of this remark represents a tensor, and transforms accordingly. In expression (F.21), the covariant derivative needs to be explicitly invoked. In the other two expressions, it does not.

Thus, the Christoffel symbol is introduced to allow the differentiation of tensor components in curvilinear coordinates to be such that the result of differentiation is also a tensor. As a consequence, the derivative is to measure changes within a tensor field, regardless of coordinates in which this field is expressed. If a given tensor is of the first rank or the coordinates are Cartesian, there is no need to invoke the covariant derivative, explicitly.

The covariant derivatives satisfy the Leibniz integral rule, discussed in Remark C.1. However, they do not commute, which means that they do not satisfy the equality of mixed partial derivatives.

Let us return to, and examine, expression (F.20). The vector field in Figure F.1 has a magnitude that is linearly proportional to the distance from the origin, and no component orthogonal to the radius; $\nabla_r r = 1$

due to the linear increase of magnitude with the distance from the origin; $\partial V^\theta / \partial\theta = 0$ due to the rotational symmetry of the coordinate system, which is also the symmetry of the vector field. However, $\nabla_\theta V^\theta \neq 0$, since the orientation of the vector field does vary; this field does not consist of parallel arrows. According to the covariant derivative, the description of the field shown in Figure F.1 depends on two coordinates, as expected.

The Christoffel symbol, the metric tensor and the covariant derivative are closely related to each other. Notably, the symbol itself can be derived as the condition for the covariant derivative of the metric to be zero, which means that the changes of values of the metric tensor due to the curvilinearity of coordinates have no effect on the covariant derivative, which is the purpose of that covariance.

Example F.3. Let us use expression (F.14) to obtain the divergence of the vector field in Figure F.1. In $\mathbb{R}^2$, the general form of the divergence is

$$\nabla \cdot V = \sum_{i=1}^{2} \nabla_i V^i ,$$

which, using expression (F.14), we write explicitly as

$$\nabla_1 V^1 = \frac{\partial V^1}{\partial \xi^1} + \Gamma^1_{11} V^1 + \Gamma^1_{12} V^2$$

and

$$\nabla_2 V^2 = \frac{\partial V^2}{\partial \xi^2} + \Gamma^2_{21} V^1 + \Gamma^2_{22} V^2 .$$

Since—in Cartesian Coordinates—the Christoffel symbols are zero, we have

$$\nabla \cdot V = \frac{\partial V^1}{\partial x^1} + \frac{\partial V^2}{\partial x^2} = \frac{\partial x^1}{\partial x^1} + \frac{\partial x^2}{\partial x^2} = 1 + 1 = 2 ,$$

where $x^1 \equiv x$ and $x^2 \equiv y$, in the notation of Example F.1. In polar coordinates, following expressions (F.7), (F.9), (F.11), (F.12), we obtain

$$\begin{aligned}\nabla \cdot V &= \nabla_1 V^1 + \nabla_2 V^2 = \frac{\partial r}{\partial r} + \Gamma^r_{rr}\, r + \Gamma^r_{r\theta}\, 0 + \frac{\partial 0}{\partial \theta} + \Gamma^\theta_{\theta r}\, r + \Gamma^\theta_{\theta\theta}\, 0 \\ &= 1 + 0\, r + 0 + 0 + \frac{1}{r} r + 0 = 2 .\end{aligned}$$

The two results agree, as required.

Example F.4. Let us derive expression (F.6) by requiring the covariant derivative of a metric to be zero,

$$\nabla_i g_{k\ell} = \frac{\partial g_{k\ell}}{\partial \xi^i} - \sum_{j=1}^{n} \Gamma^j_{ik} g_{j\ell} - \sum_{j=1}^{n} \Gamma^j_{i\ell} g_{kj} = 0 , \qquad i, k, \ell = 1, \ldots, n , \quad \text{(F.22)}$$

where n is the dimension of the space. In expression (F.22), in contrast to expressions (F.15)–(F.18), the Christoffel symbols are subtracted, not added. This is a consequence of the covariant transformation properties, discussed in Appendix E.2, of the components of tensor $g_{k\ell}$, in contrast to components of vector V, in expressions (F.15)–(F.18), which are contravariant, as discussed in Appendix E.1. The general form of expression (F.22) is expression (F.29), below.

To solve equation (F.22) for the Christoffel symbol, we permute its indices to write

$$\nabla_k g_{\ell i} = \frac{\partial g_{\ell i}}{\partial \xi^k} - \sum_{j=1}^{n} \Gamma^j_{k\ell} g_{ji} - \sum_{j=1}^{n} \Gamma^j_{ki} g_{\ell j} = 0\,, \qquad i,k,\ell = 1,\ldots,n\,,$$

and

$$\nabla_\ell g_{ik} = \frac{\partial g_{ik}}{\partial \xi^\ell} - \sum_{j=1}^{n} \Gamma^j_{\ell i} g_{jk} - \sum_{j=1}^{n} \Gamma^j_{\ell k} g_{ij} = 0\,, \qquad i,k,\ell = 1,\ldots,n\,.$$

Subtracting the two latter expressions from the first one and using the symmetry of the lower indices of $\Gamma^j_{k\ell}$, we obtain

$$\frac{\partial g_{k\ell}}{\partial \xi^i} - \frac{\partial g_{\ell i}}{\partial \xi^k} - \frac{\partial g_{ik}}{\partial \xi^\ell} + 2\sum_{j=1}^{n} \Gamma^j_{k\ell} g_{ji} = 0\,, \qquad i,k,\ell = 1,\ldots,n\,,$$

which, to solve for $\Gamma^j_{k\ell}$, we rewrite as

$$\sum_{j=1}^{n} \Gamma^j_{k\ell} g_{ji} = \frac{1}{2}\left(\frac{\partial g_{\ell i}}{\partial \xi^k} + \frac{\partial g_{ik}}{\partial \xi^\ell} - \frac{\partial g_{k\ell}}{\partial \xi^i}\right), \qquad i,k,\ell = 1,\ldots,n\,. \tag{F.23}$$

Multiplying both sides by g^{mi}, which is the inverse of the metric tensor, we write

$$\sum_{i=1}^{n}\sum_{j=1}^{n} \Gamma^j_{k\ell} g_{ji} g^{mi} = \frac{1}{2}\sum_{i=1}^{n} g^{mi}\left(\frac{\partial g_{\ell i}}{\partial \xi^k} + \frac{\partial g_{ik}}{\partial \xi^\ell} - \frac{\partial g_{k\ell}}{\partial \xi^i}\right), \quad k,\ell,m = 1,\ldots,n\,,$$

which results in

$$\sum_{j=1}^{n} \Gamma^j_{k\ell} \delta^m_j = \frac{1}{2}\sum_{i=1}^{n} g^{mi}\left(\frac{\partial g_{\ell i}}{\partial \xi^k} + \frac{\partial g_{ik}}{\partial \xi^\ell} - \frac{\partial g_{k\ell}}{\partial \xi^i}\right), \qquad k,\ell,m = 1,\ldots,n\,.$$

Using the property of the Kronecker delta, we obtain

$$\Gamma^m_{k\ell} = \frac{1}{2}\sum_{i=1}^{n} g^{mi}\left(\frac{\partial g_{\ell i}}{\partial \xi^k} + \frac{\partial g_{ik}}{\partial \xi^\ell} - \frac{\partial g_{k\ell}}{\partial \xi^i}\right), \qquad k,\ell,m = 1,\ldots,n\,,$$

which is expression (F.6), as expected.

Remark F.3. The Christoffel symbol used in this appendix is of the second kind. The Christoffel symbol of the first kind, which is used to study geometric properties of a metric, is

$$\Gamma_{ijk} = \sum_{\ell=1}^{n} g_{\ell k}\Gamma_{ij}^{\ell}\,, \qquad i,j,k = 1,\ldots,n\,, \tag{F.24}$$

which is also expression (F.23). To obtain this explicit expression from definition (F.24), we invoke expression (F.6), and write the right-hand side of the definition as

$$\frac{1}{2}\sum_{\ell=1}^{n}\sum_{m=1}^{n} g_{\ell k}g^{\ell m}\left(\frac{\partial g_{mi}}{\partial \xi^{j}} + \frac{\partial g_{mj}}{\partial \xi^{i}} - \frac{\partial g_{ij}}{\partial \xi^{m}}\right), \qquad i,j,k = 1,\ldots,n\,.$$

Using the reciprocity of the metric tensor and its inverse, we write

$$\frac{1}{2}\sum_{m=1}^{n} \delta_{k}^{m}\left(\frac{\partial g_{mi}}{\partial \xi^{j}} + \frac{\partial g_{mj}}{\partial \xi^{i}} - \frac{\partial g_{ij}}{\partial \xi^{m}}\right), \qquad i,j,k = 1,\ldots,n\,.$$

Using the property of the Kronecker delta, we write

$$\Gamma_{ijk} = \frac{1}{2}\left(\frac{\partial g_{ik}}{\partial \xi^{j}} + \frac{\partial g_{jk}}{\partial \xi^{i}} - \frac{\partial g_{ij}}{\partial \xi^{k}}\right), \qquad i,j,k = 1,\ldots,n\,;$$

it also follows that $\Gamma_{ijk} = \Gamma_{jik}$.

Example F.5. Let us consider infinitesimal strains in curvilinear coordinates in $\mathbb{R}^3$. To do so, we generalize expression (A.76), which represents quantities expressed in Cartesian coordinates, and write it as[5]

$$\varepsilon_{j}^{i} = \frac{1}{2}\left(\frac{\partial u^{i}}{\partial x^{j}} + \frac{\partial u^{j}}{\partial x^{i}}\right), \qquad i,j = 1,2,3\,,$$

where ε is a second-rank tensor of the $(1,1)$ type (Carroll, 2004, Expression 3.5), discussed in Appendix E.3. Herein, for consistency with definition (A.76), we use u to denote the displacement vector, which is tantamount to vector V, used in this appendix.

In terms of covariant derivatives,

$$\varepsilon_{j}^{i} = \tfrac{1}{2}\left(\nabla_{j}u^{i} + \nabla_{i}u^{j}\right), \qquad i,j = 1,2,3\,.$$

Remark F.4. In a similar manner, we can express the dilatation, stated in expression (5.3), on page 183, as

$$\varphi = \sum_{i=1}^{3}\nabla_{i}u^{i}\,,$$

which is the divergence expressed in terms of covariant derivatives.

[5] *see also*: Slawinski (2015, Expression 1.4.6)

In view of discussions in Appendix E.1, we infer that u, which is a displacement vector, is a contravariant entity. Hence, in accordance with expression (F.26), below, we write explicitly,

$$\varepsilon_1^1 = \tfrac{1}{2}\left(\nabla_1 u^1 + \nabla_1 u^1\right) = \frac{1}{2}\left(\frac{\partial u^1}{\partial \xi^1} + \sum_{k=1}^{3}\Gamma_{1k}^1 u^k + \frac{\partial u^1}{\partial \xi^1} + \sum_{k=1}^{3}\Gamma_{1k}^1 u^k\right)$$
$$= \frac{\partial u^1}{\partial \xi^1} + \sum_{k=1}^{3}\Gamma_{1k}^1 u^k .$$

Following the same pattern,

$$\varepsilon_2^1 = \varepsilon_1^2 = \frac{1}{2}\left(\frac{\partial u^1}{\partial \xi^2} + \sum_{k=1}^{3}\Gamma_{2k}^1 u^k + \frac{\partial u^2}{\partial \xi^1} + \sum_{k=1}^{3}\Gamma_{1k}^2 u^k\right),$$

$$\varepsilon_3^1 = \varepsilon_1^3 = \frac{1}{2}\left(\frac{\partial u^1}{\partial \xi^3} + \sum_{k=1}^{3}\Gamma_{3k}^1 u^k + \frac{\partial u^3}{\partial \xi^1} + \sum_{k=1}^{3}\Gamma_{1k}^3 u^k\right),$$

$$\varepsilon_2^2 = \frac{\partial u^2}{\partial \xi^2} + \sum_{k=1}^{3}\Gamma_{2k}^2 u^k ,$$

$$\varepsilon_3^2 = \varepsilon_2^3 = \frac{1}{2}\left(\frac{\partial u^2}{\partial \xi^3} + \sum_{k=1}^{3}\Gamma_{3k}^2 u^k + \frac{\partial u^3}{\partial \xi^2} + \sum_{k=1}^{3}\Gamma_{2k}^3 u^k\right),$$

and

$$\varepsilon_3^3 = \frac{\partial u^3}{\partial \xi^3} + \sum_{k=1}^{3}\Gamma_{3k}^3 u^k ,$$

where ξ^1, ξ^2, ξ^3 are the curvilinear coordinates.

Remark F.5. If, subsequently, one is to apply the covariant derivative to ε_j^i, where $i, j = 1, 2, 3$—as opposed to expressing ε_j^i, itself, in terms of such derivatives—one needs to use formula (F.30), below.

Let us continue Example F.5, by considering the strain tensor in $\mathbb{R}^2$. We write explicitly

$$\varepsilon_1^1 = \frac{\partial u^1}{\partial \xi^1} + \Gamma_{11}^1 u^1 + \Gamma_{12}^1 u^2 ,$$

$$\varepsilon_2^1 = \varepsilon_1^2 = \frac{1}{2}\left(\frac{\partial u^1}{\partial \xi^2} + \Gamma_{21}^1 u^1 + \Gamma_{22}^1 u^2 + \frac{\partial u^2}{\partial \xi^1} + \Gamma_{11}^2 u^1 + \Gamma_{12}^2 u^2\right),$$

$$\varepsilon_2^2 = \frac{\partial u^2}{\partial \xi^2} + \Gamma_{21}^2 u^1 + \Gamma_{22}^2 u^2 .$$

Let us express this result in polar coordinates, where indices 1 and 2 correspond to r and θ, respectively,

$$\varepsilon_r^r = \frac{\partial u^r}{\partial \xi^r} + \Gamma_{rr}^r u^r + \Gamma_{r\theta}^r u^\theta = \frac{\partial u^r}{\partial \xi^r} ,$$

$$\begin{aligned}\varepsilon_\theta^r = \varepsilon_r^\theta &= \frac{1}{2}\left(\frac{\partial u^r}{\partial \xi^\theta} + \Gamma_{\theta r}^r u^r + \Gamma_{\theta\theta}^r u^\theta + \frac{\partial u^\theta}{\partial \xi^r} + \Gamma_{rr}^\theta u^r + \Gamma_{r\theta}^\theta u^\theta\right)\\ &= \frac{1}{2}\left(\frac{\partial u^r}{\partial \xi^\theta} - r u^\theta + \frac{\partial u^\theta}{\partial \xi^r} + \frac{1}{r} u^\theta\right) ,\end{aligned}$$

$$\varepsilon_\theta^\theta = \frac{\partial u^\theta}{\partial \xi^\theta} + \Gamma_{\theta r}^\theta u^r + \Gamma_{\theta\theta}^\theta u^\theta = \frac{\partial u^\theta}{\partial \xi^\theta} + \frac{1}{r} u^r ;$$

the final form of each component is obtained using expressions (F.7)–(F.12).

Closing remarks

Using the covariant derivative, we examine changes in properties intrinsic to an object described by a function; the value of a covariant derivative of that function is not affected by properties of the coordinates. Covariant derivatives—with respect to curvilinear coordinates, $\xi^1, \ldots, \xi^n$—can be applied to a tensor of any rank. For a scalar field, they are reduced to

$$\nabla_i \varphi := \frac{\partial \varphi}{\partial \xi^i} \qquad i = 1, \ldots, n , \tag{F.25}$$

which is the standard gradient operation, and the origin of symbol ∇ to denote covariant derivatives. For a contravariant first-rank-tensor field, illustrated in Example F.1,

$$\nabla_i V^j := \frac{\partial V^j}{\partial \xi^i} + \sum_{k=1}^{n} \Gamma_{ik}^j V^k , \qquad i, j = 1, \ldots, n . \tag{F.26}$$

For a covariant first-rank-tensor field,

$$\nabla_i V_j := \frac{\partial V_j}{\partial \xi^i} - \sum_{k=1}^{n} \Gamma_{ij}^k V_k , \qquad i, j = 1, \ldots, n . \tag{F.27}$$

For a contravariant second-rank-tensor tensor field,

$$\nabla_i T^{jk} := \frac{\partial T^{jk}}{\partial \xi^i} + \sum_{\ell=1}^{n} \Gamma_{i\ell}^j T^{\ell k} + \sum_{\ell=1}^{n} \Gamma_{i\ell}^k T^{j\ell} , \quad i, j, k = 1, \ldots, n . \tag{F.28}$$

For a covariant second-rank-tensor field, illustrated in Example F.4,

$$\nabla_i T_{jk} := \frac{\partial T_{jk}}{\partial \xi^i} - \sum_{\ell=1}^{n} \Gamma_{ij}^{\ell} T_{\ell k} - \sum_{\ell=1}^{n} \Gamma_{ik}^{\ell} T_{j\ell}\,, \quad i,j,k = 1,\ldots,n\,. \tag{F.29}$$

The number of sums is the rank of a tensor; it is zero in expression (F.25), one in expressions (F.26) and (F.27), two in expressions (F.28) and (F.29). The number of terms, n, is the dimension of the space; it is two in Example F.1, n in Example F.4, and three and two in Example F.5. Also, contravariant entities result in the operation of addition of the corresponding sum and the covariant entities result in its subtraction. For instance—for a second-rank tensor of the $(1,1)$ type, which consists of both contravariant and covariant entities—we have

$$\nabla_i T^j_k := \frac{\partial T^j_k}{\partial \xi^i} + \sum_{\ell=1}^{n} \Gamma_{i\ell}^{j} T^{\ell}_k - \sum_{\ell=1}^{n} \Gamma_{ik}^{\ell} T^j_{\ell}\,, \quad i,j,k = 1,\ldots,n\,. \tag{F.30}$$

In each case, the Christoffel symbol ensures that the covariant derivative provides intrinsic information about a given field, regardless of the adopted coordinate system.

The concept of a covariant derivative agrees with the *covariance principle* in physics. As stated by Schutz (2003), in the context of general relativity, but applicable to continuum mechanics,

> the principle of covariance says that the field equations must be able to be used in any *coordinate system*, no matter how peculiar.

Since coordinates do not exist in nature, they cannot be intrinsic to physical laws; Mount Everest exists regardless of its latitude and longitude, which are useful, convenient and consistent—but arbitrarily selected—coordinates. For quantitative studies, mathematical entities representing physical properties and processes must transform unambiguously among coordinate systems.

Notably, *covariant*, in the context of the derivative, refers to the covariance principle in physics; it implies the agreement in expressing properties regardless of the coordinate system. It is not related explicitly to the distinction between the *covariant* and *contravariant* transformations, discussed in Appendix E.

Mathematical physics invokes many generalizations of the derivative concept besides the covariant derivative, among them the *Lie derivative* and *exterior derivative*.[6] The latter is named in honour of a Norwegian mathematician Sophus Lie.[7]

In a conventional setting, the covariant derivative—as is the case with many generalizations—is reduced to the standard partial derivative. Herein, the constancy of the metric tensor for Cartesian coordinates results in their Christoffel symbol being zero.

[6]Readers interested in the Lie derivative might refer to Epstein (2010, Section 5.6).
[7]Readers interested in a biography of Sophus Lie might refer to Stubhaug (2002).

Appendix G

List of symbols

Il y a plusieurs points de vue sur l'axiomatisation d'une théorie physique. On peut les schématiser en parlant du point de vue du logicien, du point de vue du mathématicien et du point de vue du physicien. [...] L'accent du logicien est concentré sur les méta-théoremes non sur les théoremes de la théorie [...] L'accent du physicien est concentré sur le contenu physique d'une théorie et non sur sa structure formelle. Le mathématicien a une position intermédiaire. [...] C'est la théorie mathématique qui fournit la précision définitive des concepts de la physique.[1]

Walter Noll (1974c)

G.1 Mathematical relations and operations

Distinctions between different meanings of the same symbol are clear from subtle notation differences and the context.

ι : $\sqrt{-1}$

$=$: equality

$\equiv$: equivalence

$:=$: definition

$\propto$: proportionality

[1] *There are several points of view on axiomatization of a physical theory. One can structure it speaking from the point of view of a logician, the point of view of a mathematician and the point of view of a physicist. [...] The emphasis of a logician is focussed on metatheorems, not on the theorems of the theory. [...] The emphasis of a physicist is focussed on the physical content of a theory and not on its formal structure. A mathematician has an intermediate position. [...] It is the mathematical theory that provides the definite precision of physical concepts.*

$\sim$: similarity—see page 312

$\approx$: approximation—see page 161

$\cdot$: scalar product—see page 40

$:$: double-dot product—see page 447

$\times$: vector product—see page 386

$\otimes$: tensor product—see page 365

δ_{ij} : Kronecker delta—see page 65

ϵ_{ijk} : Levi-Civita symbol—see page 387

∇ : gradient—see page 313

$\nabla\cdot$: divergence—see page 30

$\nabla\times$: curl—see page 187

$\nabla^2 := \nabla\cdot\nabla$: Laplacian—see page 225

$\nabla_{\mathbf{X}}$: deformation-gradient operator—see page 364

$\nabla_{\mathbf{x}}$: inverse deformation-gradient operator—see page 364

∇_i : covariant-derivative operator—see page 482

$\Box_P$: P-wave d'Alembertian—see page 187

$\Box_S$: S-wave d'Alembertian—see page 188

$\phi_*[\,\cdot\,]$: pushforward—see page 385

$\phi^*[\,\cdot\,]$: pullback—see page 385

J : Jacobian—see page 387

tr : trace—see page 382

det : determinant—see page 382

^{T} : transpose—see page 382

G.2 Physical quantities

G.2.1 *Greek letters*

Distinctions between different meanings of the same symbol are clear from the context.

α : P-wave speed—see page 226

β : S-wave speed—see page 226

$\varepsilon_{k\ell}$: infinitesimal-strain tensor—see page 43

η : entropy—see page 431

$\eta_{\alpha\beta}$: Minkowski metric—see page 319

θ : ray angle—see page 166

ϑ : wavefront angle—see page 16

κ : wave number—see page 239

κ : thermal diffusivity—see page 325

κ : curvature—see page 276

λ : stretch ratio—see page 389

λ : Lamé parameter—see page 183

μ : Lamé parameter—see page 183

ρ : electric-charge density—see page 31

ρ : mass density—see page 30

σ_{ij} : Cauchy-stress-tensor components—see page 43

Σ_{ij} : total-stress-tensor components—see page 310

φ : dilatation—see page 183

$\mathbf{\Psi}$: rotation vector—see page 183

ψ : Helmholtz free energy—see page 434

ψ : eikonal function—see page 470

ω : angular frequency—see page 195

Ω : Earth's angular frequency—see page 320

G.2.2 *Roman letters*

Distinctions between different meanings of the same symbol are clear from the context.

b : left Cauchy-Green strain tensor—see page 380

c : speed of light—see page 322

c : specific heat—see page 326

c_0 : specific heat at zero stress—see page 327

$\mathbf{c}$: elasticity tensor in $\mathbb{R}^3$—see page 63

$c_{ijk\ell}$: elasticity-tensor components in $\mathbb{R}^3$—see page 43

C : elasticity tensor in $\mathbb{R}^6$—see page 65

C_{mn} : elasticity-tensor components in $\mathbb{R}^6$—see page 229

C : right Cauchy-Green strain tensor—see page 377

C_{IJ} : right-Cauchy-Green-strain-tensor components—see page 377

e_0 : internal energy—see page 440

e : Almansi strain tensor—see page 383

E : Green-Lagrange strain tensor—see page 383

F : deformation-gradient tensor—see page 365

g : acceleration due to gravity—see page 303

g : symmetry-group element—see page 63

g : Gibbs free energy—see page 441

g_{ij} : metric-tensor components—see page 479

$\mathbf{g}$: gravitational force—see page 311

G : symmetry group—see page 64

$G = 6.674 \times 10^{-11}\,\mathrm{Nm}^2/\mathrm{kg}^2$: gravitational constant—see page 312

h_0 : enthalpy—see page 441

h_{ij} : wave-metric-tensor components—see page 319

H : Hamiltonian (ray theory)—see page 279

H : Hamiltonian (mechanics)—see page 299

J : electric-current density—see page 31

k : spring constant—see page 292

$\mathbf{k}$: wave vector—see page 240

k : thermal conductivity—see page 437

K : bulk modulus—see page 82

K : thermal-conductivity tensor—see page 437

L : Lagrangian (ray theory)—see page 279

L : Lagrangian (mechanics)—see page 290

$\mathcal{L}$: Lagrangian density—see page 291

P : First Piola-Kirchhoff stress tensor—see page 420

$\mathcal{P}$: scalar potential—see page 225

$\mathcal{M}$: material point—see page 25

$s_{ijk\ell}$: compliance-tensor components in $\mathbb{R}^3$—see page 44

S : Second Piola-Kirchhoff stress tensor—see page 422

$\mathcal{S}$: spatial location—see page 25

$\mathcal{S}$: surface—see page 51

$\boldsymbol{\mathcal{S}}$: vector potential—see page 225

T : period—see page 315

T : temperature—see page 325

T : kinetic energy—see page 290

U : potential energy—see page 290

U : right Stretch tensor—see page 391

$\mathbf{u}$: displacement vector—see page 30

v : left Stretch tensor—see page 396

v : wavefront speed—see pages 229 and 314

v_P : P-wave speed—see page 15

v_S : S-wave speed—see page 15

V : ray velocity—see page 15

W : strain energy—see page 454

Bibliography

Abbott, B. P., Abbott, R., Abbott, T. D. and LIGO Scientific Collaboration (2016). Observation of gravitational waves from a binary black hole merger, *Phys. Rev. Lett.* **116**, 6, doi:10.1103/PhysRevLett.116.061102.

Achenbach, J. D. (1973). *Wave propagation in elastic solids* (North-Holland).

Adamus, F. P., Slawinski, M. A. and Stanoev, T. (2018). On effects of inhomogeneity on anisotropy in Backus average, *arXiv* **[physics.geo-ph]**, 1802.04075.

Aki, K. and Richards, P. G. (2002). *Quantitative seismology*, 2nd edn. (University Science Books).

Anderson, D. L. (1961). Elastic wave propagation in layered anisotropic media, *J. Geophys. Res.* **66**, pp. 2953–2964.

Arfken, G., Weber, H. and Harris, F. (2013). *Mathematical methods for physicists: A comprehensive guide*, 7th edn. (Elsevier/Academic Press).

Arnold, V. I. (1989). *Mathematical methods of classical mechanics*, 2nd edn. (Springer-Verlag).

Arnold, V. I. (2004). *Lectures on partial differential equations* (Springer).

Auld, B. A. (1990). *Acoustic fields and waves in solids*, 2nd edn. (Krieger Publishing Company).

Ayer, A. (1972). *Probability and evidence* (Columbia University Press).

Babich, V. M. (1994). Ray method of calculating the intensity of wavefronts in the case of a heterogeneous, anisotropic, elastic medium (translation from 1961), *Geophysical Journal International* **118**, pp. 379–383.

Backus, G. E. (1962). Long-wave elastic anisotropy produced by horizontal layering, *J. Geophys. Res.* **67**, 11, pp. 4427–4440.

Barton, G. (1989). *Elements of Green's functions and propagation: Potentials, diffusion, and waves* (Oxford University Press).

Basdevant, J.-L. (2005). *Principes variationnels et dynamique* (Vuibert).

Berdichevsky, V. L. (2009). *Variational principles of continuum mechanics: I. Fundamentals*, Vol. 1 (Springer-Verlag).

Berryman, J. G. (1997). Range of the P-wave anisotropy parameter for finely layered VTI media, *Stanford Exploration Project* **93**, pp. 179–192.

Bóna, A., Bucataru, I. and Slawinski, M. A. (2004a). Characterization of elasticity-tensor symmetries using $SU(2)$, *Journal of Elasticity* **75**, 3, pp. 267–289.

Bóna, A., Bucataru, I. and Slawinski, M. A. (2004b). Material symmetries of elasticity tensors, *The Quarterly Journal of Mechanics and Applied Mathematics* **57**, 4, pp. 583–598.

Bóna, A., Bucataru, I. and Slawinski, M. A. (2007a). Coordinate-free characterization of the symmetry classes of elasticity tensors, *Journal of Elasticity* **87**, pp. 109–132.

Bóna, A., Bucataru, I. and Slawinski, M. A. (2007b). Material symmetries versus wavefront symmetries, *The Quarterly Journal of Mechanics and Applied Mathematics* **60**, 2, pp. 73–84.

Bóna, A., Bucataru, I. and Slawinski, M. A. (2008). Space of SO(3)-orbits of elasticity tensors, *Archives of Mechanics* **60**, 2, pp. 121–136.

Bóna, A., Diner, Ç., Kochetov, M. and Slawinski, M. A. (2010). On symmetries of elasticity tensors and Christoffel matrices, *arXiv* **[physics.geo-ph]**, 1011.4975v2.

Bóna, A. and Slawinski, M. A. (2003). Fermat's principle for seismic rays in elastic media, *Journal of Applied Geophysics* **54**, pp. 445–451.

Bóna, A. and Slawinski, M. A. (2015). *Wavefronts and rays as characteristics and asymptotics*, 2nd edn. (World Scientific).

Bonet, J. and Wood, R. D. (2008). *Nonlinear continuum mechanics for finite element analysis*, 2nd edn. (Cambridge University Press).

Borcherdt, R. D. (2009). *Viscoelastic waves in layered media* (Cambridge University Press).

Born, M. and Wolf, E. (1999). *Principles of optics*, 7th edn. (Cambridge University Press).

Bos, L. (2003). *Introductory remarks on PDEs*, Lecture notes.

Bos, L., Dalton, D. R. and Slawinski, M. A. (2017a). On commutativity and near commutativity of translational and rotational averages: Analytical proofs and numerical examinations, *arXiv* **[physics.geo-ph]**, 1704.05541.

Bos, L., Dalton, D. R., Slawinski, M. A. and Stanoev, T. (2017b). On Backus average for generally anisotropic layers, *Journal of Elasticity* **127**, 2, pp. 179–196.

Bos, L., Danek, T., Slawinski, M. A. and Stanoev, T. (2017c). Statistical and numerical considerations of Backus-average product approximation, *Journal of Elasticity* , DOI: 10.1007/s10659-017-9659-9.

Bos, L. and Slawinski, M. A. (2010). Elastodynamic equations: Characteristics, wavefronts and rays, *The Quarterly Journal of Mechanics and Applied Mathematics* **63**, 1, pp. 23–37.

Bos, L. and Slawinski, M. A. (2015). 2-norm effective isotropic Hookean solids, *Journal of Elasticity* **120**, 1, pp. 1–22.

Brekhovskikh, L. M. and Goncharov, V. (1994). *Mechanics of continua and wave dynamics*, 2nd edn. (Springer-Verlag).

Brisco, C. (2014). *Anisotropy vs. inhomogeneity: Algorithm formulation, coding and modelling*, Honours thesis, Memorial University.

Brown, J. R. (2001). *Who rules in science: An opinionated guide to the wars* (Harvard University Press).

Brown, J. R. and Slawinski, M. A. (2017). *On foundations of seismology: Bringing idealizations down to Earth* (World Scientific).

Brush, S. G. (1980). Discovery of the Earth's core, *Am. J. Phys.* **48**, 9, pp. 705–724.

Brush, S. G. (1996). *Nebulous Earth: The origin of the Solar System and the core of the Earth from Laplace to Jeffreys* (Cambridge University Press).

Bucataru, I. and Slawinski, M. A. (2009a). Invariant properties for finding distance in space of elasticity tensors, *Journal of Elasticity* **94**, pp. 97–114.

Bucataru, I. and Slawinski, M. A. (2009b). On convexity and detachment of innermost wavefront-slowness sheet, *Geophysics* **74**, 5, pp. 63–66.

Bullen, K. and Bolt, B. (1987). *An introduction to the theory of seismology*, 4th edn. (Cambridge University Press).

Bunge, M. (1967). *Foundations of physics* (Springer-Verlag).

Bunge, M. (ed.) (1999). *Critical approaches to science and philosophy* (Transaction Publishers).

Bunge, M. (2014). *Emergence and convergence: Qualitative novelty and the unity of knowledge* (University of Toronto Press).

Cannell, D. M. (2001). *George Green: mathematician and physicist 1793-1841: The background to his life and work* (SIAM).

Carroll, S. (2004). *Spacetime and geometry: An introduction to General Relativity* (Addison-Wesley).

Chadwick, P. (1976). *Continuum mechanics: Concise theory and problems* (Dover).

Chadwick, P. and Jarvis, D. A. (1979). Surface waves in a pre-stressed elastic body, *Proceedings of the Royal Society of London. Series A, Mathematical, Physical and Engineering Sciences* **366**, 1727, pp. 517–536.

Chadwick, P. and Smith, G. D. (1977). Foundations of the theory of surface waves in anisotropic elastic materials, *Advances in Applied Mechanics* **17**, pp. 303–376.

Chapman, A. (2005). *England's Leonardo: Robert Hooke and the seventeenth-century scientific revolution* (IOP Publishing Ltd).

Chapman, C. H. (2004). *Fundamentals of seismic wave propagation* (Cambridge University Press).

Chou, P. C. and Pagano, N. J. (1992). *Elasticity: Tensor, dyadic, and engineering approaches* (Dover).

Coleman, B. D. and Noll, W. (1974). The thermodynamics of elastic materials with heat conduction and viscosity, in W. Noll (ed.), *The Foundations of Mechanics and Thermodynamics: Selected papers by W. Noll* (Springer).

Cosserat, E. and Cosserat, F. (1909). Théorie des corps déformables, Hermann et Fils.

Curie, P. (1894). Sur la symétrie dans les phénomènes physiques, symétrie d'un champ électrique et d'un champ magnétique, *Journal de Physique Théorie Appliquée* **3**, 1, pp. 393–415.

Červený, V. (2001). *Seismic ray theory* (Cambridge University Press).

Červený, V. (2002). Fermat's variational principle for anisotropic inhomogeneous media, *Studia geophysica and geodetica* **46**, pp. 567–588.

Červený, V. and Ravindra, R. (1971). *Theory of seismic head waves* (University of Toronto Press).

Dahlen, F. A. and Tromp, J. (1998). *Theoretical global seismology* (Princeton University Press).

Dalton, D. R. and Slawinski, M. A. (2016). On Backus average for oblique incidence, *arXiv* [**physics.geo-ph**], 1601.02966.

Dalton, D. R., Slawinski, M. A., Stachura, P. and Stanoev, T. (2017). Sensitivity of Love and quasi-Rayleigh waves to model parameters, *Quarterly Journal of Mechanics and Applied Mathematics* .

Danek, T., Kochetov, M. and Slawinski, M. A. (2013). Uncertainty analysis of effective elasticity tensors using quaternion-based global optimization and Monte-Carlo method, *The Quarterly Journal of Mechanics and Applied Mathematics* **66**, 2, pp. 253–272.

Danek, T., Kochetov, M. and Slawinski, M. A. (2015). Effective elasticity tensors in the context of random errors, *Journal of Elasticity* **121**, 1, pp. 55–67.

Danek, T., Noseworthy, A. and Slawinski, M. A. (2018). Effects of norms on general Hookean solids for their isotropic counterparts, *Dolomites Research Notes on Approximation* **11**, pp. 15–28.

Danek, T. and Slawinski, M. A. (2012). Bayesian inversion of VSP traveltimes for linear inhomogeneity and elliptical anisotropy, *Geophysics* **77**, 6, pp. 239–243.

Danek, T. and Slawinski, M. A. (2014). On effective transversely isotropic elasticity tensors of Frobenius and L_2 operator norms, *Dolomites Research Notes on Approximation* **7**.

Descartes, R. (1992). *Discours de la méthode* (Flammarion).

Destrade, M. and Saccomandi, G. (eds.) (2007). *Waves in nonlinear pre-stressed materials* (Springer).

Dewangan, P. and Grechka, V. (2003). Inversion of multicomponent, multiazimuth, walkaway VSP data for the stiffness tensor, *Geophysics* **68**, 3, pp. 1022–1031.

Dey, S. and Mukherjee, S. P. (1984). Rayleigh waves in an initially stressed layer over a half-space under gravity, *Acta Geophysica Polonica* **32**, 1, pp. 81–90.

Diner, Ç., Kochetov, M. and Slawinski, M. A. (2011a). Identifying symmetry classes of elasticity tensors using monoclinic distance function, *Journal of Elasticity* **102**, 2, pp. 175–190.

Diner, Ç., Kochetov, M. and Slawinski, M. A. (2011b). On choosing effective symmetry classes for elasticity tensors, *The Quarterly Journal of Mechanics and Applied Mathematics* **64**, 1, pp. 57–74.

Dong, H. and Hovem, J. M. (2011). Interface waves, waves in fluids and solids, URL `http://www.intechopen.com/books/waves-in-fluidsand-solids/interface-waves`.

Dorfmann, L. and Ogden, R. W. (2014). *Nonlinear theory of electroelastic and magnetoelastic interactions* (Springer).

Dyson, F. J. (1964). Mathematics in the physical sciences, *Scientific American* **211**, 3, pp. 129–146.

Dyson, F. J. (1969). Seismic response of the Earth to a gravitational wave in the 1-Hz band, *The Astrophysical Journal* **156**, pp. 529–540.

Dyson, F. J. (1996a). Missed opportunities, in *Selected papers of Freeman Dyson with commentary* (American Mathematical Society), pp. 169–186.

Dyson, F. J. (1996b). Seismic response of the Earth to a gravitational wave in the 1-Hz band, in *Selected papers of Freeman Dyson with commentary* (American Mathematical Society), pp. 485–496.

Dyson, F. J. (1996c). *Selected papers of Freeman Dyson with commentary* (American Mathematical Society).

Dyson, F. J. (2001). Homage to George Green: How physics looked in the nineteen-forties, in *George Green, mathematician and physicist: The background to his life and work*, 2nd edn. (SIAM), pp. 232–247.

Dyson, F. J. (2007). *Advanced quantum mechanics* (World Scientific).

Dyson, F. J. (2011). Birds and frogs, in M. Pitici (ed.), *The best writings on mathematics 2010* (Princeton University Press), pp. 57–78.

Dyson, F. J. (2012). Science on the rampage, *New York Review of Books* **59**, 6, pp. 38–39.

Dziewoński, A. M. and Anderson, D. L. (1981). Preliminary reference Earth model, *Physics of the Earth and Planetary Interiors* **25**, 4, pp. 297–356.

Einstein, A. (1916). Näherungsweise Integration der Feldgleichungen der Gravitation, *Sitzungsberichte der Preussischen Akademie der Wissenschaften zu Berlin*

Ekeland, I. (2000). *Le meilleur des mondes possibles: Mathématiques et destinée* (Editions du Seuil).

Epstein, M. (2010). *The geometrical language of continuum mechanics* (Cambridge University Press).

Epstein, M. and Slawinski, M. A. (1998). On some aspects of the continuum-mechanics context, *Revue de l'Institut Français du Pétrole* **53**, 5, pp. 669–677.

Epstein, M. and Slawinski, M. A. (1999). On rays and ray parameters in inhomogeneous-isotropic media, *Canadian Journal of Exploration Geophysics* **35**, 1/2, pp. 7–19.

Epstein, M. and Śniatycki, J. (1992). Fermat's principle in elastodynamics, *Journal of Elasticity* **27**, pp. 45–56.

Ewing, W. M., Jardetzky, W. S. and Press, F. (1957). *Elastic waves in layered media*, International series in the Earth Sciences (McGraw-Hill, Inc.).

Ferry, L. (2013). *Descartes-Spinoza-Leibniz: L'œvre philosophique expliquée* (Frémeaux et Associés).

Feynman, R. P. (2006). *QED: The strange theory of light and matter* (Princeton University Press).

Feynman, R. P., Leighton, R. B. and Sands, M. L. (1964). *The Feynman lectures on physics* (Addison-Wesley).

Forte, S. and Vianello, M. (1996). Symmetry classes for elasticity tensors, *Journal of Elasticity* **43**, 2, pp. 81–108.

Fowler, A. C. (1997). *Mathematical models in the applied sciences* (Cambridge University Press).

Freegarde, T. (2013). *Introduction to the physics of waves* (Cambridge University Press).

Gardner, G., Gardner, L. and Gregory, A. (1974). Formation velocity and density—the diagnostic basics for stratigraphic traps, *Geophysics* **39**, pp. 770–780.

Gazis, D. C., Tadjbakhsh, I. and Toupin, R. A. (1963). The elastic tensor of given symmetry nearest to an anisotropic elastic tensor, *Acta Crystallographica* **16**, 9, pp. 917–922.

Goldstein, H. (1980). *Classical mechanics*, 2nd edn. (Addison-Wesley).

Grant, F. S. and West, G. F. (1965). *Interpretation theory in applied geophysics*, International series in the Earth Sciences (McGraw-Hill, Inc.).

Green, G. (1828). An essay on the application of mathematical analysis to the theories of electricity and magnetism,

Gurtin, M. E. (1981). *An introduction to continuum mechanics* (Academic Press).

Gurtin, M. E., Fried, E. and Anand, L. (2010). *The mechanics and thermodynamics of continua* (Cambridge University Press).

Hadamard, J. (1903). *Leçons sur la propagation des ondes et les équations de l'hydrodynamique* (Broché).

Hadamard, J. (1932). *Le problème de Cauchy et les équations aux dérivées partielles linéaires hyperboliques* (Broché).

Hanyga, A. (ed.) (1984). *Seismic wave propagation in the Earth*, Physics and the evolution of the Earth interior (PWN-Elsevier).

Haskell, N. (1953). The dispersion of surface waves on multilayered media, *Bull. Seism. Soc. Am.* **43**, pp. 17–34.

Helbig, K. (1958). Elastische Wellen in anisotropen Medien, *Getlands Beitr. Geophys.* **67**, 256-288.

Helbig, K. (1994). *Foundations of anisotropy for exploration seismics* (Pergamon Press).

Helbig, K. (1998). Layer-induced anisotropy: Forward relations between between constituent parameters and compound parameters, *Revista Brasileira de Geofísica* **16**, 2–3, pp. 103–114.

Helbig, K. (2000). Inversion of compound parameters to constituent parameters, *Revista Brasileira de Geofísica* **18**, 2, pp. 173–185.

Hempel, C. G. (2000). *Selected philosophical essays* (Cambridge University Press).

Herman, B. (1945). Some theorems of the theory of anisotropic media, *Comptes rendus de l'Académie des Sciences de l'URSS* **48**, 2, pp. 89–92.

Hertel, P. (2012). *Continuum Physics* (Springer).

Hesse, M. B. (2005). *Forces and fields: The concept of action at a distance in the history of physics* (Dover).

Hildebrandt, S. and Tromba, A. (1996). *The parsimonious Universe: Shape and form in the Natural World* (Copernicus).

Hoger, A. (1986). On the determination of residual stress in an elastic body, *Journal of Elasticity* **16**, pp. 303–324.

Holland, J. H. (2014). *Complexity: A very short introduction* (Oxford University Press).

Holzapfel, G. A. (2000). *Nonlinear solid mechanics: A continuum approach for engineering* (John Wiley & Sons).

Hume, D. (2007). *An enquiry concerning human understanding* (Oxford University Press).

Irgens, F. (2008). *Continuum Mechanics* (Springer-Verlag).

James, R. D. (2015). Continuum mechanics, in N. J. Higham (ed.), *The Princeton companion to applied mathematics* (Princeton University Press), pp. 446–458.

Jaynes, E. T. (1980). *Bayesian methods: General background, An introductory tutorial*, St. John's College and Cavendish Laboratory, Cambridge, England.

Jaynes, E. T. (2003). *Probability theory: The logic of science* (Cambridge University Press).

Jeffreys, H. (1924). *The Earth: Its origin, history, and physical constitution*, 1st edn. (Cambridge University Press).

Jeffreys, H. (1925). On certain approximate solutions of linear differential equations of the second order, *Proceedings of the London Mathematical Society* **s2-23**, 1, pp. 428–436.

Jeffreys, H. (1961a). *Cartesian tensors* (Cambridge University Press).

Jeffreys, H. (1961b). *Theory of probability*, 3rd edn., Oxford Classic Texts in the Physical Sciences (Oxford University Press).

Kelly, P. L., Rodney, S. A. and Treu, T. (2015). Multiple images of a highly magnified supernova formed by an early-type cluster galaxy lens, *Science* **347**, 6226, pp. 1123–1126.

Kochetov, M. and Slawinski, M. A. (2009a). Estimating effective elasticity tensors from Christoffel equations, *Geophysics* **74**, 5, pp. WB67–WB73.

Kochetov, M. and Slawinski, M. A. (2009b). On obtaining effective orthotropic elasticity tensors, *The Quarterly Journal of Mechanics and Applied Mathematics* **62**, 2, pp. 149–166.

Kochetov, M. and Slawinski, M. A. (2009c). On obtaining effective transversely isotropic elasticity tensors, *Journal of Elasticity* **94**, pp. 1–13.

Kołakowski, L. (1990). *Modernity on endless trial* (The University of Chicago Press).

Krebes, E. S. (2004). *Seismic theory and methods*, Lecture notes.

Krebes, E. S. and Slawinski, M. A. (1991). On raytracing in an elastic-anelastic medium, *Bulletin of the Seismological Society of America* **81**, 2, pp. 667–686.

Kuhn, T. S. (1996). *The structure of scientific revolutions*, 3rd edn. (University of Chicago Press).

Lagrange, J.-L. (1788). *Mécanique analytique* (Éditions Jacques Gabay).

Landau, L. D. and Lifshitz, E. M. (1975). *The classical theory of fields, Course of theoretical physics*, Vol. 2, 4th edn. (Elsevier).

Landau, L. D. and Lifshitz, E. M. (1976). *Mechanics, Course of theoretical physics*, Vol. 1, 3rd edn. (Elsevier).

Landau, L. D. and Lifshitz, E. M. (1986). *Theory of Elasticity, Course of theoretical physics*, Vol. 7, 3rd edn. (Elsevier).

Leng, M. (2011). Creation and discovery in mathematics, in J. Polkinghorne (ed.), *Meaning in mathematics*, chap. 6 (Oxford University Press), pp. 61–69.

Lewin, K. (1952). *Field theory in social science: Selected theoretical papers by Kurt Lewin* (Tavistock).

Logan, R. K. (2010). *The poetry of physics and the physics of poetry* (World Scientific).

Long, C. (1967). On the completeness of the Lamé potentials, *Acta Mechanica* **3**, pp. 371–375.

Love, A. (1892). *A treatise on the mathematical theory of elasticity* (Cambridge University Press).

Love, A. (1911). *Some problems of geodynamics* (Cambridge University Press).

Love, A. E. H. (1944). *A treatise on the mathematical theory of elasticity*, 4th edn. (Dover).

Mach, E. (1989). *The science of mechanics*, 6th edn. (Open Court Publishing Company).

Marsden, J. E. and Hughes, T. J. R. (1983). *Mathematical foundations of elasticity* (Prentice-Hall).

Mase, G. E. (1970). *Theory and problems of continuum mechanics* (McGraw-Hill, Inc.).

Maugin, G. A. (1993). *Material inhomogeneities in elasticity* (Chapman and Hall).

Maugin, G. A. (2013). *Continuum mechanics through the twentieth century: A concise historical perspective*, Vol. 2 (Springer).

Maugin, G. A. (2014). *Continuum mechanics through the eighteenth and nineteenth centuries: Historical perspectives from John Bernoulli (1727) to Ernst Hellinger (1914)*, Vol. 1 (Springer).

McGrayne, S. B. (2011). *The theory that would not die: How Bayes' rule cracked the Enigma code, hunted down Russian submarines, and emerged triumphant from two centuries of controversy* (Yale University Press).

McLellan, A. G. (1980). *The classical thermodynamics of deformable materials* (Cambridge University Press).

Misner, C. W., Thorne, K. S. and Wheeler, J. A. (1973). *Gravitation* (W.H. Freeman and Company).

Moakher, M. and Norris, A. N. (2006). The closest elastic tensor of arbitrary symmetry to an elastic tensor of lower symmetry, *Journal of Elasticity* **85**, 3, pp. 215–263.

Morse, P. M. (1964). *Thermal physics* (W. A. Benjamin, Inc.).

Morse, P. M. and Feshbach, H. (1953). *Methods of theoretical physics* (McGraw-Hill, Inc.).

Müller, W. H. and Weiss, W. (2016). *The state of deformation in Earth-like self-gravitating objects* (Springer).

Musgrave, M. (1970). *Crystal acoustics: Introduction to the study of elastic waves and vibrations in crystals* (Holden-Day).

Noether, E. (1918). Invariante variationsprobleme, *Nachrichten von der Gesellschaft der Wissenschaften zu Göttingen* **1918**, pp. 235–257.

Noll, W. (1974a). The foundations of classical mechanics in the light of recent advances in continuum mechanics, in W. Noll (ed.), *The Foundations of Mechanics and Thermodynamics: Selected papers by W. Noll* (Springer).

Noll, W. (ed.) (1974b). *The Foundations of Mechanics and Thermodynamics: Selected papers by by W. Noll* (Springer).

Noll, W. (1974c). La mécanique classique, basée sur un axiome d'objectivité, in W. Noll (ed.), *The Foundations of Mechanics and Thermodynamics: Selected papers by W. Noll* (Springer).

Norris, A. N. (2006). Elastic moduli approximation of higher symmetry for the acoustical properties of an anisotropic material, *J. Acoust. Soc. Am.* **119**, 4, pp. 2114–2121.

Nowacki, W. (1965). Dynamiczne zagadnienia termosprężystości, *Journal of Theoretical and Applied Mechanics* **3**, 3, pp. 3–49.

Nowacki, W. (1986). *Thermoelasticity*, 2nd edn. (Elsevier).

Nye, J. F. (1987). *Physical properties of crystals: Their representation by tensors and matrices* (Oxford University Press).

Ogden, R. W. (1997). *Non-linear elastic deformations* (Dover).

Ogden, R. W. (2001). Elements of the theory of finite elasticity, in Y. B. Fu and R. W. Ogden (eds.), *Nonlinear elasticity: Theory and applications* (Cambridge University Press), pp. 1–58.

Okasha, S. (2002). *Philosophy of science: A very short introduction* (Cambridge University Press).

Oldroyd, J. G. (1950). On the formulation of rheological equations of state, *Proceedings of the Royal Society of London. Series A, Mathematical and Physical Sciences* **200**, 1063, pp. 523–541.

Peirce, C. S. (1955). *Philosophical writings of Peirce* (Dover).

Peirce, C. S. (1998). *The essential Peirce: Selected philosophical writings (1893-1913)*, Vol. 2 (Indiana University Press).

Penrose, R. (1997). *The large, the small and the human mind* (Cambridge University Press).

Penrose, R. (2004). *The road to reality: A complete guide to the laws of the Universe* (Jonathan Cape).

Penrose, R. (2016). *Fashion, faith, and fantasy in the new physics of the universe* (Princeton University Press).

Polkinghorne, J. C. (1986). *The quantum world* (Pelikan Books).

Popper, K. (1963). *Conjectures and refutations: The growth of scientific knowledge* (Routledge).

Popper, K. (1979). *Objective knowledge: An evolutionary approach*, revised edn. (Oxford University Press).

Postma, G. (1955). Wave propagation in a stratified medium, *Geophysics* **20**, pp. 780–806.

Proust, M. (1992). *A la recherche du temps perdu*, Vol. 6 (Gallimard).

Rayleigh, J. (1945). *The theory of sound*, 2nd edn. (Dover).

Reid, C. (1998). *Neyman* (Springer).

Riznichenko, Y. V. (1949). On seismic anisotropy, *Invest. Akad. Nauk SSSR, Ser. Geograf. i Geofiz.* **13**, pp. 518–544.

Roberts, A. J. (1994). *A one-dimensional introduction to continuum mechanics* (World Scientific).

Rochester, M. G. (2010). Note on the necessary conditions for P and S wave propagation in a homogeneous isotropic elastic solid, *Journal of Elasticity* **98**, 1, pp. 111–114.

Romano, A. and Marasco, A. (2014). *Continuum mechanics using Mathematica: Fundamentals, methods, and applications*, 2nd edn. (Birkhäuser).

Rota, G.-C. (1997). *Indiscrete thoughts* (Birkhäuser/Kindle).

Rudzki, M. P. (1911). Parametrische Darstellung der elastischen Wellen in anisotropischen Medien, *Bull. Acad. de Cracovie*, pp. 503–536.

Rudzki, M. P. (1912). Sur la propagation d'une onde élastique superficielle dans un milieu transversalement isotrope, *Bull. Acad. de Cracovie*, pp. 47–58.

Rudzki, M. P. (1913). Essai d'application du principe de Fermat aux milieux anisotrope, *Bull. Acad. de Cracovie*, 241-253.

Rudzki, M. P. (2000). On application of Fermat's principle to anisotropic media (*translation with comments*), in *Anisotropy 2000: Fractures, converted waves, case studies* (Society of Exploration Geophysicists), pp. 13–20.

Rudzki, M. P. (2003). Parametric representation of the elastic wave in anisotropic media (*translation with comments*), *Journal of Applied Geophysics* **54**, pp. 165–183.

Russell, B. (1945). *A history of western philosophy* (Simon and Schuster).

Ruta, G. (2016). Continuum models according to Piola, *Mathematics and Mechanics of Solids* **21**, 4, pp. 506–522.

Rymarz, C. (1993). *Mechanika ośrodków ciągłych* (PWN).

Santayana, G. (1983). *Reason in science* (Dover).

Schoenberg, M. and Muir, F. (1989). A calculus for finely layered anisotropic media, *Geophysics* **54**, 5, pp. 581–589.

Scholte, J. (1947). The range of existence of Rayleigh and Stoneley waves, *Geophysical Journal of the Royal Astronomical Society* **5**, pp. 120–126.

Schutz, B. (1980). *Geometrical methods of mathematical physics* (Cambridge University Press).

Schutz, B. (2003). *Gravity from the ground up: An introductory guide to gravity and general relativity* (Cambridge University Press).

Shermergor, T. (1977). *The theory of elasticity of microinhomogeneous media (in Russian)* (Nauka).

Slawinski, M. A. (1996). *On elastic-wave propagation in anisotropic media: Reflection/refraction laws, raytracing, and traveltime inversion*, Ph.D. thesis, University of Calgary.

Slawinski, M. A. (2015). *Waves and rays in elastic continua*, 3rd edn. (World Scientific).

Slawinski, M. A. and Webster, P. S. (1999). On generalized ray parameters for vertically inhomogeneous and anisotropic media, *Canadian Journal of Exploration Geophysics* **35**, 1/2, pp. 28–31.

Stacey, F. D. and Davis, P. M. (2008). *Physics of the Earth*, 4th edn. (Cambridge University Press).

Stebbing, L. S. (1937). *Philosophy and the physicists* (Dover).

Steiner, M. (1998). *The applicability of mathematics as a philosophical problem* (Harvard University Press).

Stewart, J. (1995). *Multivariable calculus*, 3rd edn. (Brooks/Cole Publishing Company).

Stillwell, J. (2008). *Naive Lie theory* (Springer).

Stone, M. and Goldbart, P. (2009). *Mathematics for Physics: A guided tour for graduate students* (Cambridge University Press).

Stubhaug, A. (2002). *The Mathematician Sophus Lie* (Springer).

Synge, J. (1937). *Geometrical optics: An introduction to Hamilton's method* (Cambridge University Press).

Tao, T. (2013). *Compactness and contradiction* (American Mathematical Society).

Tarantola, A. (2005). *Inverse problem theory and methods for model parameter estimation* (SIAM).

Tarantola, A. (2006). Popper, Bayes and the inverse problem, *Nature Physics* **2**, pp. 492–494.

Tegmark, M. (2014). *Our mathematical universe: My quest for the ultimate nature of reality* (Alfred A. Knopf).

Thomsen, L. (1986). Weak elastic anisotropy, *Geophysics* **51**, 10, pp. 1954–1966.

Thomson, W. (1890). *Mathematical and physical papers: Elasticity, heat, electromagnetism* (Cambridge University Press).

Thomson, W. T. (1950). Transmission of elastic waves through a stratified solid medium, *J. Appl. Phys.* **21**, pp. 80–93.

Thorne, K. S. and Blandford, R. D. (2017). *Modern classical physics: Optics, fluids, plasmas, elasticity, relativity, and statistical physics* (Princeton University Press).

Timoshenko, S. P. (1983). *History of strength of materials with a brief account of the history of elasticity and theory of structures* (Dover).

Timoshenko, S. P. and Goodier, J. N. (1987). *Theory of elasticity*, 3rd edn. (McGraw-Hill Publishing Company).

Tiwary, D. K. (2007). *Mathematical modeling and ultrasonic measurement of shale anisotropy and a comparison of upscaling methods from sonic to seismic*, Ph.D. thesis, University of Oklahoma.

Toretti, R. (1999). *The philosophy of physics*, The evolution of modern philosophy (Cambridge University Press).

Truesdell, C. (1966). *Six lectures on modern natural philosophy* (Springer-Verlag).

Truesdell, C. (1977). *A first course in rational continuum mechanics* (Academic Press).

Truesdell, C. and Noll, W. (2004). *The non-linear field theories of mechanics*, 3rd edn. (Springer).

Udías, A. (1999). *Principles of seismology* (Cambridge University Press).

Vinh, P. C. (2013). Scholte-wave velocity formulae, *Wave motion* **50**, pp. 180–190.

Voigt, W. (1910). *Lehrbuch der Kristallphysik* (Teubner, Leipzig).

Warner, F. W. (1983). *Foundations of differentiable manifolds and Lie groups* (Springer).

Wegener, A. (1929). *The origin of continents and oceans* (Dover).

Weinberg, S. (1987). *Towards the final laws of physics* (Cambridge University Press).

Weinberg, S. (2015). *To explain the World: The discovery of modern science* (Harper).

Weyl, H. (1949). *Philosophy of mathematics and natural sciences* (Princeton University Press).

Wigner, E. (1960). The unreasonable effectiveness of mathematics in the natural sciences, *Communications in Pure and Applied Mathematics* **13**, 1, pp. 1–14.

Will, C. M. (1986). *Was Einstein right?: Putting general relativity to the test* (Basic Books).

Winterstein, D. (1990). Velocity anisotropy terminology for geophysicists, *Geophysics* **55**, 1070-1088.

Yong-Zhong, H. and Del Piero, G. (1991). On the completeness of the crystallographic symmetries in the description of the symmetries of the elastic tensor, *Journal of Elasticity* **25**, 3, pp. 203–246.

Ziman, J. (1965). Mathematical models and physical toys, *Nature* **206**, 4990, pp. 1187–1192.

Subject Index

acceleration due to gravity, 303
  symbol, 495
adiabatic
  formulation, 324
  process, 324–328, 332, 333, 438, 439, 444, 451
æolotropy, 264
affine
  space, 409
  transformation, 410
  vector field, 368
Almansi strain tensor, **384**, 403, 407, 413
  symbol, 495
analytic geometry, 23
angular frequency, 197, 328
  Earth, 320, 328
  symbol, 495
angular momentum
  Lagrange mechanics, 301
anisotropy
  Backus average, 122, 124, 126, 138, 180
  Bayesian evaluation, 344
  continuum mechanics, 122, 142
  eikonal equation, 299
  elasticity tensor, 43
  homogeneity, 305
  Hookean solid, 92
  inhomogeneity, 143, 153, 177–179
  inner core, 12
  propagation speed, 15
  ray parameter, 296, 298
  seismic, 6
  shear-wave splitting, 344, 345
  symmetry group, 124
  tensor, 47, 92
  Thomsen parameters, 138
  weak, 142, 149, 152, 174, 175
ansatz, 226, 236, 265
approximation
  symbol, 494
Aristotelian
  physics, 3
  spacetime, 23, 24
astrophysics, xxv
asymptotic series, xxxi, 210, 282

Backus average, 126, 127, 178, 343
  Fermat's principle, 126, 354

Fourier transform, 130
isotropy, 179, 180
layers
generally anisotropic, 142
isotropic, 125
orthotropic, 142, 164
transversely isotropic, 141
weak inhomogeneity, 179
balance
of angular momentum, 29, 301
gravitational wave, 320
of linear momentum, 3, 6, 21, 31, 121, 182, 190, 336, 338
of mass
continuity equation, 336, 338
gravitational wave, 320
based science, 3
basic science, 3
Bayes Information Criterion, 323
Bayes's theorem, 343–346, 355
Bayesian
analysis, 9, 123
approach, 344
evaluation, 354
Fermat's principle, 357
geometrical optics, 357
hypothesis refutability, 353
model evaluation, 357
ray theory, 357
theory evaluation, 357
wave optics, 357
inference, 343, 344
statistics, 352
Beltrami's identity, **295**, 298, 304
bicharacteristics, 210, 279, 287, 301, 316, 355
black hole, 317
body force, 22, 182, 310, 312–314, 328
body wave, xxviii, 11, 15, 181, 182, 222, 231, 240, 252, 254, 261, 262
Hookean solid, 14
brachistochrone problem, 268, 303
Brownian motion, 374
bulk modulus, 83, 101, 327
adiabatic, 327
isothermal, 327
symbol, 496

calculus of variations
causality, 287
de Maupertuis, 188
Fermat's principle, 268, 269, 272, 277, 340
Hamilton principle, 268
history, 268, 269
optimization, xxix
teleology, 277, 289
Cartesian
plane, 23
product, 23, 24
Cauchy
elastic material, 455
equation of motion, **31**
balance of linear momentum, 33
characteristics, 291
continuum mechanics, 293
elastodynamic equation, 182, 208
Fermat's principle, 280
Hookean solid, 210

P wave, 31
ray theory, 209
S wave, 31
wave equation, 193, 208
problem, 199, 203, 287
causality, 355
wavefront healing, 355
strain tensor
strain-energy function, 454
stress quadric, 417, 418
Cauchy stress tensor, **32**, **420**
Cauchy's equation of motion, 33
corotational, 426
current configuration, 425
deformation gradient, 50, 52
elastodynamic equation, 32
finite strain, 415, 416, 426
infinitesimal deformation, 426
objectivity, 425, 461
Piola-Kirchhoff stress tensor, 57, 415, 416, 426
First, 420, 421
Second, 422
reference configuration, 56
residual stress, 56, 57, 426
strain-energy function, 454, 457, 458, 460, 461
symbol, 495
thermoelasticity, 442
Cauchy-Green strain tensor
left, **380**, 381, 382, 389, 403, 413, 461
inverse, 380
symbol, 495
right, **378**, 413, 456, 457
inverse, 378
positive-definite, 380
symbol, 495
symmetric, 379
Cayley-Hamilton theorem, 330, 331, 382
celestial mechanics, 344
characteristics, 195–198, 203, 208, 210, 279, 291, 301, 316, 317
Christoffel symbol, 475, 476, 480–482, 484–487, 490, 491
classical field theory
continuum mechanics, 464
classical mechanics
quantum mechanics, 8, 294
Clausius-Duhem inequality, 432, 433, 440, 446
Clausius-Planck inequality, 440
Clebsch-Gordan theorem, 49
coherence theory of justification, 352
comoving material coordinates, 26, 27
compliance tensor
$\mathbb{R}^3$, 44, 112
symbol, 496
condensed-matter physics
continuum mechanics, 353
conservation of angular momentum, *see* balance of angular momentum
conservation of linear momentum, *see* balance of linear momentum
conservation of mass, *see* balance of mass
constitutive equation, 20
constructive interference
Fermat's principle, 284, 286
Snell's law, 286

contact force, 22, 182, 328
continental drift, 8
continuity equation, 31, 311, 312
  balance of mass, 336
continuum, 19, 122, 123
continuum mechanics, 21
  anisotropy, 142
  atom, 33
  axiomatic approach, 21
  Backus average, 123, 127
  balance of
    angular momentum, 29
    energy, 29
    linear momentum, 29
    mass, 29
  basic science, 2, 3, 34
  body, 22
  classical field theory, 464
  condensed-matter physics, 353
  configuration, 22
  continuum, 19
  continuum physics, 464
  deductive argument, 338
  dynamical process, 181
  electromagnetism, 2
  entropy, 29
  field equation, 29
  force system, 22
  general theory, 21
  general theory of relativity, 475, 490
  hypothesis, 339
  hypotheticodeductive theory, 21
  index symmetry, 66
  inference, 34
  inhomogeneity, 142
  interpretation, 34, 353
  irreducibility, 337
  justification, 33
  macroscopic phenomenon, 19, 33
  material manifold, 23
  material symmetry, 48
  mathematical foundation, 154
  mathematics, 339, 475
  microphysics, 293
  motion, 25, 26
  nonuniqueness, 353
  observation, 339
  particle mechanics, 20
  particular theory, 21
  quantum mechanics, 3, 19, 293, 353
  reality, 22
  scale-dependent phenomenon, 122
  seismograph, 10
  seismology, 154, 354
    application, xxviii
    background, xxviii
    based science, 34, 48, 336
    explanation, xxix
    hypotheticodeductive, xxix, 3
    method, 2, 10, 12
    observation, 339
    physics, xxviii
  sensory immediacy, 337
  stress tensor, 33
  tensor analysis, 43
  tensor calculus, 374
  theory, 21
  theory of elasticity, 41
  validity, 33
  worldview, 337
continuum physics
  continuum mechanics, 464

convolution, 130, 171–173, 202, 206
coordinates, 27
  Eulerian, 27
  Lagrangian, 28
  material, 26
  spatial, 26
Coriolis effect, 328
corpuscular theory of light, 346
correspondence principle, 209
cosmological constant, 320
Cosserat medium, 211
cotangent space, *see also* tangent space, 365, 471
covariance principle, 490
covariant-derivative operator
  symbol, 494
covector, 469
covering-law model, 346
critical angle, 281–283, 304, 347, 348
crystal symmetry
  material symmetry, 20
Curie point, 9
Curie's principle, 123
curl
  symbol, 494
current configuration
  Almansi strain tensor, 385, 413
  Cauchy stress tensor, 415, 421, 425
  Cauchy-Green strain tensor
    left, 381, 382, 413
  displacement-vector field, 360
  First Piola-Kirchhoff stress tensor, 420, 421, 425
  hyperelasticity, 455
  left Stretch tensor, 396, 413
  line element, 367, 378, 389, 391
  material symmetry, 51
  objectivity, 405, 406
  polar decomposition, 392, 396
  position vector, 360
  reference configuration, 50, 359–361, 368, 383, 413, 415
  rotation, 459, 460
  Second Piola-Kirchhoff stress tensor, 422
  spatial coordinates, 361
  surface area, 388, 389
  volume element, 387
curvature
  Cartesian coordinates, 479
  Cauchy stress tensor, 32
  plane curve, 276
  ray, 276
  spacetime, 318, 322
  symbol, 495
cycloid, 269, 303

d'Alembert solution, 196–201
d'Alembertian, 187, 188, 207, 320
  symbol, 494
deformation
  equivoluminal, 36, 183, 184, 186, 189, 327, 368, 388
  gradient, *see also* deformation-gradient, xxxii, 368, 378, 379, 384
    Cauchy-Green strain tensor, 382
    compatibility, 374
    deformation matrix, 28
    determinant, 50, 388, 443

displacement-vector
gradient, 50, 364
elasticity tensor, 52
First Piola-Kirchhoff
stress tensor, 423
frame indifference, 60
homogeneous
deformation, 367
infinitesimal deformation,
59, 402, 404, 442
isochoric deformation,
368–370
material symmetry, 59,
459
noncommutativity, 403
objectivity, 406
one-dimensional
continuum, 389
polar decomposition, 394,
397
second law of
thermodynamics, 450
Second Piola-Kirchhoff
stress tensor, 423
simple shear, 36, 399
strain-energy function,
453, 454, 463
stress tensor, 51
symbol, 495
terminology, 412
thermoelasticity, 427, 430,
438, 440–442
transformation matrix,
363
homogeneous, 367, 368, 375,
377, 388, 400, 401, 403
inhomogeneous, 367, 368, 370,
400
isochoric, 36, 368, 370–372,
388
deformation matrix, 29, 35–37
deformation gradient, 28
determinant, 37
inverse, 29, 35–37
main diagonal, 37
rotation matrix, 46
deformation-gradient, *see also*
deformation gradient
operator
symbol, 494
tensor, 359, 363, **364**, 366,
451, 472, 484
symbol, 495
demarcation criterion, 335, 339,
352, 353
determinant
symbol, 494
Dido problem, 268
diffusion equation
wave equation, 193
dilatation, 183, 314, 325, 326,
331, 406, 487
symbol, 495
Dirac's delta, 129, 161, 205, 210
gravitational wave, 330
directional derivative, 276, 316,
468
displacement
field, 50
gradient, 54, 56, 57, 401, 414
tensor, 384
vector
field, 359, 360, 363
gradient, 50, 364, 406, 424
symbol, 496
divergence
symbol, 494

divergence theorem, 30, 32, 33, 38
  statement, 38
Doppler effect, 320, 321
double vector, 411
double-dot product, 430, 431
  symbol, 494
Duhamel's heat-conduction law, 437, 438

Earth's angular frequency, 320, 328
  symbol, 495
Earth's mantle, xxvi, 6, 8, 9, 92, 322, 327, 328
earthquake seismology, 1, 2, 241
effective symmetry
  material symmetry, 156
effective tensor, xxviii, 104, 106
  cubic, 117, 118
  equivalent medium, 124
  equivalent tensor, 152
  isotropic, 108, 110, 117, 118
  many-to-one map, 117
  minimization, 119, 120
  monoclinic, 71–74, 112, 118, 155
  Monte-Carlo method, 108
  norm, 112
  orthotropic, 109, 111, 113, 118
  rotation, 110, 118, 120, 121
  symmetric tensor, 153
  symmetry class, 108
  tetragonal, 114, 118
  transversely isotropic, 107, 108, 115, 118, 119
  trigonal, 115, 118
eigenvalue problem, 195
eikonal equation, 208, 279, 288, 297, 299
eikonal function, 470
  symbol, 495
Einstein Field Equation, 320, 321
Einstein tensor, 320
Einstein's relativity, 24
Einsteinian spacetime, 24, 321
elasticity tensor
  Bayesian inference, 345
  commutativity, 147
  constitutive equation, 52
  coordinate-free, 69
  effective tensor, 118, 124
  equivalent medium, 134
  four-linear map, 462
  Frobenius norm, 111, 112, 121, 157
  Hooke's law, 43
  index symmetry, 54, 65, 459
  inverse, 44
  isothermal process, 324
  isotropic, 462, 463
  Kelvin notation, 66
  Lamé parameter, 184, 333
  material symmetry, 50, 99
  mathematical entity, 44, 104
  natural coordinate system, 83
  objectivity, 409
  order relation, 63
  orthogonal transformation, 63, 68
  P wave, 183
  $\mathbb{R}^3$, 63, 66
    symbol, 495
  $\mathbb{R}^6$, 66
    symbol, 495
  S wave, 183
  space of elasticity tensor, 69

St. Venant-Kirchhoff material, 447
stability condition, 140, 214, 242
symmetry group, 59, 64, 65
tensorial property, 70, 78
thermal-conductivity tensor, 437

elastodynamic equation, **182**
anisotropy, 199
bicharacteristics, 279
body force, 328
Coriolis effect, 328
elasticity, 182
equation of motion, 22, 181
Fermat's principle, 278, 301
gravitation, 22, 182, 310
Hamilton equations, 279, 293
Hamilton mechanics, 293
Hooke's law, 298
Hookean solid, 22
hyperbolicity, 210
$JWKB$ method, 210
linearization, 183
ray theory, 293
solution, 182, 208, 209, 237
asymptotic series, 210
traction, 22
wave equation, 181, 190

electric-charge density, 31, 495

electric-current density, 31, 496

electromagnetic wave
gravitational wave, 318
vacuum, 318

electromagnetism
basic science, 2
continuum mechanics, 2

emergence, 20, 153
reduction, 20

empiricism, 351, 352

enthalpy, 441
symbol, 496

entropy, 29, 431–433, 435, 438–441, 444, 450
symbol, 494

epistemology, xxvii, 5, 13, 350, 351

equation of motion
object trajectory, 268

equivalent medium, xxviii, 126, 131, 132, 134–137, 141, 162–164, 169

equivalent tensor
effective tensor, 152

Euclidean norm, 112
Frobenius norm, 112
log-Euclidean norm, 112

Euler angle, 70–72, 74, 80, 123
quaternion, 120

Euler equation
calculus of variations, 340
Euler-Lagrange equation, 279
first-derivative test, 278
stationary action
least action, 277
stationary time
least time, 277
maximum time, 277
variational principle, 277
variational problem, 272, 274

Euler-Lagrange equation
Beltrami's identity, 295
Euler equation, 279
Fermat's principle, 278, 279
Hamilton equations
Legendre's transformation, 289, 291

Hamilton principle, 291
wave equation, 292
expansion coefficient, 325
exploration seismology, 1, 11, 254, 313, 351
exponential integral, 219
exterior derivative, 491

falsifiability, 352
Fermat traveltime, 177
Fermat's principle, xxix, 15, 177, 269, 270, 283, 286, 297
Backus average, 126, 354
Bayesian approach, 357
bicharacteristics, 355
calculus of variations, 277, 340
constructive interference, 284, 286
elastodynamic equation, 301
Euler-Lagrange equation, 292
geometrical optics, 354
global optimization, 275
Hamilton principle, 268
Huygens's principle, 267, 269, 271
justification
macroscopic, 285
microscopic, 285
Lagrange's ray equation, 16
least time, 277
Legendre's transformation, 279
mathematical analogy, 283, 355
philosophy, 271
probability amplitude, 293
proof, 267, 278–280
ray theory, 268, 290, 301
Snell's law, 270, 271, 286, 288
stationary time, 277
teleology, 287–289
units, 291
variational principle, 267
wavefront healing, 355
wavefront-slowness surface, 280
wavelength, 285, 354
field equation, 29
finite elasticity, 50–52, 54, 58–61
first integral, 294
First Piola-Kirchhoff stress tensor, **420**, 421, 423, 425, 430, 441, 472
symbol, 496
fixed material coordinates, 26, 27
fluid mechanics, 7, 21
Fourier series, 210
Fourier transform
Backus average, 130
wave equation
causality, 287, 289
solution, 201–204, 217–219
Fourier's heat-conduction law, 438
frame indifference, 51–54, 56, 60, 406, 407
Fresnel's bright spot, 285, 346, 354, 356, 357
Frobenius norm, **68**
coordinate invariance, 76, 85
distance, 76, 118, 121, 156, 333
effective isotropic tensor, 106
Euclidean norm, 112
$F21$, 157, 159
$F36$, 157, 159
formula, 156
Kelvin notation, 68
Lamé parameters, 106

log-Euclidean norm, 112
operator norm, 112
orthogonal
projection, 111
transformation, 112
positive-definite tensor, 134
product, 118
invariance, 119, 120
rotation invariance, 70, 74, 80, 86–88, 118, 121
seismology, 105
Frobenius product, 118
invariance, 120

Galilean
metric, 319
relativity, 374
spacetime, 23, 24, 27, 31, 319, 321, 374
transformation
reference frame, 26
Gardner's relation, 214
general relativity, 317, 318, 324, 490
weak field, 319
generalized momentum
ray parameter, 301
geodynamics, 7, 8, 19, 351, 414, 426
geometrical optics, 209, 280, 285, 346–348, 354, 356
Fermat's principle, 354
wave optics, 357
Gibbs free energy, 441
symbol, 495
global seismology, 1, 7, 8, 11, 328
gradient
symbol, 494
gravitation
elastodynamic equation, 22, 182
gravitational constant, 312, 320
symbol, 496
gravitational force, 311
symbol, 495
gravitational lensing, 321
gravitational potential, 311–313, 316
gravitational wave, xxix, 317, 318, 323, 328
detection, xxix, 323, 324
Love wave, 323
matter interaction, 318
Rayleigh wave, 323
Scholte wave, 323
seismic wave, 318
source, 323
catastrophic, 323
periodic, 323
Stoneley wave, 323
strain tensor, 331
theory, 324
transverse, 318, 322, 330
vacuum, 318
Green elastic material, 455
Green's function, 204
impulse-response function, 204
method of singularities, 204
potential function, 204
probability amplitude, 285
Green-Lagrange strain tensor, **384**, 403, 407, 444, 446, 455, 457
symbol, 495
groundroll, 232
guided wave, xxix, 11, 182, 221, 222, 224, 233, 234, 242, 264
properties, 262

Hamilton equations
  Cauchy's equation of motion, 293
  conserved quantity, 294
  eikonal equation, 279, 288
  equations of motion
    Legendre's transformation, 340
  Euler-Lagrange equation
    bicharacteristics, 279
    Legendre's transformation, 289, 291
  Fermat's principle, 267, 278, 279
  Lagrange equations of motion, 340
  Legendre's transformation, 268, 340
  mechanics, 291–293
  phase space, 288
  ray theory, 279, 293, 297
Hamilton mechanics, 291, 293
  Lagrange mechanics, 292
  Legendre's transformation, 268
Hamilton principle, xxix, 267, 292, 302, 340
  Fermat's principle, 268
  Lagrange mechanics, 268
  least action, 277
  mechanics, 290
  probability amplitude, 293
  teleology, 290
  variational principle, 291
Hamiltonian
  Legendre's transformation, 279
  mechanics, 268, 291, 293, 300
  ray theory, 268, 279, 291, 293, 297–299
  symbol, 496
head wave, 280–283
Heaviside's step
  gravitational wave, 330
  Green's function, 218
helioseismology, 11
Helmholtz free energy, **434**, 440–443, 447
  symbol, 495
Helmholtz's
  decomposition theorem, **186**, 189, 247, 257
  equation, 287
Hempel's paradox, 349
Hencky strain tensor, 414
Herman's theorem, xxviii, 46, 47, 61, 62, 90, 91
Hooke's law
  constitutive equation, 20
  continuum mechanics, 3
  Duhamel's heat-conduction law, 437
  elastic solid, 41
  elasticity parameters, 82
  elasticity theory, 340
  elastodynamic equation, 182, 208, 298
  equivalent medium, 136
  finite elasticity, 53
  guided wave, 238
  history, 43
  incremental stress, 44
  interface wave, 247, 258
  isotropy, 229, 230, 238, 244, 247, 258, 325, 326
  linear relation, 42, 43
  operator norm, 112
  surface wave, 229, 230, 244
  temperature, 310, 324

theory of seismology, 336
three dimensions, 6
wave equation
friction, 193
Hookean solid
accuracy, 17
effective tensor, 110
validation, 342
analogy, 336
asthenosphere, 338
Backus average, 122
Voigt formulæ, 175
Bayesian evaluation, 355
Bayesian inference, 345
best explanation, 353
bounded, 221, 222
dispersion, 237
guided wave, 262, 264
halfspace, 224, 234, 242
infinite, 222
interface wave, 262, 264
liquid, 246
Scholte wave, 253
side condition, 245
Stoneley wave, 246
surface, 229
surface wave, 262, 264
vacuum, 246
constitutive relation, 43
Cosserat medium, 211
covering-law model, 347
Curie's principle, 123
diclinic, 88–90
dynamical process, 181
effective tensor, 103, 106, 108, 110, 124
isotropy, 106, 117
norm, 112
positive-definite, 112
elastic continuum, 42
elasticity tensor, 43, 46, 47
elastodynamic equation, 22, 182, 183
hyperbolicity, 210
solution, 211
entropy, 29
equation of motion, 181
equivalent medium
Backus average, xxviii, 121–124
fourfold symmetry
fourth-rank tensor, 47, 91
Herman's theorem, 47, 91
transverse isotropy, 92
fourth-rank tensor
fourfold symmetry, 47, 91
Herman's theorem, 47, 91
generally anisotropic
isotropic, 62
gravitation effect, 315, 317, 328
gravitational wave, 318, 324, 328
Hamilton equations, 293
Herman's theorem
hexagonal, 47
pentagonal, 47
hexagonal, 90, 91
tessellation, 48, 91
transversely isotropic, 91
homogeneity
wave equation, 208
homogeneous
waves, 16
index symmetry, 66, 68
strain-energy function, 67
inference to best explanation, 349

inhomogeneity, 190
isotropy, 45
elasticity parameters, 82
general anisotropy, 62
Lamé parameters, 83
wave equation, 208
material body, 22
material symmetry, 41, 49, 62, 66, 72, 92
mathematical analogy, 92, 104, 105, 108, 154, 210, 338, 341
effective tensor, 110
physical world, 45
mathematical entity, 42
physical world, 90
physical world, 353
analogy, 42, 45, 189
measurement, 16
model, 236
Platonic analogy, 111
point symmetry, 46
Poisson's ratio, 230, 231
qP wave, 223
qS_1 wave, 223
qS_2 wave, 223
ray curvature, 305
refutation, 352
St. Venant-Kirchhoff material, 447
stability condition, 223, 231
symmetry
crystal, 48
symmetry class, 41, 47, 62, 66
diclinic, 62
temperature effect, 324, 325, 328
theory of elasticity, xxviii, 344
theory of seismology, 345
theory validation, 336
theory-independent data, 341
transversely isotropic
hexagonal, 91
wave equation, 182
solution, 194, 199, 208
waves, xxviii, 4, 6, 14, 15, 138, 223
Hulse-Taylor binary, 323
Hume's induction problem, 290
Huygens principle, 272, 285, 288
Fermat's principle, 267, 269, 271, 288
Snell's law, 269, 270
wavelength, 285
Huygens source, 282
hyperelastic material, 450, 454, 455, 460, 464
hypotheticodeductive, 336, 352

idempotent operator, 395, 396
impulse-response function
Green's function, 204
incremental stress, 44
index symmetry
Christoffel symbol, 480, 487
elasticity tensor, 43, 54, 61, 62, 65, 66, 463
fourth-rank tensor, 66–68
Hookean solid, 92
material symmetry, 65, 66
isotropy, 462
inertial reference frame, 24, 25
Aristotelian spacetime, 24
Galilean spacetime, 26
Galilean transformation, 25
inference to best explanation, 349

infinitesimal
  displacement, 362
  elasticity, 53, 59, 61
  strain, 359, 401–404, 424, 428, 446–449, 455, 457, 463, 487
infinitesimal-strain tensor, 402, 412, 414
inhomogeneity
  anisotropy, 143, 153, 177–179, 181
  Backus average, 124, 126, 127
  continuum mechanics, 142
  deformation gradient, 369
  elasticity tensor, 43
  gravitation, 316
  propagation speed, 15, 190
  thermal, 437
  thermodynamic, 439
inner core, 12, 350
instrumentalism, 351
interface wave, xxix, 11, 182, 221, 222, 224, 234, 245, 246, 262, 264
  dispersion, 253
  existence, 242, 261, 262
  properties, 261, 262
  solution, 221
  speed, 224, 229, 253
  Stoneley, 245
internal energy, 30, 427, 429, 440, 441
  symbol, 495
irreversible process, 436, 439, 450
isentropic process, 439
isochoric deformation, 36, 368–372, 388
isomorphism, 26, 67
isothermal formulation, 324
isothermal process, 324–327, 332, 333, 439, 444, 446, 451

Jacobi matrix, 28, 150, 151, 367, 421
Jacobian, 421
  symbol, 494
$JWKB$ method, *see also* $WKBJ$ method, 209, 210

Kelvin notation, 66, 68, 81, 120
kinetic energy
  symbol, 496
Kirchhoff
  formula, 204
  stress tensor, 426
Kronecker
  delta, 65, 212, 379, 458, 463, 478
    symbol, 494
  tensor, 363

Lagrange
  equations of motion
    conserved quantity, 294
    Hamilton equations, 268, 340
    Legendre's transformation, 340
  mechanics, 291, 302
    angular momentum, 301
    equation of motion, 301
    generalized momentum, 301
    Hamilton mechanics, 292
    Hamilton principle, 268
    Legendre's transformation, 268
    linear momentum, 301

ray equation
Fermat's principle, 16
Lagrangian
density, 291, 292
symbol, 496
Hamilton principle, 291
mechanics, 268, 290
symbol, 496
ray theory, 268, 279, 298, 299
symbol, 496
Lamé
parameter, **83**, 106, 126, 141, 183, 184, 230, 324, 333
compressibility, 83
rigidity, 83
symbol, 495
stress ellipsoid, 417, 418
Landau symbol, 53, 312, 319
Laplacian
geometrical meaning, 185, 189, 211, 212
physical meaning, 185, 189
polar coordinates, 217
scalar, 199
scalar potential, 225
spherical coordinates, 200
symbol, 494
vector, 184, 185, 200
vector potential, 225
least-action principle, 277
least-time principle, 277
Legendre's transformation
antisymmetry, 435, 449
Fermat's principle, 279, 280
Hamilton equations, 268, 279, 289, 340
Euler-Lagrange equation, 291
mechanics, 293
Hamilton principle
wave equation, 292
Hamiltonian
ray theory, 280
Lagrange equations of motion, 268
Hamilton equations of motion, 340
mechanics, 300, 301
partial differential equation, 268
ray theory, 268, 298, 299, 301, 470, 471
strain-energy function, 449, 450
symmetry, 435, 449
thermodynamic potential, 268, 434, 441
Leibniz integral rule, 430, 484
Levi-Civita symbol, 374–376, 387, 403
symbol, 494
Lie
algebra, 404
derivative, 490, 491
linear momentum
brachistochrone, 303
ray parameter, 305
local admissibility, 367, 387, 397
log-Euclidean norm, 112
operator norm, 112
Love wave, 237, 240, 243, 246, 262
existence, 242
gravitational wave, 323
Rayleigh wave, 241
Love's hypothesis, 311

manifest truth
scientific truth, 13

mass density
  balance of mass, 31, 443
  continuum mechanics, 30
  Earth, 315
  elastodynamic equation, 298
  gravitation, 311, 316
    wave, 321
  Hookean solid, 43
  linear momentum, 32, 310
  P wave, 82, 180, 214, 215
  Rayleigh wave, 246
  S wave, 101
  Scholte wave, 247, 253, 257
  symbol, 495
  wave equation, 190, 193, 292
material frame indifference, *see* frame indifference
material
  body, 25
  coordinates
    Cartesian coordinates, 362
    Cauchy stress tensor, 416
    comoving, 26–28, 362
    deformation gradient, 363, 364
    deformation matrix, 38
    fixed, 26–28
    Green-Lagrange strain tensor, 384
    infinitesimal deformation, 362, 402
    Lagrangian coordinates, 28
    material time derivative, 39
    motion, 362
    notation, 50
    reference configuration, 361, 362, 364
    Second Piola-Kirchhoff stress tensor, 422
    spatial coordinates, 26–28, 34, 36
  fiber, 366
  manifold, 23, 27
  point
    symbol, 496
  symmetry, 41, 45, 464
    crystal symmetry, 20
    cubic, 58, 62, 82, 117, 118
    diclinic, 88, 90
    effective symmetry, 156
    effective tensor, 119
    emergence, 20
    finite elasticity, 58
    general anisotropy, 69
    hexagonal, xxviii, 47, 48, 61, 90, 91, 154
    Hookean solid, 91
    index symmetry, 65, 66
    infinitesimal elasticity, 58
    isotropy, 83, 92, 137, 175, 176, 183
    monoclinic, 70, 74, 98, 99
    octagonal, xxviii, 61
    orthogonal transformation, 58, 63
    orthotropic, 64, 75–78, 90, 110, 113, 118, 122, 123, 155
    physical analogy, 105
    terminology, 48
    tetragonal, 74, 78–80, 84–87, 91, 114, 118
    transverse isotropy, 72, 91–93, 105, 118, 119, 174, 176, 223

trigonal, 63, 64, 74, 81, 84, 87–89, 91, 115, 118
wavefront symmetry, 123
time derivative, 32, 39
material coordinates
Lagrangian coordinates, 361
referential coordinates, 361
matrix-index algorithm, 65
Maxwell
equations, 29
model, 338
mechanics
Hamilton principle, 290
metric tensor, 319, 475, 478–481, 485, 486, 491
Cartesian, 478, 479
Galilean, 319
Minkowskian, 31, 319, 322, 332
symbol, 495
Minkowskian metric, 31, 319, 322, 332
symbol, 494
Moho discontinuity, 6
Mohr's circle, 36, 49, 94, 95
strain tensor, 49
stress tensor, 49
traction, 95
vector, 49
Monte-Carlo method
effective tensor, 108, 110
mudroll, 254

Nanson's formula, **389**, 421, 429
natural coordinate system, 69, 144, 145, 147, 148
natural form, 69, 75
Newton's law, 268, 302
first, 24
second, 190, 268
third, 29, 66
nilpotent operator, 395
Noether's theorem, 296
ray parameter, 16, 45
noninertial reference frame, 25
normal mode, xxviii, 11, 235

$O(\cdot)$, 312, 319, 425
$o(\cdot)$, 53, 424
$O(2)$, 58
$O(3)$, 51, 58, 59, 62–65, 121, 383
objectivity, 22, 405–408, 425, 461
Occam's razor, 106, 111, 352
Ohm's law, 438
one-form, xxx, 279, 465, 469, 470, 472, 473, 480
ontology, xxvii, 5, 350, 351
operator norm, 112, 159
induced, 112
orthogonal transformation
invariance, 112
seismology, 105
orthogonal group, 59, 383
orthogonal transformation
deformation gradient, 376
frame indifference, 51
Frobenius
norm, 112, 118, 121
product, 119
invariance, 74, 382, 383, 412
material symmetry, 58, 60, 61, 63
objectivity, 405, 407
reflection, 393
rotation, 100, 393
symmetry group, 64, 65, 68, 75
orthonormal
basis, 63

coordinates, 82, 370, 388, 443
matrix, 46
transformation, 100
outer core, 6, 8, 9, 12, 254, 321, 349–351

P wave
analogy, 189
Cauchy's equation of motion
balance of linear momentum, 31
coupled, 322
dispersion, 224
existence, 33, 121, 183, 187
Gardner's relation, 214
gravitation, 314
gravitational wave, 321
heat, 325, 327, 328
homogeneous halfspace, 243–245
Hookean solid, 337
unbounded, 222
irrotational, 189
liquid halfspace, 258, 261
longitudinal wave, 186
mathematical entity, 3, 4, 42, 104, 222
Rayleigh wave, 224, 226, 245, 246
scalar potential, 225
Scholte wave, 248
seismic interpretation, 4, 189, 337
selfgravitation, 328
speed, 82, 101
stability condition, 214, 242
theory-ladenness of data, 341
transverse isotropy, 81
wave equation, 183, 184
solution, 194
P-wave speed
symbol, 494, 496
paradigm shift, 13
parallel transport, 482
particle mechanics
continuum mechanics, 20
motion, 25, 26
Penrose tiling, 92
period
symbol, 496
phase space, 279, 288, 302
philosophy of science, 335
photoelasticity, 67
photoelectric effect, 374
physics
Aristotelian, 3
Piola strain tensor, 378
Piola transformation, 422
Piola-Kirchhoff stress tensor, 54, 57, 58, 415, 416, 443
First, **420**, 421, 423, 425, 426, 430, 441, 443, 446, 455, 472
two-point tensor, 425
Second, 415, **422**, 422, 423, 425, 443, 446, 456
reference configuration, 425
plate tectonics, 7, 8
point symmetry
Hookean solid, 46
Poisson
equation, 312–314, 321
ratio, 230–233
solid, 140
Poisson's
ratio, 139

polar decomposition, 359, 364, **390**, 392, 394, 397, 404, 461
  theorem, **390**
potential energy
  symbol, 496
Preliminary Reference Earth Model (PREM), 139
prestressed medium, 414, 426
principal
  strain, 386, 419
  stress, 417, 419, 423
    direction, 417
    space, 417–419
  stretch, 370, 399
probability amplitude, 283–285, 293
product solution, 194
  separation of variables, 194
pseudotensor, 387, 411, 473
pseudovector, 387, 411, 473
pullback, 368, 413, 422
  symbol, 494
pushforward, 368, 377, 413, 422
  symbol, 494

qP wave, 4–6, 15, 137, 166, 167, 174, 223, 324
  theory-ladenness of data, 341
qS wave, 16, 223
qS_1 wave, 4, 6, 15, 16, 223, 324
  S wave, 223
  shear-wave splitting, 344
  theory-ladenness of data, 341
qS_2 wave, 4, 6, 15, 16, 223, 324, 341
  S wave, 223
  shear-wave splitting, 344
qSV wave, 137, 166, 167, 174
  transverse isotropy
    shear-wave splitting, 223
quantitative seismology, xxviii, xxx, 3, 10, 12, 121, 154, 183, 339, 472, 475
quantum mechanics, 181, 339, 350
  basic science, 3
  classical mechanics, 8, 294
  continuum mechanics, 3, 19, 293, 353
  Hookean solids, 45
  linearity, 45
  probability amplitude, 283
  seismology, xxix
  variational principles, 287, 302
quasi-Rayleigh wave, *see also* Rayleigh wave, 241
quasishear wave, 223
quaternion, 120, 121
  Euler angle, 120

$\mathbb{R}$, 23, 24
$\mathbb{R}^1$, 182, 216
$\mathbb{R}^2$, 46, 48, 49, 85, 93, 94, 330, 331
$\mathbb{R}^3$
  body wave, 182
  flat space, 479
  Hooke's law, 65
  material symmetry, 61, 63, 68, 69
  pseudovector, 473
  $\mathbb{R}^6$, 101
  rotation, 69, 77, 85, 99
  vector product, 473
  wave equation, 204, 216
$\mathbb{R}^6$, 46, 69, 99, 101
  Hooke's law, 65
  material symmetry, 68
  $\mathbb{R}^3$, 101

rotation, 69
rationalism, 352
ray angle
symbol, 494
ray parameter, 16, 294, 296, 297, 303
Beltrami's identity, 295
brachistochrone, 303
conserved quantity, 298, 299
global seismology, 301
linear momentum, 301, 305
Noether's theorem, 16, 45
Snell's law, 298
translational symmetry, 296
ray theory
Bayesian approach, 357
Fermat's principle, 268, 290, 301
Hamiltonian, 301
Lagrangian, 301
mathematical analogy, 355
wavefront healing, 355
ray velocity
symbol, 496
Rayleigh wave, 224, 230–235, 237, 241, 242, 244–247, 255, 257, 262
gravitational wave, 323
Love wave, 241
quasi-Rayleigh wave, 241
speed, 241
transverse isotropy, 264
redshift, 321
reduction
emergence, 20
reference configuration
Cauchy-Green strain tensor
right, 378, 382, 393, 413
current configuration, 50, 359–361, 368, 383, 413, 415
deformation gradient, 367
First Piola-Kirchhoff stress tensor, 420, 421, 425
Green-Lagrange strain tensor, 385, 413
hyperelasticity, 455
line element, 367, 378, 386, 389, 391
material coordinates, 361, 362
objectivity, 405, 406
polar decomposition, 392, 396, 404
position vector, 360, 367
residual stress, 56
right Stretch tensor, 413
rotation, 54, 459, 460
Second Piola-Kirchhoff stress tensor, 422, 425
stress tensor, 53, 54
surface area, 388, 389
thermoelasticity, 444
volume element, 387
reference frame
inertial, 24, 25
Aristotelian spacetime, 24
Galilean spacetime, 26
Galilean transformation, 25
noninertial, 25
refutability, 352, 353
Bayesian approach, 353
refutation, xxx, 352
residual stress, 53, 55–57, 426
retrodiction, 12, 14, 17, 137, 288
reversible process, 439, 440
Reynolds transport theorem, 430

Riemannian
  manifold, 478
  metric, 478
rigid-body rotation
  deformation, 413
  deformation gradient, 364, 368, 397
  infinitesimal, 402
  material symmetry, 459
  objectivity, 407–409, 426
  polar decomposition, 399
rigidity modulus, 83, 101, 321
rotation matrix, 46, 69, 93, 94, 101, 119
  deformation matrix, 46
rotation vector
  symbol, 495

S wave
  analogy, 189
  Cauchy's equation of motion, 31
  coupled, 322
  dispersion, 224
  displacement, 138
  equivoluminal, 189
  existence, 33, 121, 183, 187, 188
  gravitational wave, 318, 321, 322, 328
  heat, 327, 329
  Hookean solid, 337
    unbounded, 222
  isotropy, 223
  liquid core, xxvi
  material property, 187
  mathematical entity, 3, 4, 42, 104, 222
  qS wave
    shear-wave splitting, 223
  Rayleigh wave, 226, 246
  seismic interpretation, 4, 189, 337
  SH wave, 223
  shear-wave splitting, 344
  speed, 82, 101
  stability condition, 214
  Stoneley wave, 246
  SV wave, 223
  theory-ladenness of data, 341
  transverse isotropy, 81
  transverse wave, 186
  vector potential, 225
  wave equation, 183, 184, 186
    solution, 194
S-wave speed
  symbol, 494, 496
scalar potential, 225
  P wave, 225
  symbol, 496
scalar product, 431
  symbol, 494
Scholte wave, 246, 254, 262, 264
  existence, 253, 262
  gravitational wave, 323
  speed, 253
scientific revolution, 3, 13
scientific truth, 12, 13
  manifest truth, 13
Second Piola-Kirchhoff stress tensor, **422**, 423, 425
  symbol, 496
seismic anisotropy, 123
  continuum mechanics, 6
seismogram, 4–6, 14, 351
seismograph, 5, 6, 9, 14, 189, 342
seismology
  abstract concept, 154
  analogy, 336

applied, 11
astrophysics, xxv
basic science, 3
branches, 1
  earthquake, 1
  exploration, 1
  global, 1
  theoretical, 1
classification, 10
  standing wave, 11
  wave propagation, 11
conserved quantities, xxix
continuum mechanics, xxviii, xxix, 12, 48, 154
  hypotheticodeductive, 3
discipline, 1
  computational science, 1
  continuum mechanics, 1
  electromagnetism, 1
  mathematical physics, 1
earthquake, 241
etymology, 2
experimental, 4, 6
exploration, 11, 313
finite strain, 414, 426
formulation, xxvi
  continuum mechanics, 34
foundation, xxx
Frobenius norm, 105
fundamental principle, 352
geodynamics, 7
geophysics, 10
global, 11, 328
gravitation, 316
heat flow, 325
helioseismology, 11
history, xxvii
Hookean solid, 338
  point symmetry, 46
hypothesis, 337
hypotheticodeductive, xxix, 183, 336, 345
interface wave, 224
mathematics, xxvi
methodology, 12
natural science, xxix
  empiricism, xxx
  realism, xxx
normal mode, xxviii
observational
  theoretical, 42
operator norm, 105
Platonic analogy, 111
preloads, 44
prestressed medium, 414
quantitative, xxx, 121, 472, 475
  continuum mechanics, 3, 339
quantum mechanics, xxix
remote measurements, 11
science, xxix, 2, 283, 335
  Bayesian inference, 345, 346
  Kuhn, 13
  Popper, 12, 13, 352
  realistic approach, 336
subjects, xxix, 2
surface wave, 224
taxonomy, 3
theoretical, 11
  continuum mechanics, 34
  observational, 42
theory
  Bayesian inference, 343, 345, 346
  continuum mechanics, 354, 374

explanation, xxv, xxix, 137, 280–282, 288, 346–349, 351, 353
prediction-explanation, xxx, 8, 347, 349, 353
underdetermination, 342
variational principles, 267, 301
wave propagation, xxviii
wavefront slowness, 468
zenoseismology, 11
selfgravitation, 310, 328
separation of variables, 194
SH wave, 137, 138, 222, 225, 240, 241, 246
S wave, 223
transverse isotropy
shear-wave splitting, 223
shear
strain, 83
stress, 83
wave, xxvi, 175, 222, 247, 253, 257, 355, 356, 403
shear-wave splitting, 223, 344
side condition
eikonal equation, 209
hyperbolic partial differential equation, 209
initial condition, 195
initial-value problem, 96
separation of variables, 195
Sturm-Liouville problem, 195
wave equation, xxix
guided, 11, 221, 222, 238, 239, 242
interface, 11, 221, 222, 247, 258, 263
solution, 201, 203, 204
surface, 11, 221, 222, 227, 230, 235, 242, 244
similarity
symbol, 493
simple shear, 36, 399–401
Snell's law, 297
anisotropy, 298
constructive interference, 286
Fermat's principle, 271, 275, 280, 286, 288
Huygens's principle, 270
ray parameter, 294–296, 298
wavelength, 285
$SO(3)$, 49, 51, 53, 57, 59, 96, 112, 114, 119, 120, 364, 461
$O(3)$, 59, 65
$SO(6)$, 69
source operator, 204
spacetime, 23
Aristotelian, 23, 24
divergence, 31
Einsteinian, 24, 321
Galilean, 23, 24, 27, 319, 321, 374
Hamilton principle, 294
manifold, 23
spacetime strain, 318
spatial coordinates
Almansi strain tensor, 384, 385
Cartesian coordinates, 362
Cauchy stress tensor, 416
Cauchy-Green tensor
left, 389
current configuration, 361, 362, 364
deformation gradient, 364
deformation matrix, 38
Eulerian coordinates, 361
infinitesimal deformation, 362, 402

material coordinates, 26–28, 34, 36
material time derivative, 39
motion, 362
notation, 50
$\mathcal{S}$, 27
Stretch tensor
left, 414
spatial location
symbol, 496
spatial manifold, 23, 27
special orthogonal group
antisymmetric tensor, 53, 57, 96
effective tensor, 120
elasticity tensor, 69
Frobenius product, 119
isotropy, 364, 461
material symmetry, 51, 59, 112, 114, 461
orthogonal group, 59, 65
special unitary group, 49
special relativity, 319, 332, 374
special unitary group, 49
specific heat, 326, 327
symbol, 495
speed of light, 322
symbol, 495
spring constant, 292
symbol, 496
St. Venant-Kirchhoff material, 447
stability condition, 134, 140, 188, 214, 223, 231, 262, 332, 333
standing wave, 11, 235, 237
stiffness tensor, *see also* elasticity tensor, 44, 112
Stokes's heat flux-theorem, 428
Stoneley wave, 246, 262
gravitational wave, 323
interface, 245
strain, *see also* strain tensor
energy, 326, 444, 455, 460
function, 54, 65–67, 382, 426, 453–455, 458–460, 463
symbol, 496
finite, xxxii, 359, 373, 403, 413–415, 425, 426, 442, 447
infinitesimal, 428, 446–448
isothermal, 428, 446–448
strain tensor
Almansi, 384, 403, 407, 413
symbol, 495
Cauchy-Green
left, 382, 389, 403, 413, 461
right, 413, 456, 457
Cayley-Hamilton theorem, 331
gravitational wave, 318, 319
Green-Lagrange, 384, 403, 407, 455, 457
symbol, 495
infinitesimal, 402
symbol, 494
left Cauchy-Green
symbol, 495
left Stretch
symbol, 496
Mohr's circle, 49
right Cauchy-Green
symbol, 495
right Stretch
symbol, 496
Stretch
left, 396, 413, 414, 461
right, 396, 413, 414, 443

stress tensor
  Cauchy, *see* Cauchy stress tensor
  continuum mechanics, 23
  definition, 33
  deformation gradient, 51
  displacement field, 50
  elastodynamic equation, 32
  First Piola-Kirchhoff, 420, 421, 423, 425, 426, 443, 446, 455
    symbol, 496
  Love's hypothesis, 311
  material symmetry, 58
  Mohr's circle, 49, 95
  Piola-Kirchhoff, 57, 58
  reference configuration, 53, 54, 56
  residual, 56
  Second Piola-Kirchhoff, 415, 422, 423, 425, 443, 446, 456
    symbol, 496
  seismology, 350
  side condition, 233, 235, 238
  wave equation, 3
stress-energy tensor, 320
stretch ratio, 383, **389**
  symbol, 495
Stretch tensor
  left, **396**, 414, 461
    symbol, 496
  right, **391**, 396, 414, 443
    symbol, 496
Sturm-Liouville theory, 195
$SU(2)$, 49
superposition principle, xxviii, 44, 208, 285, 309
surface wave, xxix, 11, 15, 182, 221, 222, 224, 230, 231, 234, 264
  properties, 261, 262
SV wave, 138, 224, 244–246
  S wave, 223
  Scholte wave, 248
symmetric tensor
  antisymmetric, 53, 54, 57, 96
  Cauchy stress tensor, 32
  deformation gradient, 375
  effective tensor, 106, 108, 153, 158
  infinitesimal deformation, 59, 61
  reflection, 59
symmetry
  group
    average, 103, 124, 143, 145, 153
    conjugation, 65
    continuous, 48
    cubic, 82, 117
    discrete, 91
    effective, 153, 155
    elasticity tensor, 64
    even-rank tensor, 65
    finite elasticity, 59–61
    Herman's theorem, 62
    infinitesimal elasticity, 59, 61
    integration, 111, 112
    isotropic, 82, 117
    material symmetry, 58, 460, 461
    monoclinic, 70, 98, 112, 113, 145, 155
    $O(2)$, 58
    $O(3)$, 58
    order relation, 64

orthogonal
transformation, 63
orthotropic, 77, 113
$SO(3)$, 112, 364
symbol, 495, 496
symmetry class, 61
tetragonal, 78, 114
transversely isotropic, 81, 115
trigonal, 81, 115
index, 66–68, 480, 487
material, 20, 45, 58, 63, 91, 119, 123
wavefront, 123

tangent space, *see also* cotangent space, 54, 365, 378, 413, 469, 471
teleology
Fermat's principle, 287–289
temperature
symbol, 496
tensor product
symbol, 494
tessellation, 48, 91, 92
theoretical seismology, 1, 11, 34
theory of elasticity
constitutive relation, 43
continuum mechanics, xxviii, 21, 41, 42
Hookean solid
anisotropic, 344
infinitesimal, 54
theory-ladenness of data, *see also* underdetermination of theory by data, 341
thermal
conductivity, 437, 438
symbol, 496
conductivity tensor
elasticity tensor, 437
symbol, 496
diffusivity, 325, 329
symbol, 494
expansion, 325
isotropy, 437
potential, 428
power, 428, 429, 431, 436
thermodynamic
inhomogeneity, 439
potential, 438, 439
Legendre's
transformation, 268
thermodynamics, xxxii, 42, 181
first law, xxxii, 428, 429, 450
second law, xxxii, 428, 431–433, 439, 440, 442, 446, 450
thermoelastic continuum, 427, 428
thermoelasticity, xxxii, 324, 427, 431, 433, 442, 443, 448, 450
Thomsen parameters, 138, 176, 177, 180
total-stress tensor
symbol, 495
trace
symbol, 494
traction, 51
elastodynamic equation, 22, 182
Mohr's circle, 95
transformation
contravariant, 466, 467, 470, 473, 479
covariant, 466, 468, 486
transition matrix, 467, 468

transpose
  symbol, 494
two-point tensor, 365, 374, 375
  deformation
    finite, 366
    gradient, 365, 366, 406, 472, 484
    infinitesimal, 366, 402
  First Piola-Kirchhoff stress tensor, 420, 425, 472
  rotation, 391, 392
  symmetry, 366

underdetermination of theory by data, *see also* theory-ladenness of data, 341
uniaxial stretch, 34, 35, 399, 400

variational principle
  Fermat's principle, 267
  Hamilton principle, 267, 291
vector potential, 225
  S wave, 225
  symbol, 496
vector product
  symbol, 494
verification, xxx
vis viva, 269

wave, *see also* wave equation, *see also* wavefront, *see also* waveguide
  body, xxviii, 11, 182, 222, 240, 261, 262
    elastodynamic equation, 181, 231
    Hookean solid, 14, 15
    speed, 252, 254
  electromagnetic, 318
  gravitational, xxix, 317, 318, 323, 328, 331
    detection, 324
    speed, 318
  guided, xxix, 11, 182, 221, 222, 224, 234, 264
    existence, 233, 234, 242
    properties, 262
  interface, xxix, 11, 182, 221, 222, 224, 234, 245, 246, 262, 264
    dispersion, 253
    existence, 242, 261, 262
    properties, 261, 262
    solution, 221
    speed, 224, 229, 253
    Stoneley, 245
  Love, 237, 240, 241, 243, 246, 323
    existence, 242
    speed, 262
  metric, 319, 320, 328
    symbol, 496
  number, 195, 197, 201, 239
    symbol, 494
  optics
    geometrical optics, 357
  quasi-Rayleigh, 241
  Rayleigh, 224, 230–235, 237, 241, 242, 244–247, 255, 257, 262, 323
    existence, 233
    transverse isotropy, 264
  Scholte, 246, 264, 323
    existence, 253, 262
    speed, 253, 254, 262
  SH, 240, 241
  standing, 11, 235, 237
  Stoneley, 246, 323

existence, 262
speed, 262
surface, xxix, 11, 15, 182, 221, 222, 224, 230, 231, 234, 264
properties, 261, 262
vector, 240
symbol, 496
wave equation
body wave, xxviii, 11, 222
Cauchy problem, 288
continuum mechanics, 3
diffusion equation, 193
displacement potential, 225, 226
Fourier transform, 287, 289
friction, 193
guided wave, xxix, 11, 234
Hamilton principle, 287, 292
homogeneous continuum, 190, 192
Hookean solid, 42
inhomogeneous continuum, 190
interface wave, xxix, 11, 257, 258
SH wave, 263
isotropic continuum, 3, 190
Love wave, 238
P wave, 183, 184, 186, 187, 189
S wave, 183, 184, 186–188
Scholte wave, 247, 248
side condition, xxix, 222
solution, xxviii, 181, 194, 208, 221
d'Alembert, 196, 197, 199, 200
Fourier transform, 201, 203, 204, 217
friction, 193
Green's function, 204, 207, 208
product, 194
surface wave, xxix, 11
P wave, 243
wavefront
angle
symbol, 494
healing, 355
slowness
convexity, 280
Fermat's principle, 280
speed
symbol, 496
symmetry, 123
waveguide, 234, 237, 262, 320
wavenumber, *see* wave number
WKB method, *see also* $JWKB$ method, 209
$WKBJ$ method, *see also* $JWKB$ method, 209
work conjugate, 447

Zipf's law, 414

Name Index

Achenbach, Jan Drewes (1935–), 225, 230–232, 237, 245, 262

Aki, Keiiti (1930–2005), 48

Albertoni, Giovanni (1806–1887), 273

Almansi, Emilio (1869–1948), 366, 384

Anderson, Don Lynn (1933–2014), 127, 139

Arago, François Jean Dominique (1786–1853), 285

Aristotle (384 BC–322 BC), 22

Arnold, Vladimir Igorevich (1937–2010), 203

Backus, George Edward (1930–), 122–127, 129, 130, 134–137, 141, 154

Bacon, Francis (1561–1626), 8

Bayes, Thomas (1701–1761), 344

Beltrami, Eugenio (1835–1899), 296, 366, 373

Bernard D. Coleman (1929–), 21

Bernoulli, Jacob (1655–1705), 344, 366, 370

Bernoulli, Johannes (1667–1748), 268

Biot, Jean-Baptiste (1774–1862), 438

Borcherdt, Roger D. (1941–), 237

Brillouin, Louis Marcel (1854–1948), 209

Bullen, Keith Edward (1906–1976), 9

Bunge, Mario Augusto (1919–), v, xxv, xxvi, 3, 21, 45, 154, 293, 337, 353

Byron, George Gordon (1788–1824), 13

Cauchy, Augustin-Louis (1789–1857), 3, 366

Chadwick, Peter (1931–), 364, 398, 412, 414, 437

Clairaut, Alexis Claude (1713–1765), 188

Clausius, Rudolf Julius Emanuel (1822–1888), 366, 432, 433, 440, 446

Cosserat, Eugène Maurice Pierre (1866–1931), 211

Cosserat, François Nicolas (1852–1914), 211
Curie, Pierre (1859–1906), 9, 123

d'Alembert, Jean-Baptiste le Rond (1717–1783), 3, 187, 196, 203
Dahlen, Francis Anthony (1942–2007), 11, 19, 55, 209, 254, 312, 325, 414, 426
Darwin, George Howard (1845–1912), 9
Democritus (*ca.* 460 BC–370 BC), 22
Descartes, René (1596–1650), 23, 351
du Toit, Alexander Logie (1878–1948), 8
Duhamel, Jean-Marie Constant (1797–1872), 437, 438
Duhem, Pierre Maurice Marie (1861–1916), 366, 432, 433, 440, 446
Dyson, Freeman John (1923–), xxvii, 209, 285, 321
Dziewoński, Adam Marian (1936–2016), 139

Einstein, Albert (1879–1955), 317, 339, 374
Epstein, Marcelo (1944–), 23, 30, 34, 275, 276, 406, 442, 451, 453
Euler, Leonhard (1707–1783), 197, 203, 269, 287
Ewing, William Maurice (1906–1974), 225, 237, 245

Fermat, Pierre de (1607–1665), 15, 267, 269
Fermi, Enrico (1901–1954), 10
Feshbach, Herman (1917–2000), 193, 209
Feynman, Richard Phillips (1918-1988), 17, 287
Forte, Sandra (1953–), 62
Fourier, Jean-Baptiste Joseph (1768–1830), 438
Fresnel, Augustin (1788–1827), 285
Frobenius, Ferdinand Georg (1849–1917), 68

Galilei, Galileo (1564–1642), 309, 374
Gazis, Denos C. (1930–2004), 112, 124, 134, 154
Gibbs, Josiah Willard (1839–1903), 366, 441, 449
Golitsyn, Boris Borisovich (1862–1916), 5, 6
Grant, Fraser S. (1926–1984), 221, 225, 237, 245, 246, 264
Green, George (1793–1841), 4, 204, 205, 366
Gurtin, Morton E. (1934–), 359, 415

Halley, Edmond (1656–1742), 2, 3, 5
Hamilton, William Rowan (1805–1865), 267, 269, 287
Heisenberg, Werner Karl (1901–1976), 339
Helbig, Klaus (1927–), 81, 127, 137, 154
Helmholtz, Hermann Ludwig Ferdinand von (1821–1894), 186, 189, 247, 257, 287, 366,

434, 440, 441, 449
Hempel, Carl Gustav (1905–1997), 347, 349, 353
Hesse, Mary Brenda (1924–2016), 103, 351
Hooke, Robert (1635–1703), 3, 4, 42, 43
Hulse, Russell A. (1950–), 317
Hume, David (1711–1776), 289, 290
Huygens, Christiaan (1629–1695), 43, 267, 269, 272

Jacobi, Carl Gustav Jacob (1804–1851), 28, 269
Jardetzky, Wenceslas S. (1896–1962), 225, 237, 245
Jeffreys, Harold (1891–1989), xxvi, xxvii, 9, 209, 324, 326, 344, 472, 473

Kepler, Johannes (1571–1630), 268
Kirchhoff, Gustav Robert (1824–1887), 204, 366, 447
Kołakowski, Leszek (1927–2009), 351
Kramers, Hendrik Anthony (1894–1952), 209
Kronecker, Leopold (1823 – 1891), 65
Kuhn, Thomas Samuel (1922–1996), 13, 342

Lagrange, Joseph-Louis (1736–1813), 269, 273, 366
Lamé, Gabriel Léon Jean Baptiste (1795–1870), 83, 106, 140, 366, 417, 463
Landau, Lev Davidovich (1908–1968), 53, 184, 291, 312, 319, 322, 324, 340
Laplace, Pierre-Simon de (1749–1827), 184, 344
Legendre, Adrien-Marie (1752–1833), 268
Leibniz, Gottfried Wilhelm (1646–1716), 269
Levi-Civita, Tullio (1873–1941), 374
Lewin, Kurt (1890–1947), 11
Lie, Sophus (1842 –1899), 49, 491
Lifshitz, Evgeny Mikhailovich (1915–1985), 184, 291, 322, 324, 340
Lord Kelvin (1824–1907) *see also* Thomson, William
Love, Augustus Edward Hough (1863–1940), 237, 311

Mach, Ernst (1838–1916), 350
Maclaurin, Colin (1698–1746), 188
Maugin, Gérard A. (1944–2016), 20, 365, 366
Maupertuis, Pierre-Louis Moreau de (1698–1759), 188, 269, 287
Maxwell, James Clerk (1831–1879), 5, 19, 337
McLellan, Alister George (1919–2012), 427
Minkowski, Hermann (1864–1909), 31
Mintrop, Ludger (1880–1956), 9, 10, 280, 281, 348
Mohorovičić, Andrija (1857–1936), 6

Mohr, Christian Otto (1835–1918), 49, 366
Morse, Philip McCord (1903–1985), 193, 209, 427, 438
Muir, Francis (1926–), 137

Nanson, Edward John (1850–1936), 366, 389
Neumann, Carl Gottfried (1832–1925), 204
Noether, Emmy (1882–1935), 16, 45, 296
Noll, Walter (1925–2017), 20, 21, 25, 32, 51, 154, 181, 493
Nowacki, Witold (1911–1986), 427

Occam, William of (*ca.* 1288–*ca.* 1348), 106
Ogden, Raymond William (1943–), 36, 363, 364, 368, 398, 399, 412, 414, 426, 460
Ohm, Georg Simon (1789–1854), 438
Oldham, Richard Dixon (1858–1936), 4–6, 121, 189
Oldroyd, James Gardner (1921–1982), 51

Peirce, Charles Sanders (1839–1914), xxv, 122
Penrose, Roger (1931–), 92
Piola, Gabrio (1794–1850), 54–56, 366
Planck, Max Karl Ernst Ludwig (1858–1947), 366, 440
Plato (*ca.* 425 BC–*ca.* 347 BC), 350
Poisson, Siméon Denis (1781–1840), 4, 139, 140, 230, 231, 285, 312–314, 321
Pontecorvo, Bruno (1913–1993), 9, 10
Pope, Alexander (1688–1744), 290
Popper, Karl Raimund (1902–1994), 13, 335, 343, 352
Press, Frank (1924–), 225, 237, 245
Ptolemy (*ca.* 90–*ca.* 168), 268

Rayleigh, John William Strutt (1842–1919), 224
Ricci-Curbastro, Gregorio (1853–1925), 374
Richards, Paul G. (1943–), 48
Riemann, Georg Friedrich Bernhard (1826–1866), 478
Rochester, Michael Grant (1933–), 183, 184, 310
Rudzki, Maurycy Pius (1862–1916), 6, 7, 127, 154, 264
Russell, Bertrand (1872–1970), 329

Santayana, George (1863–1952), xxx
Schoenberg, Michael (1939–2008), 137
Schrödinger, Erwin Rudolf Josef Alexander (1887–1961), 339
Snel van Royen, Willebrord (1580–1626), 271
Snider-Pellegrini, Antonio (1802–1885), 8

St. Venant, Adhémar Jean Claude Barré de (1797–1886), 366, 447
Stacey, Frank Donald (1929–), 241, 315, 325, 328
Stebbing, Lizzie Susan (1885–1943), 45
Stokes, George Gabriel (1819–1903), 188
Stoneley, Robert (1894–1976), 245

Tadjbakhsh, Iradj G. (1928–1993), 112, 124, 134, 154
Tarantola, Albert (1949–2009), 12, 23, 343, 345
Taylor, Jr., Joseph H. (1941–), 317
Thomson, William (Lord Kelvin) (1824–1907), 338
Timoshenko, Stephen Prokopovich (1878–1972), 386, 417
Toupin, Richard A. (1926–), 112, 124, 134, 154
Tromp, Jeroen (1966–), 11, 19, 23, 55, 209, 254, 312, 325, 414, 426
Truesdell, Clifford Ambrose III (1919–2000), 51, 154, 338

Udías Vallina, Agustín (1935–), 19, 34, 42, 241, 314

Vela, Vincenzo (1820–1891), 55
Vianello, Maurizio Stefano (1952–), 49, 62
Voigt, Woldemar (1850–1919), 6, 106, 110, 117, 138

Wegener, Alfred Lothar (1880–1930), 8
Wentzel, Gregor (1898–1978), 209
West, Gordon Fox (1933–), 221, 225, 237, 245, 246, 264
Wiechert, Emil Johann (1861–1928), 6
Wigner, Eugene (1902–1995), 34, 189, 224

Zipf, George Kingsley (1902–1950), 414

About the Author

photo by Massimo Cervetti

Michael A. Slawinski is a professor in the Department of Earth Sciences at Memorial University in St. John's, Newfoundland.

He studied at University of Warsaw, Université de Paris and University of Calgary, where he obtained a doctoral degree.

Between 1986 and 1997, Slawinski worked for several years as a geophysicist in the petroleum industry. In 1998, he joined the Department of Mechanical Engineering at University of Calgary. In 2001, he joined Memorial University as a Research Chair in Applied Seismology.

Since 1998, Slawinski has been the director of The Geomechanics Project, a theoretical-research group whose focus is on extending seismic theory within the realm of continuum mechanics and the language of differential geometry. Together with his collaborators, he has worked on the group-theory underpinning of material symmetries of Hookean solids, and on wave phenomena in effective and equivalent media; these media are formulated, respectively, in terms of distances in the space of tensors, and their weighted averages. Also, his research focus includes finite elasticity, to examine theoretical foundations of seismology, as well as numerical analysis, to solve nonlinear problems, and perturbation methods, to examine stability and error-sensitivity of solutions.

In pursuit of aforementioned studies, he has long-term research collaborations with Department of Mathematics at Politecnico di Milano, Department of Geosciences at Princeton University, Department of Philosophy at University of Toronto, and Department of Geoinformatics and Applied Computer Science at University of Science and Technology in Kraków.

He has published three other books with World Scientific. "Waves and Rays in Elastic Continua" is already in its third edition. "Wavefronts and Rays as Characteristics and Asymptotics", with Andrej Bóna, is in its second edition. "On foundations of seismology: Bringing idealizations down to Earth", with James Robert Brown, is in its first edition.

www.ingramcontent.com/pod-product-compliance
Lightning Source LLC
LaVergne TN
LVHW010538300926
849162LV00002B/173

* 9 7 8 9 8 1 3 2 3 9 8 7 6 *